Resilience
The Science of Adaptation to Climate Change

Resilience
The Science of Adaptation to Climate Change

Edited by

Zinta Zommers
Mercy Corps, London, United Kingdom

Keith Alverson
International Environmental Technology Center, UN Environment, Osaka, Japan

ELSEVIER

Elsevier
Radarweg 29, PO Box 211, 1000 AE Amsterdam, Netherlands
The Boulevard, Langford Lane, Kidlington, Oxford OX5 1GB, United Kingdom
50 Hampshire Street, 5th Floor, Cambridge, MA 02139, United States

Notices
Knowledge and best practice in this field are constantly changing. As new research and experience broaden our understanding, changes in research methods, professional practices, or medical treatment may become necessary.

Practitioners and researchers must always rely on their own experience and knowledge in evaluating and using any information, methods, compounds, or experiments described herein. In using such information or methods they should be mindful of their own safety and the safety of others, including parties for whom they have a professional responsibility.

To the fullest extent of the law, neither the Publisher nor the authors, contributors, or editors, assume any liability for any injury and/or damage to persons or property as a matter of products liability, negligence or otherwise, or from any use or operation of any methods, products, instructions, or ideas contained in the material herein.

British Library Cataloguing-in-Publication Data
A catalogue record for this book is available from the British Library

Library of Congress Cataloging-in-Publication Data
A catalog record for this book is available from the Library of Congress

ISBN: **978-0-12-811891-7**

For Information on all Elsevier publications
visit our website at https://www.elsevier.com/books-and-journals

Working together
to grow libraries in
developing countries

www.elsevier.com • www.bookaid.org

Publisher: Candice Janco
Acquisition Editor: Laura Kelleher
Editorial Project Manager: Tasha Frank
Production Project Manager: Nilesh Kumar Shah
Cover Designer: Christian Bilbow

Typeset by MPS Limited, Chennai, India

Cover Image: IrinaK/Shutterstock - Working traffic lights over streets flooded by Hurricane Harvey in Houston, Texas, September 2017

Contents

Section II
Adaptation Actions—Hazards, Ecosystems, Sectors

8. Measuring Drought Resilience Through Community Capitals

Andries J. Jordaan, Dusan M. Sakulski, Curtis Mashimbye and Fumiso Mayumbe

9. Community-Based Adaptation: Alaska Native Communities Design a Relocation Process to Protect Their Human Rights

Robin Bronen

10. California: It's Complicated: Drought, Drinking Water, and Drylands

Gillisann Harootunian

11. Advancing Coastal Climate Resilience: Inclusive Data and Decision-Making for Small Island Communities

Roger-Mark De Souza and Judi Clarke

12. Building Urban Resilience to Address Urbanization and Climate Change

Julie Greenwalt, Nina Raasakka and Keith Alverson

18. Supporting Farmers Facing Drought: Lessons from a Climate Service in Jamaica

John Furlow, James Buizer, Simon J. Mason and Glenroy Brown

19. Forecast-Based Financing and Climate Change Adaptation: Uganda Makes History Using Science to Prepare for Floods

Eddie Wasswa Jjemba, Brian Kanaahe Mwebaze, Julie Arrighi, Erin Coughlan de Perez and Meghan Bailey

20. Managing Risks from Climate Change on the African Continent: The African Risk Capacity (ARC) as an Innovative Risk Financing Mechanism

Ekhosuehi Iyahen and Joanna Syroka

21. Climate Change Adaptation in Ethiopia: Developing a Method to Assess Program Options

Karyn M. Fox, Suzanne Nelson, Timothy R. Frankenberger and Mark Langworthy

List of Contributors

Daniel Abrahams, University of South Carolina, Columbia, SC, United States

Keith Alverson, International Environmental Technology Center, UN Environment, Osaka, Japan

Julie Arrighi, Red Cross/Red Crescent Climate Centre, The Hague, The Netherlands

Meghan Bailey, Red Cross/Red Crescent Climate Centre, The Hague, The Netherlands; School of Geography and the Environment, University of Oxford, Oxford, United Kingdom

Jane Barr, Environmental Pulse Institute, Trent, SD, United States

Amanda Bourne, Conservation South Africa, Cape Town, South Africa

Kerry W. Bowman, University of Toronto, Toronto, ON, Canada

Robin Bronen, Alaska Institute for Justice, Anchorage, AK, United States

Glenroy Brown, Meteorological Service of Jamaica, Kingston, Jamaica

James Buizer, Institute of the Environment, University of Arizona, Tucson, AZ, United States

Alvin Chandra, The University of Queensland, Brisbane, QLD, Australia

Judi Clarke, J F Clarke Consulting Inc., Bridgetown, Barbados

Ethan D. Coffel, Columbia University, New York, NY, United States

Cosmin Corendea, United Nations University – Institute for Environment and Human Security, Bonn, Germany

Erin Coughlan de Perez, Red Cross/Red Crescent Climate Centre, The Hague, The Netherlands; VU University Amsterdam, Amsterdam, The Netherlands; Columbia University, Palisades, NY, United States

Alex de Sherbinin, Columbia University, New York, NY, United States

Roger-Mark De Souza, Sister Cities International, Washington, DC, United States

Barney Dickson, UN Environment, Nairobi, Kenya

Ebinezer R. Florano, Center for Policy and Executive Development, National College of Public Administration and Governance, University of the Philippines, Quezon City, Philippines

Karyn M. Fox, TANGO International, Washington, DC, United States

Timothy R. Frankenberger, TANGO International, Tucson, AZ, United States

John Furlow, International Research Institute for Climate and Society, Columbia University, New York, NY, United States

Corey Gabriel, Department of Environmental Sciences, Rutgers University, New Brunswick, NJ, United States; Scripps Institution of Oceanography, La Jolla, CA, United States

Julie Greenwalt, Cities Alliance, Brussels, Belgium

H. Gyde Lund, Environmental Pulse Institute, Trent, SD, United States

Gillisann Harootunian, California State University, Fresno, CA, United States; University of Maryland, University College, Adelphi, MD, United States

Sarah Henly-Shepard, Mercy Corps, Washington, DC, United States

Radley M. Horton, Columbia University, New York, NY, United States; Lamont Doherty Earth Observatory, Palisades, NY, United States

Saleemul Huq, International Centre for Climate Change and Development, Dhaka, Bangladesh

Ekhosuehi Iyahen, Policy & Advisory Services, African Risk Capacity, Johannesburg, South Africa

Eddie Wasswa Jjemba, Red Cross/Red Crescent Climate Centre, The Hague, The Netherlands

Andries J. Jordaan, University of the Free State, Bloemfontein, South Africa

Siobhan E. Kerr, University of Maryland College Park, College Park, MD, United States

Stefan Kienberger, University of Salzburg, Salzburg, Austria

Kathryn Lane, New York City Department of Health & Mental Hygiene, New York, NY, United States

Mark Langworthy, TANGO International, Tucson, AZ, United States

Eliot Levine, Mercy Corps, Portland, Oregon, United States

Filipe D.F. Lúcio, World Meteorological Organization, Geneva, Switzerland

Darren Lumbroso, HR Wallingford, Wallingford, United Kingdom

Yousuf Mahid, International Centre for Climate Change and Development, Dhaka, Bangladesh

Tanvi Mani, The Graduate Institute of International and Development Studies, Geneva, Switzerland

Yousef M. Manialawy, University of Toronto, Toronto, ON, Canada

Curtis Mashimbye, University of the Free State, Bloemfontein, South Africa

Simon J. Mason, International Research Institute for Climate and Society, Columbia University, New York, NY, United States

Fumiso Mayumbe, University of the Free State, Bloemfontein, South Africa

Karen E. McNamara, The University of Queensland, Brisbane, QLD, Australia

Rachelle Meyer, Australian-German Climate & Energy College, University of Melbourne, Parkville, Victoria, Australia; Faculty of Veterinary and Agricultural Sciences, University of Melbourne, Parkville, Victoria, Australia

Halcyone Muller, Conservation South Africa, Cape Town, South Africa

Brian Kanaahe Mwebaze, Uganda Red Cross Society, Kampala Uganda

Suzanne Nelson, TANGO International, Tucson, AZ, United States

Robert J. Nicholls, University of Southampton, Southampton, United Kingdom

Anne Olhoff, UNEP DTU Partnership, Technical University of Denmark, Copenhagen, Denmark

Victor Orindi, National Drought Management Authority, Nairobi, Kenya

Janak Pathak, World Meteorological Organization, Geneva, Switzerland

Karen Podvin, Regional Office for South America, Quito, Ecuador

Nina Raasakka, UN Environment, Bangkok, Thailand

David Ramsbottom, HR Wallingford, Wallingford, United Kingdom

Hannah Reid, International Institute for Environment and Development, London, United Kingdom

Alan Robock, Department of Environmental Sciences, Rutgers University, New Brunswick, NJ, United States

Dusan M. Sakulski, University of the Free State, Bloemfontein, South Africa

Sarshen Scorgie, Conservation South Africa, Cape Town, South Africa

Ashbindu Singh, Environmental Pulse Institute, Trent, SD, United States

Adam H. Sobel, Columbia University, New York, NY, United States

Nadine Suliman, International Centre for Climate Change and Development, Dhaka, Bangladesh; The University of Alberta, Alberta, Canada

Joanna Syroka, Research & Development, African Risk Capacity, Johannesburg, South Africa

Michael K. Tippett, Columbia University, New York, NY, United States

Christopher H. Trisos, National Socio-Environmental Synthesis Center (SESYNC), University of Maryland, Annapolis, MD, United States

Elina Väänänen, UN Environment, Nairobi, Kenya

Elisabeth Vogel, Australian-German Climate & Energy College, University of Melbourne, Parkville, Victoria, Australia; ARC Centre of Excellence for Climate System Science, University of Melbourne, Parkville, Victoria, Australia

Alan Warner, University of Toronto, Toronto, ON, Canada

Olga Wilhelmi, National Center for Atmospheric Research, Boulder, CO, United States

Lili Xia, Department of Environmental Sciences, Rutgers University, New Brunswick, NJ, United States

Zinta Zommers, Food and Agriculture Organization of the United Nations, Freetown, Sierra Leone

Preface

"Today we must set the world on a new course." These were the words of the UN Secretary-General Ban Ki Moon at the opening Ceremony of Climate Summit in 2014. Four years later, these sentiments ring ever more true.

Over the course of that September day in 2014, from behind the lectern of the UN General Assembly, I watched over 100 Heads of State make new and significant climate change commitments. More than 800 representatives from business, finance, and civil society joined national and local leaders. People showed their solidarity in the streets of more than 100 cities around the world, with over 400,000 in New York alone.

Climate change is perhaps the greatest defining issue of the 21st century. The human, environmental, and financial costs are unbearable. It is time for definitive action.

We have come a long way since 2014. The Paris Agreement provides a collective vision for action on climate change. At the time of writing, 174 parties have ratified the Agreement. When the Trump Administration announced its intent to withdraw, countries around the world, as well as businesses, states, cities, and citizens across the United States rallied to it even more strongly. Now we need to make sure we have the tools to achieve our collective goals.

As the Dean of the School of Public Policy at the University of Maryland, I am determined to ensure research plays a critical role informing public policy and action. The climate research community has been highly engaged over the past few decades in detecting global climate change and attributing this change, including extreme events, to anthropogenic causes. However there has been less scientific and political focus on how to adapt to these climatic changes.

According to the 2017 *Adaptation Gap Report*, produced by UN Environment, current national monitoring and evaluation systems mainly focus on monitoring adaptation. Few countries have undertaken an evaluation of actual adaptation progress. There are currently no agreed-upon methods, indicators, metrics, or frameworks designed for an assessment of progress toward the global goal on adaptation. This is a critical gap, especially given that tens of billions of dollars of public finance is spent globally on climate change adaptation each year.

This book makes a significant contribution to understanding these issues. Chapters from leading experts analyze ongoing adaptation interventions. The contributions all assess, from different disciplinary and geographical perspectives, how to adapt to climate change impacts such as sea level rise and changing hydrological variability. Chapters critically examine tools and methods for adaptation, including early warnings and forecast-based finance, risk transfer mechanisms including insurance, community-based adaptation, and even the possible impacts of geo-engineering. They identify gaps in knowledge and highlight areas where further research and innovation is needed to build resilience. Ultimately, this book promotes an evidence-based approach to adaptation with the scientific community engaged in evaluating, learning, and improving on existing resilience building actions.

This book takes a valuable step toward building a sustained conversation between climate experts, leaders, and citizens. We have never faced such a challenge. Nor have we encountered such great opportunity. We must ensure that a low-carbon, climate resilient future will be a better future. Each day we must all recommit to taking this new course.

Robert C. Orr

Dean, School of Public Policy, University of Maryland

Special Advisor, UN Secretary-General on Climate Change

Acknowledgments

The editors would like to thank Peter Gilruth and Mette Wilkie for their support, comments, and contributions. Your assistance was greatly appreciated. We also thank friends and family for your encouragement and patience during the writing process.

Numerous chapters in this book were presented at the IAPSO-IAMAS-IAGA Good Hope for Earth Sciences Joint Assembly in August 2017. The session "Resilience: The Science of Adapting to Climate Change" allowed authors to interact, discuss, and further refine their ideas and chapters. We would like to thank UN Environment for extending its support to this process.

Introduction

Keith Alverson and Zinta Zommers

As this introduction was being written, the news headline on CNN was "Catastrophic Maria churns toward Irma battered Islands." The story of these two massive hurricanes, their short-term, and as yet unknown longer-term, consequences on islands in the Caribbean captures much about resilience and adaptation to climate change. In his visit to the scene of the aftermath, Antonio Guterres, the UN Secretary General, stated that "the link between climate change and the devastation we are witnessing is clear" (UN News). At around the same time, Kerry Emmanuel, Professor of Atmospheric Sciences at M.I.T. stated that "Climate change, if unimpeded, will greatly increase the probability of extreme events" (MIT News).

The first section of the book focuses on the need for adaptation. Adam Sobel and Michael Tippett (Chapter 1: Extreme Events: Trends and Risk Assessment Methodologies) have a detailed look at the state of the science on attributing extreme events such as hurricanes to anthropogenic climate change, and conclude with a reminder that irrespective of technical debates, the need to build resilience in the face of such storms remains. The rest of the volume follows suit, with little or no attempt to weigh in on the overly politicized nuances of detecting or attributing anthropogenic causality; instead, the chapters directly tackle the vital business of applying scientific rigor to the process of building resilience to climate impacts and extremes. Robert Nicholls (Chapter 2: Adapting to Sea-Level Rise) provides a global overview of sea-level rise. Elisabeth Vogel and Rachelle Meyer discuss the immense challenge of global food security (Chapter 3: Climate Change, Climate Extremes, and Global Food Production—Adaptation in the Agricultural Sector), while Barney Dickson et al. (Chapter 4: Tracking Adaptation Progress at the Global Level: Key Issues and Priorities) weigh in on tracking progress on the ground, and Saleem Huq et al. (Chapter 5: Evolution of Climate Change Adaptation Policy and Negotiation) evaluate global negotiations and national legislation. Both note that while the need for adaptation is growing, so too is its prominence in the political and policy sphere.

DEFINITIONS

Vulnerability. Adaptation to climate change has often been built on the notion of reducing vulnerability. There have been many definitions of vulnerability in the context of climate change, with IPCC's consensus effort being the degree to which geophysical, biological, and socioeconomic systems are susceptible to, or unable to cope with, adverse impacts of climate change. Adaptation projects are often designed to start with understanding and mapping vulnerabilities in order to prioritize actions. One common index of vulnerability combines exposure and sensitivity to estimate impact and then compares impact with existing adaptive capacity, to provide an estimate of vulnerability to a given threat. It is important to note that reducing vulnerability is not the only way to adapt to climate change, and is certainly not the same as building resilience. For example, the city of New Orleans decreased its vulnerability to sea-level rise and storm surges by building levees and dikes, but this reduced vulnerability may have led to a sense of complacency and overdevelopment in low-lying areas, such that when the defences were breached by hurricane Katrina in 2005, the city found it was unable to quickly bounce back from the devastation. Arguably, the city's resilience even decreased as a result of protections aimed at decreasing vulnerability. One illustration of this concept is that of American Football helmets. Players started wearing helmets to reduce their vulnerability to catastrophic head damage due to high speed collisions, but the unintended consequence was decreased resilience since players tended not to avoid head collisions when wearing helmets. One recent study found 110 out of 111 NFL players whose brains were inspected had suffered chronic traumatic encephalopathy—an extremely debilitating condition associated with memory loss, impulsivity, and suicide—due to repeated head trauma.

Resilience. The concept of resilience has been used in the context of many challenges including climate change, as well as a wide range of issues from food security to sustainable economic growth. The IPCC defines resilience as the ability of a system or community that is exposed to hazards to resist, absorb, accommodate, and recover in a timely and efficient manner. As noted by several chapters in this book, there are critical capacities for resilience—adaptive capacity, absorptive capacity, and transformative capacity.

Clearly increasing resilience is a much broader task than decreasing vulnerability, and in that sense seems a more appropriate outcome to seek from adaptation actions. On the other side of the coin, it is of course much more difficult to quantify resilience—nobody has been so bold as to publish indices or maps of resilience as they have for vulnerability in order to prioritize adaptation actions on the ground. Consider the response of small islands in the Caribbean to hurricanes. As of this writing, the entire population of Barbuda had evacuated, mostly to nearby Antigua (CNN, September 15, 2017). Even before the hurricanes, more Puerto Ricans were living in the continental United States than on the island due to economic shocks (The Atlantic, July 6, 2017), a trend that is only likely to increase with most of the island unable to regain electricity supply for many months after the storms. Migration out of vulnerable areas is clearly part of resilience building. On the other hand, the population in Southern Florida grew from 217,438 homes in 1940 to 8,648,404 homes in 2017, mostly concentrated along the coasts (Sneed, 2017). Demographic changes, including migration, both drive and respond to environmental extremes and is only one of many examples of why it is inherently difficult to assess or quantify the concept of resilience.

Adaptation. The IPCC defines adaptation as an adjustment in natural or human systems in response to actual or expected climatic stimuli or their effects (which moderates harm or exploits beneficial opportunities). It is important to note that the portion of this definition in brackets restricts the definition to "successful adaptation" in that it inherently assumes success. In fact, many adaptation actions have delivered no measurable outcomes, while others may have even backfired due to unintended negative consequences, particularly in certain segments of society. Adaptation actions may even contribute to increasing climate change, exacerbating the problem they seek to solve. Air-conditioning, e.g., certainly moderates harm due to heat stress (Coffel et al., this volume), but it is also usually a substantial contributor to greenhouse gas emissions. In this context, it is important to keep in mind that the most effective global adaptation action is to reduce greenhouse gas emissions, making the impacts of global warming easier to deal with at the local level. Mitigation is inherently part of any effective adaptation strategy.

SCALES OF ADAPTATION ACTION

The scales of adaptation actions are highly variable, and highly dependent on specific hazards, ecosystems, and socio-economic sectors. Given the huge diversity of challenges, the oft-touted philosophy of funding pilot demonstration projects to demonstrate success, supposedly to be followed by emulation and up-scaling is unfortunately woefully flawed. There is simply no off-the-shelf solution to anything. Adaptation actions need to occur locally and nationally but also in some cases actions must be regional or even global, e.g., in the case of shared resources such as transboundary waters. Each and every adaptation action needs to be rigorously and scientifically evaluated and adjusted in the face of future change (Zommers and Alverson, this volume).

The second section of this book breaks down some of the more common challenges specific to adaptation for different hazards, sectors, and ecosystems. After the broad discussion of sea-level rise in Chapter 2, Adapting to Sea-Level Rise, Darren Lumbroso and David Ramsbottom (Chapter 6: Flood Risk Management in the United Kingdom: Putting Climate Change Adaptation Into Practice in the Thames Estuary) takes a focused look at adaptation in one geographically miniscule, but economically vital section of coast—the Thames Estuary in London. Coffel et al. (Chapter 7: The Science of Adaptation to Extreme Heat) tackle adaptation to extreme heat, mostly from a developed country urban perspective. Drought resilience gets a look both from a rural developing country perspective by Andries Jordaan et al. (Chapter 8: Measuring Drought Resilience Through Community Capitals) and a more urban developed country perspective by Gillisann Hartoonian (Chapter 10: California: It's Complicated:Drought, Drinking Water, and Drylands). Next, a few geographical contexts with specific vulnerabilities and some common solutions get a look: Small Island Communities by Roger-Mark De Souza and Judi Clarke (Chapter 11: Advancing Coastal Climate Resilience: Inclusive Data and Decision-Making for Small Island Communities) and Urban Centers by Greenwalt et al. (Chapter 12: Urban Resilience). Robin Bronen (Chapter 9: Community-Based Adaptation: Alaska Native Communities Design a Relocation Process to Protect Their Human Rights) discusses community led efforts to adapt in the Arctic. The Arctic is a particularly vexing challenge for adaptation. Since Arctic countries are in developed countries, they are left out of all multilateral adaptation funding and assistance mechanisms. Yet, they contain communities that are in fact underdeveloped islands, and hence highly vulnerable in the face of some of the greatest climate impacts on the planet. The section ends with two sectoral overview papers—Climate Smart Agriculture by Alvin Chandra and Karen McNamara

(Chapter 13: Resilience: The Science of Climate-Smart Agriculture in Southeast Asia: Lessons From Community-Based Adaptation Programs in the Philippines and Timor-Leste) and Energy in Africa by Ashbindu Singh et al. (Chapter 14: Challenges in Building Climate-Resilient Quality Energy Infrastructure in Africa).

There will obviously be many lacunae in any such attempt to catalogue, in a series of chapters, adaptation actions by hazard and sector. Indeed, such gaps are so numerous that they cannot be filled in a short introduction either. Some are obvious—we leave out mountainous regions and their inherent challenges, such as landslides for example. Some of the gaps are less obvious but nonetheless extremely important. As one example consider highly vulnerable subgroups. The winner of the 2017 MIT climate colab "resilience" crowdsourced solutions contest (https://climatecolab.org/contests/2017/A2R-Anticipating-Climate-Hazards) was an interesting but somewhat controversial proposal to improve disaster preparedness for deaf people in the Philippines. At first glance it may be hard to understand how such an incredibly targeted proposal could have interested the judges (full disclosure, Keith Alverson was one of them). However, a few months later the decision was shown to be extremely prescient, as the state of Florida was caught without any such preparations before a hurricane. The Florida attempts to communicate evacuation orders at an official press conference clearly did not serve the deaf population. Televised evacuation instructions included a man signing gibberish alongside officials (NY Post, September 16, 2017). There are over 200,000 hearing disabled people in the state of Florida, representing nearly 2% of the state population.

THE ADAPTATION TOOLKIT

Having laid out the needs, progress achieved, and a number of case studies, the third section of the book focuses on tools and methods. Bowman et al. (Chapter 15: Ethics, Communities, and Climate Resilience: An Examination by Case Studies) remind technocrats of the need for ethical considerations and the precautionary principle in framing adaptation actions. Next, Reid et al., evaluate ecosystem-based adaptation, which seeks to harness the resilience that ecosystems have inherently built up over millions of years of climatic stress for the benefit of human populations. In the worst case scenario, ecosystem services for resilience are destroyed unthinkingly. In the best case, ecosystem preservation and protection is often employed in concert with engineering approaches in a "green-grey" holistic approach. We have two chapters focused on meteorological and climate services—making predictions useful across a wide range of socioeconomic activities. Janak Pathak and Filipe Lúcio (Chapter 17: The Global Framework for Climate Services Adaptation Programme in Africa) provide details of how climate services were strengthened in Tanzania and Malawi, while Furlow et al. (Chapter 18: Supporting Farmers Facing Drought: Lessons From a Climate Service in Jamaica) highlights the codesign of a drought information service in Jamaica.

And then there is money. According to a recent study by UN Environment, the costs of adaptation could range from USD 140 billion to USD 300 billion by 2030, and between USD 280 billion and USD 500 billion by 2050 (UNEP, 2016). These figures are far beyond the budgets available, meaning that much of the time, and particularly in developing counties, the availability of funding remains a constraint to achieving resilience. Looking at innovative ways to unlock funding is thus an important component in the adaptation toolkit. In this book we have two chapters looking at finance flows; at when and how finance can be delivered for communities and countries to take action. Jjemba et al. (Chapter 19: Forecast-Based Financing and Climate Change Adaptation: Uganda Makes History Using Science to Prepare for Floods) take a look at forecast-based financing for flood preparation in Uganda and Joanna Syroka and Ekhosuehi Iyahen (Chapter 20: Managing Risks From Climate Change on the African Continent: The African Risk Capacity (ARC) as an Innovative Risk Financing Mechanism) tackle insurance and risk transfer. Rounding out the section on approaches, Karyn Fox et al. (Chapter 21: Climate Change Adaptation in Ethiopia: Developing a Method to Assess Program Options) provides an overview of decision support tools, and Siobhan Kerr (Chapter 22: Social Capital as a Determinant of Resilience: Implications for Adaptation Policy) looks at the importance of social capital.

EMERGING ISSUES AND NEXT STEPS

The final section of the book looks at some of a few more recently acknowledged issues in adaptation, and potential future ones. Henley-Shepard et al. (Chapter 23: Climate-Resilient Development in Fragile Contexts) provide an overview of adaptation challenges in fragile and postconflict states. Trisos et al. (Chapter 24: Ecological, Agricultural, and Health Impacts of Solar Geoengineering) grapple with geoengineering, a highly dubious intervention given the enormous unknown and unanticipated consequences that would certainly result. Many of the statements in this book about applying the scientific method, using the precautionary principle, taking no regrets measures, and learning by doing, are simply not relevant in the case of global geoengineering, since interventions on this scale could lead to potentially irreversible planetary changes. Cosmin Corendea and Tanvi Mani (Chapter 25: The Progression of Climate Change, Human Rights, and Human Mobility in the Context of Transformative Resilience—A Perspective Over the Pacific) look at climate change impacts on migration and Ebinezer Florano (Chapter 26: Integrated Loss and Damage−Climate Change Adaptation−Disaster Risk Reduction Framework: the Case of the Philippines) highlights a case study from the Philippines to examine the loss and damage that occur beyond adaptation. Finally, Zinta Zommers and Keith Alverson (Chapter 27: Intelligent Tinkering in Climate Change Adaptation) conclude with the need for a greater evidence base and "intelligent tinkering" to achieve such measurable success.

CONCLUSIONS

Learning by doing. The science underpinning the detection and attribution of anthropogenic climate change is clear and convincing. Indeed this is probably one of the foremost examples of the academic research community and policy makers interacting for mutual benefit. On the other hand, the science behind climate change adaptation and resilience remains weak. It is common to involve local stakeholders in adaptation projects, but rare to involve academic researchers, be it in planning, monitoring, or evaluating actions. Building resilience will benefit immensely from the engagement of scientists, and the better integration of scientific methods and hypothesis testing, to achieve more effective adaptation, and to avoid maladaptation.

Unintended consequences. Much adaptation work is narrowly focused on measurable indices of success. Predetermined outcomes are programmed into detailed work plans and delivered in a timely manner by the end of a typical 3-year project. Evaluators praise this demonstrable success. However, in addition to planned outcomes, every intervention will also have unplanned ones. Indeed, the unplanned outcomes may even be more substantial. They could either enhance resilience, or at times lead to maladaptation. Air-conditioning, e.g., is one of the most effective methods of adaptation to extreme heat (Coffel et al., this volume). However, air-conditioning, when powered by fossil fuel-derived electricity, is also a substantial contributor to greenhouse gases, thereby contributing to enhancing both the number and degree of precisely such extreme heat events. This particular trade-off is rather obvious, but there are numerous less acknowledged unintended consequences. The use of chemicals, particularly CFCs but also many others, leads to impacts on the environment for example. Probably most importantly, the ready availability of air-conditioning was, and remains, a substantial driver of maladaptive demographic trends with enormous population growth in regions particularly susceptible to extreme events.

Extreme Events. While writing this, one of us was on a Shinkansen, also known as the bullet train, in Japan. The iPad I was typing on suddenly flashed a screen alert for a "flood tide" in Nagoya, precisely the city I was traveling through at over 300 km/h. In one of the most high tech, wealthy countries in the world the degree of preparedness and efficiency of early warning is inspiring, a real testimony to the ingenuity and resourcefulness of humankind in thinking ahead and building resilience. At the same time, Japan remains disaster prone, and even the most advanced precautions and warnings have failed to predict massive and irreversible impacts. Millions of tons of radioactive soil and groundwater from one of the two worst nuclear power disasters in history was not in a single contingency plan for tsunami preparedness and response. Although devastating tsunami waves have been known in Japan for millennia, their impact has not remained constant. Therefore adaptation measures must themselves be continuously evaluated and changed using scientific methods, or at the very least "intelligent tinkering" (Zommers and Alverson, this volume), in order to achieve resilience.

REFERENCES

CNN, September 15, 2017. <http://edition.cnn.com/2017/09/15/americas/irma-barbuda-population-trnd/index.html>.
MIT News (21 September 2017): http://news.mit.edu/2017/kerry-emanuel-hurricanes-are-taste-future-0921.
NY Post, September 16, 2017. <http://nypost.com/2017/09/16/deaf-community-outraged-after-interpreter-signed-gibberish-before-irma/>.
Sneed, A., September 12, 2017. Hurricane Irma: Florida's overdevelopment has created a ticking time bomb. Sci. Am.
The Atlantic, July 2017. <https://www.theatlantic.com/politics/archive/2015/07/due-to-crisis-more-puerto-ricans-now-live-in-the-us-than-on-the-island/432246/>.
UNEP, 2016. The Adaptation Finance Gap Report 2016. United Nations Environment Programme (UNEP), Nairobi, <http://www.unep.org/climate-change/adaptation/gapreport2016/>.
UN News (8 October 2017): https://medium.com/we-the-peoples/devastating-hurricanes-could-become-the-new-normal-44cba2b6fef1.

Adaptation Needs

Chapter 1

Extreme Events: Trends and Risk Assessment Methodologies

Adam H. Sobel and Michael K. Tippett

Columbia University, New York, NY, United States

Chapter Outline

1.1 IMPACT OF CLIMATE CHANGE ON EXTREMES

There is no simple statement which accurately describes the state of the science on how extreme weather events respond to climate change. Statements such as "climate change is making weather more extreme" are oversimplifications with the potential to be misleading. The statement in quotations above does carry two truths, however. First, much of the damage from climate change will be felt through changes in extreme events. Second, for many kinds of events, despite the uncertainties, current science justifies a legitimate concern that the *risk*—a probabilistic concept that can include scientific uncertainty as well as other forms—is increasing. But the dependence of extreme events on climate is substantially different for different types of events, as is our degree of scientific knowledge and understanding.

In this section we give a brief summary of the current understandings regarding a subset of extreme event types. The recent review by IPCC (2012) remains relevant and can be consulted for more details. The more recent National Academy (2016) report addresses individual extreme event attribution—the science of making quantitative statements about how different causes, including anthropogenic global warming, contribute to specific individual events—but contains a wider-ranging discussion illustrating the broader point that our confidence in our understanding of the human influence on any event type is affected by many factors, including the quality and length of observational records, the quality of numerical models in simulating and predicting that event type, etc. Fig. 1.1, adapted from that report, provides a graphical assessment of the state of attribution science for different types of extreme events, as explained briefly in the caption and in more detail in the report itself.

The National Academy (2016) report emphasizes that our degree of scientific understanding of different types of events' relation to warming is greatest for events most closely related to atmospheric temperature, since temperature is the variable in which greenhouse gas influence is first and most directly felt. Thus the human influence is most clear on heat waves and cold snaps, as indicated by their position above and to the right of other event types in Fig. 1.1.

1.1.1 Heat Waves

Heat waves are now occurring frequently with a magnitude that was very rare until the last few decades. Dynamical climate models are able to simulate this trend with, but not without anthropogenic greenhouse gas emissions

Resilience. DOI: https://doi.org/10.1016/B978-0-12-811891-7.00001-3

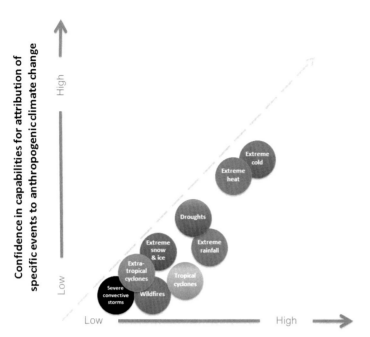

FIGURE 1.1 Schematic depiction of the state of attribution science for different event types. The horizontal position of each event type reflects an assessment of the level of understanding of the effect of climate change on the event type. The vertical position of each event type indicates an assessment of scientific confidence in current capabilities for attribution of specific events to anthropogenic climate change for that event type. A position below the 1:1 line indicates an assessment that there is potential for improvement in attribution capability through technical progress alone (such as improved modeling, or the recovery of additional historical data), which would move the symbol upward. A position above the 1:1 line is not possible because this would indicate confident attribution in the absence of adequate understanding. In all cases, there is potential to increase event attribution confidence by overcoming remaining challenges that limit the current level of understanding, as indicated by the blank space in the upper right corner. Source: *Adapted from National Academies of Sciences, Engineering, and Medicine, 2016. Attribution of Extreme Weather Events in the Context of Climate Change. Washington, DC: The National Academies Press. http://dx.doi.org/10.17226/21852, where more detailed explanation can be found.*

(e.g., Arblaster et al., 2014). The magnitude of the anthropogenic influence can depend, however, on whether one considers the change in probability of exceeding a fixed temperature threshold or the relative magnitude (in degrees) of the anthropogenic component compared to the natural variability component; the latter can be small while the former is large (Otto et al., 2012).

Heat waves illustrate, in a manner relevant to other event types as well, issues around the different roles of thermodynamic and dynamic effects, and different levels of uncertainty associated with them. The probability of a heat wave—defined as temperature exceeding some specified threshold for some specified duration—can be thought of as being influenced by the climatological mean temperature of the location of interest at a given time of year plus a fluctuating "weather" component related to the variable atmospheric circulation. In particular, a strong heat wave is generally associated with a persistent high pressure system. Changes in climatological mean temperature at most locations on earth increase roughly in sync with global mean temperature under greenhouse gas forcing (though at different rates, e.g., due to polar amplification) while the extent to which the frequency or intensity of high pressure systems may change in response to warming is much less clear. (In addition, many human populations are experiencing warming at a faster rate than the global mean due to the urban heat island effect as well as the fact that land is warming faster than ocean.)

This is an instance of the more general fact that, under climate change, all aspects of atmospheric circulation change are much more uncertain than are changes in temperature or other thermodynamic quantities (e.g., water vapor) that are closely coupled to temperature (e.g., Trenberth et al., 2015). A simple null hypothesis is that mean temperature changes while circulation does not (e.g., Held and Soden, 2006), and thus that the entire probability distribution of temperatures shifts to higher values while its shape remains unchanged. Though this hypothesis is not exactly true, it is useful as a starting point for understanding. That is, the global mean temperature increase is almost certainly the dominant driver of increasing heat wave frequency and intensity, with circulation modifying that trend quantitatively on a regional basis, but not qualitatively on the global scale.

1.1.2 Extreme Precipitation Events

Changes in regional mean precipitation over the past century are documented and statistically significant in many regions (e.g., Walsh et al., 2014), and it is plausible to expect that extremes should change as well. Consistent with this expectation, observations of precipitation show increasing trends in many regions in the intensity of rain falling in the heaviest events (e.g., Groisman et al., 2005), where "heaviest events" are typically defined as those exceeding some high percentile, say 99%, of the distribution from some set of earlier years. An example for the United States is shown in Fig. 1.2, where the threshold is defined by the 2-day precipitation total occurring on average once every 5 years over

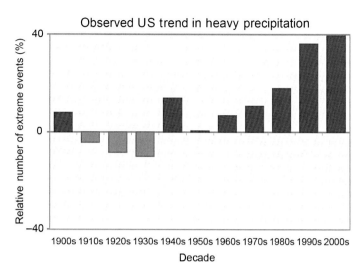

FIGURE 1.2 One measure of a heavy precipitation event is a 2-day precipitation total that is exceeded on average only once in a 5-year period, also known as a once-in-5-year event. As this extreme precipitation index for the United States during 1901—2012 shows, the occurrence of such events has become much more common in recent decades. Changes are compared to the period 1901—60, and do not include Alaska or Hawai'i. The 2000s decade (far right bar) includes 2001—12. Source: *Adapted from Walsh, J., D. Wuebbles, K. Hayhoe, J. Kossin, K. Kunkel, G. Stephens, et al., 2014. Ch. 2: Our Changing Climate. Climate Change Impacts in the United States: The Third National Climate Assessment, J.M. Melillo, Terese (T.C.) Richmond, and G.W. Yohe, Eds., U.S. Global Change Research Program, 19—67. http://dx.doi.org/10.7930/J0KW5CXT, originally from Kunkel, K.E., coauthors, 2013. Monitoring and understanding trends in extreme storms: State of knowledge. Bull. Amer. Meteor. Soc., doi: 10.1175/BAMS-D-11-00262.1 (Kunkel et al., 2013).*

a reference historical period. This too is predicted by numerical models as a consequence of warming (e.g., Kharin et al., 2007), and is supported by physical understanding: more water can be in the vapor phase at higher temperature (the Clausius—Clapeyron relation). Observations show increasing specific humidity on a global scale, and models predict with great consistency that this should occur as relative humidity changes remain small, a prediction that is also supported by dynamical understanding (Sherwood et al., 2010). While changes in global mean precipitation are controlled by radiative processes, extreme precipitation events are both largely unconstrained by such global budget considerations and limited by available moisture in the atmosphere, so that the amount of rain that falls in such events is more closely coupled to atmospheric water vapor content than is global (or perhaps even regional) mean precipitation (Trenberth, 1999; Allen and Ingram, 2002).

A simple hypothesis is that precipitation extremes should scale with surface temperature as specific humidity does, approximately following the Clausius—Clapeyron relation and increasing approximately 7% per degree Celsius. Storm dynamics can also change, however, in response to the increased convective heating and other environmental changes associated with warming, leading to changes greater or less than this naïve estimate. Climate models uniformly show precipitation extremes increasing in magnitude with warming, but at different rates in different models—especially in the tropics—suggesting that dynamical feedbacks are both significant and uncertain (O'Gorman and Schneider, 2009; Sugiyama et al., 2010).

1.1.3 Droughts

Droughts are multifaceted events, related to climate in multiple ways. Meteorological drought refers to deficits of precipitation (compared to historical climatology) over an extended period. It is inherently related to atmospheric circulation, and changes expected under warming are in principle quite uncertain. There is robust consensus among models that meteorological droughts should become more prevalent in some specific regions in a warming climate, however, and some dynamical understanding of those changes, leading to greater confidence in those regions. A highly visible example is the Middle East, where increasing drought is strongly projected by models and recent droughts have been found difficult to explain in terms of natural climate variability, leading to the inference of an anthropogenic role (Kelley et al., 2015). Generally, though, meteorological droughts are subject to large low-frequency internal climate variability. Very long and severe "megadroughts" are apparent in the historical and paleoclimate records, and difficult to explain in terms of radiative forcings (Coats et al., 2016). This makes it more difficult to make strong statements about the role of human influence on present droughts in many cases, even in other regions where that influence is expected to be strong in future such as southwestern North America.

Hydrologic drought, on the other hand—defined as a deficit in surface water reservoirs, including snow and soil moisture—is influenced not just by precipitation, but also by temperature, through temperature's control on surface evaporation. Thus arguments for a human influence on hydrologic drought are compelling: the absence of precipitation may in many cases be largely natural, but the surface water then becomes still more depleted than it would otherwise be for the same precipitation deficit, due to warming, evaporation increase, and snowpack loss as found to be the case for the recent drought in the US state of California (Diffenbaugh et al., 2015; Williams et al., 2015; Hartoonian, 2018, this volume).

1.1.4 Tropical Cyclones

The influence of global warming on tropical cyclones is a complex subject. Our understanding has evolved rapidly over the last dozen years or so, as summarized in recent reviews (Knutson et al., 2010; Walsh et al., 2015). The degree of agreement within the field is greatest for projections of the future, when we expect the influence of greenhouse gases to be larger than at present. The expectation is that warming will lead to "fewer but stronger" tropical cyclones. Increases in tropical cyclone intensity are expected with considerable confidence, supported by theoretical understanding via the theory of potential intensity as well as by numerical model results. The projection of decreasing tropical cyclone number, on the other hand, is primarily a result from global "high-resolution" (20−50 km horizontal grid spacing) models, and we lack a solid theoretical understanding of it. The most likely explanation involves increasing saturation deficit (difference between actual water vapor content and the maximum possible) in a warming atmosphere in which relative humidity changes are small (Emanuel, 2010; Camargo et al., 2014), but this is not yet a very well developed or scrutinized argument. It also applies primarily at the global scale; tropical cyclone frequency changes in individual basins are likely to be dominated by changes in regional climate and circulation, and are subject to all the uncertainties that go along with those. In particular, many models project a shift to an "El Nino-like" state in the Pacific, associated with a weakened Walker circulation, and this yields patterns in tropical cyclone activity typically associated with El Nino, with increases in the Pacific and decreases in the Atlantic (e.g., Vecchi and Soden, 2007). But this projection results from a cancellation between competing processes in the atmosphere and ocean, and it remains possible that it could be wrong (DiNezio et al., 2009).

Perhaps the most confident projections we can make about tropical cyclones involve their hydrological aspects. Tropical cyclone precipitation is almost certain to increase—essentially for the same reason as other precipitation extremes, namely increased water vapor in a warmer atmosphere. The risk of storm surge-driven coastal flooding is certain to increase in many regions as well, due to sea level rise. Even if statistics of storm frequency, intensity, and surge don't change, the higher baseline sea level increases the chance of a given water level's occurrence, relative to a fixed datum. It is theoretically possible that sufficiently large decreases in storm frequency could compensate, but that is highly unlikely, given the range of plausible estimates of sea level rise.

Attribution of changes in the recent historical record is more difficult due to the limitations of the data record and, especially, strong low-frequency natural variability. Many studies find statistically significant increases in intensity over the last few decades, but others find that these results depend to some extent on the data set and analysis method used (e.g., Kossin et al., 2013). Aerosol cooling has also likely compensated to a significant extent for the greenhouse warming, inhibiting tropical cyclone intensity increases (e.g., Sobel et al., 2016), though this compensation has already weakened, and will weaken further in future as greenhouse gas concentrations will almost certainly continue to increase while aerosol concentrations are likely to remain level or decrease. Thus we expect the projected greenhouse gas-driven increases in TC intensity to emerge more clearly in future as warming proceeds.

1.1.5 Severe Convection

Severe convective storms (thunderstorms producing tornadoes, large hail, or damaging straight-line winds) are relatively small and short-lived. Their expected behavior under climate change, as well as observed trends, has tended to be more uncertain than that of other extreme weather events (IPCC, 2012; Tippett et al., 2015). Although severe thunderstorms occur around the world, observational records in much of the world are limited and incomplete, and climate analysis of even US storm reports is difficult because of changing reporting practices (Verbout et al., 2006). To date, there is no strong evidence of trends in US storm reports due to climate change, despite signs of increased clustering and variability (Brooks et al., 2014; Tippett et al., 2016). Projection of future severe convective storm activity is challenging because the spatial resolution of numerical models commonly used in climate change projections is not adequate to resolve thunderstorms. The best current understanding of how severe convective storm activity will change in the future comes from looking at changes in conditions that are favorable for severe convective storms. Climate projections show that the number of days with favorable environments will increase in the United States, Australia, and Europe (Diffenbaugh et al., 2013; Allen et al., 2014; Púčik et al., 2017). These increases are primarily due to increases in convective available potential energy (CAPE), which is understood to increase with warmer surface temperatures and enhanced low-level moisture (Seeley and Romps, 2016). However, an important caveat when interpreting such findings is that favorable environments are not the same as storms' occurrence. Recent work suggests that increases in storm frequency and intensity in a warmer climate might be less than that indicated by changes in favorable environments (Trapp and Hoogewind, 2016).

While hail occurrence and size might be expected to increase with increasing CAPE, warming temperatures increase the height of the freezing level and are expected to cause smaller hailstones to melt before reaching the ground (Dessens et al., 2014; Mahoney et al., 2012). On the other hand, larger hail is less affected, so that hail frequency has been projected to decrease along with increases in maximum hail size (Brimelow et al., 2017). Decreases in the number of days with hail in China have been related to changes in the height of the freezing level (Li et al., 2016), and increases in hail kinetic energy have been observed in France (Berthet et al., 2011).

1.1.6　Human Impacts

This section has summarized our current understanding of the effect of global warming on extreme events of different types considering only changes in the meteorological events themselves. The risks to human populations from these events, however, are also strong functions of social and economic variables, including adaptation options themselves. The same meteorological event may have very different human impacts if it occurs in two different locations where infrastructure and societal vulnerabilities are different. These aspects are considered in detail from a range of perspectives in the rest of this book. The following section, on modeling tools, continues our focus here on the meteorological component of risk, though with some brief consideration of the vulnerability component.

1.2　CATASTROPHE MODELING AND RISK ASSESSMENT FOR ADAPTATION

The impacts of climate change are expected to occur, to a large extent, through changes in the frequency, intensity, or other characteristics of extreme events. Thus rational approaches to climate adaptation should include assessments of extreme event risk over the timescale being considered. In this section we consider the methodologies available for doing such risk assessments. Our description is by no means all-inclusive, but aims to give a broad sense of the types of tools available and their strengths and weaknesses, particularly those related to their representation of the extreme weather events themselves.

Risk is commonly defined as the product of hazard—the probability that a "natural" event of some given characteristics will occur—and the impacts to human society that would follow from such an event (fatalities, financial losses, health impacts, infrastructure damage, etc.), so that risk as a whole refers to the probabilities of those impacts. Some of the methods described below apply only to the meteorological hazard, while others—particularly catastrophe models—also include representations of some kinds of vulnerability, and thus can be said to model risk.

1.2.1　Historical Observations and Extreme Value Theory

For some purposes and some types of events, it is common to estimate hazard directly from historical observations. Records are often too short to characterize the extreme events that are of greatest interest by direct means, however—i.e., by simply counting how many of the events of interest have occurred over a given time period. The distribution of damage from natural disasters is generally found to be "fat-tailed," meaning that in a long-time average, a disproportionately large fraction of it comes from the rarest and largest magnitude events (Muir-Wood, 2016). Let us say that "rarest" here means, for specificity, annual probabilities of 1/100 or less. To estimate the "200-year" event, e.g.—the one with an annual probability of 1/200—reliably and directly from historical data, one needs a data record at least several times longer than 200 years. Good historical weather data are often not available for periods of even 100 years, however. One obviously cannot estimate the 200-year flood directly from, say, 50 years of data. Perhaps the most commonly used method to address this problem directly and empirically—i.e., without constructing explicit physical models—is extreme value theory (e.g., Coles, 2001; Embrechts et al., 1999).

Consider a random process at a point, represented by a single time series. If the events represented by the data satisfy some assumptions, then extreme value theory says that the statistics of the extremes—represented either by "block maxima," e.g., the set of annual maxima, or "peaks over threshold," the set of all values in the data exceeding some specified threshold value—can be approximated asymptotically by general distributions with only small sets of free parameters that can, in principle, be estimated even from a time series that is short compared to the return periods of interest. Knowing those parameters, the shape of the tail can be determined and the magnitude of an event of any given frequency can be estimated, including those more rare and extreme than are present in the data.

Due to low-frequency climate variability, however, meteorological variables cannot be assumed to be truly satisfy the assumptions of extreme value theory over periods of decades to centuries. In particular, observations from one epoch may not be representative, and return periods computed from records even several decades to a century long may not accurately reflect the present or future hazard (Jain and Lall, 2001), even without considering nonstationarity due to anthropogenic global warming (which only compounds this problem).

In addition, extreme value theory in its standard form assumes a time series which is populated at a regular interval by physically meaningful values (including zeros), and it considers only point processes. These assumptions are problematic for some events of interest. Continuous time series are available for variables like temperature or precipitation, but not for specific types of rare events, such as tropical cyclones, which are absent nearly all the time. In addition, many (really, all) real meteorological events have spatiotemporal structures which are not captured by standard extreme value theory, but which are important to the events' impacts. In the case of a tropical cyclone, extratropical wind storm, or major flood event, e.g., the damage-inducing extreme values of meteorological variables (wind, precipitation, storm surge-induced flooding, etc.) are often distributed over a wide area. The damage at different spatial locations within that footprint is thus highly correlated. That correlation will not be captured by independent applications of extreme value theory at nearby locations, but is terribly important to assessing the overall risk.[1] While it is possible to generalize extreme value theory to account for such correlation, it makes more sense in many applications to move to models which have explicit knowledge of the spatiotemporal structures of the events of interest.

1.2.2 Catastrophe Models

The approach used in the insurance industry—and to some extent in other arenas—involves "catastrophe models." These are used to estimate the risk of insured financial losses from extreme weather events (as well as other natural and, to some extent, human-made disasters).

Catastrophe models used in insurance have three components: a hazard module, which estimates the probability of an event with given physical characteristics in the atmosphere, ocean, or land surface; a vulnerability module, which contains data on the assets at risk (i.e., buildings or other physical structures) and "vulnerability curves" which predict the fraction of their value that would be destroyed if a given physical variable (e.g., wind speed or flood water depth) were to reach a given threshold; and a financial module, which estimates the insured loss that would result from such damage.

The strength of catastrophe models is that they are integrated tools that assess risk, rather than just hazard. The different components are ideally developed in tandem, and evaluated together. The desired risk is that of a loss of a given magnitude, and ideally data on losses from past events are available to calibrate the model.

The existing catastrophe models used in insurance have several limitations, however, that may limit their application to climate adaptation (though they are not particularly problematic for their traditional use in insurance, that being of course the reason they have developed as they have).

First, they are not open source, and the science going into them is not fully documented in the peer-reviewed literature, or even visible to their users. The models used most widely are commercial products provided by catastrophe modeling firms whose business models require some degree of proprietariness. Open-source models are only recently being developed (e.g., Bresch, 2014), and are not the standard in industry.

Second, catastrophe models developed in the insurance industry do not generally address impacts other than insured financial losses, such as loss of life or livelihood, or even financial losses in regions (such as much of the developing world) where insurance penetration is limited. Catastrophe models have begun to be adapted more widely for a range of problems in international development finance and disaster risk reduction (e.g., Cummins and Mahul, 2009; Joyette et al., 2015; Linnerooth-Bayer Hochrainer-Stigler, 2015; Souvignet et al., 2016; Bresch, 2016); these applications are for the most part similar to those in industry, focusing on financial loss, but considering a wider range of assets and in some cases considering risk transfer mechanisms different than traditional insurance. Some models are explicitly designed to consider the impacts of specific adaptation actions (e.g., Souvignet et al., 2016), though any model which includes the vulnerability of physical assets can in principle represent such actions through changes in the representation of those assets and their vulnerabilities.

Third, and of greatest interest here, the hazard components of standard catastrophe models are based closely on historical observations, and incorporate little if any of the physics that relates extreme weather events to the large-scale climate. This limits the models' utility for assessment of changing risks under climate change. Some important facets of climate change can be handled relatively straightforwardly—e.g., sea level rise can be incorporated into coastal flood risk calculations, as those are generally treated by physical models for storm surge and inland flooding (e.g., Hallegatte

---

1. The standard theory used to set insurance premiums, e.g., assumes that individual claims are uncorrelated, something that cannot be assumed about property damage claims in regions prone to natural hazards (Kunreuther et al., 2013).

et al., 2011, 2013). Other facets of climate change cannot be so easily incorporated. For example, capturing changes in the frequency and intensity of tropical cyclones or extreme precipitation events requires some degree of physical modeling, as one is attempting to predict the behavior of the climate system outside the regime in which the historical observations were taken. Hybrid statistical-dynamical approaches which generate large sets of synthetic events cheaply, in the spirit of traditional catastrophe models, but using enough physics to tackle the climate change problem, are beginning to be developed, pioneered by Emanuel (2006) with his statistical-dynamical model for TC hazard. Another TC hazard model of this type, to our knowledge the second (after Emanuel's), has been developed by Lee et al. (2018).

1.2.3 Dynamical Models

The primary tools for making predictive statements about climate change are dynamical models, also known as "climate models" or "earth system models." These are renderings of the deterministic physical (and, to some extent, chemical and even biological) laws governing the atmosphere, ocean, and other components of the climate system into discrete mathematical equations that can then be solved on computers. The models often represent the whole globe, as is appropriate for representing climate change. Regional models can be used if higher resolution is needed—this is known as "dynamical downscaling"—using lateral boundary conditions from another source, generally a global model. The choice of domain and the influence of the boundary data invoke distinct sets of potentially vexing problems, however, both technical and scientific. Global models with variable resolution, "zoomed in" over some region of interest, offer an intermediate approach which is now being explored with greater intensity (Harris and Lin, 2013; Duvel et al., 2017).

Dynamical models have the great advantage that they can, in principle, represent behavior outside the historical record, since the laws governing the system will presumably remain the same in the future as in the past even as greenhouse gas concentrations and other "forcings" change. They have the disadvantage that their representations of the climate inevitably contain "biases," or persistent errors, whose magnitudes may or may not be significant enough to compromise their utility in risk assessment. As a class, dynamical models generally represent some kinds of extreme events well and others poorly. The difference depends in large part on whether the physics of the event in question critically involves fluid-dynamical processes at scales smaller than the models can resolve (that resolved scale being, generally, hundreds of km, at best tens of km). Heat waves, e.g., being a result of large-scale weather systems, are represented quite well. Tropical cyclones and heavy rain events, on the other hand, are represented much more poorly. Some higher-resolution models exist which perform much better than earlier generations on these latter types of events (e.g., Shaevitz et al., 2014; Van der Wiel et al., 2016), but these require great computational power and are, at present, not accessible to many researchers. Some kinds of events—tornadoes come to mind—occur at such small scales that they remain inaccessible to any climate model, and can only be represented by some form of statistical downscaling which relates the events to larger-scale environmental conditions that the models can simulate (e.g., Tippett et al., 2012; Diffenbaugh et al., 2013).

1.3 DIFFERENT QUESTIONS

We can identify several types of inquiry regarding extreme events and their relation to climate. What question one asks might influence which scientific tools are most appropriate to the problem and how they might best be used.

1. Risk assessment. If we are doing risk assessment for the present, and our results need not be valid far into the future (say, because we will redo the analysis every year, as in the reinsurance industry where contracts are written annually) we arguably do not need to consider human influence explicitly at all for some kinds of events, and can base our analysis on the historical record, either directly or using some empirical or semiempirical method, such as extreme value theory or a catastrophe model, to extrapolate from it to generate probabilities for rare events. One could argue that even there, however, inasmuch as the present is already different from the past due to human influence, one should consider that change explicitly, and use a method that recognizes the difference between the earlier and later parts of the historical record. For some kinds of events (e.g., heat waves), where the change in the mean is large enough compared to the width of the historical range of natural variability, this could make some difference. For others (e.g., tropical cyclones), where natural variability is large and anthropogenic trends have not yet been conclusively detected, it may or may not.

2. Detection and Attribution (e.g., Stott et al., 2010). Detection of trends, almost by definition, involves observational data alone. Attribution, however compares the detected trends to what would have been expected with and without human influence, and that must involve a model of some kind. Attribution of trends in extreme events is generally done using some form of climate model, and thus has been done more (and with greater confidence) for the events which are most amenable to simulation in such models, such as heat waves. The issues around attribution of individual events (National Academy, 2016) are similar.

3. Projection of future change, given climate forcing scenarios. Here one is interested explicitly in the change between the future and the present, rather than the present and the past. Since there are no observations of the future, predictive models become yet more critical. For events which are not handled well by global dynamical models, one could imagine using a catastrophe model designed for such events, if it had appropriate sensitivity to the climate state. The climate state could be obtained from global models, using a downscaling approach. This has been done for tropical cyclones and storm surge (Lin et al., 2010, 2012) with the downscaling model of Emanuel (2006).

For climate adaptation, which activity is most needed, and therefore which tools? The term "adaptation" is often taken to imply an adjustment to a change in external circumstances. This would mean that climate adaptation studies should focus on measures needed in order to deal with the effects of anthropogenic climate change, and thus that the tools needed to do so should be those associated with climate projection (or perhaps with attribution, if we consider the anthropogenic change that has already occurred). This implicitly assumes, however, that the systems being studied were already "adapted" to the historical climate. When we consider the most damaging extreme events, this is often not the case. Because the most damaging events are also quite rare—having return periods longer than a human lifetime, and much longer than the time in office of a typical political leader—much physical infrastructure and human settlement is very vulnerable to such events, having been constructed under the assumption (conscious or not) that such extreme events will not occur (e.g., Sobel, 2014; Muir-Wood, 2016).

We advocate that climate adaptation studies take a pragmatic view in which anthropogenic global warming is considered as one factor influencing future extreme event risk, but is not assumed a priori to be the dominant one. Whether it is important or not, and what tools should be used for extreme event risk assessment in any given application, depend on the time horizon and the extreme event types being considered. If one is planning decades ahead and considering heat extremes, global warming is almost certainly important and dynamical models may be practical. If one is considering tropical cyclone risk for a single year (as in writing reinsurance contracts) then standard catastrophe models, without climate change, may be adequate. When deciding whether to rebuild a facility or community after a disaster, event attribution might be relevant; a finding that climate change played a significant role, implying continued increasing risk in the future, could inform a decision not to rebuild (in itself a possible form of adaptation). For other purposes, we need to make nuanced judgments, and the right tools may not yet exist in some cases.

On virtually any timescale, however, risk assessment for climate adaptation should be probabilistic, considering the full range of possibilities, and considering internal climate variability as well as human-induced climate change (e.g., Goddard et al., 2009). While accurate, high-resolution, robust decadal predictions, e.g., would certainly be very valuable and constitute a worthy target for research (e.g., Meehl et al., 2009; Shukla et al., 2009), adaptation planning need not wait for them. Where present vulnerabilities to historical levels of hazard are large, an obvious and conservative adaptation strategy is simply to build resilience to a wide range of plausible extreme events. As a starting point, this range can be inferred from the historical record, interpreted through methods such as extreme value theory or catastrophe modeling to correct for the shortness of that record (so that the full range of plausible extreme events may not have been sampled). Changes in hazard due to climate change should by all means be considered to the fullest extent allowed by current science, but science's perfection should not become the enemy of adaptation's good.

REFERENCES

Allen, J.T., Karoly, D.J., Walsh, K.J., 2014. Future Australian severe thunderstorm environments. Part II: The influence of a strongly warming climate on convective environments. J. Climate 27, 3848−3868. Available from: https://doi.org/10.1175/JCLI-D-13-00426.1.

Allen, M., Ingram, W., 2002. Constraints on future changes in climate and the hydrologic cycle. Nature 419, 224−232.

Arblaster, J.M., Lim, E.-P., Hendon, H.H., Trewin, B.C., Wheeler, M.C., Liu, G., et al., 2014. Understanding Australia's hottest September on record [in "Explaining Extremes of 2013 from a Climate Perspective"]. Bull. Amer. Meteor. Soc. 95 (9), S37−S41.

Berthet, C., Dessens, J., Sanchez, J.L., 2011. Regional and yearly variations of hail frequency and intensity in France. Atmos. Res. 100, 391−400. Available from: https://doi.org/10.1016/j.atmosres.2010.08.008.

Bresch, D.N., 2014. Climada − the open-source economics of climate adaptation (ECA) tool. https://github.com/davidnbresch/climada.

Bresch, D.N., 2016. Shaping climate resilient development: economics of climate adaptation. In: N. Salzmann et al. (eds.), *Climate Change Adaptation Strategies − An Upstream-downstream Perspective*, 241−254. Available from: https://doi.org/10.1007/978-3-319-40773-9_13.

Brimelow, J.C., Burrows, W.R., Hanesiak, J.M., 2017. The changing hail threat over North America in response to anthropogenic climate change. Nature Clim. Change 7, 516−522. Available from: https://doi.org/10.1038/nclimate3321.

Brooks, H.E., Carbin, G.W., Marsh, P.T., 2014. Increased variability of tornado occurrence in the United States. Science 346, 349−352.

Camargo, S.J., Tippett, M.K., Vecchi, G.A., Zhao, M., 2014. Testing the performance of tropical cyclone genesis indices in future climates using the HIRAM model. J. Climate 27, 9171−9196.

Coats, S., Smerdon, J.E., Cook, B.I., Seager, R., Cook, E.R., Anchukaitis, K.J., 2016. Internal ocean-atmosphere variability drives megadroughts in Western North America. Geophys. Res. Lett. 43, 9886−9894.

Coles, S., 2001. *An Introduction to Statistical Modeling of Extreme Values.* Springer, London.

Cummins, D., Mahul, O., 2009. Catastrophe Risk Financing in Developing Countries. World Bank, Washington, DC.

Dessens, J., Berthet, C., Sanchez, J., 2014. Change in hailstone size distributions with an increase in the melting level height. Atmos. Res. Available from: https://doi.org/10.1016/j.atmosres.2014.07.004.

Diffenbaugh, N.S., Scherer, M., Trapp, R.J., 2013. Robust increases in severe thunderstorm environments in response to greenhouse forcing. Proc. Natl. Acad. Sci. USA 110, 16361−16366. Available from: https://doi.org/10.1073/pnas.1307758110.

Diffenbaugh, N.S., Swain, D.L., Touma, D., 2015. Anthropogenic warming has increased drought risk in California. Proc. Natl. Acad. Sci. USA 112 (13), 3931−3936. Available from: https://doi.org/10.1073/pnas.1422385112.

DiNezio, P., Clement, A.C., Vecchi, G.A., Soden, B.J., Kirtman, B.P., Lee, S.-K., 2009. Climate response of the equatorial pacific to global warming. J. Climate 22, 4873−4892.

Duvel, J.-P., Camargo, S.J., Sobel, A.H., 2017. Role of the convection scheme in modeling initiation and intensification of tropical depressions over the North Atlantic. Mon. Wea. Rev. 145. Available from: https://doi.org/10.1175/MWR-D-16-0201.1.

Emanuel, K.A., 2006. Climate and tropical cyclone activity: a new model downscaling approach. J. Climate 19, 4797−4802. Available from: https://doi.org/10.1175/JCLI3908.1.

Emanuel, K.A., 2010. Tropical cyclone activity downscaled from NOAA-CIRES reanalysis, 1908−1958. J. Adv. Model. Earth Syst. 2 (1). Available from: https://doi.org/10.3894/JAMES.2010.2.1.

Embrechts, P., Resnick, S.I., Samorodnitsky, G., 1999. Extreme value theory as a risk management tool. N Am. Act. J. 3, 30−41.

Goddard, L., Baethgen, W., Kirtman, B., Meehl, G., 2009. The urgent need for improved climate models and predictions. EOS 90 (39), 343.

Groisman, P.Y., Knight, R.W., Easterling, D.R., Karl, T.R., Hegerl, G.C., Razuvaev, V.N., 2005. Trends in intense precipitation in the climate record. J. Climate 18, 1326−1350.

Hallegatte, S., et al., 2011. Assessing climate change impacts, sea level rise and storm surge risk in port cities: a case study on Copenhagen. Climatic Change 104, 113137.

Hallegatte, S., Green, C., Nicholls, R.J., Corfee-Morlot, J., 2013. Future flood losses in major coastal cities. Nat. Climate Change 3, 802−806. Available from: https://doi.org/10.1038/NCLIMATE1979.

Harris, L.M., Lin, S.-J., 2013. A two-way nested global-regional dynamical core on the cubed-sphere grid. Mon. Wea. Rev. 141, 283−306.

Hartoonian, 2018. California: it's complicated: a case study of the San Joaquin Valley. This volume.

Held, I.M., Soden, B.J., 2006. Robust responses of the hydrological cycle to global warming. J. Climate 19 (21), 5686−5699. Available from: https://doi.org/10.1175/Jcli3990.1.

IPCC., 2012. Managing the risks of extreme events and disasters to advance climate change adaptation. In: Field, C.B., Barros, V., Stocker, T.F., Qin, D., Dokken, D.J., Ebi, K.L., et al., (Eds.), A Special Report of Working Groups I and II of the Intergovernmental Panel on Climate Change. Cambridge University Press, Cambridge, UK, and New York, NY, USA.

Jain, S., Lall, U., 2001. Floods in a changing climate: does the past represent the future? Water Resour. Res. 37 (12), 3193−3205. Available from: https://doi.org/10.1029/2001wr000495.

Joyette, A.R.T., Nurse, L.A., Pulwarty, R.S., 2015. Disaster risk insurance and catastrophe models in risk-prone small Caribbean islands. Disasters 39, 467−492. Available from: https://doi.org/10.1111/disa.12118.

Kelley, C.P., Mohtadi, S., Cane, M.A., Seager, R., Kushnir, Y., 2015. Climate change in the Fertile Crescent and implications of the recent Syrian drought. Proc. Nat. Acad. Sci 112, 3241−3246.

Kharin, V.V., Zwiers, F.W., Zhang, X., Hegerl, G.C., 2007. Changes in temperature and precipitation extremes in the IPCC ensemble of global coupled model simulations. J. Climate 20, 1419−1444.

Knutson, T.R., McBride, J.L., Chan, J., Emanuel, K., Holland, G., Landsea, C., et al., 2010. Tropical cyclones and climate change. Nat. Geosci. 3, 157−163. Available from: https://doi.org/10.1038/ngeo779.

Kossin, J.P., Olander, T.L., Knapp, K.R., 2013. Trend analysis with a new global record of tropical cyclone intensity. J. Climate 26, 9960−9976.

Kunkel, K.E., coauthors, 2013. Monitoring and understanding trends in extreme storms: state of knowledge. Bull. Amer. Meteor. Soc. Available from: https://doi.org/10.1175/BAMS-D-11-00262.1.

Kunreuther, H.C., Pauly, M.V., McMorrow, S., 2013. Insurance and Behavioral Economics: Improving Decisions in the Most Misunderstood Industry. Cambridge University Press, New York.

Lee, C., Tippett, M.K., Sobel, A.H., Camargo, S.J., 2018. An environmentally forced tropical cyclone hazard model. J. Adv. Model. Earth Syst. 10. Available from: https://doi.org/10.1002/2017MS001186.

Li, M., Zhang, Q., Zhang, F., 2016. Hail day frequency trends and associated atmospheric circulation patterns over China during 1960−2012. J. Climate 29, 7027−7044. Available from: https://doi.org/10.1175/JCLI-D-15-0500.1.

Lin, N., Emanuel, K., Oppenheimer, M., Vanmarcke, E., 2012. Physically based assessment of hurricane surge threat under climate change. Nat. Climate Change 2 (6), 462−467.

Lin, N., Emanuel, K.A., Smith, J.A., Vanmarcke, E., 2010. Risk assessment of hurricane storm surge for New York City. J. Geophys. Res. 115, D18121.

Linnerooth-Bayer, J., Hochrainer-Stigler, S., 2015. Financial instruments for disaster risk management and climate change adaptation. Climatic Change. 133, 85−100.

Mahoney, K., Alexander, M.A., Thompson, G., Barsugli, J.J., Scott, J.D., 2012. Changes in hail and flood risk in high-resolution simulations over Colorado's mountains. Nat. Climate Change 2, 125−131. Available from: https://doi.org/10.1038/nclimate1344.

Meehl, G.A., coauthors, 2009. Decadal prediction: can it be skillful? Bull. Amer. Meteor. Soc. 90, 1467−1485.

Muir-Wood, R., 2016. The Cure for Catastrophe: How We Can Stop Manufacturing Natural Disasters. Basic Books, New York.

National Academies of Sciences, Engineering, and Medicine, 2016. Attribution of Extreme Weather Events in the Context of Climate Change. The National Academies Press, Washington, DC. Available from: http://dx.doi.org/10.17226/21852.

O'Gorman, P.A., Schneider, T., 2009. The physical basis for increases in precipitation extremes in simulations of 21st-century climate change. Proc. Natl. Acad. Sci. USA 106, 14773—14777.

Otto, F.E.L., Massey, N., van Oldenborgh, G.J., Jones, R.G., Allen, M.R., 2012. Reconciling two approaches to attribution of the 2010 Russian heat wave. Geophys. Res. Lett. 39. Available from: https://doi.org/10.1029/2011gl050422.

Púčik, T., coauthors, 2017. Future changes in European severe convection environments in a regional climate model ensemble. J. Climate 30, 6771—6794. Available from: https://doi.org/10.1175/JCLI-D-16-0777.1.

Seeley, J.T., Romps, D.M., 2016. Why does tropical convective available potential energy (CAPE) increase with warming? Geophys. Res. Lett. 42, 10429—10437.

Shaevitz, D.A., coauthors, 2014. Characteristics of tropical cyclones in high-resolution models in the present climate. J. Adv. Model. Earth Sys. Available from: https://doi.org/10.1002/2014MS000372.

Sherwood, S.C., Roca, R., Weckwerth, T.M., Andronova, N.G., 2010. Tropospheric water vapor, convection, and climate. Rev. Geophys. 48, RG2001. Available from: https://doi.org/10.1029/2009RG000301.

Shukla, J., Hagedorn, R., Hoskins, B., Kinter, J., Marotzke, J., Miller, M., et al., 2009. Revolution in climate prediction is both necessary and possible. Bull. Amer. Meteor. Soc. 90, 175—178.

Sobel, A.H., 2014. Storm Surge: Hurricane Sandy, Our Changing Climate, and Extreme Weather of the Past and Future. Harper-Collins, New York.

Sobel, A.H., Camargo, S.J., Hall, T., Lee, C.-Y., Tippett, M.K., Wing, A.A., 2016. Human influence on tropical cyclone intensity. Science 353, 242—246. Available from: https://doi.org/10.1126/science.aaf6574.

Souvignet, M., Wieneke, F., Mueller, L., Bresch, D.N., 2016. Economics of Climate Adaptation (ECA): Guidebook for Practitioners. KfW Group, KfW Development Bank, Frankfurt am Main.

Stott, P.A., Gillett, N.P., Hegerl, G.C., Karoly, D.J., Stone, D.A., Zhang, X., et al., 2010. Detection and attribution of climate change: a regional perspective. WIREs Clim. Chg. 1, 192—211. Available from: https://doi.org/10.1002/wcc.34.

Sugiyama, M., Shiogama, H., Emori, S., 2010. Precipitation extreme changes exceeding moisture content increases in MIROC and IPCC climate models. Proc. Nat. Acad. Sci. 107, 571—575. Available from: https://doi.org/10.1073/pnas.0903186107.

Tippett, M.K., Sobel, A.H., Camargo, S.J., 2012. Association of U.S. tornado occurrence with monthly environmental parameters. Geophys. Res. Lett. 39. Available from: https://doi.org/10.1029/2011GL050368.

Tippett, M.K., Allen, J.T., Gensini, V.A., Brooks, H.E., 2015. Climate and hazardous convective weather. Curr. Clim. Change Rep. 1, 60—73. Available from: https://doi.org/10.1007/s40641-015-0006-6.

Tippett, M.K., Lepore, C., Cohen, J.E., 2016. More tornadoes in the most extreme U.S. tornado outbreaks. Science 354, 1419—1423. Available from: https://doi.org/10.1126/science.aah7393.

Trapp, R.J., Hoogewind, K.A., 2016. The realization of extreme tornadic storm events under future anthropogenic climate change. J. Climate 29, 5251—5265. Available from: https://doi.org/10.1175/JCLI-D-15-0623.1.

Trenberth, K.E., 1999. Conceptual framework for changes of extremes of the hydrological cycle with climate change. Clim. Change 42, 327—339.

Trenberth, K.E., Fasullo, J.T., Shepherd, T.G., 2015. Attribution of climate extreme events. Nat. Climate Change 5 (8), 725—730. Available from: https://doi.org/10.1038/Nclimate2657.

Van der Wiel, K., Kapnick, S.B., Vecchi, G.A., Cooke, W.F., Delworth, T.L., Jia, L., et al., 2016. The resolution dependence of contiguous U.S. precipitation extremes in response to CO2 forcing. J. Climate 29, 7991—8012. Available from: https://doi.org/10.1175/JCLI-D-16-0307.1.

Vecchi, G.A., Soden, B.J., 2007. Increased tropical Atlantic wind shear in model projections of global warming. Geophys. Res. Lett. 34, L08702. Available from: https://doi.org/10.1029/2006GL028905.

Verbout, S.M., Brooks, H.E., Leslie, L.M., Schultz, D.M., 2006. Evolution of the U.S. tornado database: 1954-2003. Wea. Forecasting 21, 86—93.

Walsh, J., Wuebbles, D., Hayhoe, K., Kossin, J., Kunkel, K., Stephens, G., et al., 2014. In: Melillo, J.M., Richmond, Terese (T.C.), Yohe, G.W. (Eds.), Our Changing Climate. Climate Change Impacts in the United States: The Third National Climate Assessment. U.S. Global Change Research Program, pp. 19—67. Ch. 2 . Available from: http://dx.doi.org/10.7930/J0KW5CXT.

Walsh, K.J.E., McBride, J.L., Klotzbach, P.J., Balachandran, S., Camargo, S.J., Holland, G., et al., 2015. Tropical cyclones and climate change. WIREs Climate Change. Available from: https://doi.org/10.1002/wcc.371.

Williams, A.P., Seager, R., Abatzoglou, J.T., Cook, B.I., Smerdon, J.E., Cook, E.R., 2015. Contribution of anthropogenic warming to California drought during 2012—2014. Geophys. Res. Lett. 42 (16), 6819—6828. Available from: https://doi.org/10.1002/2015GL064924.

FURTHER READING

Michel-Kerjan, E., Hochrainer-Stigler, S., Kunreuther, H., Linnerooth-Bayer, J., Mechler, R., Muir-Wood, R., et al., 2013. Catastrophe risk models for evaluating disaster risk reduction investments in developing countries. Risk Analysis 33, 984—999. Available from: https://doi.org/10.1111/j.1539-6924.2012.01928.x.

Tippett, M.K., Camargo, S.J., Sobel, A.H., 2011. A poisson regression index for tropical cyclone genesis and the role of large-scale vorticity in genesis. J. Climate 24 (9), 2335—2357. Available from: https://doi.org/10.1175/2010JCLI3811.1.

Chapter 2

Adapting to Sea-Level Rise

Robert J. Nicholls
University of Southampton, Southampton, United Kingdom

Chapter Outline

2.1 INTRODUCTION

Sea-level rise (SLR) has been recognized as a major threat to low-lying coastal areas since the 1980s (e.g., Barth and Titus, 1984; Milliman et al., 1989; Tsyban et al., 1990). There is an ever growing literature demonstrating the large potential impacts of SLR. Hence, interest in coastal adaptation is also increasing (Linham and Nicholls, 2010; Moser et al., 2012; Wong et al., 2014). Although SLR only directly impacts coastal areas, these are the most densely-populated and economically active land areas on Earth. More than 600 million people live below 10 m elevation in the Low Elevation Coastal Zone (McGranahan et al., 2007; Neumann et al., 2015), and coastal urban areas are expanding rapidly (Hanson et al., 2011). These people and assets are exposed to multiple meteorological and geophysical hazards, including storms and storm-induced flooding (Kron, 2013). Many low-lying coastal areas already depend on various flood risk adaptation strategies, be it natural and/or artificial flood defences and drainage or flood resilient construction methods. Recent major events such as Hurricane Katrina, 2005 (New Orleans and environs, USA), Cyclone Nagris, 2008 (Irrawaddy delta, Myanmar), Superstorm Sandy, 2012 (New York and environs), Typhoon Haiyan, 2013 (the Philippines), or the 2017 Atlantic hurricane season in the Caribbean and US Gulf Coasts demonstrate the present vulnerability of low-lying coastal areas to floods during storms. SLR and potentially more intense storms have the potential to exacerbate these risks significantly unless we adapt (Wong et al., 2014). As well as the human environment, coastal areas also support important and productive ecosystems that are sensitive to SLR (Crossland et al., 2005).

This chapter focuses on adaptation to SLR. This can be defined as reducing the impacts of SLR via behavioral changes. This includes a range of actions from individuals/households to collective coastal management policy, such as upgraded defence systems, warning systems and land management approaches. SLR is a pervasive long-term problem that will continue for centuries. Hence, the chapter focuses on collective actions as only at this level of response can the long-term challenge of SLR be met. Coastal adaptation to SLR has been considered for the last 20−30 years (Barth and Titus, 1984; Dronkers et al., 1990; Bijlsma et al., 1996), building on the large experience of coastal adaptation to extremes and other stresses such as coastal subsidence. Despite this experience, uncertainties about the success or failure of adaptation to SLR remains large, contributing significant uncertainty to the overall consequences of SLR for society (Nicholls et al., 2014a; Nicholls, 2014). Hence, the chapter reviews and evaluates current efforts in coastal adaptation to SLR.

The chapter is structured as follows. First the coast as a system is elaborated. This provides an appropriate framework to analyze coasts, SLR and adaptation. Second, climate change and SLR are considered in more detail, including

Resilience. DOI: https://doi.org/10.1016/B978-0-12-811891-7.00002-5

the important distinction between global-mean and relative SLR. Then the impacts of SLR are briefly considered from both a biophysical and a socioeconomic perspective, including drawing on experience from subsiding coasts. This is followed by a more detailed consideration of adaptation. This demonstrates the complexity of adaptation and the multiple factors that need to be considered. A discussion/conclusion ends the chapter, including consideration of success and failure and how can best practice be defined.

2.2 COASTAL SYSTEMS

Global SLR and the need to adapt does not happen in isolation: coasts are changing significantly due to numerous other factors such as urbanization and changing water/sediment inputs due to river regulation and watershed land use and coastal land cover change (Crossland et al., 2005; Valiela, 2006; Bianchi, 2016). Table 2.1 summarizes the key trends of which SLR is one important and pervasive factor. This requires a systems approach to analyze the full range of interacting drivers, including feedbacks of which adaptation is an example. Fig. 2.1 presents a simplified systems model of the impacts of SLR and other drivers on the coastal zone. Such a conceptual model highlights the varying implicit and explicit assumptions, simplifications, and limitations of any assessment of coastal impacts. The overall coastal system is characterized as interacting natural and socioeconomic systems. Both systems can be characterized by key system properties such as their exposure, sensitivity and adaptive capacity, both due to SLR, related climate change, and other nonclimate stresses (see Klein and Nicholls, 1999). These conceptual models have been considered and developed by the Intergovernmental Panel on Climate Change (IPCC) assessments.

TABLE 2.1 Key Coastal Trends at the Global Scale

- Population
  - Growing coastal population (double global trends)
  - Urbanizing coastal zone (new residents are urban)
  - Increasing tourism, recreation and retirement
- Subsiding, densely-populated deltas, especially in urban areas
- Globalization of trade and international shipping routes
- Increasingly costly coastal disasters
- Climate change and sea-level rise
- A reactive approach to adaptation
- Degrading coastal habitats and declining ecosystem services

Source: From Dawson, R.J., Nicholls, R.J., Day, S.A., 2015. The challenge for coastal management during the third millenium. In R.J. Nicholls, R.J. Dawson, S.A. Day (Eds.), Broad Scale Coastal Simulation: New Techniques to Understand and Manage Shorelines in the Third Millennium. Springer, Dordrecht, NL, pp. 1—78 (Dawson et al., 2015).

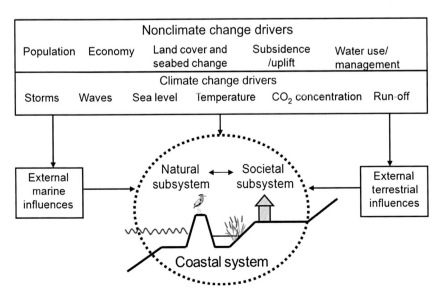

FIGURE 2.1 The Coastal System comprises interacting natural and human subsystems, and external terrestrial and marine influences. A range of nonclimate and climate drivers can act directly or indirectly on the Coastal System, including sea-level rise. *Adapted by Nicholls (2014) from Figure 6.1 in IPCC 4th Assessment Report (2007). Nicholls, R.J., Stive, M.J.F., Tol, R.S.J., 2014a. Coping with coastal change. In: Masselink, G., Gehrels, R. (Eds.), Coastal Environments and GlobalChange. Wiley, Chichester.*

A range of drivers may influence the coastal zone, either directly or via terrestrial or oceanic influences (Fig. 2.1). SLR is one aspect of climate change and these drivers interact with nonclimate stresses, often exacerbating impacts (see Table 2.2). The socioeconomic system is not passive and influences the natural system by deliberate changes such as construction of sea dykes, destruction of wetlands, and building of port and harbor works, as well as unintended changes such as reductions of sediment and water fluxes due to the building of dams. Hence, the socioeconomic system

TABLE 2.2 The Main Natural System Effects of Relative Sea-Level Rise and Examples of Adaptation Options. Potential Interacting Factors Which Could Offset or Exacerbate These Impacts are Also Shown. Some Interacting Factors (e.g., Sediment Supply) Appear Twice as They can be Influenced both by Climate and Nonclimate Factors. Adaptation Options are Coded: At, Attack; Ph and Ps, Hard or Soft Protection, respectively; Ac, Accommodation; R—Retreat

Natural System Effect		Possible Interacting Factors		Possible Adaptation Options
		Climate	Nonclimate	
1. Inundation/ flooding	a. Surge (flooding from the sea)	Wave/storm climate, Erosion, Sediment supply.	Sediment supply, Flood management, Erosion.	Land claim [At] Dikes/surge barriers/closure dams [Ph], Nourishment, including dune construction [Ps], Ecosystem-based barriers (e.g., mangrove buffers) [Ps], Building codes/flood-proof buildings [Ac], Land use planning/hazard mapping [Ac/R], Planned migration [R].
	b. Backwater effect (flooding from rivers)	Run-off.	Catchment management and land use.	
2. Wetland loss (and change)		CO_2 fertilization, Sediment supply, Migration space.	Sediment supply, Migration space, Land claim (i.e., direct destruction).	Gabions/breakwaters [Ph], Nourishment/sediment management [Ps], Wetland planting [Ps], Land use planning [Ac/R], Managed realignment/ forbid hard defences [R].
3. Erosion (of "soft" morphology)		Sediment supply, Wave/storm climate.	Sediment supply.	Land claim [At] Coastal defences/seawalls [Ph], Ecosystem-based barriers (e.g., mangroves) [Ps], Nourishment [Ps], Building setbacks/rolling easements [R].
4. Saltwater Intrusion	a. Surface Waters	Run-off.	Catchment management (water extraction/ diversion), Land use.	Saltwater intrusion barriers [P], Desalination [Ac], Move water abstraction upstream [R].
	b. Groundwater	Rainfall.	Land use, Aquifer utilization.	Impermeable groundwater barriers [P] Freshwater injection [P], Desalination [Ac], Change water abstraction [Ac/R].
5. Impeded drainage/ rising water tables		Rainfall, Run-off.	Land use, Aquifer utilization, Catchment management.	Drainage systems/polders [Ph], Change land use/crop type [Ac], Land use planning/hazard delineation [Ac/R].

Source: Adapted from Nicholls, R.J., 2010. Impacts of and responses to sea-level rise. In: Church, J.A., Woodworth, P.L. Aarup T., Wilson, S. (Eds.), Understanding Sea-Level Rise and Variability. Wiley-Blackwell, 2010, pp. 17—51; Nicholls, R.J., 2014. Adapting to sea level rise. In J.T. Ellis, D.J. Sherman (Eds.), Coastal and Marine Hazards, Risks and Disasters. London, GB: Elsevier, pp. 243—270.

constrains the natural system, and vice versa. This raises the prospect of the coast as a coevolving system where the natural system shapes the socioeconomic system and vice versa, with adaptation being an important feedback. It raises a new way of thinking about the future of coasts, which is shedding new insights on coastal evolution and the role of adaptation (Lazarus et al., 2016; Welch et al., 2017).

2.3 GLOBAL-MEAN AND RELATIVE SEA-LEVEL CHANGE

Climate-induced SLR is mainly due to (1) thermal expansion of seawater as it warms, and (2) the melting/destabilization of land-based ice, comprising components from (a) small glaciers, (b) the Greenland ice sheet, and (c) the West Antarctic ice sheet (Church et al., 2010, 2013). The global SLR trend was 1.1 ± 0.3 mm/year from 1900 to 1990 and 3.1 ± 1.4 mm/year from 1993 to 2012 showing a significant recent acceleration (Dangendorf et al., 2017). There is a large uncertainty about future SLR, and hence adaptation needs which hinders action. Over the 21st century a rise of 1 m or more is plausible if the major ice sheets make a large positive contribution to sea level with larger changes beyond 2100 (Church et al., 2013). Future SLR depends in part on future greenhouse gas emissions. Even if we achieve the Paris Agreement targets (United Nations, 2015), there will still be a significant rise due to the strong inertia of SLR. For unmitigated emissions versus 2.0°C temperature stabilization, Nicholls et al. (2017) made a median estimate of global SLR of 0.72 and 0.49 m by 2100, and 3.65 and 1.17 m by 2300 (relative to 1985−2005 mean), respectively. Hence, while aggressive climate mitigation significantly reduces the rise, a growing need for adaptation remains. As such SLR adaptation seems essential under all plausible futures, but the quantitative need is uncertain (cf. Nicholls et al., 2007; Wong et al., 2014).

When analyzing SLR impacts and adaptation responses, it is fundamental that impacts are a product of *relative* (or local) RSLR rather than global changes alone (Nicholls et al., 2014b). Relative sea-level change considers the sum of global, regional, and local components of sea-level change: the underlying drivers of these components are (1) climate change, as already discussed, and changing ocean dynamics, and (2) nonclimate land level change (i.e., uplift/subsidence) processes such as tectonics, glacial-isostatic adjustment, and natural and human-induced subsidence. Gravitational effects due to mass redistribution of melting ice also need to be considered. Hence, relative sea-level change is only partly a response to climate change and varies in space (Fig. 2.2). Many populated deltaic areas and alluvial plains are experiencing enhanced subsidence (Ericson et al., 2006; Syvitski et al., 2009; Chaussard et al., 2013) and RSLR exceeds the global rise, as at Grand Isle in the Mississippi delta (Fig. 2.2). Most dramatically, subsidence can be enhanced by drainage and withdrawal of groundwater in susceptible soils. Multimeter RSLR has been observed in a number of coastal cities built on deltas and alluvial plains over the last 100 years due to this cause, such as Tokyo, Osaka, Shanghai, and Bangkok (Figs. 2.2 and 2.3). Human-induced subsidence can be mitigated by stopping shallow subsurface fluid withdrawals and managing water levels, but natural "background" rates of subsidence typical of deltas (1−5 mm/year and maybe more) will continue and RSLR will still exceed global trends in these areas. The four cities mentioned above have all successively implemented subsidence mitigation policies (Kaneko and Toyota, 2011), combined with the provision of improved flood defence and pumped drainage systems to avoid submergence and/or frequent flooding. However, other cities such as Jakarta and Metro Manila are still subsiding, suggesting that the

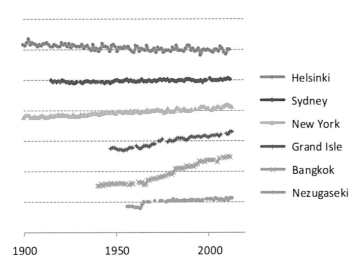

FIGURE 2.2 Selected relative sea-level observations since 1900, illustrating different trends (offset for display purposes with meter divisions). Helsinki shows a falling trend (−2.0 mm/year) as the land is rising, Sydney shows a gradual rise (0.9 mm/year), New York is subsiding slowly (3.1 mm/year), Grand Isle is on a subsiding delta (9.1 mm/ycar), Bangkok (Station: Fort Phrachula Chomklao) is also on a delta and includes the additional effects of human-induced subsidence (18.9 mm/year from 1962 to 2012), and Nezugaseki shows an abrupt 0.15−0.20 mm rise due to an earthquake. *Data from Holgate, S.J., Matthews, A., Woodworth, P.L., Rickards, L.J., Tamisiea, M.E., Bradshaw, E., et al., 2013. New data systems and products at the permanent service for mean sea level. J. Coastal Res. 29, 493−504. doi:10.2112/ JCOASTRES-D-12-00175.1 (Holgate et al., 2013); Permanent Service for Mean Sea Level (PSMSL), 2014. Tide Gauge Data. Retrieved 30 Jun 2014 from http://www.psmsl.org/data/obtaining/ (PSMSL, 2014).*

Helsinki

Sydney

New York

Grand Isle

Bangkok

Nezugaseki

1900   1950   2000

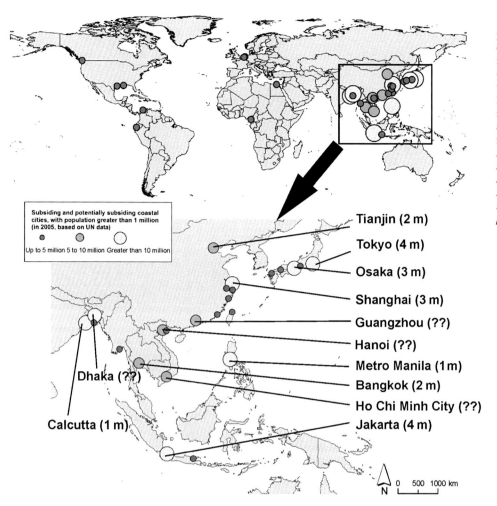

FIGURE 2.3 Subsiding and potentially subsiding coastal cities due to human influence. The maximum observed subsidence (in meters) is shown for cities with populations exceeding 5 million people, where known. The maximum subsidence is reported as data on average subsidence is not available. *Reproduced from Nicholls, R.J., 2014. Adapting to sea level rise. In J.T. Ellis, D.J. Sherman (Eds.), Coastal and Marine Hazards, Risks and Disasters. London, GB: Elsevier, pp. 243–270.*

Subsiding and potentially subsiding coastal cities, with population greater than 1 million (in 2005, based on UN data)

Up to 5 million   5 to 10 million   Greater than 10 million

Tianjin (2 m)
Tokyo (4 m)
Osaka (3 m)
Shanghai (3 m)
Guangzhou (??)
Hanoi (??)
Metro Manila (1 m)
Bangkok (2 m)
Ho Chi Minh City (??)
Jakarta (4 m)
Dhaka (??)
Calcutta (1 m)

0   500   1000 km
N

problem of enhanced subsidence are likely to be widely repeated in other susceptible coastal cities through the 21st century. It is important to emphasize that only some cities are susceptible to this problem: of the 136 large coastal cities considered by Hallegatte et al. (2013), 32 (or 24%) have an appropriate geological setting to experience enhanced subsidence (Fig. 2.3). Note the concentration of large cities in south, south-east, or east Asia, giving this issue a strong regional dimension.

More appreciation of the importance of subsidence is urgently needed to promote appropriate responses and evidence-based adaptation needs to distinguish climate-induced and subsidence-induced RSLR as a first step to finding solutions. New measurement systems will permit analysis and quantification of subsidence (Chatterjee et al., 2006; Teatini et al., 2005), supporting such analysis.

2.4   IMPACTS OF SEA-LEVEL RISE

RSLR causes more effects than simple submergence (the "bath-tub" effect); the five main effects are summarized in Table 2.2. Flooding/submergence, ecosystem change, and erosion have received significantly more assessment than salinization and rising water tables. Along with rising sea levels, there are changes to all processes that operate around the coast. The immediate effect is submergence and increased flooding of coastal land, as well as saltwater intrusion into surface waters. Longer term effects also occur as the coast adjusts to the new environmental conditions, including ecosystem change and loss, erosion of beaches and soft cliffs, and saltwater intrusion into groundwater. These lagged changes interact with the immediate effects of SLR and generally exacerbate them. For instance, erosion of saltmarshes, mangroves, sand dunes, and coral reefs degrade or remove natural protection from waves and storms and increase the likelihood of coastal flooding.

TABLE 2.3 Summary of Sea-Level Rise Impacts on Socioeconomic Sectors in Coastal Zones. These Impacts Are Overwhelmingly Negative

Coastal Socioeconomic Sector	Sea-Level Rise Natural System Effect (Table 2.2)				
	Inundation/ Flooding	Wetland Loss	Erosion	Saltwater Intrusion	Impeded Drainage
Freshwater resources	X	x	−	X	X
Agriculture and forestry	X	x	−	X	X
Fisheries and Aquaculture	X	X	x	X	−
Health	X	X	−	X	x
Recreation and tourism	X	X	X	−	−
Biodiversity	X	X	X	X	X
Settlements/ infrastructure	X	X	X	X	X

X, strong; x, weak; −, negligible or not established.
Source: Reproduced from Nicholls, R.J., 2010. Impacts of and responses to sea-level rise. In: Church, J.A., Woodworth, P.L. Aarup T., Wilson, S. (Eds.), Understanding Sea-Level Rise and Variability. Wiley-Blackwell, 2010, pp. 17−51.

A rise in mean sea level also raises extreme water levels. Changes in storm characteristics could also influence extreme water levels. For example, an increase in the intensity of tropical cyclones will generally raise extreme water levels in the areas affected (Church et al., 2013). Extratropical storms may also intensify in some regions, although this effect is uncertain. An improved understanding of these changes is an important research topic to support impact and adaptation assessments (e.g., Wahl et al., 2017).

Changes in natural systems resulting from SLR have many important direct socioeconomic impacts on a range of sectors, with these impacts being overwhelmingly negative (Table 2.3). For instance, flooding can damage coastal infrastructure, ports and industry, the built environment, and agricultural areas. In the worst case, coastal flooding can lead to significant mortality. There were 1200−1800 deaths due to Hurricane Katrina (USA) in 2005, about 138,000 deaths due Cyclone Nargis (Myanmar) in 2008, and at least 6000 deaths due to Typhoon Haiyan (the Philippines) in 2013. Erosion can lead to the loss of beachfront/cliff-top buildings and other infrastructure, and have adverse consequences for sectors such as tourism and recreation. In addition to these direct impacts, there are potential indirect impacts such as mental health problems triggered by floods, or economic effects that cascade through the whole economy (Nicholls and Kebede, 2012; Hallegatte, 2012). These indirect impacts are less understood and appreciated, but can be significant. Thus, SLR has the potential to trigger a cascade of direct and indirect human impacts.

Potential interactions of nonclimate drivers with SLR need to considered. They are indicated in Table 2.2 (column entitled "Potential Interacting Factors"). For instance, a coast with a positive sediment budget may not erode given SLR and vice versa. Hence, coastal change ideally requires an integrated assessment approach to analyze the full range of interacting drivers, including the feedback of policy interventions (i.e., adaptation).

2.5 RECENT IMPACTS OF SEA-LEVEL RISE

Since 1900, global sea level rose in round terms 10−20 cm following the analysis of Dangendorf et al. (2017). While this change may seem small, it has had many significant coastal effects, such as reducing the return periods of extreme sea levels (Zhang et al., 2000; Menendez and Woodworth, 2010), development of "blue sky" chronic flooding where floods become a regular rather than a rare occurrence (Sweet et al., 2014), and promoting an erosive tendency for coasts (Bird, 1985, 2000). However, linking SLR quantitatively to impacts is difficult due to the multiple drivers of change already discussed (see Nicholls et al., 2015a). Good data on rising sea levels has only been measured in a few locations, and growing coastal populations and infrastructure have significantly increased the exposure available to damage. At the same time, natural defences have been widely degraded by urban expansion and erosion trends. In addition, adaptation has occurred and, e.g., artificial flood defences have often been upgraded substantially, reducing the incidence of floods despite higher sea levels (e.g., Ruocco et al., 2011). Most of these defence upgrades coincide with expanding populations and wealth in the coastal flood plain and growing risk aversion. Hence, RSLR may not have been explicitly considered in design. Equally, impacts can be promoted by processes other than SLR (Table 2.2), including subsidence. As another example, widespread human reduction in sediment supply to the coast must contribute to the observed erosional changes around the world and this probably dominates erosion in many locations (Bird, 1985; Syvitski et al., 2009). Hence, while global SLR is a pervasive process, other processes obscure its link to impacts, except in some special cases; most coastal change in the 20th century was a response to multiple drivers of change.

There have certainly been impacts from RSLR resulting from subsidence (Nicholls, 2010). Notable sites include increased floods in the iconic city of Venice, which will shortly be protected by the MOSES storm surge barriers and the Mississippi delta where thousands of square kilometers of intertidal coastal marshes and adjacent lands were converted to open water in the last 50 years. There are also significant impacts of RSLR associated with subsiding coastal cities (e.g., Fig. 2.3), in terms of increased waterlogging, flooding and submergence, and the resulting need for adaptation/management responses.

These empirical observations also provide lessons for adaptation. Subsiding areas with a low population density were often abandoned, such as around Galveston Bay, Texas and south of Bangkok, Thailand. However, most of the major developed areas that were impacted by RSLR have been defended and continue to experience population and economic growth (Nicholls, 2010). This includes areas where the change in RSLR was rapid—several meters over several decades. However, there are exceptions such as New Orleans where population declines seem linked to defence failure (Grossi and Muir Wood, 2006). The city population peaked in 1965 at more than 625,000 immediately before the Hurricane Betsy floods, and was 500,000 before Katrina in 2005. Subsequently, the population has yet to recover to pre-Katrina levels, although US$15 billion has been invested to significantly upgrade defences (completed 2011). The future of New Orleans in terms of flood occurrence, socioeconomic changes and adaptation will be instructive to monitor.

Observations since 1900 reinforce the importance of understanding the impacts of SLR in the context of multiple drivers of change; this will remain the case under more rapid rises in sea level. RSLR due to human-induced subsidence is of particular interest, but this remains relatively unstudied. Observations also emphasize the ability to protect against RSLR, especially for the most densely-populated areas, such as the subsiding Asian megacities or urban areas around the southern North Sea.

2.6 FUTURE IMPACTS OF SEA-LEVEL RISE

The future impacts of SLR will depend on a range of factors, including: (1) the magnitude of SLR; (2) the coastal physiographic setting; (3) the nature of coastal development; and (4) the success (or failure) of adaptation. Assessments of the future impacts of SLR have taken place on a range of scales from local to global. They all demonstrate large potential impacts consistent with Table 2.2, especially increases in inundation, flooding, and storm damage. Recent studies of flood risk (i.e., expected annual damages) under SLR all emphasize catastrophic impacts assuming no adaptation (e.g., Hallegatte et al., 2013; Hinkel et al., 2014). However, if defences are upgraded and other adaptation takes place, flood impacts are more limited and possibly almost totally avoided. This raises the question of the long-term implications of such an adaptation pathway.

In absolute numbers East, South-East, and East Asia and Africa appear to be most threatened by SLR (Fig. 2.4). Vietnam and Bangladesh appear especially threatened due to large absolute and relative populations in low-lying deltaic plains. There are also large absolute threatened populations in India and China. In Africa, Egypt (the Nile Delta) and Mozambique are two potential hotspots for impacts due to SLR. Impact hotspots also exist outside these regions, such as Guyana, Suriname, and French Guiana in South America. There will be significant residual risk in other coastal areas of the world, such as around the southern North Sea, and major flood disasters are possible in many coastal regions. Small island regions in the Pacific, Indian Ocean, and Caribbean stand out as being especially vulnerable to SLR impacts (Nurse et al., 2014). The populations of low-lying island nations, founded on atolls, such as the Maldives, Kiribati, or Tuvalu face the real prospect of increased flooding, submergence, salinization, and even forced abandonment due to SLR.

2.7 ADAPTATION TO SEA-LEVEL RISE

Adaptation to SLR involves responding to both mean and extreme rise. It is a complex process with multiple dimensions which are characterized differently across the literature. The overall field of climate adaptation is evolving rapidly (e.g., Klein et al., 2014), and this is influencing coastal adaptation, even though coastal adaptation is one of the more mature sectors in climate adaptation. It is useful to recognize different dimensions of adaptation, such as (1) autonomous (or spontaneous) adaptation versus planned adaptation, (2) proactive versus reactive

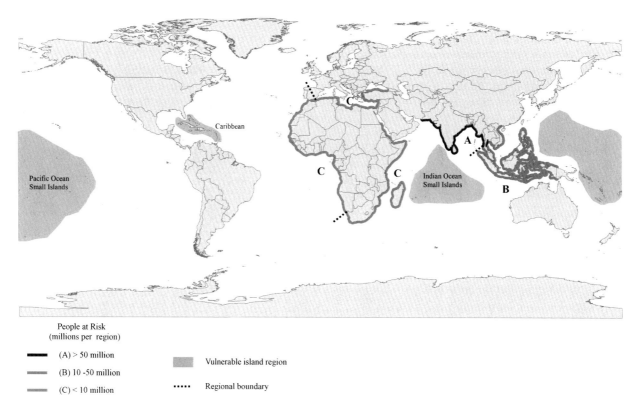

FIGURE 2.4 Regions most vulnerable to coastal flooding and sea-level rise. At highest risk are coastal zones with dense populations, low elevations, appreciable rates of subsidence, and/or inadequate adaptive capacity. *From Nicholls, R.J., Cazenave, A., 2010. Sea level rise and its impact on coastal zones. Science, 328, 1517−1520 (Nicholls and Cazenave, 2010).*

planned adaptation, and (3) individual/household versus collective adaptation. Given the large and rapidly growing concentration of people and activity in the coastal zone, autonomous adaptation processes alone will not be able to cope with SLR. Further, adaptation in the coastal context is widely seen as a public rather than a private responsibility, meaning that government is expected to develop and facilitate adaptation (Klein et al., 2000; Tribba and Moser, 2008).

Historically, society has tended to react to coastal trends and disasters rather than anticipate them. For SLR there are significant benefits in promoting proactive planning of adaptation, as the demand for adaptation can be expected to grow. Further, many adaptation decisions have long-term (10−100 years) implications (e.g., Hallegatte, 2009). Examples of proactive adaptation in coastal zones include upgraded flood defences and drainage systems, higher elevation designs for new coastal infrastructure such as coastal bridges, building standards/regulations to promote flood proofing and resilience, and building setbacks to prevent development in areas threatened by erosion and flooding.

This section considers adaptation strategies and options which are the building blocks of adaptation. It then considers adaptation processes and frameworks and how adaptation strategies and options are put together in space and time. The selection of adaptation strategies and options is considered, followed by experience of adaptation.

2.7.1   Adaptation Strategies and Options

Coastal adaptation can be classified in various ways: one of the most widely followed approaches is the IPCC typology of retreat, accommodation, and protection (Dronkers et al., 1990; Bijlsma et al., 1996). The concept of "attack" has been suggested as an additional adaptation strategy against SLR (e.g., RIBA and ICE, 2010). These can be linked together as shown in Fig. 2.5 and defined below:

- *(Planned) Retreat*—all natural system effects are allowed to occur and human impacts are minimized by pulling back from the coast via land use planning, development controls, planned migration, etc. (e.g., Fig. 2.6);

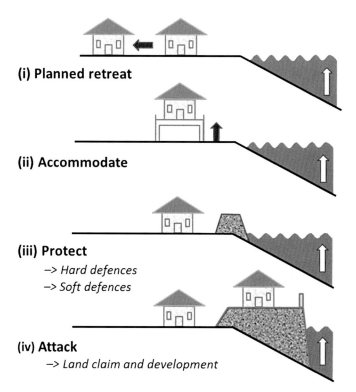

(i) Planned retreat

(ii) Accommodate

(iii) Protect
    –> *Hard defences*
    –> *Soft defences*

(iv) Attack
    –> *Land claim and development*

FIGURE 2.5 Generic adaptation approaches for sea-level rise, modified to include attack (via land claim and coastal development). *Modified from Dronkers, J., Gilbert, J.T.E., Butler, L.W., Carey, J.J., Campbell, J., James, E., et al., 1990. Strategies for adaption to sea level rise. Report of the Coastal Zone Management Subgroup, Response Strategies Working Group of the Intergovernmental Panel on Climate Change. The Hague, The Netherlands, Ministry of Transport, Public Works and Water Management.*

(A)

FIGURE 2.6 An example of a retreat option: Managed realignment at Medmerry, West Sussex, UK. The defence line (a shingle barrier beach) was breached allowing the low-lying flood plain behind to be inundated creating new intertidal habitats. To landward, a new (longer) defence line was constructed. (A) Aerial view from the east and (B) the breach from the west (photograph). (A) UK Environment Agency Copyright, reprinted with permission (B) Photograph by Hachem Kassem.

(B)

FIGURE 2.7   An example of accommodation in a coastal flood plain in the United Kingdom. The older property on the left is built at grade, while the adjoining property has been raised to enhance flood resilience—the design elevation considers present extreme water levels plus an allowance for future sea-level rise.

FIGURE 2.8   An example of a protection option: a new sea dike which has been raised as part of the phase 1 upgrade to the coastal flood defences in North Portsea Island, Portsmouth. The design height includes an allowance for sea-level rise.

- *Accommodation*—all natural system effects are allowed to occur and human impacts are minimized by adjusting human use of the coastal zone via changing land use/crop types, applying flood resilience measures, etc. (e.g., Fig. 2.7);
- *Protection*—natural system effects are controlled by soft or hard barriers (e.g., nourished beaches and dunes, or seawalls), reducing human impacts in the zone that would be impacted without protection (e.g., Fig. 2.8);
- *Attack*—build seaward and upwards, such as claiming new land to promote economic development.

Individually, there are a large set of potential adaptation options. Examples of adaptation options linked to each natural system impact of SLR are provided in Table 2.2.

Attack is consistent with land claim and advancing seawards (Linham and Nicholls, 2010). Land claim has a long history in coastal areas, including around most large coastal cities due to space constraints and high land values (e.g., Seasholes, 2003). Land claim is an active strategy in many coastal countries such as Singapore, Hong Kong, Dubai, and the Maldives to expand land area for coastal activities. In the context of Attack, SLR must be considered as shown in Fig. 2.5. As an extreme option, floating islands have been suggested as a response to SLR, although this remains untested in practise (Marris, 2017). Attack will be costly and it is most likely where it promotes increased economic activity, which can finance this investment.

Information measures such as disaster preparedness, hazard mapping, and flood warning/evacuation are growing in importance, and are cross-cutting and complementary of the approaches above. There is also growing interest in nature-based approaches, which have the advantage of being self-sustaining and providing multiple benefits (Borsje et al., 2011; Temmerman et al., 2013; Pontee et al., 2016). However, the uncertainties about their future state and function is much higher than engineered defences hindering application. In the future there is likely to be widespread potential for hybrid approaches which combine nature-based and more traditional engineering options. For example, flood protection could combine ecosystem buffers in front of artificial defences, reducing the required defence size (or further reduce risk). Building elevation via controlled sedimentation is another innovation option for coastal areas with large sediment supplies, especially deltas (Darby et al., 2018).

Where protection is used, we need to evaluate the residual risk that inevitably remains (but is often neglected). This suggests that protection must be combined with forecast/warning systems. Adaptation for one sector may also exacerbate impacts elsewhere: a good example is coastal squeeze of intertidal and shallow coastal habitats where onshore migration of habitats due to rising sea levels is prevented by attack and hard protection (Jones et al., 2011). In contrast, retreat and accommodation options allow habitat migration. Coastal management needs to consider the balance between protecting socioeconomic activity/human safety and the habitats and ecological functioning of the coastal zone under rising sea levels. While the 20th century saw large losses of coastal habitats due to direct and indirect destruction, most coastal countries now aspire to protect these areas and their ecosystem services: SLR threatens such initiatives and requires additional action to achieve these desired goals.

2.7.2 Adaptation Processes and Frameworks

As well as the adaptation measures, the adaptation process needs to be considered in terms of when adaptation will be implemented, how options might be combined and the long-term trajectory of adaptation. As already noted, SLR is expected to continue (at a slower rate) even with aggressive climate mitigation and hence adaptation needs are growing into the future under all plausible climate scenarios.

While adaptation to SLR is relatively new, there is considerable experience of adapting to climate and sea-level extremes and other coastal problems. This experience can inform decision making under a changing climate. Importantly, adaptation to coastal problems is a multistage *process*, which can be characterized as (1) information and awareness building, (2) planning and design, (3) evaluation, and (4) monitoring and evaluation operating within multiple policy cycles (e.g., Klein et al., 2000; Hay, 2009). There are also constraints on the approach to adaptation due to broader policy and development goals. Once implemented, monitoring and evaluation of adaptation measures is a critical step and yet easily ignored. This is essential given the large uncertainties associated with SLR and other future conditions, adaptation performance and coastal management in general. Monitoring and evaluation allows learning and improved future adaptation and this needs to be built into coastal management and adaptation.

A range of adaptation frameworks are apparent in the literature, with a diverse range of experience. For example, Integrated Coastal Zone Management (ICZM) was strongly advocated as the response to SLR in the early 1990s (e.g., Dronkers et al., 1990; Bijlsma et al., 1996), recognizing that SLR and climate change occur in a multistressed situation. However, it remains unproven as an effective response approach (Wong et al., 2014). Adaptive management is also advocated and has important merits in dealing with the uncertainties of SLR. Community-based adaptation (CBA) is widely advocated as a bottom-up development focused approach (e.g., Huq and Reid, 2004; Rawlani and Sovacool, 2011). These address community concerns, but there is concern that the focus may not be strategic. For example, will the 1 in 100 or 1 in 1000 event be considered, when they are beyond what most if not all the community have experienced? In the worse case, could CBA promote people to stay in increasingly hazardous locations—this suggests CBA should be practised within a broader science-based framework, including consideration of extremes and warning systems. Linking climate adaptation with Disaster Risk Reduction (Smith, 2013), which can be seen in part as adapting to climate variability, is also receiving increasing interest.

Shoreline management planning (SMP) has emerged in England and Wales over the last two decades as a government-led response to coastal erosion and flood risk management (Nicholls et al., 2013). It provides a framework in applying future coastal interventions and management over long timescales, including the nonlocal effects of management. The English/Welsh coast has been divided into 22 SMPs and further about 2000 management units and three time epochs are considered. Four generic responses are considered for each management unit and epoch without considering the technical detail: (1) advance the existing defence line; (2) hold the existing defence line; (3) managed realignment; and (4) no active intervention. Options 1 and 2 can be considered as generic protection (or even attack for option 1), while options 3 and 4 can be considered retreat. Note that accommodation is also being implemented in the United Kingdom for flood management purposes, as demonstrated by Fig. 2.7, but this is implemented at the property level, and hence at a submanagement unit scale. Supporting the SMP process are national monitoring systems. This high level approach could be applied widely around the world's coast, including consideration of SLR.

In parallel with this, there has also been recognition that while we need to adapt to SLR, there is great uncertainty about timing and an opportunity to learn. Hence, while we can see different qualitative directions of travel (or possible adaptation pathways), we are not sure how fast we need to travel along each pathway as the magnitude of future SLR is uncertain. Hence, we can define adaptation pathways and take actions that preserve these options without spending the large sums that are needed to realize them, until required. Adaptation pathways, combined with monitoring and learning, are an appropriate approach for SLR, especially in cities where significant adaptation will be needed. This approach has been adopted in the Thames Estuary 2100 Project addressing future flood protection for London (Ranger et al., 2013; Tarrant and Sayers, 2013; Chapter 6: Flood Risk Management in the United Kingdom, Putting Climate Change Adaptation Into Practice in the Thames Estuary).

2.7.3 Choosing Between Adaptation Measures/Options

Retreat is often argued as the best response to SLR (e.g., Pilkey and Young, 2009). However, benefit—cost models that compare protection with retreat generally suggest that it is worth investing in widespread protection as populated coastal areas have high economic value (e.g., Fankhauser, 1995; Anthoff et al., 2010; Nicholls et al., 2014a). This does not mean that we *should* protect. Rather the main insight is that these results suggest that significant resources are available for adapting to SLR, and further that protection can be expected to be a significant part of the portfolio of responses. With or without protection, small island and deltaic areas stand out as more vulnerable in these analyses and the impacts fall disproportionately on poorer countries. Even though optimal in a benefit—cost sense, protection costs may overwhelm the capacity of local economies to fund them, especially when they are small such as islands (Fankhauser and Tol, 2005; Nicholls and Tol, 2006). While adaptation is essentially a local activity, these funding challenges are an issue of international concern due to the shared responsibility for climate-induced SLR.

The existing state of coastal adaptation and any associated "adaptation deficit" is an important concern—the adaptation deficit is the cost of adapting to today's situation, before we consider adapting to future SLR and other change (Burton, 2004; Parry et al., 2009). There is a lot of evidence to support an adaptation deficit on coasts, but it has not been systematically quantified. For example, Hallegatte et al. (2013) showed that US coastal cities have much higher expected damage costs than European coastal cities today reflecting much lower protection standards. Equally, less developed and rapidly growing regions are likely to have a significant and growing adaptation deficit due to explosive coastal development in increasingly hazardous areas (e.g., Hinkel et al., 2011; Mycoo and Chadwick, 2012).

Global cost estimates of coastal adaptation normally focus on the incremental costs of upgrading defence infrastructure, assuming no adaptation deficit, as this is consistent with the United Nations Framework Convention on Climate Change. There are several cost estimates for protection (Nicholls et al., 2014a). These are lower than conventional instinct suggests, reflecting the high benefit—cost estimates already mentioned. For example, recent global protection costs for flooding were estimated to rise to US\$20 and \$70 billion/year over the 21st century (Hinkel et al., 2014). For 136 coastal cities, Hallegatte et al. (2013) argued that adaptation to SLR would be about US\$50 billion/year globally to 2050. Considering beach erosion, global adaptation costs for SLR only via nourishment estimated costs in 2100 of US \$1.5—5 billion/year (Hinkel et al., 2013). These costs all seem affordable, but as the cost of the adaptation deficit is not addressed they must be considered as minima. Further, a protection pathway raises questions of where it takes the coastal system in the long term: key concerns with protection are lock-in and residual risk.

The adaptation deficit reflects that many countries have a limited capacity to address today's coastal problems, let alone consider tomorrow's problems, including SLR. Therefore, promoting coastal adaptation should include developing coastal management capacity and institutions, as already widely recommended (USAID, 2009; Moser et al., 2012) and integrating it with wider coastal development.

2.7.4 Adaptation Experience

Existing adaptation experience mainly reflects adaptation to extreme events or land subsidence, which is good analogue for climate-induced SLR. Through human history, developing technology has increased the range of adaptation options in the face of coastal hazards, and there has been a move from retreat and accommodation to hard protection and active seaward advance via attack and land claim as exemplified by the Netherlands (Van Koningsveld et al., 2008). Rising sea level is one factor calling widespread reliance on protection into question, and the appropriate mixture of protection, accommodation and retreat, and the whole philosophy of coastal adaptation is under debate, as already discussed (Wong et al., 2014).

While there is growing awareness of the need to adapt to SLR, only a few countries or locations are preparing for this challenge that recognizes the huge uncertainty around the magnitude and timing of SLR, and the inevitability of some ongoing SLR even under aggressive climate mitigation policies. Examples include London (Lavery and Donovan, 2005; Tarrant and Sayers, 2013; Chapter 6: Flood Risk Management in the United Kingdom, Putting Climate Change Adaptation Into Practice in the Thames Estuary) and the Netherlands (Kabat et al., 2009; Stive et al., 2011). Both these sites considered a wide range of SLR scenarios, including scenarios of up to 5 and 4 m, respectively, addressing post-2100 challenges. Importantly they considered adaptation as a process and London in particular focused on adaptation pathways as a function of SLR rather than time. Both analyses demonstrate that there are feasible protection options available for large rises in sea level, and progressive upgrade is feasible over a 100/200 year time frame. This is an effective way to deal with the uncertainty of future SLR and implies an adaptive approach where higher protection is provided when needed, avoiding over or under adapting (Ranger et al., 2013; Haasnoot et al., 2013; Chapter 6: Flood Risk Management in the United Kingdom, Putting Climate Change Adaptation Into Practice in the Thames Estuary). It is worth noting that these activities to date were more about enhancing planning, process, and capacity than actual responses. For instance, the Netherlands has created the new governance institution of the Delta Commission to manage the national Delta Plan and develop strategic policy processes and model tools to support this process to facilitate more strategic planning and investment (Van Alphen, 2015). The Delta Commission enjoys widespread cross-party political support and hence funds are guaranteed beyond a political cycle allowing a longer term perspective. This approach is being exported to other deltas such as Bangladesh.

In other locations such as New York City, adaptation is also being carefully considered (Rosenzweig and Solecki, 2010), but the timing of implementation is less clear. In this case, the major event of Superstorm Sandy has accelerated action and a number of projects are planned. But a strategic response as in London or the Netherlands is not yet apparent. In Singapore new land claim will be raised by approximately a meter to allow for SLR. In general coastal cities are expected to be a major focus for these efforts given the concentration of people and assets, and their ability to fund large investments (Hallegatte et al., 2013; Aerts et al., 2014). If all coastal cities followed an assessment process similar to Thames Estuary 2100 in London, it would give them a better basis for understanding potential impacts and provide a basis for more systematic planning of adaptation (Nicholls et al., 2015b). Similar arguments could be made in vulnerable deltas and islands, and ultimately all populated coasts might make such an assessment.

2.8 DISCUSSION/CONCLUSIONS

The chapter illustrates that adapting to SLR and enhancing resilience is a multidimensional problem that crosses many disciplines and embraces natural, social, and engineering sciences, as well as engaging stakeholders, policy, and governance. Importantly, significant SLR seems inevitable under all plausible scenarios and hence adaptation is essential: the uncertainty is the magnitude and timing. The actual outcome will also depend on our responses, both in terms of climate mitigation and adaptation and their success or failure. For adaptation in general, and protection in particular, there are widely divergent views on the likely success or failure, and this strongly influences how the issue of SLR is considered in public policy. Much current thinking can be characterized into pessimist and optimist camps (Nicholls and Tol, 2006; Anthoff et al., 2010; Nicholls et al., 2014a). "Pessimists" tend to focus on high rises in sea level, extreme events with large impacts like Katrina, and view our ability to adapt to SLR as limited, resulting in widespread human migration away from coastal areas. In contrast, "optimists" tend to focus on lower rises in sea level and stress the growing technical ability to forecast, warn, and protect and the high benefit−cost ratios in developed areas leading to widespread protection. Hence to optimists a major consequence of SLR is the diversion of investment to coastal adaptation in general, and protection in particular. More work is required to reconcile these opposing perspectives, which will influence the relative role of attack, protect, accommodate, and retreat in our adaptation responses.

SLR is clearly a threat that demands a response, with adaptation being an essential activity. There is a need to better understand these threats, including the implications of different mixtures of climate mitigation and adaptation, and different portfolios of adaptation and adaptation pathways (Nicholls et al., 2007; Wong et al., 2014). To provide the basic data, coastal monitoring and coastal climate services are increasingly required (Nicholls et al., 2013; Le Cozannet et al., 2017). As the coast is a coupled system, it will be important to examine different scenarios of SLR and climate change, socioeconomic changes, and how adaptation coevolves with the wider coastal system. There is also a need to engage with and inform the coastal and climate policy process.

Assessing the success or failure of adaptation to SLR will be challenging given the long timescale of the issue. We can assess the success or failure of adaptation to present hazards such as flooding and erosion. Equally, we can learn from experience adapting to human-induced subsidence. Research aspects and operational aspects of adaptation need to be separated, as both are required in terms of developing general guidance and more detailed analysis and assessment of specific sites. Adaptation inventories provide one foundational step to both types of analysis (e.g., Tompkins et al., 2017). Similarly assessing best practise will be challenging. Research is required at all scales from local to global, but much will be learnt about adaptation in practise (Wong et al., 2014). This will promote more appropriate adaptation options and the opportunity to learn from experience.

ACKNOWLEDGMENTS

Dr. Abiy Kebede assisted with preparation of Fig. 2.5. Dr. Hachem Kassem took Fig. 2.6b. This chapter is a contribution to the WRCP Grand Challenge on Regional Sea Level Change and Coastal Impacts, Working Group 5 "Sea level science for coastal zone management."

REFERENCES

Aerts, J.C.J.H., Botzen, W.J.W., Emanuel, K., Lin, N., de Moel, H., Michel-Kerjan, E.O., 2014. Evaluating flood resilience strategies for coastal megacities. Science 344, 473−475.

Anthoff, D., Nicholls, R.J., Tol, R.S.J., 2010. The economic impact of substantial sea-level rise. Mitig. Adapt. Strategies Glob. Chang. 15, 321−335.

Barth, M.C., Titus, J.G. (Eds.), 1984. Greenhouse Effect and Sea Level Rise: A Challenge for This Generation. Van Nostrand Reinhold, New York.

Bianchi, T., 2016. Deltas and Humans. Oxford University Press, New York.

Bijlsma, L., Ehler, C.N., Klein, R.J.T., Kulshrestha, S.M., McLean, R.F., Mimura, N., et al., 1996. Coastal zones and small islands. In: Watson, R.T., Zinyowera, M.C., Moss, R.H. (Eds.), Climate Change 1995: Impacts, Adaptations, and Mitigation of Climate Change: Scientific-Technical Analyses. Contribution of Working Group II to the Second Assessment Report of the Intergovernmental Panel on Climate Change. Cambridge University Press, Cambridge and New York, pp. 289−324.

Bird, E.C.F., 1985. Coastline Changes: A Global Review. John Wiley and Sons, New York, p. 219.

Bird, E.C.F., 2000. Coastal Geomorphology: AN introduction. Wiley and Sons, Chichester.

Borsje, B.W., van Wesenbeeck, B.K., Dekker, F., Paalvast, P., Bouma, T.J., van Katwijk, M.M., et al., 2011. How ecological engineering can serve in coastal protection. Ecol. Eng. 37, 113−122.

Burton, I., 2004. Climate change and the adaptation deficit. In: French, A., et al., (Eds.), Climate Change: Building the Adaptive Capacity, Meteorological Service of Canada. Environment Canada, Gatineau, Quebec, pp. 25−33.

Chatterjee, R.S., Fruneau, B., Rudant, J.P., Roy, P.S., Frison, P., Lakhera, R.C., et al., 2006. Subsidence of Kolkata (Calcutta) City, India during the 1990s as observed from space by Differential Synthetic Aperture Radar Interferometry (D-InSAR) technique. Remote Sens. Environ. 102, 176−185.

Chaussard, E., Amelung, F., Abidin, H., Hong, S.-H., 2013. Sinking cities in Indonesia: ALOS PALSAR detects rapid subsidence due to groundwater and gas extraction. Remote Sens. Environ. 128, 150−161.

Church, J.A., Clark, P.U., Cazenave, A., Gregory, J.M., Jevrejeva, S., Levermann, A., et al., 2013. Sea Level Change. In IPCC Working Group I, Fifth Assessment Report.

Church, J.A., Woodworth, P.L., Aarup, T., Wilson, W.S. (Eds.), 2010. Understanding Sea-Level Rise and Variability. Wiley-Blackwell, Hoboken, NJ.

Crossland, C.J., Kremer, H.H., Lindeboom, H.J., Marshall Crossland, J.I., Le Tissier, M.D.A. (Eds.), 2005. Coastal Fluxes in the Anthropocene. Springer, Berlin.

Dangendorf, S., Marcos, M., Wöppelmann, G., Conrad, C.P., Frederiksee, T., Riva, R., 2017. Reassessment of 20th century global mean sea level rise. Proc. Natl. Acad. Sci. 114, 5946−5951. Available from: https://doi.org/10.1073/pnas.1616007114.

Darby, S.E., Nicholls, R.J., Rahman, Md.M., Brown, S., Karim, Md.R., 2018. A sustainable future supply of fluvial sediment for the Ganges-Brahmaputra Delta. In: Nicholls, R.J., et al., (Eds.), Ecosystem Services For Well-Being In Deltas: Integrated Assessment For Policy Analysis. Palgrave, forthcoming.

Dawson, R.J., Nicholls, R.J., Day, S.A., 2015. The challenge for coastal management during the third millenium. In: Nicholls, R.J., Dawson, R.J., Day, S.A. (Eds.), Broad Scale Coastal Simulation: New Techniques to Understand and Manage Shorelines in the Third Millennium. Springer, Dordrecht, NL, pp. 1−78.

Dronkers, J., Gilbert, J.T.E., Butler, L.W., Carey, J.J., Campbell, J., James, E., et al., 1990. Strategies for Adaption to Sea Level Rise. Report of the Coastal Zone Management Subgroup, Response Strategies Working Group of the Intergovernmental Panel on Climate Change. The Hague, The Netherlands, Ministry of Transport, Public Works and Water Management.

Ericson, J.P., Vorosmarty, C.J., Dingman, S.L., Ward, L.G., Meybeck, M., 2006. Effective sea-level rise and deltas: causes of change and human dimension implications. Glob. Planet. Change 50, 63−82.

Fankhauser, S., 1995. Protection versus retreat: estimating the costs of sea-level rise. Environ. Plan. A 27, 299−319.

Fankhauser, S., Tol, R.S.J., 2005. On climate change and economic growth. Resour. Energy Econ. 27, 1−17.

Grossi, P., Muir-Wood, R., 2006. Flood Risk in New Orleans: Implications for Future Management and Insurability. Risk Management Solutions (RMS), London.

Haasnoot, M., Kwakkel, J.H., Walker, W.E., ter Maat, J., 2013. Dynamic adaptive policy pathways: a method for crafting robust decisions for a deeply uncertain world. Glob. Environ. Change 23, 485−498.

Hallegatte, S., 2009. Strategies to adapt to an uncertain climate change. Glob. Environ. Change 19, 240−247.

Hallegatte, S., 2012. A framework to investigate the economic growth impact of sea level rise. Environ. Res. Lett. 7 (1). Available from: https://doi.org/10.1088/1748-9326/7/1/015604.

Hallegatte, S., Green, C., Nicholls, R.J., Corfee-Morlot, J., 2013. Future flood losses in major coastal cities. Nat. Climate Change 3, 802−806. Available from: https://doi.org/10.1038/nclimate1979.

Hanson, S., Nicholls, R.J., Patmore, N., Hallegatte, S., Corfee-Morlot, J., Herweijer, C., et al., 2011. A global ranking of port cities with high exposure to climate extremes. Climatic Change 140 (1), 89−111. Available from: https://doi.org/10.1007/s10584-010-9977-4.

Hay, J.E., 2009. Institutional and Policy Analysis of Disaster Risk Reduction and Climate Change Adaptation in Pacific Island Countries. United Nations International System for Disaster Reduction (UNISDR) and the United Nations Development Programme (UNDP), Suva, Fiji.

Hinkel, J., Brown, S., Exner, L., Nicholls, R.J., Vafeidis, A.T., Kebede, A.S., 2011. Sea-level rise impacts on Africa and the effects of mitigation and adaptation: an application of DIVA. Reg. Environ. Change. Available from: https://doi.org/10.1007/s10113-011-0249-2.

Hinkel, J., Lincke, D., Vafeidis, A.T., Perrette, M., Nicholls, R.J., Tol, R.S.J., et al., 2014. Coastal flood damage and adaptation costs under 21st century sea-level rise. Proc. Natl. Acad. Sci. Available from: https://doi.org/10.1073/pnas.1222469111.

Hinkel, J., Nicholls, R.J., Tol, R.S.J., Wang, Z.B., Hamilton, J.B., Boot, G., et al., 2013. A global analysis of erosion of sandy beaches and sea-level rise: an application of DIVA. Glob. Plan. Change 111, 150−158. Available from: https://doi.org/10.1016/j.gloplacha.2013.09.002.

Holgate, S.J., Matthews, A., Woodworth, P.L., Rickards, L.J., Tamisiea, M.E., Bradshaw, E., et al., 2013. New data systems and products at the permanent service for mean sea level. J. Coastal Res. 29, 493−504. Available from: https://doi.org/10.2112/JCOASTRES-D-12-00175.1.

Huq, S., Reid, H., 2004. Mainstreaming adaptation in development. IDS Bull. 35, 15−21. Available from: https://doi.org/10.1111/j.1759-5436.2004.tb00129.x.

Jones, L., Angus, S., Cooper, A., Doody, P., Everard, M., Garbutt, A., et al., 2011. Coastal margins. UK National Ecosystem Assessment Technical Report. United Nations Environment Programme World Conservation Monitoring Centre, Cambridge, GB (UNEP-WCMC).

Kabat, P., Fresco, L.O., Stive, M.J.F., Veerman, C.P., van Alphen, J., Parmet, B., et al., 2009. Dutch coasts in transition. Nat. Geosci. 2, 450−452.

Kaneko, S., Toyota, T., 2011. Long-term urbanization and land subsidence in Asian megacities: an indicators system approach. In: M. Taniguchi (ed.), Groundwater and Subsurface Environments: Human Impacts in Asian Coastal Cities, DOI 10.1007/978-4-431-53904-9_13, pp. 249−270.

Klein, R.J.T., Aston, J., Buckley, E.N., Capobianco, M., Mizutani, N., Nicholls, R.J., et al., 2000. Coastal adaptation. In: Metz, B., Davidson, O.R., Martens, J.W., Van Rooijen, S.N.M., Van Wie McGrory, L.L. (Eds.), IPCC Special Report on Methodological and Technological Issues in Technology Transfer. Cambridge University Press, Cambridge.

Klein, R.J.T., Midgley, G.F., Preston, B.L. Alam, M., Berkhout, F.G.H., Dow, K. et al., 2014. Adaptation Opportunities, Constraints, and Limits. In IPCC Working Group II, Fifth Assessment Report.

Klein, R.J.T., Nicholls, R.J., 1999. Assessment of coastal vulnerability to climate change. Ambio 28, 182−187.

Kron, W., 2013. Coasts: the high-risk areas of the world. Natl. Hazards 66, 1363−1382.

Lavery, S., Donovan, B., 2005. Flood risk management in the Thames Estuary looking ahead 100 years. Philos. Trans. Royal Soc. A 363, 1455−1474. Available from: https://doi.org/10.1098/rsta.2005.1579.

Lazarus, E.D., Ellis, M.A., Murray, A.B., Hall, D.M., 2016. An evolving research agenda for human-coastal systems. Geomorphology 256, 81−90.

Le Cozannet, G., Nicholls, R.J., Hinkel, J., Sweet, W.V., McInnes, K.L., Van de Wal, R.S.W., et al., 2017. Sea level change and coastal climate services: the way forward. J. Marine Sci. Eng. 5 (4). Available from: https://doi.org/10.3390/jmse5040049.

Linham, M.M., Nicholls, R.J., 2010. Technologies for Climate Change Adaptation: Coastal Erosion and Flooding. TNA Guidebook Series, UNEP Risø Centre on Energy, Climate and Sustainable Development, Roskilde, Denmark, pp. 150.

Marris, E., 2017. Why fake islands might be a real boon for science. Nature 550, 22−24.

McGranahan, G., Balk, D., Anderson, B., 2007. The rising tide: assessing the risks of climate change and human settlements in low elevation coastal zones. Environ. Urban. 19, 17−37.

Menéndez, M., Woodworth, P.L., 2010. Changes in extreme high water levels based on a quasi-global tide-gauge dataset. J. Geophys. Res. 115, C10011.

Milliman, J.D., Broadus, J.M., Gable, F., 1989. Environmental and economic implications of rising sea level and subsiding deltas: the Nile and Bengal examples. Ambio 18, 340−345.

Moser, S.C., Williams, S.J., Boesch, D.F., 2012. Wicked challenges at land's end: managing coastal vulnerability under climate change. Annu. Rev. Environ. Resour. 37, 51−78.

Mycoo, M., Chadwick, A., 2012. Adaptation to climate change: the coastal zone of Barbados. Maritime Eng. 165 (4), 159−168.

Neumann, B., Vafeidis, A.T., Zimmermann, J., Nicholls, R.J., 2015. Future coastal population growth and exposure to sea-level rise and coastal flooding - a global assessment. PLoS One 10 (6), e0131375. Available from: https://doi.org/10.1371/journal.pone.0118571.

Nicholls, R.J., 2010. Impacts of and responses to sea-level rise. In: Church, J.A., Woodworth, P.L., Aarup, T., Wilson, S. (Eds.), Understanding Sea-Level Rise and Variability, 2010. Wiley-Blackwell, Chichester, pp. 17−51.

Nicholls, R.J., 2014. Adapting to sea level rise. In: Ellis, J.T., Sherman, D.J. (Eds.), Coastal and Marine Hazards, Risks and Disasters. Elsevier, London, pp. 243−270.

Nicholls, R.J., Brown, S., Goodwin, P., Wahl, T., Lowe, J., Solan, M., et al., 2017. Stabilisation of global temperature at 1.5°C and 2.0°C: implications for coastal areas. Philos. Trans. Royal Soc. Accepted.

Nicholls, R.J., Cazenave, A., 2010. Sea level rise and its impact on coastal zones. Science 328, 1517−1520.

Nicholls, R.J., Hanson, S.E., Lowe, J.A., Warrick, R.A., Lu, X., Long, A.J., 2014b. Sea-level scenarios for evaluating coastal impacts. Wiley Inter. Rev. Climate Change 5 (1), 129−150.

Nicholls, R.J., Kebede, A.S., 2012. Indirect impacts of coastal climate change and sea-level rise: the UK example. Climate Policy 12, S28−S52.

Nicholls, R.J., Reeder, T., Brown, S., Haigh, I.D., 2015b. The risks of sea-level rise for coastal cities. In: King, D., Schrag, D., Dadi, Z., Ye, Q., Ghosh, A. (Eds.), Climate Change: A Risk Assessment. Foreign and Commonwealth Office, London, pp. 94−98.

Nicholls, R.J., Stive, M.J.F., Tol, R.S.J., 2014a. Coping with coastal change. In: Masselink, G., Gehrels, R. (Eds.), Coastal Environments and Global Change. Wiley, Chichester.

Nicholls, R.J., Tol, R.S.J., 2006. Impacts and responses to sea-level rise: a global analysis of the SRES scenarios over the twenty-first century. Philos. Trans. Royal Soc. A 364, 1073−1095.

Nicholls, R.J., Townend, I.H., Bradbury, A., Ramsbottom, D., Day, S., 2013. Planning for long-term coastal change: experiences from England and Wales. Ocean Eng. 71, 3−16. Available from: https://doi.org/10.1016/j.oceaneng.2013.01.025.

Nicholls, R.J., Wong, P.P., Burkett, V.R., Codignotto, J.O., Hay, J.E., McLean, R.F., et al., 2007. Coastal systems and low-lying areas. In: Parry, M.L., Canziani, O.F., Palutikof, J.P., Van Der Linden, P., Hanson, C.E. (Eds.), Climate Change2007: Impacts, Adaptation and Vulnerability. Cambridge University Press, Cambridge.

Nicholls, R.J., Woodroffe, C.D., Burkett, V.R., 2015a. Coastline degradation as an indicator of global change. In: Letcher, T.M. (Ed.), Climate Change: Observed Impacts on Planet Earth, 2nd ed. Elsevier, Amsterdam, pp. 309−324.

Nurse, L., McLean, R., Agard, J., Briguglio, L.P., Duvat, V., Pelesikoti, N., et al., 2014. Small Islands. In IPCC Working Group II, Fifth Assessment Report.

Parry M.L., Arnell N.W., Berry P.M., Dodman D., Fankhauser S., Hope C., et al., 2009. Assessing the costs of adaptation to climate change: a review of the UNFCCC and other recent estimates. International Institute for Environment and Development and Grantham Institute for Climate Change, London. 111 pp.

Permanent Service for Mean Sea Level (PSMSL), 2014. Tide Gauge Data. Retrieved 30 Jun 2014 from http://www.psmsl.org/data/obtaining/.

Pilkey, O.H., Young, R., 2009. The Rising Sea. Island Press/Shearwater Books, Washington, DC, p. 203.

Pontee, N., Narayan, S., Beck, M.W., Hosking, A.H., 2016. Nature-based solutions: lessons from around the world. Maritime Eng. 169, 29−36.

Ranger, N., Reeder, T., Lowe, J.A., 2013. Addressing 'deep' uncertainty over long-term climate in major infrastructure projects: four innovations of the Thames Estuary 2100 Project. Eur. J. Decis. Proces. 1, 233−262.

Rawlani, A.K., Sovacool, B.J., 2011. Building responsiveness to climate change through community based adaptation in Bangladesh. Mitig. Adapt. Strategies Glob. Chang. 16, 845−863.

RIBA and ICE, 2010. Facing up to rising sea levels. Retreat? Defend? Attack? RIBA (Royal Institute of British Architects) and ICE (Institution of Civil Engineers), London. http://www.buildingfutures.org.uk/assets/downloads/Facing_Up_To_Rising_Sea_Levels.pdf.

Climate change adaptation in New York City: building a risk management response. In: Rosenzweig, C., Solecki, W. (Eds.), Ann. NY Acad. Sci., 1196. pp. 1−354.

Ruocco, A.C., Nicholls, R.J., Haigh, I.D., Wadey, M.P., 2011. Reconstructing coastal flood occurrence combining sea level and media sources: a case study of the Solent, UK since 1935. Nat. Hazards 59 (3), 1773−1796. Available from: https://doi.org/10.1007/s11069-011-9868-7.

Seasholes, N.C., 2003. Gaining Ground: A History of Landmaking in Boston. MIT Press, Cambridge, MA.

Smith, K., 2013. Environmental Hazards: Assessing Risk and Reducing Disaster. Routledge, London.

Stive, M.J.C., Fresco, L.O., Kabat, P., Parmet, B.W.A.H., Veerman, C.P., 2011. How the Dutch plan to stay dry over the next century. Proc. Instit. Civil Eng. 164, 114−121.

Sweet, W., Park, J., Marra, J., Zervas, C., Gill, S., 2014. Sea-Level Rise and Nuisance Flood Frequency Changes around the United States. National Oceanic and Atmospheric Administration. NOAA Technical Report NOS CO-OPS 073.

Syvitski, J.P.M., Kettner, A.J., Overeem, I., Hutton, E.W.H., Hannon, M.T., Brakenridge, G.R., et al., 2009. Sinking deltas. Nat. Geosci. 2, 681−689.

Tarrant, O., Sayers, P.B., 2013. Managing flood risk in the Thames Estuary -- the development of a long-term robust and flexible strategy. In: Sayers, P.B. (Ed.), Flood Risk: Planning, Design and Management of Flood Defence Infrastructure. ICE Publishing, London, pp. 303−326.

Teatini, P., Tosi, L., Strozzi, T., Carbognin, L., Wegmüller, U., Rizzetto, F., 2005. Mapping regional land displacements in the Venice coastland by an integrated monitoring system. Remote Sens. Environ. 98 (4), 403−413.

Temmerman, S., Meire, P., Bouma, T.J., Herman, P.M.J., Ysebaert, T., De Vriend, H.J., 2013. Ecosystem-based coastal defence in the face of global change. Nature 504, 79−83.

Tompkins, E.L., Suckall, N., Vincent, K., Rahman, R., Mensah, A., Ghosh, T., et al., 2017. Observed adaptation in deltas. DECCMA Working Paper, Deltas, Vulnerability and Climate Change: Migration and Adaptation, IDRC Project Number 107642. Available online at: www.deccma.com, date accessed 20 October2017.

Tribbia, J., Moser, S.C., 2008. More than information: what coastal managers need to plan for climate change. Environ. Sci. Policy 11, 315−328.

Tsyban, A., Everett, J., Titus, J., 1990. World oceans and coastal zones. In: Tegart, W.J. Mc.G., Sheldon, G.W., Griffiths, D.C. (Eds.), Climate Change: The IPCC Impacts Assessment. Australian Government Publishing Service, Canberra, pp. 6-1−6-28.

United Nations, 2015. Paris Agreement. http://unfccc.int/files/essential_background/convention/application/pdf/english_paris_agreement.pdf.

USAID, 2009. Adapting to Coastal Climate Change: A Guidebook for Development Planners. USAID, Rhode Island. Available from: www.crc.uri.edu/download/CoastalAdaptationGuide.pdf [Accessed: 07/10/10].

Valiela, I., 2006. Global Coastal Change. Blackwell, Malden, MA.

Van Alphen, J., 2015. The Delta Programme and updated flood risk management policies in the Netherlands. J. Flood Risk Manage. Available from: https://doi.org/10.1111/jfr3.12183.

Van Koningsveld, M., Mulder, J.P.M., Stive, M.J.F., Van Der Valk, L., Van Der Weck, A.W., 2008. Living with sea-level rise and climate change: a case study of the Netherlands. J. Coastal Res. 24, 367−379.

Wahl, T., Haigh, I.D., Nicholls, R.J., Arns, A., Dangendorf, S., Hinkel, J., et al., 2017. Understanding extreme sea levels for broad-scale coastal impact and adaptation analysis. Nat. Commun. 8 (16075). Available from: https://doi.org/10.1038/ncomms16075.

Welch, A.C., Nicholls, R.J., Lazar, A.N., 2017. Evolving deltas: coevolution with engineered interventions. Elementa Sci. Anthrop. 5 (49). Available from: https://doi.org/10.1525/elementa.128.

Wong, P.P., Losada, I.J., Gattuso, J., Hinkel, J., Khattabi, A., McInnes, K., et al. 2014. Coastal systems and low-lying areas. In IPCC Working Group II, Fifth Assessment Report.

Zhang, K.Q., Douglas, B.C., Leatherman, S.P., 2000. Twentieth-century storm activity along the US east coast. J. Climate 13, 1748−1761.

FURTHER READING

Dang, V.K., Doubre, C., Weber, C., Gourmelen, N., Masson, F., 2014. Recent land subsidence caused by rapid urban development in the Hanoi region (Vietnam) using ALOS InSAR data. Nat. Hazards Earth Syst. Sci. 14, 657−674.

Gornitz, V., 2013. Rising Seas: Past, Present and Future. Columbia University Press, New York, p. 344.

Nicholls, R.J., Hanson, S., Herweijer, C., Patmore, N., Hallegatte, S., Corfee-Morlot, J., et al., 2008. Ranking port cities with high exposure and vulnerability to climate extremes—exposure estimates. Environmental Working Paper No. 1. Paris Organisation for Economic Co-operation and Development (OECD).

Phien-Wej, N., Giao, P.H., Nutalaya, P., 2006. Land subsidence in Bangkok, Thailand. Eng. Geol. 82, 187−201.

Pugh, D., Woodworth, P., 2014. Sea-Level Science: Understanding Tides, Surges, Tsunamis and Mean Sea-Level Change. Cambridge University Press, Cambridge, p. 395.

Rodolfo, K.S., Siringan, F.P., 2006. Global sea-level rise is recognised, but flooding from anthropogenic land subsidence is ignored around northern Manila Bay, Philippines. Disaster Manage. 30, 118−139.

Chapter 3

Climate Change, Climate Extremes, and Global Food Production—Adaptation in the Agricultural Sector

Elisabeth Vogel[1,2] and Rachelle Meyer[1,3]

[1]Australian-German Climate & Energy College, University of Melbourne, Parkville, Victoria, Australia,

[2]ARC Centre of Excellence for Climate System Science, University of Melbourne, Parkville, Victoria, Australia, [3]Faculty of Veterinary and Agricultural Sciences, University of Melbourne, Parkville, Victoria, Australia

Chapter Outline

3.1 INTRODUCTION

Climate change poses major risks for world agriculture in the 21st century, with implications for the livelihoods of farmers and the food security of communities across the globe. The agricultural sector employs close to one billion people worldwide—or about one-third of the global workforce (ILOSTAT, 2016). In developing countries, the share of people who depend on agriculture for their livelihoods is even higher, putting the greatest risks on the most vulnerable. In addition, the global food system is faced with increasing pressures from demographic change—by 2050, the world population is estimated to reach 9.7 billion people. Furthermore, rising income and urbanization levels are projected to lead to higher per capita consumption and increasing demands for animal products. The United Nations (UN) Food and Agriculture Organization (FAO) estimates that food production has to increase by 70% by 2050 to meet future demands (FAO, 2009), while mitigating climate change and adapting to the effects of it.

This chapter reviews the biggest challenges and risks from climate change, including climate extreme events, for global agricultural production based on a bibliographic network analysis of the existing literature. Bibliographic networks built from publications and their citations allow visualization and analysis of the connections within the body of literature, to map out research directions, and to objectively identify the most influential publications that built the foundation of current knowledge (Knutas et al., 2015). Based on the findings of the bibliographic network analysis, this chapter highlights the most important needs, opportunities, and constraints for climate change adaptation in the agricultural sector. The focus of this chapter is on crop production, as a comprehensive analysis of impacts on all food sectors (including livestock and dairy,

Resilience. DOI: https://doi.org/10.1016/B978-0-12-811891-7.00003-7

fisheries, horticulture, and viticulture) were beyond the scope of this analysis. However, several challenges and recommendations for the adaptation options discussed in this chapter are applicable to other food sectors as well.

3.2 BIBLIOGRAPHIC NETWORK ANALYSIS

A bibliographic network analysis is a type of literature review that allows visualization of the main structure of the body of literature in a given field and to identify leading authors and publications. Here, it was used to analyze the body of literature focusing on climate change and agriculture and determine the most influential publications that build the foundation of our current knowledge on climate change impacts and adaptation options in the agricultural sector.

A bibliographic network analysis is related to a systematic quantitative review (Pickering and Byrne, 2014) in that it systematically queries the existing literature based on objective search criteria that allow for reproducible results. Both types of literature review provide an overview of bibliometric characteristics (e.g., number of articles, geographic distribution of authors, time trends in publications) and can be used to identify gaps in the literature (Pickering and Byrne, 2014). However, there are differences in aims and approaches: A systematic quantitative review aims to evaluate the existing literature to answer specified research questions (e.g., "Does A have an effect on B?") and to provide a quantitative overview of methodological approaches and findings in the literature. It is generally based on a relatively narrow subsection of the literature, as it requires each publication included in the review to be (partially) read and its content to be documented in a database. A bibliographic network analysis, on the other hand, provides an overview of a whole research field and can be built upon a comparatively large number of publications. A bibliographic network analysis has not yet been applied to the literature focusing on climate change and agriculture, therefore, it was used here to visualize the body of literature, establish subclusters, and identify the most influential authors and publications.

3.2.1 Methodology

A bibliographic network analysis consists of (1) a systematic query of the existing literature, (2) the development of a bibliographic network based on publications and their citation links, and (3) identification of the most influential publications using an importance ranking algorithm (Knutas et al., 2015). An overview of the methodological approach is shown in Fig. 3.1; and each step is described in the following sections.

3.2.1.1 Systematic Query of the Literature

The Web of Science literature database, created by Thomson Reuters (WoS, 2017), was queried to systematically identify all relevant publications (peer-reviewed articles, proceedings papers, reviews, or book chapters) that focus on

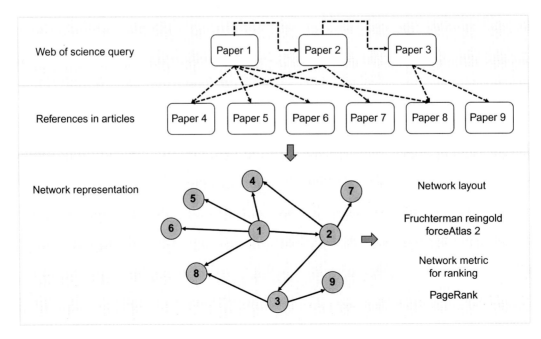

FIGURE 3.1 Overview of bibliographic network analysis.

the impacts of climate change and climate extreme events on agriculture and were published in the years 1990–2017. The search scope was limited to include documents that focus, directly or indirectly, on the four major crops—rice, wheat, maize, and soy—due to their relevance as staple crops. Combined, these four crops account for nearly two-thirds of consumed calories worldwide (Ray et al., 2013).

The following search terms were used:

- *In title*: (agricultur* OR crop OR (harvest* area) OR ("crop product*") OR ("agricultur* product*") OR "crop yield*" OR ("agricultur* yield*") OR food)
- *In topic (i.e., title, abstract or keywords)*: ("climate change" OR "high-impact weather" OR disaster OR "natural hazard" OR drought OR flood OR hail OR cyclone OR frost OR (extreme AND (weather OR climate OR precipitation OR temperature OR "soil moisture" OR dry OR wet OR cold OR hot OR warm OR indicator))) AND (maize OR rice OR wheat OR soy* OR corn OR "major crop*")

The query results, including their references, were downloaded in a standardized format provided as text files by Web of Science. The search yielded 3480 sources (2900 peer-reviewed articles, 383 proceedings papers, 354 reviews, and 170 book chapters, with overlapping classifications). In a following step, the references of each publication were extracted as individual references and—if not yet included within the search results—added to the literature database. Each document was identified by its DOI, where available, or a unique combination of first author name, year of publication, and journal. The data preparation of the Web of Science data was carried out in R using the "nails: Network Analysis Interface for Literature Studies" (Knutas et al., 2015; Knutas, 2017).

3.2.1.2 Creation of a Bibliographic Network

A bibliographic network was built using the complete set of publications and their citations. A network consists of a set of nodes that are pairwise connected via edges. In a bibliographic network, the references in the literature dataset are the nodes and their citations are directed edges. The direction of the edge implies the citation relationship, e.g., an edge from A to B means publication A referenced publication B.

Node and edge tables were created from the literature dataset and imported into Gephi, a network visualization and analysis tool (Bastian et al., 2009). After creating and importing the network structure, the nodes and edges were spatially organized using a combination of the Fruchterman Reingold (Fruchterman and Reingold, 1991) and ForceAtlas 2 (Jacomy et al., 2014) algorithms to identify clusters of documents that are highly connected.

3.2.1.3 Identification of the Most Influential Publications Using PageRank

The PageRank metric (Page et al., 1999) was calculated for all nodes to identify the most influential publications focusing on climate and agriculture. PageRank is a ranking algorithm that underpins the Google web search engine and ranks the relevance of publications by taking into account the number of publications that reference a given document, weighted recursively by their own PageRank values. This way, the PageRank metric reveals articles with particularly high relevance within a body of literature.

3.2.2 Overview of the Literature

3.2.2.1 Considerable Increase in Publications Focusing on Climate Change and Agriculture in Recent Decades

Research activities focusing on the links between climate change or climate extremes and agriculture have considerably intensified since 1990. The number of articles published per year increased by nearly 400% in the last decade, from 90 articles in 2005 to 445 articles in 2015 (Fig. 3.2). This increase reflects a growing interest from the research community and stakeholders in climate-related risks to food security.

3.2.2.2 Most Influential Authors Affiliated With US and European Institutions

Fig. 3.3 shows the 20 most important journals (A) and leading authors (B) based on articles that were found as part of the Web of Science query, ranked by the number of total citations. The author ranking reveals a dominance of US and European universities, which may reflect a (relative) underrepresentation of research perspectives from other geographic regions, particularly developing countries as well as a number of important food producing countries, such as China, India, and Russia.

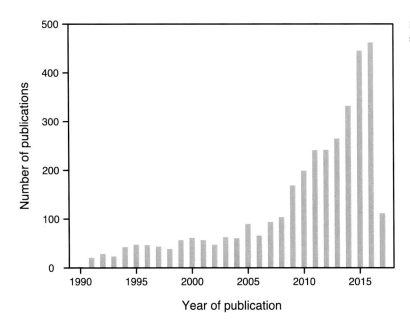

FIGURE 3.2 Publication count per year, 1990—2017, showing an increase by nearly 400% in the last decade.

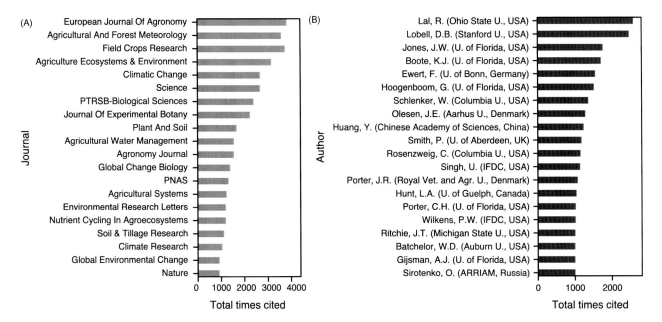

FIGURE 3.3 Most influential journals and authors. (A) shows journals, (B) authors, ranked by number of total number of citations among the articles found as part of Web of Science literature search. Among those journals with more than one article, the highest average citations per article are found in *Science* (1140 citations per article) and *Nature* (252 citations per article). Authors with the most citations are predominantly from US and European institutions. Please note, citations here refer to citations within the body of literature included in this literature review, not absolute number of citations.

3.2.2.3 Bibliographic Network Structure

The bibliographic network consists of 128,264 edges (publications originating from Web of Science query plus their references) and 189,788 directed links, i.e., their citations (Fig. 3.4). The network was filtered to only show publications that have been cited at least five times by publications within the network, and therefore, only 3.16% of nodes and 4.59% of edges are presented.

The network shows one main cluster of relevant publications with highly dense connections in the centre, but no clearly distinguishable secondary clusters. The concentric cloud of nodes around the main cluster, which are not connected to the centre, are publications that have been referenced at least five times, however, the referencing articles

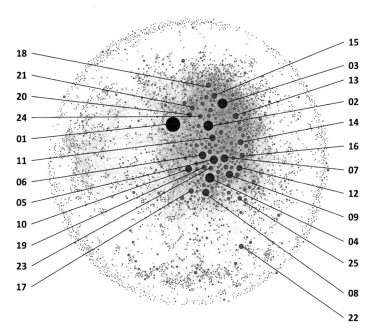

01 Allen RG, et al. (1998) Crop evapotranspiration-Guidelines for computing crop water requirements-FAO Irrigation and drainage paper 56.

02 Rosenzweig C, Parry ML (1994) Potential impact of climate change on world food supply. Nature 367(6459):133–138.

03 Jones JW, et al. (2003) The DSSAT cropping system model. Eur J Agron 18(3):235–265.

04 Lobell DB, et al. (2008) Prioritizing Climate Change Adaptation Needs for Food Security in 2030. Science 319(5863):607–610.

05 Parry M, et al. (2004) Effects of climate change on global food production under SRES emissions and socio-economic scenarios. Glob Environ Change 14(1):53–67.

06 Long SP, et al. (2006) Food for thought: lower-than-expected crop yield stimulation with rising CO2 concentrations. Science 312(5782):1918–1921.

07 Lobell, David B., and Christopher B. Field. "Global scale climate–crop yield relationships and the impacts of recent warming." Environmental research letters 2.1 (2007): 014002.

08 Peng S, et al. (2004) Rice yields decline with higher night temperature from global warming. Proc Natl Acad Sci U S A 101(27):9971–9975.

09 Lobell DB, Schlenker W, Costa-Roberts J (2011) Climate Trends and Global Crop Production Since 1980. Science 333(6042):616–620.

10 Keating BA, et al. (2003) An overview of APSIM, a model designed for farming systems simulation. Eur J Agron 18(3):267–288.

11 Schlenker W, Roberts MJ (2009) Nonlinear temperature effects indicate severe damages to U.S. crop yields under climate change. Proc Natl Acad Sci 106(37):15594–15598.

12 Jones PG, Thornton PK (2003) The potential impacts of climate change on maize production in Africa and Latin America in 2055. Glob Environ Change 13(1):51–59.

13 Olesen JE, Bindi M (2002) Consequences of climate change for European agricultural productivity, land use and policy. Eur J Agron 16(4):239–262.

14 Solomon S (2007) Climate change 2007-the physical science basis: Working group I contribution to the fourth assessment report of the IPCC (Cambridge University Press)

15 Mendelsohn R, Nordhaus WD, Shaw D (1994) The impact of global warming on agriculture: a Ricardian analysis. Am Econ Rev:753–771.

16 Asseng S, et al. (2013) Uncertainty in simulating wheat yields under climate change. Nat Clim Change 3(9):827–832.

17 Godfray HCJ, et al. (2010) Food Security: The Challenge of Feeding 9 Billion People. Science 327(5967):812–818.

18 Adams RM, et al. (1990) Global climate change and US agriculture. Nature 345(6272):219.

19 Lobell DB, Asner GP (2003) Climate and management contributions to recent trends in US agricultural yields. Science 299(5609):1032–1032.

20 Kimball BA (1983) Carbon dioxide and agricultural yield: an assemblage and analysis of 430 prior observations. Agron J 75(5):779–788.

21 Jones CA, Kiniry JR, Dyke PT (1986) CERES-Maize: A simulation model of maize growth and development (Texas A& M University Press)

22 Lal R (2004) Soil Carbon Sequestration Impacts on Global Climate Change and Food Security. Science 304(5677):1623–1627.

23 Christensen JH, Christensen OB (2007) A summary of the PRUDENCE model projections of changes in European climate by the end of this century. Clim Change 81:7–30.

24 Stöckle CO, Donatelli M, Nelson R (2003) CropSyst, a cropping systems simulation model. Eur J Agron 18(3):289–307.

25 Monfreda C, Ramankutty N, Foley JA (2008) Farming the planet: 2. Geographic distribution of crop areas, yields, physiological types, and net primary production in the year 2000. Glob Biogeochem Cycles 22(1):GB1022.

FIGURE 3.4 Bibliographic network of the scientific literature focusing on climate change/climate extremes and agriculture. *Dots* represent publications; *lines* represent citations (each line is curved in clockwise direction from referencing paper to cited paper). The size and color of each dot represents its importance based on the PageRank metric—the larger and darker the dot, the higher the PageRank value. The 25 most relevant papers are named and highlighted within the plot.

themselves were cited less than five times and are therefore not visible and do not connect those publications with the main body of literature. These nodes represent either articles that are less influential or central within the present subset of literature and therefore less connected or more recent publications that may be influential, but are not yet "weaved" into the body of scientific knowledge through citations.

The network structure, consisting of only one main cluster of dense connections, suggests that the scientific literature focusing on climate impacts on agriculture is highly interconnected, without any clearly separated subthemes or group of publications, and research findings are well integrated into the body of literature. This may be due to global efforts that help to facilitate the communication within the global agricultural research community, data and knowledge exchange, and the integration and alignment of research activities (e.g., by institutions or global programmes including the Food and Agriculture Organization, the Agricultural Model Intercomparison and Improvement Project, the International Food Policy Research Institute, CGIAR, and many others). However, it may also point toward a lack of diversity in the research field and a missing variety in research approaches. In view of the finding that the main influential authors are located at US and European institutions only (Section 3.2.2.2), it may also indicate an underrepresentation of researchers from certain geographic regions and a lack of diversity in perspectives.

3.2.2.4 Most Influential Publications

The most influential publications within the complete literature dataset, estimated via PageRank, are highlighted in Fig. 3.4. The most significant paper is a technical report entitled "Crop evapotranspiration-Guidelines for computing crop water requirements-FAO Irrigation and drainage paper 56" (Allen et al., 1998), which summarizes equations used for the calculation of evapotranspiration, particularly the Penman−Monteith equation. It has been cited 175 times within the literature database of this study (more than 19,000 times in total according to Google Scholar), among which there are many model description or benchmarking papers. Furthermore, among the 20 most influential articles within this study, four are model description papers, introducing the crop models DSSAT (Jones et al., 2003), CERES-maize (Jones et al., 1986), CropSyst (Stöckle et al., 2003), and APSIM (Keating et al., 2003). The article introducing DSSAT (Jones et al., 2003) is the third most influential paper. This highlights the importance of physical and technical foundations for the analysis of climate change impacts on agriculture. The second most influential paper dates back to 1994 and is one of the first major studies to systematically assess the implications of climate change for global food supply (Rosenzweig and Parry, 1994). This paper cautioned that the greatest share of impacts will be burdened by developing countries. The fourth most important paper discusses climate change adaption in the agricultural sector and outlines hotspots of adaptation needs worldwide (Lobell et al., 2008).

The distribution of articles examining the link between climate and agriculture in the past and future is equal. Six of the 20 most important articles focus on the impacts of temperature, precipitation, or increased CO_2 on agricultural yields or production using historical observations or experimental data (Kimball, 1983; Lobell et al., 2011; Lobell and Field, 2007; Long et al., 2006; Peng et al., 2004; Schlenker and Roberts, 2009). Similarly, six articles assess future climate change impacts on agriculture using model simulations (Adams et al., 1990; Asseng et al., 2013; Jones and Thornton, 2003; Olesen and Bindi, 2002; Parry et al., 2004; Rosenzweig and Parry, 1994).

Based on the bibliographic network developed here, the following sections will summarize the literature and review challenges for food security in the coming decades, impacts of climate change and climate extreme events on agriculture, and summarize options and challenges for adaptation in the agricultural sector.

3.3 THE CONTEXT: MAIN CHALLENGES FOR FOOD SECURITY IN THE 21ST CENTURY

3.3.1 Population Growth and Changes in Consumption Patterns Increase Food Demand

The global food system faces various challenges in the coming decades: A growing world population and changing consumption patterns mean that food demand will increase considerably in the next several decades. According to UN estimates, the global population will grow from 7.5 billion in 2017 to 9.7 billion people in 2050 (U.N. DESA, 2015). Most of this population growth will take place in developing countries, particularly in sub-Saharan Africa, which is going to exacerbate existing pressures on local food production. Some countries that are already food insecure today are going to increase their population by a factor of up to ∼4 (e.g., Niger, Yemen, and Uganda are estimated to grow in population by factors of 4.1, 2.8, and 3.0, respectively, until 2050 relative to 2006) (Alexandratos and Bruinsma, 2012). Additionally, human diets are changing. Due to growing income levels, the global average calorie intake is projected to rise, especially in developing countries, and consumers' preferences are changing toward higher shares of animal products—including meat, eggs, and dairy (Alexandratos and Bruinsma, 2012). The FAO estimates that food production must increase by 70% by 2050 relative to 2005/07 to meet increasing demands. In the developing countries, production has to nearly double (FAO, 2009).

3.3.2 Limitations in Land Area Available for Agricultural Production

There are constraints with respect to how much agricultural land can expand to meet increasing food demands. Arable land is limited and competing demands for land area (e.g., for biofuel production, bioenergy with carbon capture and storage, afforestation, or protection of ecosystems) put further constraints on the expansion of agricultural areas. According to estimates, most of the production growth (90% of the growth globally, 80% in developing countries) needs to be met by increases in yields and cropping intensity (FAO, 2009). Additionally, measures to reduce food losses and food waste and changes in diets, especially in developed countries, can help reduce the demand for food in the future (Fahrenkamp-Uppenbrink, 2016; Gustavsson et al., 2011; Lipinski et al., 2013).

3.3.3   Yield Trends and Yield Gaps

Current yield trends are insufficient to meet future production demands to ensure food security (Ray et al., 2013). Across 24%−39% of the cropping areas of the four major crops (wheat, maize, rice, soy), yields are not increasing— they are either stagnating, have peaked in the past and are currently on a downwards trend, or have never improved in the past (Ray et al., 2012). This is particularly the case for rice and wheat, where yield is not increasing in 37% and 39% of global cropping areas, respectively. Some of the stagnation or collapse in yields are due to socioeconomic factors (e.g., financial resources of farmers) or due to management (Mueller et al., 2012). However, in some regions, yield stagnation or collapse is related to adverse impacts from climate change (Brisson et al., 2010; Hochman et al., 2017; Lobell et al., 2012; Peng et al., 2004), which are predicted to worsen in many regions in the future (Rosenzweig et al., 2013). Ensuring the food security of an ever-growing world population, while adapting to an already changing climate and at the same time transitioning all sectors and aspects of human life (including agriculture) to a carbon-neutral future will pose significant challenges (Godfray et al., 2010). The next section will explore the impacts of climate change and climate extreme events on agriculture in more detail.

3.4   IMPACTS OF CLIMATE CHANGE AND CLIMATE EXTREME EVENTS ON CROP PRODUCTION

3.4.1   Introduction

Climate change has an impact on agroecological variables and hence the suitability of land areas for cropping and crop yields (Schmidhuber and Tubiello, 2007). Agricultural yields and production are significantly influenced by climate conditions during the growing season. Approximately one-third of the variance in yield anomalies at the global scale can be explained by interannual changes in precipitation and temperature (Ray et al., 2015), with explained variances of more than 50% in many major crop producing countries (Frieler et al., 2017). Plant growth processes are driven by climate variables (temperature, precipitation, solar radiation), soil physical and chemical processes and parameters (e.g., soil moisture, temperature, acidity, bulk density and nutrient concentrations), atmospheric carbon dioxide and ozone concentrations, and microbial processes (Schauberger et al., 2016), which are all directly or indirectly affected by climate change.

3.4.2   Climate Trends and Agricultural Production

The global mean temperature increased by $\sim 0.86°C$ between 1880 and 2012 (IPCC, 2013). In relation to this, the growing seasons of the most important producing countries of the major crops almost all experienced significant temperature increases that exceeded one standard deviation of the past interannual variability (with the exception of the United States) (Lobell et al., 2011). The effect of climate change on precipitation is more complex and spatially heterogeneous. Many high-latitude regions are expected to become wetter, while arid and semiarid regions in the subtropics and mid-latitudes are expected to become drier (IPCC, 2013).

Increases in temperature lead to accelerated crop growth and shorten the time for plants to absorb radiation and nutrients, thereby decreasing yields and yield quality (Howden et al., 2003). Temperature is a key driver of evapotranspiration and warming can reduce the soil water content, which affects processes that strongly depend on water availability, such as photosynthesis and carbon allocation, nutrient uptake, and reproduction (McGregor and Dawson, 2016; Schauberger et al., 2016). Direct effects of temperature on crop yields result from impacts on enzyme production and gene expression (McGregor and Dawson, 2016; Reynolds et al., 2010). Previous research has identified detrimental temperature thresholds for various crops (McGregor and Dawson, 2016; Porter and Gawith, 1999; Wheeler et al., 2000). For example, temperatures exceeding about 35°C lead to sterility in rice, and temperatures beyond 36°C are associated with reduced pollen viability in maize (Gornall et al., 2010; McGregor and Dawson, 2016). Nonlinearities in yield responses to increasing temperatures (Schlenker and Roberts, 2009) mean that negative effects of high temperatures will become more harmful with increasing temperatures.

The trends in climate in the last decades already had a negative impact on global crop yields for several major crops (Lobell and Field, 2007) and have in some regions even suppressed yield growth (Hochman et al., 2017; Lobell and Field, 2007). Wheat and maize production were particularly affected by climate trends, with aggregated climate-related production losses of 3.8% and 5.5%, respectively, over the period 1980−2008 (Lobell et al., 2011). Impacts of climate on global agricultural production trends are primarily related to temperature trends, since these trends are more homogeneous and significant at the global scale (Lobell and Field, 2007).

3.4.3 Changes in the Variability of Climate and Climate Extreme Events

In addition to global mean trends, changes in climate variability, including the frequency and severity of extreme events, pose significant threats to agricultural production in the future (Thornton et al., 2014). It is these types of changes that are the most challenging to adapt to due to their larger uncertainties (Bryant et al., 2000; Rosenzweig et al., 2001). The warming of the atmosphere has led to warming trends in both hot and cold temperature extremes, with an increase in anomalous warm days and a reduction in cold days (IPCC, 2012, 2013). Heat waves have become more frequent and their duration has increased in most regions across the globe (IPCC, 2013). Climate change alters the shape of the distribution of precipitation and leads to an increase in the spatial and seasonal variability of rainfall (IPCC, 2013), which has many implications, including for soil water availability in cropping systems. Projected changes in precipitation extremes are more spatially heterogeneous than changes in temperature extremes. In the Mediterranean and West Africa, the frequency and intensity of droughts have increased, while in central North America and northwest Australia, the number and intensity of droughts have decreased since 1950 (IPCC, 2012, 2013).

Lesk et al. (2016) analyzed the impacts of four types of climate related extreme events (drought, flood, hot temperature, and cold temperature extremes) on crop production between 1964 and 2007. Their research suggested that drought and extreme heat events significantly reduced national agricultural production by 9%−10% compared to expected production. Crops are particularly vulnerable to a combination of different climate extremes, particularly drought and high temperature extremes. The opening and closure of leaf stomata regulate two key physiological processes—transpiration and photosynthesis—leading to critical trade-offs between water and temperature regulation and carbon uptake. Stomatal closure as a coping strategy in response to water stress can lead to additional heat stress and reductions in yields, as transpiration and photosynthesis are reduced (Davies et al., 2005).

3.4.4 Impacts of Increased CO_2 Concentrations

Increased atmospheric CO_2 concentrations have a positive effect on crop yields and may offset some of the negative effects related to climate change (Tubiello et al., 2007a). However, the magnitude of this effect is highly uncertain and the subject of ongoing research (Ainsworth et al., 2008; Long et al., 2006; Parry et al., 2004; Rosenzweig et al., 2013; Tubiello et al., 2007b). The CO_2 effect on plant growth and water use efficiency varies considerably between regions as it depends on crop type and interactions with climate, soil properties, nutrient availability, pests and diseases, air pollutants, and other factors. Hence, beneficial CO_2 effects may not be experienced equally in all regions across the globe (Deryng et al., 2016; McGrath and Lobell, 2013; Tubiello et al., 2007a).

Additionally, findings from free-air carbon dioxide enrichment experiments (Ainsworth and Long, 2005; McLeod and Long, 1999) indicate that elevated CO_2 concentrations will reduce the levels of vital nutrients, such as zinc, iron, and protein, in major crops (Ainsworth and McGrath, 2010; DaMatta et al., 2010; Erda et al., 2005; Myers et al., 2014). This has important health implications for communities worldwide, especially in developing countries in which a large share of the population depends on these crops as their main source of nutrients. Iron deficiency increases mortality from infectious diseases and poses significant health risks for pregnant women (Allen, 2000), while zinc deficiency is related to increased, maternal morbidity, child mortality, and growth impairment (Arlette, 1982; Black, 1998, 2003; Prasad, 1996). These findings highlight that, although the CO_2 fertilization effect may offset yield losses or even increase yields in some regions, the dilution of nutrients in major food crops counteracts these positive effects and poses a significant health risk for populations depending on cereals for their nutrient intake.

3.4.5 Other Impacts of Climate Change on Food Production

Climate change further affects food security via the following, indirect impact pathways:

- *Pests and diseases:* Many crop pests and diseases are influenced by climate conditions and hence their distribution patterns are predicted to change in a warming climate (Howden et al., 2003; Rosenzweig et al., 2001; Tubiello et al., 2007a).
- *Sea level rise:* Sea level rise due to climate change leads to salt-water intrusion into aquifers and increases the frequency of storm surges, with negative consequences for food production in coastal areas and atoll islands (McGregor and Dawson, 2016).
- *Impacts on farm operations:* Climate extreme events, such as heavy precipitation events and flooding, can lead to waterlogging and soil erosion, which both may negatively affect farm operations.
- *Impacts on food transport, storage, and cooling:* Extreme events, such as flooding or heatwaves, can increase the risks of food losses and waste along the transport chain. For example, flooding events can block important transport routes, while heatwaves may increase food losses in regions where sufficient cooling is not available (Gustavsson et al., 2011; Jowel et al., 2017).
- *Impacts on health of farm labor:* Climate extreme events, particularly heat extremes and drought, have negative effects on the health of farm labor in regions where agriculture is based primarily on manual labor (McGregor and Dawson, 2016). Furthermore, heat waves and droughts, and related production losses, have negative impacts on the mental health of farm managers and have been associated with increased rates of suicide among farmers (Carleton, 2017; Hanigan et al., 2012).

In summary, adapting the global food system to a changing climate is critical to increasing agricultural productivity to ensure food security under more uncertain and variable climate conditions in the future. The following section will highlight adaptation options and the main challenges to adaptation in the agricultural sector.

3.5 ADAPTING GLOBAL CROP PRODUCTION TO CLIMATE CHANGE

Adaptation is defined as "reductions in risk and vulnerability through the actions of adjusting practices, processes, and capital in response to the actuality or threat of climate change" (Porter et al., 2014). Adaptation is context-specific and depends on local conditions, including regional historical climate and future climate projections, biophysical constraints, traditional agricultural practices, socioeconomic conditions, as well as political and institutional settings (Belliveau et al., 2006). This section provides an overview of the types of adaptation and adaptation options in the agricultural sector, and outlines the main challenges.

3.5.1 Types of Adaptation

3.5.1.1 Incremental, Systems and Transformational Adaptation

The literature differentiates adaptation based on the scale and complexity of change, ranging from incremental adaptation to more profound systemic and transformational adaptation (Howden et al., 2010; Kates et al., 2012; Park et al., 2012; Porter et al., 2014; Rickards and Howden, 2012). The majority of adaptation options considered by research as well as by farmers concern short-term incremental changes to an existing agricultural system (Park et al., 2012). However, under business-as-usual scenarios of climate change (Meinshausen et al., 2011), such changes may not be sufficient and more profound, transformational changes may be required (Kates et al., 2012). Incremental adaptation comprises short-term measures that maintain existing food production systems (Park et al., 2012) and usually enhance existing coping strategies and behaviors, e.g., those to deal with natural climate variability (Kates et al., 2012). Examples are changes in sowing and planting dates, additional irrigation, or planting of more heat- or drought-tolerant crop varieties (Rickards and Howden, 2012). Systems adaptation lies in between incremental and transformational adaptation in terms of degree of change and complexity and includes more fundamental modifications of the system in place. Examples are introducing climate-sensitive precision agriculture or diversification of agricultural productions and risk management (Rickards and Howden, 2012).

Transformational adaptations are more profound changes to the agricultural production landscape and include measures that are either (1) implemented at a larger scale or intensity than known processes, (2) measures that have not previously been applied in a specific location, or (3) measures that transform a whole region or relocate production systems to new regions (Kates et al., 2012; Rickards and Howden, 2012). An example of transformational adaptation in agriculture is the relocation of wine production to more suitable regions under future climate conditions, as viticulture depends on very narrow climate niches (Moriondo et al., 2013).

3.5.2 Adaptation Options

Adaptation options depend strongly on the agricultural system, the region and climate scenario considered (Rosenzweig and Tubiello, 2007). Adaptation options are fundamentally different for crop production, livestock, viticulture, horticulture, and fishing. For regions which are projected to become warmer and drier, the breeding and sowing of drought-tolerant crop varieties or species and earlier planting during the colder season will be considered adaptation options. In contrast, in regions that are predicted to become wetter, the focus may be on erosion mitigation measures and adaptation to increased pathogen risks. Table 3.1 provides an overview of examples of adaptation options for crop production, some of which will be described in the following sections.

TABLE 3.1 Examples of Adaptation Options for Crop Production

Area	Adaptation Options
Farm management	• Changes in planting and harvesting times (e.g., earlier planting to benefit from lower frost risk in cooler months) • Optimization of row spacing, plant population
Change of crop cultivars and species	• Breeding and planting of more adapted crop cultivars or crop species (e.g., slower maturing varieties, drought-tolerant crops) • Support of breeding activities, research, and development
Water conversation and erosion reduction	• Adopting zero- or minimal-tillage and reducing soil disturbance to increase infiltration of water and reduce soil erosion • Contour banking, maintaining crop cover, and retaining crop residues to reduce erosion • Water and soil moisture conservation measures, e.g., extension of fallows, raised bed technologies • Adding irrigation, reducing irrigation losses
Nutrient management	• Altering crop rotations, e.g., including pulses to offset nitrogen losses • Optimizing fertilizer applications based on predictions and monitoring of climate data, soil moisture stores and soil nutrient data • Breeding and use of crop varieties with higher nutrient content
Pests and diseases	• Plant breeding and genetic modification to produce more resistant crop varieties • Altering crop rotations, physical measures to reduce transmission of diseases • Application of pesticides • Predictive modeling of pest and disease risks
Seasonal forecasting and observations	• Seasonal predictions and on-the-ground monitoring of climate data, soil water storage and/or pest and disease risks
Financial options and risk management	• Implementing or adopting crop insurance • Changes of lending policies of financial institutions to reflect changes in risks • Diversifying income, e.g., mixed crop–livestock farming • Purchasing farm land in regions that may become more suitable for given agricultural sector in the future
Research and education	• Supporting breeding programmes • Facilitating and integrating research into cultivar development, breeding, management, and technology • Knowledge exchange and community learning processes among farmers
Governmental programs	• Financial assistance for adaptation efforts, e.g., investments in irrigation infrastructure, investments into research and development • Cross-policy integration of adaptation policies in a range of sectors (agriculture, water, energy, etc.)

Sources: Benhin, J.K., 2006. Climate change and South African agriculture: impacts and adaptation options, CEEPA discussion paper. Available from: <https://www.researchgate.net/profile/James_Benhin/publication/265074602_CLIMATE_CHANGE_AND_SOUTH_AFRICAN_AGRICULTURE_IMPACTS_AND_ADAPTATION_OPTIONS/links/54b930910cf2d11571a31c86/CLIMATE-CHANGE-AND-SOUTH-AFRICAN-AGRICULTURE-IMPACTS-AND-ADAPTATION-OPTIONS.pdf> (accessed 15.10.17.) (Benhin, 2006); Easterling, W.E., 1996. Adapting North American agriculture to climate change in review. Agric. For. Meteorol. 80(1), 1–53, doi:10.1016/0168-1923(95) 02315-1 (Easterling, 1996); Howden, M.S., Ash, A., Barlow, S., Booth, T., Charles, S., Cechet, B., et al., 2003. An Overview of the Adaptive Capacity of the Australian Agricultural Sector to Climate Change: Options, Costs and Benefits. Australian Greenhouse Office, Canberra, ACT (Howden et al., 2003); Howden, S.M., Soussana, J.-F., Tubiello, F.N., Chhetri, N., Dunlop, M., Meinke, H., 2007. Adapting agriculture to climate change. Proc. Natl. Acad. Sci. 104(50), 19691–19696 (Howden et al., 2007); Kates, R.W., Travis, W.R., Wilbanks, T.J., 2012. Transformational adaptation when incremental adaptations to climate change are insufficient. Proc. Natl. Acad. Sci. 109(19), 7156–7161, doi:10.1073/pnas.1115521109 (Kates et al., 2012); Rosenzweig, C., Tubiello, F.N., 2007. Adaptation and mitigation strategies in agriculture: an analysis of potential synergies. Mitig. Adapt. Strateg. Glob. Change 12(5), 855–873, doi:10.1007/s11027-007-9103-8 (Rosenzweig and Tubiello, 2007).

3.5.2.1 Adjustments in Farm Management

Many short-term adaptations are already employed in many regions to cope with the impacts of natural climate variability (Howden et al., 2003), such as the El Nino Southern Oscillation (Iizumi et al., 2014). Farmers are already adjusting their farming operations to changes in precipitation and temperature. For example, they may optimize planting dates toward earlier sowing to benefit from reduced frost risks during the cold season and to avoid negative impacts from extreme heat during later crop growth stages (Howden et al., 2003, 2007). Reduced nutrient contents due to CO_2 fertilization may be offset by changes in crop rotations (i.e., inclusion of legumes to increase nitrogen contents) or by optimized fertilizer applications (Howden et al., 2003, 2007).

3.5.2.2 Change of Crop Cultivars and Species

Over the long term, one of the most important adaptation measures will be the breeding and selection of crop varieties that are better adapted to future climate conditions. Under higher temperatures, phenological stages will be shorter, resulting in reduced yields. These may be offset using slower-maturing crop varieties (Howden et al., 2003). In regions with higher risks of drought or high temperature extremes, cultivars with increased drought tolerance or lower temperature sensitivity will help mitigate negative effects (Bänziger et al., 2006; Cattivelli et al., 2008; Challinor et al., 2016; Howden et al., 2003, 2007; Kumar et al., 2008).

3.5.2.3 Water Conversation and Erosion Reduction

Changes in hydrological conditions require improvements in water and soil erosion management. For example, in regions that become drier, an important focus is to increase the retention of water by increasing the soil cover (e.g., using crop residues in combination with zero- or minimum-tillage), extending fallows or applying raised-bed techniques (Howden et al., 2003, 2007). Where rainfall intensities are projected to increase, adaptation will have to focus on soil erosion reduction to maintain soil fertility. Measures include increasing soil cover (e.g., by using crop residues or cover crops), reducing soil disturbance (e.g., minimum- or zero-tillage), and implementing contour buffer strips or vegetated waterways (Howden et al., 2003, 2007; Vogel et al., 2016).

3.5.2.4 Seasonal Forecasting and Observations

Climate information tailored to the needs of the agricultural sector provide a useful tool for farmers to adapt to impacts of climate change (Ziervogel et al., 2010). Seasonal forecasts of climate and hydrological variables are particularly important to help farmers reduce impacts from climate extreme events (e.g., heatwaves or drought). These forecasts provide the advanced warning required for optimized irrigation planning, fertilizer and pesticide applications, and the scheduling of planting and harvest dates (Cantelaube and Terres, 2005; Hansen, 2002; Hansen et al., 2011; Howden et al., 2003, 2007; Klopper et al., 2006) and can help reducing some of the negative impacts of climate extreme events.

3.5.2.5 Financial Options and Risk Management

More systemic or transformational adaptation options include the adjustment of agricultural businesses and diversification of risks. For example, farmers may switch to mixed crop–livestock farming systems (Bell and Moore, 2012; Ryschawy et al., 2012; Schiere and Kater, 2001; Thornton and Herrero, 2015), or relocate farm enterprises to regions that will become more suitable to the specific industry in the future (Anderson et al., 2008; Cabré et al., 2016).

3.5.3 Challenges and Opportunities

Adaptation is generally considered in reaction to a changing climate or climate shocks and framed as avoidance of negative impacts. Given that a reactive approach leads to harm before implementing changes, which in itself can reduce the adaptive capacity of communities or nations, it is preferable to take up an anticipatory approach and focus on the benefits of adaptation. A more positive framing focuses on the chance for communities to improve existing systems and adopt more sustainable and desirable farming practices. It can help overcome existing barriers that hinder the implementation of adaptation measures, particularly the uncertainties and costs related to such measures and institutional and behavioral barriers (Kates et al., 2012). There are several other factors that may slow down adaptation in the food sector, including the costs of adaptation, uncertainties of impacts, the adaptive capacity of communities, and institutional barriers.

3.5.3.1 Costs of Adaptation

Rickards and Howden (2012) summarize the costs of adaptation and differentiate the following categories:

- *Transaction costs:* These costs include economic costs, but also tolls on mental, emotional, and social resources. Transaction costs are an important source of reluctance among farmers for adapting to future climate change, as understanding climate impacts and the uncertainty of such impacts may exceed the time and resources available to the farm manager.
- *Opportunity costs:* These costs relate to the path-dependency of adaptation and reflect that some measures may limit options available in the future (Inderberg and Eikeland, 2009). Ideally, adaptation measures are no-regret options, that provide benefits besides avoiding negative climate change impacts; an example is the increase in soil organic carbon, which has positive effects on soil properties more generally (Lal, 2004a,b; Meyer et al., 2015). Costs related to the avoidance of climate change adaptation fall into this category as well.
- *Unintended consequences:* These costs include long-term, unintended costs (e.g., increases in greenhouse gas emissions) related to adaptation measures.
- *Residual costs:* Since perfect adaptation will never be possible (due to limited resources and uncertainties in climate predictions), these costs include costs related to the incompleteness of adaptation, i.e., residual climate change impacts, which were not avoided.

While the benefits of adaptation usually outweigh the costs of adaptation, the initial investments may be beyond the financial capacity of countries or individual food producers. Therefore, these barriers need to be accounted for and financial support mechanisms put in place.

3.5.3.2 Uncertainties of Impacts

One of the most important obstacles is the uncertainty in climate projections, particularly at the regional and local scale (Knutti and Sedláček, 2012; Olesen et al., 2007). As an example, projections of the frequency of precipitation extremes at the local or regional scale show high uncertainties and large model spread. When aggregated over larger regions, however, projections become more robust (Fischer et al., 2013). Therefore, adaptation planning should be matched with the spatial and temporal scale at which climate projections are available and reliable, i.e., by creating policies that are applicable over larger geographic areas (i.e., at the national or state level) and diversifying risks over a larger geographic region or longer time frames.

3.5.3.3 Adaptive Capacity

Adaptation, particularly transformational, large-scale measures, can have requirements that may be beyond the adaptive capacity of communities and farm managers. The definition of adaptive capacity is generally based on sustainable livelihoods theory and includes five types of resources or assets—social, human, physical, natural, and financial capital (Ellis, 2000; Nelson et al., 2005, 2010; Smit and Wandel, 2006). One of the most important barriers to adaptation can be the costs of it (Kates et al., 2012), in cases where economic resources are limited and other more near-term needs have to be met. This can be the case at the individual farm level as well as at the national/governmental scale. Adaptive capacity also includes the willingness to adapt, which depends on many factors including the perceived benefits and legitimacy of the adaptation process (Brooks and Adger, 2005; Jones et al., 2010; Rickards and Howden, 2012).

3.5.3.4 Institutional Barriers

Institutional barriers can hinder effective climate change adaptation (Inderberg and Eikeland, 2009). Oberlack (2017) identified 28 common institutional constraints to climate change adaptation; among those barriers are:

- *Eligibility and responsibility for adaptation decision-making*: e.g., unclear, fragmented, or overlapping responsibilities for adaptation processes (resulting in high transaction costs and inefficiencies in decision-making processes); or clear responsibilities, but within a too rigid framework that does not reflect the biophysical and socioeconomic reality;
- *Control over adaptation outcomes*: e.g., vested interests of decision-makers and institutions, or limited control in governance systems where power is distributed between different levels of government or actors;

- *Social connectivity and conflict presentation*: e.g., mal-coordination (including lack of coordination, siloing) between agencies, mistrust or competition among organizational actors, ineffective conflict prevention or resolution mechanisms;
- *Social learning:* limited mechanisms for knowledge exchange and learning, limited exchange between science and policy;
- *Accountability:* lack of accountability for adaptation outcomes or inefficient processes for enforcing responsibilities;
- *Inherent attributes of institutions:* e.g., temporal mismatches between decision-making processes and incentives in governmental institutions and the timescale of climate change processes; too high or too low institutional adaptiveness (rigidity or instability of governmental structures).

Successful adaptation requires an integrated approach that cuts across a number of governmental institutions and policy themes. Adapting agriculture to climate change has linkages with other policy areas, including water management (in particular drought and flood preparedness), economic policies (e.g., subsidies, trade policies), climate change mitigation policies (Box 3.1), as well as infrastructure, health, and other sectors (Howden et al., 2007). More broadly, climate change adaptation relies on and should be embedded into policies that foster sustainable development and increase resilience of communities as a whole (Schipper, 2007).

BOX 3.1 Integrating Climate Change Mitigation and Adaptation in Agriculture

The food sector is responsible for approximately a third of global greenhouse gas emissions (Gilbert, 2012; Vermeulen et al., 2012) and hence agriculture plays an important role in avoiding dangerous climate change. Mitigation measures, i.e., "interventions to reduce the sources or enhance the sinks of greenhouse gases" (IPCC, 2014), will reduce the extent to which the agricultural sector will need to adapt to climate change (Field et al., 2014; Watkiss et al., 2015).

Co-benefits between mitigation and adaptation are common in the agricultural sector. For instance, a review of climate change project designs for the forestry and agricultural sectors found that, although not typically made explicit, all mitigation projects had potential to contribute to adaptation and approximately 90% of the analyzed adaptation projects had potential mitigation co-benefits (Kongsager et al., 2016). Smith and Olesen (2010) identified six categories of potential mitigation co-benefits from agricultural adaptation: (1) reduction of soil erosion; (2) reduction of nitrogen and phosphorus losses; (3) conservation of soil moisture; (4) increased diversity of crop rotations; (5) positive changes to the microclimate; and (6) positive effects on land use. Resource use efficiency and practices that increase soil carbon are often mentioned as providing both adaptation and mitigation benefits (Rosenzweig and Tubiello, 2007; Stokes and Howden, 2010). In the case of soil carbon sequestration, co-benefits are temporally separated with mitigation potential declining and adaptive benefits increasing as soil carbon increases (Meyer et al., 2016). Effective adaptation to climate change increases yields and thereby reduces the need to convert land for agricultural purposes. Lobell et al. (2013) found that investments in agricultural adaptation of about 225 billion USD by 2050 would lead to yield increases that reduce land requirements by 61 Mha and greenhouse gas emissions by 15 $GtCO_2eq$ over the coming decades. This is equivalent to 0.35 $GtCO_2eq$ avoided emissions per year and 15 USD per avoided $GtCO_2eq$ (Lobell et al., 2013). Mitigation or adaptation options, however, can also negatively affect one another, leading to trade-offs which need to be accounted for. Examples of agricultural adaptation negatively affecting climate change mitigation include the installation of irrigation systems to address increasing drought risks or the additional application of nitrogen fertilizers to offset nutrient reductions or yield losses. Both approaches lead to substantial increases in energy use and therefore may increase greenhouse gas emissions, depending on the energy mix (Cullen and Eckard, 2011; Eckard and Cullen, 2011; Olesen, 2006; Stokes and Howden, 2010). An example of mitigation efforts with negative implications for adaptation is the use of biofuels, which affect land and water availability and limit the adaptive capacity of food systems (Watkiss et al., 2015). Explicit trade-offs between mitigation and adaptation exist, for instance, when both efforts have the same funding source (Klein et al., 2007). Climate change mitigation and adaptation differ in several respects that can lead to their interconnections and contradictions being overlooked (Berry et al., 2015; Locatelli et al., 2015). These differences include the spatial and temporal scale of the solutions and the sectors concerned (Harvey et al., 2014; Klein et al., 2007; van Vuuren et al., 2011). Such discrepancies can contribute to a lack of integration of mitigation and adaptation planning resulting in unintended negative side effects (Locatelli et al., 2015), higher costs of mitigation and adaptation (Kongsager et al., 2016), and underestimation of co-benefits (Suckall et al., 2015). In summary, an integrated approach to adaptation and mitigation in the agricultural sector is preferable, as it maximizes co-benefits and reduces negative trade-offs.

3.6 CONCLUSIONS

A bibliographic network analysis was used to provide an overview of the research landscape focusing on climate change and agriculture and to identify most influential publications. The results indicate that the body of literature is relatively integrated, without clear subclusters, but instead a well-connected network of publications. The network structure suggests a high connectivity and communication within the agricultural research community, but may also indicate a lack of diversity in research approaches and perspectives. The most influential authors are primarily from institutions in the United States and Europe, suggesting an underrepresentation of other geographic regions.

Recent climate change and associated temperature trends have already had negative impacts on crop yields. Depending on human mitigation efforts and the magnitude of climate change in the future, these negative impacts are projected to worsen in many regions across the globe. Ensuring food security for nine billion people by 2050 under these conditions is a complex challenge and requires strategic, widespread adaptation efforts.

There are many uncertainties in climate projections, particularly at the regional scale. However, a lack of certainty should not deter planners from assessing risks and implementing adaptation options in a proactive way, as a lack of early action may decrease adaptive capacity in the future. Incremental adaptation may not be sufficient to respond to climate impacts. Depending on the magnitude of the changes large-scale transformational adaptation efforts may be needed. These efforts will require a broad, integrated, intersectoral and interdisciplinary approach that is suited to local food production systems, biophysical and socioeconomic conditions, and that is robust enough to withstand a variety of climate scenarios and regional model projections.

ACKNOWLEDGMENTS

The authors thank the University of Melbourne and the Australian-German Climate & Energy College for their support, David Karoly and Sebastian Rattansen for comments on drafts of the chapter, and Sebastian Rattansen for providing input on the methodology. EV acknowledges the support of the Australian Research Council (ARC) Centre of Excellence for Climate System Science. EV and RM were supported by an Australian Government Research Training Award.

REFERENCES

Adams, R.M., Rosenzweig, C., Peart, R., Ritchie, J.T., McCarl, B.A., et al., 1990. Global climate change and US agriculture. Nature 345 (6272), 219.

Ainsworth, E.A., Long, S.P., 2005. What have we learned from 15 years of free-air CO_2 enrichment (FACE)? A meta-analytic review of the responses of photosynthesis, canopy properties and plant production to rising CO_2. New Phytol. 165 (2), 351−372.

Ainsworth, E.A., McGrath, J.M., 2010. Direct effects of rising atmospheric carbon dioxide and ozone on crop yields. Climate Change and Food Security. Springer, Dordrecht.

Ainsworth, E.A., Leakey, A.D.B., Ort, D.R., Long, S.P., 2008. FACE-ing the facts: inconsistencies and interdependence among field, chamber and modeling studies of elevated [CO_2] impacts on crop yield and food supply. New Phytol. 179 (1), 5−9. Available from: https://doi.org/10.1111/j.1469-8137.2008.02500.x.

Alexandratos, N., Bruinsma, J., 2012. Available from: <http://large.stanford.edu/courses/2014/ph240/yuan2/docs/ap106e.pdf> (accessed 15.05.17.) World Agriculture Towards 2030/2050: The 2012 Revision, ESA Working paper No 12-03. FAO, Rome.

Allen, L.H., 2000. Anemia and iron deficiency: effects on pregnancy outcome. Am. J. Clin. Nutr. 71 (5), 1280s−1284s.

Allen, R.G., Pereira, L.S., Raes, D., Smith, M. et al., 1998. Crop Evapotranspiration-Guidelines for Computing Crop Water Requirements-FAO Irrigation and drainage paper 56, FAO Rome, 300(9), D05109.

Anderson, K., Findlay, C., Fuentes, S., Tyerman, S., 2008. Viticulture, wine and climate change. Garnaut Clim. Change Rev. 1−22.

Arlette, J.P., 1982. Zinc deficiency in children. Int. J. Dermatol. 21 (8), 447−448. Available from: https://doi.org/10.1111/j.1365-4362.1982.tb03179.x.

Asseng, S., Ewert, F., Rosenzweig, C., Jones, J.W., Hatfield, J.L., Ruane, A.C., et al., 2013. Uncertainty in simulating wheat yields under climate change. Nat. Clim. Change 3 (9), 827−832. Available from: https://doi.org/10.1038/nclimate1916.

Bänziger, M., Setimela, P.S., Hodson, D., Vivek, B., 2006. Breeding for improved abiotic stress tolerance in maize adapted to southern Africa, Agric. Water Manag. 80 (1), 212−224. Available from: https://doi.org/10.1016/j.agwat.2005.07.014.

Bastian, M., Heymann, S., Jacomy, M., et al., 2009. Gephi: an open source software for exploring and manipulating networks. ICWSM 8, 361−362.

Bell, L.W., Moore, A.D., 2012. Integrated crop−livestock systems in Australian agriculture: trends, drivers and implications. Agric. Syst. 111 (Suppl. C), 1−12. Available from: https://doi.org/10.1016/j.agsy.2012.04.003.

Belliveau, S., Smit, B., Bradshaw, B., 2006. Multiple exposures and dynamic vulnerability: evidence from the grape industry in the Okanagan Valley, Canada. Glob. Environ. Change 16 (4), 364−378. Available from: https://doi.org/10.1016/j.gloenvcha.2006.03.003.

Benhin, J.K., 2006. Climate change and South African agriculture: impacts and adaptation options, CEEPA discussion paper. Available from: <https://www.researchgate.net/profile/James_Benhin/publication/265074602_CLIMATE_CHANGE_AND_SOUTH_AFRICAN_AGRICULTURE_IMPACTS_AND_ADAPTATION_OPTIONS/links/54b930910cf2d11571a31c86/CLIMATE-CHANGE-AND-SOUTH-AFRICAN-AGRICULTURE-IMPACTS-AND-ADAPTATION-OPTIONS.pdf> (accessed 15.10.17.).

Berry, P.M., Brown, S., Chen, M., Kontogianni, A., Rowlands, O., Simpson, G., et al., 2015. Cross-sectoral interactions of adaptation and mitigation measures. Clim. Change 128 (3−4), 381−393. Available from: https://doi.org/10.1007/s10584-014-1214-0.

Black, M.M., 1998. Zinc deficiency and child development. Am. J. Clin. Nutr. 68 (2), 464S−469S.

Black, R.E., 2003. Zinc deficiency, infectious disease and mortality in the developing world. J. Nutr. 133 (5), 1485S−1489S.

Brisson, N., Gate, P., Gouache, D., Charmet, G., Oury, F.-X., Huard, F., 2010. Why are wheat yields stagnating in Europe? A comprehensive data analysis for France. Field Crops Res. 119 (1), 201−212. Available from: https://doi.org/10.1016/j.fcr.2010.07.012.

Brooks, N., Adger, W.N., 2005. Assessing and enhancing adaptive capacity. Adapt. Policy Framew. Clim. Change Dev. Strateg. Policies Meas. 165−181.

Bryant, C.R., Smit, B., Brklacich, M., Johnston, T.R., Smithers, J., Chjotti, Q., et al., 2000. Adaptation in Canadian agriculture to climatic variability and change. Clim. Change 45 (1), 181−201. Available from: https://doi.org/10.1023/A:1005653320241.

Cabré, M.F., Quénol, H., Nuñez, M., 2016. Regional climate change scenarios applied to viticultural zoning in Mendoza, Argentina. Int. J. Biometeorol. 60 (9), 1325−1340. Available from: https://doi.org/10.1007/s00484-015-1126-3.

Cantelaube, P., Terres, J.-M., 2005. Seasonal weather forecasts for crop yield modelling in Europe. Tellus A 57 (3), 476−487. Available from: https://doi.org/10.1111/j.1600-0870.2005.00125.x.

Carleton, T.A., 2017. Crop-damaging temperatures increase suicide rates in India. Proc. Natl. Acad. Sci. 114 (33), 8746−8751. Available from: https://doi.org/10.1073/pnas.1701354114.

Cattivelli, L., Rizza, F., Badeck, F.-W., Mazzucotelli, E., Mastrangelo, A.M., Francia, E., et al., 2008. Drought tolerance improvement in crop plants: an integrated view from breeding to genomics. Field Crops Res. 105 (1), 1−14. Available from: https://doi.org/10.1016/j.fcr.2007.07.004.

Challinor, A.J., Koehler, A.-K., Ramirez-Villegas, J., Whitfield, S., Das, B., 2016. Current warming will reduce yields unless maize breeding and seed systems adapt immediately. Nat. Clim. Change 6 (10), 954−958. Available from: https://doi.org/10.1038/nclimate3061.

Cullen, B.R., Eckard, R.J., 2011. Impacts of future climate scenarios on the balance between productivity and total greenhouse gas emissions from pasture based dairy systems in south-eastern Australia. Anim. Feed Sci. Technol. 166−167 (Suppl. C), 721−735. Available from: https://doi.org/10.1016/j.anifeedsci.2011.04.051.

DaMatta, F.M., Grandis, A., Arenque, B.C., Buckeridge, M.S., 2010. Impacts of climate changes on crop physiology and food quality. Food Res. Int. 43 (7), 1814−1823. Available from: https://doi.org/10.1016/j.foodres.2009.11.001.

Davies, W.J., Kudoyarova, G., Hartung, W., 2005. Long-distance ABA signaling and its relation to other signaling pathways in the detection of soil drying and the mediation of the plant's response to drought. J. Plant Growth Regul. 24 (4), 285. Available from: https://doi.org/10.1007/s00344-005-0103-1.

Deryng, D., Elliott, J., Folberth, C., Müller, C., Pugh, T.A.M., Boote, K.J., et al., 2016. Regional disparities in the beneficial effects of rising CO_2 concentrations on crop water productivity. Nat. Clim. Change 6 (8), 786−790. Available from: https://doi.org/10.1038/nclimate2995.

Easterling, W.E., 1996. Adapting North American agriculture to climate change in review. Agric. For. Meteorol. 80 (1), 1−53. Available from: https://doi.org/10.1016/0168-1923(95)02315-1.

Eckard, R.J., Cullen, B.R., 2011. Impacts of future climate scenarios on nitrous oxide emissions from pasture based dairy systems in south eastern Australia. Anim. Feed Sci. Technol. 166−167 (Suppl. C), 736−748. Available from: https://doi.org/10.1016/j.anifeedsci.2011.04.052.

Ellis, F., 2000. Rural Livelihoods and Diversity in Developing Countries. Oxford University Press, Oxford.

Erda, L., Wei, X., Hui, J., Yinlong, X., Yue, L., Liping, B., et al., 2005. Climate change impacts on crop yield and quality with CO_2 fertilization in China. Philos. Trans. R. Soc. Lond. B Biol. Sci. 360 (1463), 2149−2154. Available from: https://doi.org/10.1098/rstb.2005.1743.

Fahrenkamp-Uppenbrink, J., 2016. Reducing food loss and waste. Science 352 (6284), 424−426. Available from: https://doi.org/10.1126/science.352.6284.424-p.

FAO, 2009. Available from: <http://www.fao.org/fileadmin/templates/wsfs/docs/Issues_papers/HLEF2050_Global_Agriculture.pdf> How to Feed the World in 2050. High-Level Expert Forum, Rome.

Field, C.B., Barros, V.R., Mach, K., Mastrandrea, M., 2014. Climate Change 2014: Impacts, Adaptation, and Vulnerability. Cambridge University Press, Cambridge and New York, NY.

Fischer, E.M., Beyerle, U., Knutti, R., 2013. Robust spatially aggregated projections of climate extremes. Nat. Clim. Change 3 (12), 1033−1038. Available from: https://doi.org/10.1038/nclimate2051.

Frieler, K., Schauberger, B., Arneth, A., Balkovič, J., Chryssanthacopoulos, J., Deryng, D., et al., 2017. Understanding the weather signal in national crop-yield variability. Earths Future 5 (6), 605−616. Available from: https://doi.org/10.1002/2016EF000525.

Fruchterman, T.M.J., Reingold, E.M., 1991. Graph drawing by force-directed placement. Softw. Pract. Exp. 21 (11), 1129−1164. Available from: https://doi.org/10.1002/spe.4380211102.

Gilbert, N., 2012. One-third of our greenhouse gas emissions come from agriculture. Nat. News. Available from: https://doi.org/10.1038/nature.2012.11708.

Godfray, H.C.J., Beddington, J.R., Crute, I.R., Haddad, L., Lawrence, D., Muir, J.F., et al., 2010. Food security: the challenge of feeding 9 billion people. Science 327 (5967), 812−818. Available from: https://doi.org/10.1126/science.1185383.

Gornall, J., Betts, R., Burke, E., Clark, R., Camp, J., Willett, K., et al., 2010. Implications of climate change for agricultural productivity in the early twenty-first century. Philos. Trans. R. Soc. Lond. B Biol. Sci. 365 (1554), 2973−2989. Available from: https://doi.org/10.1098/rstb.2010.0158.

Gustavsson, J., Cederberg, C., Sonesson, U., Van Otterdijk, R., Meybeck, A., 2011. Global Food Losses and Food Waste: Extent, Causes and Prevention. Food and Agriculture Organization of the United Nations, Rome.

Hanigan, I.C., Butler, C.D., Kokic, P.N., Hutchinson, M.F., 2012. Suicide and drought in New South Wales, Australia, 1970−2007. Proc. Natl. Acad. Sci. 109 (35), 13950−13955. Available from: https://doi.org/10.1073/pnas.1112965109.

Hansen, J.W., 2002. Realizing the potential benefits of climate prediction to agriculture: issues, approaches, challenges. Agric. Syst. 74 (3), 309−330. Available from: https://doi.org/10.1016/S0308-521X(02)00043-4.

Hansen, J.W., Mason, S.J., Sun, L., Tall, A., 2011. Review of seasonal climate forecasting for agriculture in Sub-Saharan Africa. Exp. Agric. 47 (Special Issue 02), 205−240. Available from: https://doi.org/10.1017/S0014479710000876.

Harvey, C.A., Chacón, M., Donatti, C.I., Garen, E., Hannah, L., Andrade, A., et al., 2014. Climate-smart landscapes: opportunities and challenges for integrating adaptation and mitigation in tropical agriculture. Conserv. Lett. 7 (2), 77−90. Available from: https://doi.org/10.1111/conl.12066.

Hochman, Z., Gobbett, D.L., Horan, H., 2017. Climate trends account for stalled wheat yields in Australia since 1990. Glob. Change Biol. . Available from: https://doi.org/10.1111/gcb.13604.

Howden, M.S., Ash, A., Barlow, S., Booth, T., Charles, S., Cechet, B., et al., 2003. An Overview of the Adaptive Capacity of the Australian Agricultural Sector to Climate Change: Options, Costs and Benefits. Australian Greenhouse Office, Canberra, ACT.

Howden, S.M., Soussana, J.-F., Tubiello, F.N., Chhetri, N., Dunlop, M., Meinke, H., 2007. Adapting agriculture to climate change. Proc. Natl. Acad. Sci. 104 (50), 19691−19696.

Howden, S.M., Crimp, S.J., Nelson, R. et al., 2010. Australian agriculture in a climate of change, in Managing Climate Change. In: Jubb, I., Holper, P., and Cai, W. (Eds.), Papers from the Greenhouse 2009 Conference, CSIRO Publishing, Collingwood, Australia, pp. 101−111.

Iizumi, T., Luo, J.-J., Challinor, A.J., Sakurai, G., Yokozawa, M., Sakuma, H., et al., 2014. Impacts of El Niño Southern Oscillation on the global yields of major crops. Nat. Commun. 5. Available from: https://doi.org/10.1038/ncomms4712.

ILOSTAT, 2016. Employment by sector-ILO modeled estimates, Nov. 2016.Available from: <http://www.ilo.org/ilostat/faces/oracle/webcenter/porta-lapp/pagehierarchy/Page3.jspx?MBI_ID=33> (accessed 15.05.17.).

Inderberg, T.H., Eikeland, P.O., 2009. Limits to Adaptation: Analysing Institutional Constraints in: Adger, WN., Lorenzoni, I., and O'Brien, K. L. (Eds.), Adapting to Climate Change. Thresholds, Values, Governance, pp. 433−447. Cambridge University Press, Cambridge, UK, and New York, NY, USA.

IPCC, 2012. Managing the risks of extreme events and disasters to advance climate change adaptation. In: Field, C.B., Barros, V., Stocker, T.F., Qin, D., Dokken, D.J., Ebi, K.L., et al., A Special Report of Working Groups I and II of the Intergovernmental Panel on Climate Change. Cambridge University Press, Cambridge, UK, and New York, NY, USA.

IPCC: Climate Change, 2013. The physical science basis. In: Stocker, T., Qin, D., Plattner, G.-K., Tignor, M., Allen, S.K., Boschung, J., et al. (Eds.), Contribution of Working Group I to the Fifth Assessment Report of the Intergovernmental Panel on Climate Change, 2013, Cambridge University Press, Cambridge, and New York, NY, USA.

IPCC, Climate Change, 2014. Mitigation of climate change. In: Edenhofer, O., Pichs-Madruga, R., Sokona, Y., Farahani, E., Kadner, S., Seyboth, K., et al. (Eds.), Contribution of Working Group III to the Fifth Assessment Report of the Intergovernmental Panel on Climate Change, 2014, Cambridge University Press, Cambridge, and New York, NY, USA.

Jacomy, M., Venturini, T., Heymann, S., Bastian, M., 2014. ForceAtlas2, a continuous graph layout algorithm for handy network visualization designed for the Gephi Software, edited by M.R. Muldoon. PLoS One 9 (6), e98679. Available from: https://doi.org/10.1371/journal.pone.0098679.

Jones, C.A., Kiniry, J.R., Dyke, P.T., 1986. Available from: <https://espace.library.uq.edu/view/UQ:388362> (accessed 10.05. 17.) CERES-Maize: A simulation Model of Maize Growth and Development. Texas A&M University Press.

Jones, J.W., Hoogenboom, G., Porter, C.H., Boote, K.J., Batchelor, W.D., Hunt, L.A., et al., 2003. The DSSAT cropping system model. Eur. J. Agron. 18 (3), 235−265.

Jones, L., Ludi, E., Levine, S., 2010. Towards a Characterisation of Adaptive Capacity: A Framework for Analysing Adaptive Capacity at the Local Level. Overseas Development Institute.

Jones, P.G., Thornton, P.K., 2003. The potential impacts of climate change on maize production in Africa and Latin America in 2055. Glob. Environ. Change 13 (1), 51−59. Available from: https://doi.org/10.1016/S0959-3780(02)00090-0.

Jowel, C., Claire, D., Rebecca, M., Jessica, F., 2017. Climate change and variability: what are the risks for nutrition, diets, and food systems? Intl. Food Policy Res. Inst.

Kates, R.W., Travis, W.R., Wilbanks, T.J., 2012. Transformational adaptation when incremental adaptations to climate change are insufficient. Proc. Natl. Acad. Sci. 109 (19), 7156−7161. Available from: https://doi.org/10.1073/pnas.1115521109.

Keating, B.A., Carberry, P.S., Hammer, G.L., Probert, M.E., Robertson, M.J., Holzworth, D., et al., 2003. An overview of APSIM, a model designed for farming systems simulation. Eur. J. Agron. 18 (3), 267−288.

Kimball, B.A., 1983. Carbon dioxide and agricultural yield: an assemblage and analysis of 430 prior observations. Agron. J. 75 (5), 779−788.

Klein, R.J., Huq, S., Denton, F., Downing, T.E., Richels, R.G., Robinson, J.B., et al., 2007. Inter-relationships Between Adaptation and Mitigation.

Klopper, E., Vogel, C.H., Landman, W.A., 2006. Seasonal climate forecasts − potential agricultural-risk management tools? Clim. Change 76 (1−2), 73−90. Available from: https://doi.org/10.1007/s10584-005-9019-9.

Knutas, A., 2017. nails: Network analysis interface for literature studies. Available from: <https://github.com/aknutas/nails>.

Knutas, A., Hajikhani, A., Salminen, J., Ikonen, J., Porras, J., 2015. Cloud-based bibliometric analysis service for systematic mapping studies. In: Proceedings of the 16th International Conference on Computer Systems and Technologies, ACM, New York, NY, pp. 184−191.

Knutti, R., Sedláček, J., 2012. Robustness and uncertainties in the new CMIP5 climate model projections. Nat. Clim. Change 3 (4), 369−373. Available from: https://doi.org/10.1038/nclimate1716.

Kongsager, R., Locatelli, B., Chazarin, F., 2016. Addressing climate change mitigation and adaptation together: a global assessment of agriculture and forestry projects. Environ. Manage. 57 (2), 271−282. Available from: https://doi.org/10.1007/s00267-015-0605-y.

Kumar, A., Bernier, J., Verulkar, S., Lafitte, H.R., Atlin, G.N., 2008. Breeding for drought tolerance: direct selection for yield, response to selection and use of drought-tolerant donors in upland and lowland-adapted populations. Field Crops Res. 107 (3), 221−231. Available from: https://doi.org/10.1016/j.fcr.2008.02.007.

Lal, R., 2004a. Soil carbon sequestration impacts on global climate change and food security. Science 304 (5677), 1623−1627. Available from: https://doi.org/10.1126/science.1097396.

Lal, R., 2004b. Soil carbon sequestration to mitigate climate change. Geoderma 123 (1−2), 1−22. Available from: https://doi.org/10.1016/j.geoderma.2004.01.032.

Lesk, C., Rowhani, P., Ramankutty, N., 2016. Influence of extreme weather disasters on global crop production. Nature 529 (7584), 84−87. Available from: https://doi.org/10.1038/nature16467.

Lipinski, B., Hanson, C., Lomax, J., Kitinoja, L., Waite, R., Searchinger, T., 2013. Available from: <http://www.worldresourcesreport.org> (accessed 15.05.17.) Reducing Food Loss and Waste. World Resources Institute, Washington, DC.

Lobell, D.B., Field, C.B., 2007. Global scale climate−crop yield relationships and the impacts of recent warming. Environ. Res. Lett. 2 (1), 014002. Available from: https://doi.org/10.1088/1748-9326/2/1/014002.

Lobell, D.B., Burke, M.B., Tebaldi, C., Mastrandrea, M.D., Falcon, W.P., Naylor, R.L., 2008. Prioritizing climate change adaptation needs for food security in 2030. Science 319 (5863), 607−610. Available from: https://doi.org/10.1126/science.1152339.

Lobell, D.B., Schlenker, W., Costa-Roberts, J., 2011. Climate trends and global crop production since 1980. Science 333 (6042), 616−620. Available from: https://doi.org/10.1126/science.1204531.

Lobell, D.B., Sibley, A., Ivan Ortiz-Monasterio, J., 2012. Extreme heat effects on wheat senescence in India. Nat. Clim. Change 2 (3), 186−189. Available from: https://doi.org/10.1038/nclimate1356.

Lobell, D.B., Baldos, U.L.C., Hertel, T.W., 2013. Climate adaptation as mitigation: the case of agricultural investments. Environ. Res. Lett. 8 (1), 015012. Available from: https://doi.org/10.1088/1748-9326/8/1/015012.

Locatelli, B., Pavageau, C., Pramova, E., Di Gregorio, M., 2015. Integrating climate change mitigation and adaptation in agriculture and forestry: opportunities and trade-offs. Wiley Interdiscip. Rev. Clim. Change 6 (6), 585−598. Available from: https://doi.org/10.1002/wcc.357.

Long, S.P., Ainsworth, E.A., Leakey, A.D., Nösberger, J., Ort, D.R., 2006. Food for thought: lower-than-expected crop yield stimulation with rising CO_2 concentrations. Science 312 (5782), 1918−1921.

McGrath, J.M., Lobell, D.B., 2013. Regional disparities in the CO_2 fertilization effect and implications for crop yields. Environ. Res. Lett. 8 (1), 014054. Available from: https://doi.org/10.1088/1748-9326/8/1/014054.

McGregor, A., Dawson, B., 2016. Vulnerability of Pacific Island Agriculture and Forestry to Climate Change. SPC.

McLeod, A.R., Long, S.P., 1999. Free-air Carbon Dioxide Enrichment (FACE) in global change research: a review. In: Fitter, A.H., Raffaelli, D. (Eds.), Advances in Ecological Research, vol. 28. Academic Press.

Meinshausen, M., Smith, S.J., Calvin, K., Daniel, J.S., Kainuma, M.L.T., Lamarque, J.-F., et al., 2011. The RCP greenhouse gas concentrations and their extensions from 1765 to 2300. Clim. Change 109 (1−2), 213. Available from: https://doi.org/10.1007/s10584-011-0156-z.

Meyer, R., Cullen, B.R., Johnson, I.R., Eckard, R.J., 2015. Process modelling to assess the sequestration and productivity benefits of soil carbon for pasture. Agric. Ecosyst. Environ. 213 (Suppl. C), 272−280. Available from: https://doi.org/10.1016/j.agee.2015.07.024.

Meyer, R., Cullen, B.R., Eckard, R.J., 2016. Modelling the influence of soil carbon on net greenhouse gas emissions from grazed pastures. Anim. Prod. Sci. 56 (3), 585−593. Available from: https://doi.org/10.1071/AN15508.

Moriondo, M., Jones, G.V., Bois, B., Dibari, C., Ferrise, R., Trombi, G., et al., 2013. Projected shifts of wine regions in response to climate change. Clim. Change 119 (3−4), 825−839. Available from: https://doi.org/10.1007/s10584-013-0739-y.

Mueller, N.D., Gerber, J.S., Johnston, M., Ray, D.K., Ramankutty, N., Foley, J.A., 2012. Closing yield gaps through nutrient and water management. Nature 490 (7419), 254−257. Available from: https://doi.org/10.1038/nature11420.

Myers, S.S., Zanobetti, A., Kloog, I., Huybers, P., Leakey, A.D.B., Bloom, A.J., et al., 2014. Increasing CO_2 threatens human nutrition. Nature 510 (7503), 139−142. Available from: https://doi.org/10.1038/nature13179.

Nelson, R., Kokic, P., Elliston, L., King, J.-A., 2005. Structural adjustment: a vulnerability index for Australian broadacre agriculture. Aust. Commod. Forecasts Issues 12 (1), 171.

Nelson, R., Kokic, P., Crimp, S., Meinke, H., Howden, S.M., 2010. The vulnerability of Australian rural communities to climate variability and change: Part I—Conceptualising and measuring vulnerability. Environ. Sci. Policy 13 (1), 8−17. Available from: https://doi.org/10.1016/j.envsci.2009.09.006.

Oberlack, C., 2017. Diagnosing institutional barriers and opportunities for adaptation to climate change. Mitig. Adapt. Strateg. Glob. Change 22 (5), 805−838. Available from: https://doi.org/10.1007/s11027-015-9699-z.

Olesen, J.E., 2006. Reconciling adaptation and mitigation to climate change in agriculture. J. Phys. IV Proc 139, 403−411. Available from: https://doi.org/10.1051/jp4:2006139026.

Olesen, J.E., Bindi, M., 2002. Consequences of climate change for European agricultural productivity, land use and policy. Eur. J. Agron. 16 (4), 239−262. Available from: https://doi.org/10.1016/S1161-0301(02)00004-7.

Olesen, J.E., Carter, T.R., Díaz-Ambrona, C.H., Fronzek, S., Heidmann, T., Hickler, T., et al., 2007. Uncertainties in projected impacts of climate change on European agriculture and terrestrial ecosystems based on scenarios from regional climate models. Clim. Change 81 (1), 123–143. Available from: https://doi.org/10.1007/s10584-006-9216-1.

Page, L., Brin, S., Motwani, R., Winograd, T., 1999. The PageRank citation ranking: bringing order to the web, Stanford InfoLab. Available from: <http://p8090-ilpubs.stanford.edu/422> (accessed 11.05.17.).

Park, S.E., Marshall, N.A., Jakku, E., Dowd, A.M., Howden, S.M., Mendham, E., et al., 2012. Informing adaptation responses to climate change through theories of transformation. Glob. Environ. Change 22 (1), 115–126. Available from: https://doi.org/10.1016/j.gloenvcha.2011.10.003.

Parry, M., Rosenzweig, C., Iglesias, A., Livermore, M., Fischer, G., 2004. Effects of climate change on global food production under SRES emissions and socio-economic scenarios. Glob. Environ. Change 14 (1), 53–67. Available from: https://doi.org/10.1016/j.gloenvcha.2003.10.008.

Peng, S., Huang, J., Sheehy, J.E., Laza, R.C., Visperas, R.M., Zhong, X., et al., 2004. Rice yields decline with higher night temperature from global warming. Proc. Natl. Acad. Sci. U.S.A. 101 (27), 9971–9975. Available from: https://doi.org/10.1073/pnas.0403720101.

Pickering, C., Byrne, J., 2014. The benefits of publishing systematic quantitative literature reviews for PhD candidates and other early-career researchers. High. Educ. Res. Dev. 33 (3), 534–548. Available from: https://doi.org/10.1080/07294360.2013.841651.

Porter, J.R., Gawith, M., 1999. Temperatures and the growth and development of wheat: a review. Eur. J. Agron. 10 (1), 23–36.

Porter, J.R., Xie, L., Challinor, A.J., Cochrane, K., Howden, S.M., Iqbal, M.M., et al., 2014. Food security and food production systems. In: Field, C. B., Barros, V.R., Dokken, D.J., Mach, K.J., Mastrandrea, M.D., Bilir, T.E., Chatterjee, M., Ebi, K.L., et al.,Climate Change 2014: Impacts, Adaptation, and Vulnerability. Part A: Global and Sectoral Aspects. Contribution of Working Group II to the Fifth Assessment Report of the Intergovernmental Panel on Climate Change. Cambridge University Press, Cambridge, UK, and New York, NY, USA.

Prasad, A.S., 1996. Zinc deficiency in women, infants and children. J. Am. Coll. Nutr. 15 (2), 113–120. Available from: https://doi.org/10.1080/07315724.1996.10718575.

Ray, D.K., Ramankutty, N., Mueller, N.D., West, P.C., Foley, J.A., 2012. Recent patterns of crop yield growth and stagnation. Nat. Commun. 3, 1293. Available from: https://doi.org/10.1038/ncomms2296.

Ray, D.K., Mueller, N.D., West, P.C., Foley, J.A., 2013. Yield trends are insufficient to double global crop production by 2050. PLoS One 8 (6), e66428. Available from: https://doi.org/10.1371/journal.pone.0066428.

Ray, D.K., Gerber, J.S., MacDonald, G.K., West, P.C., 2015. Climate variation explains a third of global crop yield variability. Nat. Commun. 6. Available from: https://doi.org/10.1038/ncomms6989.

Reynolds, M.P., Hays, D., Chapman, S., 2010. Breeding for adaptation to heat and drought stress. Clim. Change Crop Prod. 1, 71–91.

Rickards, L., Howden, S.M., 2012. Transformational adaptation: agriculture and climate change. Crop Pasture Sci. 63 (3), 240–250. Available from: https://doi.org/10.1071/CP11172.

Rosenzweig, C., Parry, M.L., 1994. Potential impact of climate change on world food supply. Nature 367 (6459), 133–138.

Rosenzweig, C., Tubiello, F.N., 2007. Adaptation and mitigation strategies in agriculture: an analysis of potential synergies. Mitig. Adapt. Strateg. Glob. Change 12 (5), 855–873. Available from: https://doi.org/10.1007/s11027-007-9103-8.

Rosenzweig, C., Iglesias, A., Yang, X.B., Epstein, P.R., Chivian, E., 2001. Climate change and extreme weather events; implications for food production, plant diseases, and pests. Glob. Change Hum. Health 2 (2), 90–104. Available from: https://doi.org/10.1023/A:1015086831467.

Rosenzweig, C., Elliott, J., Deryng, D., Ruane, A.C., Müller, C., Arneth, A., et al., 2013. Assessing agricultural risks of climate change in the 21st century in a global gridded crop model intercomparison. Proc. Natl. Acad. Sci. Available from: https://doi.org/10.1073/pnas.1222463110201222463.

Ryschawy, J., Choisis, N., Choisis, J.P., Joannon, A., Gibon, A., 2012. Mixed crop-livestock systems: an economic and environmental-friendly way of farming? Animal. Available from: https://doi.org/10.1017/S1751731112000675.

Schauberger, B., Rolinski, S., Müller, C., 2016. A network-based approach for semi-quantitative knowledge mining and its application to yield variability. Environ. Res. Lett. 11 (12), 123001. Available from: https://doi.org/10.1088/1748-9326/11/12/123001.

Schiere, H., Kater, L., 2001. Mixed crop-livestock farming. A review of traditional technologies based on literature and field experience, FAO. Available from: <http://agris.fao.org/agris-search/search.do?recordID = XF2003410307> (accessed 15.10.17.).

Schipper, E.L.F., 2007. Climate change adaptation and development: exploring the linkages. Tyndall Cent. Clim. Change Res. Work. Pap. 107, 13.

Schlenker, W., Roberts, M.J., 2009. Nonlinear temperature effects indicate severe damages to U.S. crop yields under climate change. Proc. Natl. Acad. Sci. 106 (37), 15594–15598. Available from: https://doi.org/10.1073/pnas.0906865106.

Schmidhuber, J., Tubiello, F.N., 2007. Global food security under climate change. Proc. Natl. Acad. Sci. 104 (50), 19703–19708. Available from: https://doi.org/10.1073/pnas.0701976104.

Smit, B., Wandel, J., 2006. Adaptation, adaptive capacity and vulnerability. Glob. Environ. Change 16 (3), 282–292. Available from: https://doi.org/10.1016/j.gloenvcha.2006.03.008.

Smith, P., Olesen, J.E., 2010. Synergies between the mitigation of, and adaptation to, climate change in agriculture. J. Agric. Sci. 148 (05), 543–552. Available from: https://doi.org/10.1017/S0021859610000341.

Stöckle, C.O., Donatelli, M., Nelson, R., 2003. CropSyst, a cropping systems simulation model. Eur. J. Agron. 18 (3), 289–307.

Stokes, C., Howden, 2010. Adaptation-Mitigation Interactions in Agriculture – Identifying Synergies and Conflict.

Suckall, N., Stringer, L.C., Tompkins, E.L., 2015. Presenting triple-wins? Assessing projects that deliver adaptation, mitigation and development co-benefits in rural sub-Saharan Africa. AMBIO 44 (1), 34–41. Available from: https://doi.org/10.1007/s13280-014-0520-0.

Thornton, P.K., Herrero, M., 2015. Adapting to climate change in the mixed crop and livestock farming systems in sub-Saharan Africa. Nat. Clim. Change 5 (9), 830–836. Available from: https://doi.org/10.1038/nclimate2754.

Thornton, P.K., Ericksen, P.J., Herrero, M., Challinor, A.J., 2014. Climate variability and vulnerability to climate change: a review. Glob. Change Biol. 20 (11), 3313−3328. Available from: https://doi.org/10.1111/gcb.12581.

Tubiello, F.N., Soussana, J.-F., Howden, S.M., 2007a. Crop and pasture response to climate change. Proc. Natl. Acad. Sci. 104 (50), 19686−19690. Available from: https://doi.org/10.1073/pnas.0701728104.

Tubiello, F.N., Amthor, J.S., Boote, K.J., Donatelli, M., Easterling, W., Fischer, G., et al., 2007b. Crop response to elevated CO_2 and world food supply: a comment on "Food for Thought..." by Long et al., Science 312:1918−1921, 2006. Eur. J. Agron. 26 (3), 215−223. Available from: https://doi.org/10.1016/j.eja.2006.10.002.

U.N. DESA, 2015. World population prospects: the 2015 revision, key findings and advance tables, United Nations, Department of Economic and Social Affairs, Population Division.

Vermeulen, S.J., Campbell, B.M., Ingram, J.S.I., 2012. Climate change and food systems. Annu. Rev. Environ. Resour. 37 (1), 195−222. Available from: https://doi.org/10.1146/annurev-environ-020411-130608.

Vogel, E., Deumlich, D., Kaupenjohann, M., 2016. Bioenergy maize and soil erosion—risk assessment and erosion control concepts. Geoderma 261, 80−92. Available from: https://doi.org/10.1016/j.geoderma.2015.06.020.

van Vuuren, D.P., Isaac, M., Kundzewicz, Z.W., Arnell, N., Barker, T., Criqui, P., et al., 2011. The use of scenarios as the basis for combined assessment of climate change mitigation and adaptation. Glob. Environ. Change 21 (2), 575−591. Available from: https://doi.org/10.1016/j.gloenvcha.2010.11.003.

Watkiss, P., Benzie, M., Klein, R.J.T., 2015. The complementarity and comparability of climate change adaptation and mitigation. Wiley Interdiscip. Rev. Clim. Change 6 (6), 541−557. Available from: https://doi.org/10.1002/wcc.368.

Wheeler, T.R., Craufurd, P.Q., Ellis, R.H., Porter, J.R., Prasad, P.V., 2000. Temperature variability and the yield of annual crops. Agric. Ecosyst. Environ. 82 (1), 159−167.

WoS: Web of Science, 2017. Available from: <http://apps.webofknowledge.com> (accessed 29.05.17.).

Ziervogel, G., Johnston, P., Matthew, M., Mukheibir, P., 2010. Using climate information for supporting climate change adaptation in water resource management in South Africa. Clim. Change 103 (3−4), 537−554. Available from: https://doi.org/10.1007/s10584-009-9771-3.

Tracking Adaptation Progress at the Global Level: Key Issues and Priorities

Anne Olhoff[1], Elina Väänänen[2] and Barney Dickson[2]

[1]UNEP DTU Partnership, Technical University of Denmark, Copenhagen, Denmark, [2]UN Environment, Nairobi, Kenya

Chapter Outline

4.1 INTRODUCTION

Until recently, policy, research and practitioner communities have primarily focused on adaptation at local to national scale. This focus has been motivated by the need for adaptation action to be designed and implemented in the light of the expected climate impacts in a given location and the specific features that shape vulnerability to those impacts. However, in recent years, and particularly since the 2015 Paris Agreement and the adoption of the global adaptation goal of "enhancing adaptive capacity, strengthening resilience and reducing vulnerability to climate change" (UNFCCC, 2015), there is increasing emphasis on the global dimensions of adaptation. This has generated a new dynamic between local, national, and global perspectives on adaptation.

This dynamic is particularly visible in the field of metrics for adaptation. The increasing policy priority and funding flows to adaptation have generated a demand for systematic approaches and metrics to assess the different dimensions of adaptation, including needs, progress, and effectiveness. At the national scale, it has motivated the development and implementation of Monitoring, Reporting, and Evaluation systems for adaptation in a number of countries. At the global level, attention has turned to the Paris Agreement's Global Stocktake and reviewing the effectiveness of adaptation and adaptation support as well as the overall progress in achieving the global adaptation goal.

This chapter examines the emerging global perspective on adaptation. It begins by considering why adaptation is often perceived as local and the factors that have led to the increased attention to global dimensions of adaptation. It then discusses how this relates to the question of measuring adaptation progress. It examines the challenges of developing appropriate approaches and metrics for this purpose and suggests some ways forward, focusing on three priority areas of action for advancing on a global adaptation framework under the United Nations Framework Convention on Climate Change (UNFCCC): clearly defining the global goal and achieving conceptual clarity, identifying tracking criteria, and ensuring availability of necessary data.

4.2 THE EMERGING GLOBAL PERSPECTIVE ON ADAPTATION

There has been a tendency to perceive adaptation planning and implementation in a local and national context (Ford and Berrang-Ford, 2016; Magnan and Ribera, 2016; Ford et al., 2015; IPCC, 2014a,b). It is certainly true that local and national actions and actors are key to adaptation, as both adaptation costs and benefits, such as reduced effects of climate change impacts, enhanced adaptive capacity, and more resilient ecosystems, are generally incurred at those levels, and because development decisions are mostly made in national and subnational contexts (Magnan et al., 2015).

Resilience. DOI: https://doi.org/10.1016/B978-0-12-811891-7.00004-9

However, several factors have led to a growing global perspective on adaptation. Firstly, there is increasing awareness among policy makers, practitioners, donors, and researchers of the need for broad-scale systematic, coherent, consistent, and comparable insights to guide adaptation at national and global level (Lesnikowski et al., 2016; Leiter, 2015). A global perspective on adaptation facilitates capturing the current state of adaptation across nations; provides the basis for characterizing and evaluating adaptations taking place in different settings; and underpins the monitoring of change in adaptation over time (Ford and Berrang-Ford, 2016).

Secondly, there is increasing recognition that adaptation action and inaction can have implications that transcend national contexts. National frameworks may ignore potential transboundary benefits of domestic adaptation, as well as transboundary risks of inadequate or maladaptive adaptation responses at the national level that can compound global aggregate impacts (Magnan and Ribera, 2016; Magnan et al., 2015; IPCC, 2014a,b). Examples include transboundary externalities of river basin management, inadequate adaptation action leading to or exacerbating unplanned migration and violent conflict, and inadequate adaptation action to manage the climate risks to agricultural production. Transboundary and global aggregate impacts thus indicate that global action and coordination of adaptation is required. A global framework for adaptation can facilitate addressing such impacts, their interlinkages with sustainable development goals, and potential political barriers.

Thirdly, the fact that the UNFCCC has played a key role in catalyzing adaptation planning and implementation, particularly in developing countries, has also contributed to the emergence of a global perspective on adaptation. Countries have planned their adaptation efforts through National Adaptation Programmes of Action (NAPAs) and increasingly through National Adaptation Plans (NAPs) and are therefore approaching their national adaptation challenges through instruments and processes that have been articulated through a global body. While the first guidelines for NAPAs date back to 2001, subsequent UNFCCC support on NAPAs and NAPs has extended beyond technical guidelines to workshops and expert meetings, training, regional exchanges and syntheses of experiences and best practices (Wiseman, 2016). This has inevitably encouraged common approaches and cross-country learning and has been accompanied by various bodies outside the UNFCCC providing generic adaptation planning tools to be applied in different countries and contexts.

However, probably most important for the emergence of a global perspective has been the increased prominence of adaptation within the Paris Agreement. Following pressure from developing country parties to place adaptation on par with mitigation, the Agreement established the global adaptation goal of "enhancing adaptive capacity, strengthening resilience and reducing vulnerability to climate change." It also set out the "global stocktake" to regularly assess progress towards the purpose of the Agreement, including the adaptation goal. The Paris Agreement thus provides a new impetus for establishing and reinforcing a global framework for adaptation. To be successful, such a framework needs to build on a common understanding and operational definition of what adaptation means. A global framework for adaptation can facilitate an increased understanding of the status of adaptation, and is a requisite for tracking adaptation progress and outcomes across countries and over time (Ford and Berrang-Ford, 2016; Lesnikowski et al., 2016; Magnan and Ribera, 2016; Ngwadla and El-Bakri, 2016; Singh et al., 2016; Ford et al., 2015).

4.3 TRACKING PROGRESS IN ADAPTATION AT THE GLOBAL LEVEL

Implementation of the adaptation provisions under the Paris Agreement calls for systematic approaches for tracking adaptation progress across Parties to the UNFCCC (Ford et al., 2015).[1] The global stocktakes, which are the means of tracking progress, comprise four adaptation-related components. In addition to reviewing the overall progress towards the global goal on adaptation, these include a review of the adequacy and effectiveness of adaptation and support provided for adaptation, recognition of adaptation efforts of developing country parties, and enhancing the implementation of adaptation action (UNFCCC, 2015). The text of the Paris Agreement does not specify reporting requirements and the procedures and modalities for reporting are still under discussion (Kato and Ellis, 2016). The Paris Agreement also stresses that adaptation reporting needs to avoid creating an "undue" additional burden for developing country Parties. While some argue for maximum flexibility for countries, others press for the adoption of common reporting frameworks and associated metrics that would allow for comparison and aggregation across countries and the substantive assessment of progress toward the global goal. The extent to which the process under the UNFCCC leads to clarity, consistency, and comparability of reporting will have significant implications for the usefulness of the global stocktake for global adaptation tracking purposes.

---

1. Notably Article 7.14, Article 13.5 on transparency, and Article 14 on the global stocktakes to be undertaken at 5-year intervals.

The Paris Agreement's global stocktake is however not the first instance that the challenge of measuring global adaptation progress has been raised. UN Environment has tackled related questions through its work on the Adaptation Gap at the global level, examining the difference between the actual level of adaptation action and the level required to achieve a societal goal, including finance, technology, and knowledge considerations (UNEP, 2014, 2016). More recently, the UN Climate Resilience Initiative has undertaken initial work on assessing the global state of progress on three key capacities for climate resilience: anticipating climate hazards, absorbing climate shocks and stresses, and reshaping development pathways (UN Climate Resilience Initiative, 2017). Nevertheless, it is the Paris Agreement that has provided the major push for the increasing interest in measuring adaptation progress at the global level. Since Paris, the Parties have been working to address this question under UNFCCC bodies and processes, including in the Subsidiary Body for Scientific and Technological Advice and the Adaptation Committee.

Does the existing scientific literature provide any support for these efforts under the UNFCCC? The adaptation literature has, to a large extent, focused on examining a small number of cases to analyze whether adaptation is taking place and identify determinants of success, reinforcing a general argument that the scope and effectiveness of adaptation measures is location and context specific (Swart et al., 2014; Adger et al., 2007). In addition, most assessments of adaptation have so far only covered impacts, vulnerability, and adaptation planning. Very few have assessed the processes of implementation and evaluation of actual adaptation actions (Noble et al., 2014) and fewer still have attempted to compare experiences across countries (Ford and Berrang-Ford, 2016). As a consequence, our understanding of the global state of adaptation, adaptation progress, and factors explaining differences in progress across programmes, sectors, regions, and countries is currently partial and fragmented (Ford and Berrang-Ford, 2016; Mimura et al., 2014; Noble et al., 2014; Ford et al., 2011; Gagnon-Lebrun and Agrawala, 2007).

The contrast with mitigation is striking. There are well-established approaches to tracking mitigation progress. Tracking of mitigation at global and national scale is facilitated by a clear global goal of staying well below global average warming of 2 or 1.5°C in 2100 compared to preindustrial levels, a single common metric of tonnes of carbon dioxide equivalent, and well-established methodologies and data sources. For adaptation a clear goal, a baseline, a single common metric and well-established methodologies and data sources are lacking. The Paris Agreement's global goal of "enhancing adaptive capacity, strengthening resilience and reducing vulnerability to climate change" gives limited guidance in all of these respects: By how much should we enhance capacity, strengthen resilience, and reduce vulnerability? By when and compared to what? What do we mean by adaptive capacity, resilience, and vulnerability and how do we provide operational definitions of them? How do we assess status and track progress—conceptually, methodologically, and in terms of metrics and data? Establishing a meaningful global framework for adaptation involves tackling these issues.

4.4 PRIORITIES FOR TRACKING ADAPTATION UNDER THE UNFCCC

Tackling the issues outlined in the previous section translates into three priority action areas:

1. Agreeing on what to track: Establishing conceptual clarity on the global goal on adaptation and other aspects of adaptation that should be tracked globally.
2. Agreeing on how to track it: Identifying methodologies, metrics, and indicators.
3. Addressing information and data challenges.

Each of these priority areas are discussed in the following, acknowledging their political, academic, and institutional dimensions and realities.

4.4.1 Agreeing on What to Track

Agreeing on what to track involves identifying and establishing conceptual and operational clarity on the aspects of adaptation to be tracked nationally and globally. Aspects specified in the Paris Agreement include the adequacy and effectiveness of adaptation and support for adaptation, in addition to assessing collective progress on the global goal on adaptation. All of these aspects are conceptually and operationally ambiguous. They indicate the variety of questions that different stakeholders at global, regional, national, and local levels seek to answer through tracking of adaptation, including are actions consistent with the risks posed by climate change? How can adaptation funds be most effectively invested? Is progress being made to meet adaptation objectives? (Ford and Berrang-Ford, 2016).

Following the Paris Agreement, there has been significant attention to adaptation metrics and methodologies. For example, the Adaptation Committee highlights the need for strengthening methodologies and identifying adaptation indicators that can be compared and aggregated across scales (Adaptation Committee, 2017). Adaptation metrics and methodologies featured prominently in the Adaptation Futures conference in 2016 (Adaptation Futures, 2016), and the Scientific Committee of COP22 organized Adaptation Metrics Conferences in Morocco in 2016 and 2017 (UNFCCC, 2016a). Surprisingly, there has been relatively less attention to discussing and clarifying how the global goal on adaptation and aspects of adequacy and effectiveness can be made operational under the UNFCCC. However, this will largely determine the need for and appropriateness of methodologies, metrics, and indicators, as a single methodology or set of indicators and metrics cannot accommodate tracking all adaptation aspects highlighted above. In the following, some of the potential ways forward for advancing the conceptual and operational clarity on what to track are outlined.

In theory, collective progress towards the global goal on adaptation could be tracked against the three dimensions comprised in the goal: adaptive capacity, resilience, and vulnerability. There are several examples of indices for these three dimensions (Noble et al., 2014). Box 4.1 provides examples of applied indices for vulnerability, illustrating how these are often combined with resilience and adaptive capacity dimensions. However, studies show that there is low agreement in terms of ranking of countries based on indices (Leiter et al., 2017; UNEP, 2017; Eriksen and Kelly, 2007), which suggests that it could be difficult to obtain agreement on their use for global tracking purposes. The indices furthermore indicate the methodological challenges that would be associated with separating the three dimensions of the global goal causally from each other. For example, action to build adaptive capacity will have positive implications for vulnerability and resilience. Further, distinguishing the effects of other policies and processes would be equally challenging due to their contributive effect, as is illustrated by the example of policies tackling underlying determinants of vulnerability, such as investments in education, poverty alleviation, and health care (Ford et al., 2015).

Others propose to define and operationalize the global goal on adaptation in terms of ensuring human security (Magnan and Ribera, 2016). Saved wealth and saved health is also proposed in the literature and involves quantifying adaptation benefits in terms of avoided economic damages, illnesses, and mortality. However, while these measures capture key elements, parties to the UNFCCC are unlikely to see them as covering all relevant dimensions of adaptation.

An alternative way forward would be to clarify and operationalize the global goal on adaptation in terms of key sectoral and cross-cutting priorities for increased adaptive capacity and resilience and reduced vulnerability. Such an approach is closer aligned with the approaches adopted by many countries in their Nationally Determined Contributions, National Adaptation Plan processes, national monitoring, reporting, and evaluation frameworks, development plans, and green growth strategies. To illustrate, Sri Lanka, one of the seven countries with a National Adaptation

---

BOX 4.1 Examples of Existing Vulnerability Indices

The *University of Notre Dame Global Adaptation Initiative (ND-GAIN) country index* summarizes a country's vulnerability to climate change and other global challenges in combination with its readiness to improve resilience. Using data since 1995, it provides a ranking of 181 countries based on 36 indicators of vulnerability and nine indicators of readiness. Available at http://index.gain.org/.

The *DARA Climate Vulnerability Monitor* estimates human and economic impacts of climate change and the carbon economy for 184 countries in 2010 and 2030, across 34 indicators. Available at http://daraint.org/climate-vulnerability-monitor/climate-vulnerability-monitor-2010/.

The *GCCA + Index* aims to provide knowledge to support the reconciliation of climate change policy objectives with development goals. It allows an ex ante evaluation of key features of vulnerability to climate change based on 34 country-level indicators covering four components: natural hazards, exposure, vulnerability, capacity. Available at http://knowsdgs.jrc.ec.europa.eu/gcca/gcca-index.

The *Climate Change Vulnerability Index (CCVI)* by the global risks advisory firm Maplecroft provides a global ranking of 170 countries of the vulnerability to the impacts of climate change over the next 30 years. It aims at enabling organizations to identify areas of risk within their operations, supply chains, and investments. It evaluates 42 social, economic, and environmental factors to assess national vulnerabilities across three core areas: exposure to climate-related natural disasters and sea-level rise, human sensitivity, and future vulnerability by considering the adaptive capacity of a country's government and infrastructure. Available at https://maplecroft.com/about/news/ccvi.html.

The *Global Climate Risk Index* produced annually by Germanwatch provides an ex post indication of the level of exposure and vulnerability to extreme events of countries, rather than a comprehensive climate vulnerability scoring. The index is based on the following indicators: number of deaths, number of deaths per 100,000 inhabitants, sum of losses in US$ in purchasing power parity (PPP), and losses per unit of Gross Domestic Product (GDP). Available at https://germanwatch.org/en/12978.

Plan, has identified agriculture, fisheries, water, human health, coastal and marine environments, ecosystems and biodiversity, infrastructure, and human settlements as the most vulnerable sectors to the adverse effects of climate change. Sri Lanka's NAP provides adaptation plans for each of these sectors, and matches sectoral adaptation needs with adaptation options, actions, responsible agencies, and key performance indicators (Climate Change Secretariat of Sri Lanka, 2016).

There are a number of examples of how sectoral goals and targets can be framed in the above-mentioned plans and strategies. Ethiopia's Climate Resilient Green Economy Strategy (CRGE) provides an example of jointly addressing climate change adaptation and mitigation objectives in order to achieve a resilient economic development pathway (FDRE, 2017). The CRGE follows a sectoral approach through focus on agriculture; forestry; power; and transport, industrial sectors, and infrastructure, and is integrated into the country's 5-year National Development Plan (FDRE, 2011). More specifically the CRGE sets out the following targets: improving crop and livestock production practices for higher food security and farmer income while reducing emissions (agriculture); protecting and reestablishing forests for their economic and ecosystem services, including as carbon stocks (forestry); expanding electricity generation form renewable energy for domestic and regional markets (electricity); and leapfrogging to modern and energy efficient technologies (transport, industrial sectors, and buildings). The case of Ethiopia illustrates the potential benefits of a sectoral approach that integrates adaptation, mitigation, and development objectives under a single, coherent policy framework.

While there are many differences in specific country priorities and plans, there is also considerable convergence of countries in terms of climate risks and adaptation priority sectors. This is evident in the Nationally Determined Contributions (Olhoff et al., 2015). Fig. 4.1 provides an overview of key adaptation sectors in the adaptation components of the Nationally Determined Contributions, indicating the percentage of countries prioritizing these sectors. As the Figure illustrates, agriculture, water and health are prioritized by more than two-thirds of countries having submitted an adaptation component, followed by water), health, coastal sector, forestry, and ecosystems.

For each priority sector and cross-cutting issue agreed by parties, the UNFCCC process could identify and agree on key aspects to be tracked globally, leaving it flexible and open to parties to include additional nationally specific aspects. By disaggregating the global goal into sectoral and cross-cutting components, such an approach would furthermore facilitate the linking of assessment of the collective progress on the global goal to reporting on adequacy and effectiveness of adaptation action and support received, including finance. A sector-based approach is conceptually simpler as it avoids dealing with comparing and aggregating, e.g., progress in vulnerability reduction in coastal areas with vulnerability reduction in agriculture.

In many ways, a sectoral and cross-cutting approach would resemble the approaches adopted for the Sustainable Development Goals and Sendai Framework for disaster risk reduction frameworks, where the overall goals of

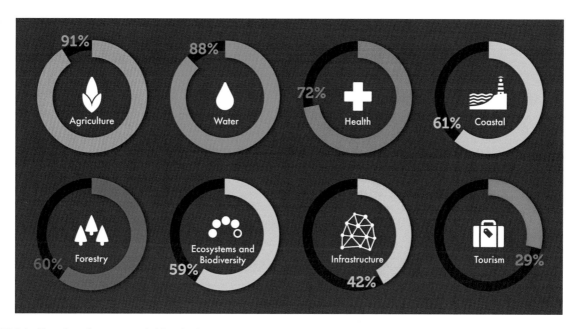

FIGURE 4.1   Key adaptation sectors prioritized in the NDCs.

sustainable development and of disaster risk reduction have been further specified in specific global targets. Table 4.1 compares key dimensions of the three frameworks (Sustainable Development Goals, Sendai Framework and Paris Agreement).

Keeping the level of complexity low while at the same time ensuring relevance and flexibility from a national perspective is likely to be critical for the political and practical feasibility of tracking progress on adaptation under the UNFCCC. The Paris Agreement specifies that an undue burden on developing countries should be avoided. It also highlights that differences between countries in adaptation priorities, preferences and starting points, current and future vulnerabilities and exposure, and institutional and economic characteristics and capacity should be acknowledged.

However, as Table 4.1 illustrates, defining and tracking progress on a limited number of key sectoral and cross-cutting adaptation dimensions globally could be combined with more detailed and specific adaptation targets, defined and adopted nationally. This would simplify the political negotiation process and honor the flexibility and sovereignty of countries.

TABLE 4.1 Comparison of Key Aspects of the Sustainable Development Goals, the Sendai Framework and the Paris Agreement

	Sustainable Development Goals	Sendai Framework	Paris Agreement
Objective of the agreement	Contributing to the achievement of sustainable development and serving as a driver for implementation and mainstreaming	A substantial reduction of disaster risk and losses in lives, livelihoods and health and in economic, physical, social, cultural, and environmental assets	Holding the increase in the global average temperature to well below 2°C and pursuing efforts to limit it to 1.5°C (mitigation); increasing the ability to adapt to the adverse impacts of climate change (adaptation); making finance flows consistent with a pathway toward low GHG emissions and climate-resilient development (Art. 2)
Quantitative goals or targets at global level	Yes, 17 global goals with several targets each Countries may define additional national targets	Yes, seven global targets. Countries may define additional national targets	Yes, for mitigation. The global goal on adaptation is qualitative. Countries define their own targets and so far no global targets have been agreed
Purpose of tracking progress	Measure global progress toward achievement of the SDG goals and targets	Measure global progress in implementation of the seven Sendai targets	Global Stocktake: "assess the collective progress towards achieving the purpose of this Agreement" (Art. 14). Transparency framework: "Clarity and tracking of progress towards achieving Parties' individual NDCs and Parties' adaptation actions" (Art. 13)
Development process	By an "Inter-Agency and Expert Group on Sustainable Development Goal Indicators," adopted by UN General Assembly	By an "open-ended intergovernmental expert working group" comprising experts nominated by States and supported by the UNISDR; adopted by UN General Assembly	Details of the Global Stocktake are still being negotiated (Art. 14). "Modalities, procedures and guidelines" for national reporting under the transparency framework (Art. 13), and details of the Adaptation Communications (Art. 7) are still to be agreed upon by the COP; Parties may develop country-specific adaptation M&E systems (Art. 7).[vii]

Source: Based on Leiter and Olivier (2017).

4.4.2   Agreeing on How to Track Progress

The second priority area involves identifying or developing appropriate methodologies, metrics, and indicators for tracking adaptation. As discussed, the appropriateness of methodologies and metrics depends on what one wants to track. It also depends on the scale at which adaptation is tracked. More specifically, consistency and comparability of what is being measured and how it is measured is required to allow comparison and aggregation across scales and over time. At present such consistency and comparability is limited, as illustrated in the UN Environment Adaptation Gap Reports (UNEP, 2017, 2016, 2014). Even in one of the most advanced fields of assessment of adaptation costs, comparison and aggregation of national and sectoral estimates is severely challenged by differences in definitions, objectives, assumptions, and methodological approaches. Aligning approaches for global tracking of adaptation with national and subnational approaches will thus be a key issue for advancing tracking adaptation under the UNFCCC.

Overall, a range of methodologies, metrics, and indicators for measuring adaptation and adaptation outcomes exist. For example, there is a large literature providing guidance and insights into the methodological challenges of tracking adaptation at various scales (Naswa et al., 2015; OECD, 2015; Bours et al., 2014; Brooks and Fisher, 2014; Dinshaw et al., 2014; Global Environment Facility, 2014; GIZ, 2014; Sanahuja, 2011; Spearman and McGray, 2011). Much of this has been developed to support monitoring and evaluation of adaptation at national and subnational level. It sheds light on the major methodological challenges of tracking adaptation, including aspects such as baselines, attribution, timelines, and the lack of adopted standard practices and methodologies.

Various compilations of quantitative and qualitative metrics and indicators for adaptation exist and provide useful starting points for sectoral, national, and global tracking efforts (see e.g., Hammill et al., 2014; WHO, 2015). Several countries have provided both qualitative and quantitative indicators in the adaptation components of their Nationally Determined Contributions, but these are far from comprehensive and comparable. It is noticeable that many of the indicators developed for the Sustainable Development Goals and the Sendai Framework for disaster risk reduction are directly relevant for adaptation and could be used to track progress on national as well as global level. Table 4.2 provides examples of such indicators. Adaptation indicators are often classified as outcome, input, process, and impact indicators. Outcome indicators demonstrate that a particular adaptation objective has been achieved (e.g., loss of lives prevented). However, as "adaptation" is not an outcome in its own right, proxies for measuring "reduced vulnerability" or "increased resilience" are required (Bours et al., 2014). Process and input indicators, such as financial resources committed to adaptation and existence of legal and policy frameworks for adaptation, are often used as proxies for progression towards the achievement of an outcome, and can be useful for tracking the complex and large-scale endeavors that adaptation involves (Bours et al., 2014). Impact indicators measure the long-term effects of outcomes of adaptation interventions, and often aim to capture the change in adaptive capacity and resilience to climate shocks of both natural systems and human communities (World Bank, 2010). Most existing efforts have focused on measuring inputs and processes, and an evidence gap remains in terms of measurable adaptation outcomes and adaptation indicators that systematically give an indication of adaptation impacts (UNFCCC, 2016b; Noble et al., 2014).

Future efforts should therefore reflect the importance of indicators that measure both process and outcomes and be sensitive to finding a balance between the need for comparability and aggregation for global purposes and the need for details and contextualization at the national level (UNFCCC, 2016b).

4.4.3   Tackling Data and Information Challenges

The third and final priority area of action highlighted in this chapter concerns addressing data and information challenges for tracking adaptation. Existing data are generally limited, too broad, and often insufficiently tailored for adaptation to allow tracking (Ford et al., 2015). These findings are supported by UN Environment adaptation gap assessments and by the recent the UN Climate Resilience assessment of the global state of progress on three key capacities for climate resilience (UN Climate Resilience Initiative, 2017; UNEP, 2017, 2016, 2014). There is a need for standardized guidelines and approaches to data collection, and data sets need to be large, comprehensive, detailed, and regularly updated and adjusted. This poses a major challenge for tracking adaptation, one that it is highly unlikely that the UNFCCC can address in isolation.

However, even though the global stocktakes under the Paris Agreement are likely to build on country reporting, all information and data for tracking progress does not need to come from national reporting and sources. The flexibility in reporting allowed for adaptation communications is likely to lead to differences in the content and timing of Parties' reporting to the UNFCCC, also potentially impeding the use of such information to track the progress against enhancing adaptive capacity, strengthening resilience, and reducing vulnerability, the three elements of the global adaptation goal

TABLE 4.2 Examples of Sustainable Development Goal and Sendai Framework Indicators of Relevance to Climate Change Adaptation

Sustainable Development Goal Indicators Relevant to Climate Change Adaptation

Goal		Indicators
Climate action	13.2.1	Number of countries that have formally communicated the establishment of integrated low-carbon, climate-resilient, disaster-risk reduction development strategies (e.g., a national adaptation plan process, national policies and measures to promote transition to environmentally-friendly substances and technologies)
	13.1.1	Number of deaths, missing people, injured, relocated, or evacuated due to disasters per 100,000 people
	13.3.1	Number of countries that have integrated mitigation, adaptation, impact reduction, and early warning into primary, secondary, and tertiary curricula
Zero hunger	2.4.1	Percentage of agricultural area under sustainable agricultural practices
	2.4.2	Percentage of agricultural households using irrigation systems compared to all agricultural households
Clean water and sanitation	6.4.1	Percentage change in water use efficiency over time
	6.5.1	Degree of integrated water resources management implementation (0−100)
Sustainable cities and communities	11.b.1	Percentage of cities implementing risk reduction and resilience strategies aligned with accepted international frameworks (such as the Sendai Framework)
Life on land	15.2.1	Forest cover under sustainable forest management

Sendai Framework Compound Indicators That Could Relate to Climate Change Adaptation (Outcomes and Enabling Environments)

Global Target		Indicators
Disaster mortality	A-1	Number of deaths and missing persons attributed to disasters, per 100,000 population
Affected people	B-1	Number of directly affected people attributed to disasters, per 100,000 population (including population injured or ill, whose dwelling is damaged or destroyed, and whose livelihood is disrupted or destroyed)
Economic loss	C-1	Direct economic loss attributed to disasters in relation to global gross domestic product (including losses from agriculture, housing sector, productive assets, critical infrastructure and cultural heritage damaged or destroyed)
Critical infrastructure and basic services	D-1	Damage to critical infrastructure attributed to disasters (including health and educational facilities damaged or destroyed and critical infrastructure units and facilities)
	D-5	Number of disruptions to basic services attributed to disasters (including educational, health, and other basic services)
Developing countries' support	F-1	Number of countries that adopt and implement national disaster risk reduction strategies in line with the Sendai Framework for Disaster Risk Reduction 2015−30
Early warning systems	G-1	Number of countries that have multihazard early warning systems

Source: UNEP (2017), based on IAEG-SDG (2017), Vallejo (2017), and Kato and Ellis (2017).

(Kato and Ellis, 2016). National reporting can however be combined with, based on, and assessed against science-based information and data from other sources, as is the case for mitigation. In fact, third-party sources are specifically mentioned in the Paris Agreement. Here, the international research community and particularly the IPCC can play an important role. Furthermore, UN organizations and databases can be used to gather relevant data and can be requested to collect additional information on adaptation needed under the UNFCCC. Data collection and processing systems linked to the Sustainable Development Goals and the Sendai Framework for Disaster Risk Reduction also provide relevant learning opportunities and sources of data. Furthermore, tweaking existing indicators under the Sustainable

Development Goals and the Sendai Framework may provide cost-effective ways for gathering information on adaptation in climate change impact areas that have been already agreed as global priorities.

4.5 SUMMARY

Despite the recent emphasis on exploring how to measure adaptation progress globally, tracking progress at national level will remain a key concern and driver of action for countries. A key issue will be to ensure that the global framework for adaptation under the UNFCCC acknowledges, builds on, and strengthens national level tracking frameworks and efforts. This requires compatibility between global efforts, and what countries are measuring and reporting on—and how they do it. It will be important to avoid forcing countries into using a particular set of metrics for tracking purposes that may not be appropriate to or suit their national needs. If designed properly, the adoption of common methodologies and metrics can allow for lesson learning, common guidance, and common tools that can actually make the task at the national level simpler. As efforts on and approaches to tracking adaptation mature, it remains to be seen whether national interests and global perspectives diverge or whether they can complement each other.

It would nevertheless seem that the benefits of global metrics extend to national and subnational scales. Global tracking of adaptation can generate the broad-scale systematic, coherent, and comparable insights needed to guide adaptation at both national and global levels (Lesnikowski et al., 2016; Leiter, 2015). It can facilitate capturing the current state of adaptation across nations; provide the basis for characterizing and evaluating adaptations taking place in different settings, and; underpin the monitoring of change in adaptation over time (Ford and Berrang-Ford, 2016).

REFERENCES

Adaptation Committee, 2017. Synthesis of submissions from Parties and other stakeholders, and next steps for developing recommendations on methodologies for assessing adaptation needs. Eleventh meeting of the Adaptation Committee, Bonn, Germany, 7-10 Mach 2017. AC/2017/4, 4 March 2017, Agenda item 6 (b). Available at: http://unfccc.int/files/adaptation/groups_committees/adaptation_committee/application/pdf/ac11_6b_methodologies.pdf.

Adaptation Futures, 2016. Adaptation Futures 2016. Meeting Report. Available at http://edepot.wur.nl/389317.

Adger, W.N., Agrawala, S., Mirza, M.M.Q., Conde, C., O'Brien, K., Pulhin, J., et al., 2007. Assessment of adaptation practices, options, constraints and capacity. In: Parry, M.L., Canziani, O.F., Palutikof, J.P., van der Linden, P.J., Hanson, C.E. (Eds.), Climate Change 2007: Impacts, Adaptation and Vulnerability. Contribution of Working Group II to the Fourth Assessment Report of the Intergovernmental Panel on Climate Change. Cambridge University Press, Cambridge, pp. 717–743.

Bours, D., McGinn, C., Pringle, P., 2014. Guidance note 2: Selecting indicators for climate change adaptation programming, SEA Change Community of Practice and UKCIP. Available at: http://www.ukcip.org.uk/wp-content/PDFs/MandE-Guidance-Note2.pdf.

Brooks, N., Fisher, S., 2014. Tracking adaptation and measuring development: a step-by-step guide. Toolkit, IIED.

Climate Change Secretariat of Sri Lanka, 2016. National Adaptation Plan for Climate Change Impacts in Sri Lanka: 2016 − 2025. Ministry of Mahaweli Development and Environment. Available at: http://www4.unfccc.int/nap/Documents%20NAP/National%20Reports/National%20Adaptation%20Plan%20of%20Sri%20Lanka.pdf.

Dinshaw, A. et al., 2014. Monitoring and Evaluation of Climate Change Adaptation: Methodological Approaches, OECD Environment Working Papers, No. 74, OECD.

Eriksen, S.H., Kelly, P.M., 2007. Developing credible vulnerability indicators for climate adaptation policy assessment. Mitigation Adapt. Strat. Global Change 12, 495−524.

Federal Democratic Republic of Ethiopia, 2017. Intended Nationally Determined Contribution (INDC) of the Federal Democratic Republic of Ethiopia. Available at: http://www4.unfccc.int/ndcregistry/PublishedDocuments/Ethiopia%20First/INDC-Ethiopia-100615.pdf.

FDRE, 2011. Ethiopia's Climate-Resilient Green Economy: Green Economy Strategy. Federal Democratic Republic of Ethiopia, Addis Ababa.

Ford, J.D., Berrang-Ford, L., 2016. The 4Cs of adaptation tracking: consistency, comparability, comprehensiveness, coherency. Mitig. Adapt. Strat. Glob. Chang. 21 (6), 839−859. Available from: https://doi.org/10.1007/s11027-014-9627-7.

Ford, J.D., Berrang-Ford, L., Patterson, J., 2011. A systematic review of observed climate change adaptation in developed nations. Clim. Chang. Lett. 106 (2), 327−336. Available from: https://doi.org/10.1007/s10584-011-0045-5.

Ford, J.D., Berrang-Ford, L., Biesbroek, R., Araos, M., Austin, S.E., Lesnikowski, A., 2015. Adaptation tracking for a post-2015 climate agreement. Comment. Nat. Clim. Chang. 5, NOVEMBER 2015 | www.nature.com/natureclimatechange.

Gagnon-Lebrun, F., Agrawala, S., 2007. Implementing adaptation in developed countries: an analysis of progress and trends. Clim. Policy 7, 392−408.

GIZ, 2014. Monitoring and Evaluating Adaptation at Aggregated Levels: A Comparative Analysis of Ten Systems. GIZ, Bonn, Germany.

Global Environment Facility, 2014. Tracking Tool for Climate Change Adaptation Projects. The Global Environment Facility, Washington DC.

Hamill, A., Dekens, J., Leiter, T., Olivier, J., Klockemann, L., Stock, E., et al. (2014). Repository of Adaptation Indicators. Real Case Examples From National Monitoring and Evaluation Systems. GIZ and IISD. Available online at: http://www.adaptationcommunity.net/?wpfb_dl=221.

IAEG-SDGs, Inter-Agency and Expert Group on SDG Indicators (2017), Annex III – Revised list of global Sustainable Development Goal indicators, in Report of the Inter-Agency and Expert Group on Sustainable Development Goal Indicators, E/CN.3/2017/2. https://documents-ddsny.un.org/doc/UNDOC/GEN/N16/441/96/PDF/N1644196.pdf?OpenElement.

IPCC, 2014a. Summary for policymakers. In: Field, C.B., Barros, V.R., Dokken, D.J., Mach, K.J., Mastrandrea, M.D., Bilir, T.E., et al.,Climate Change 2014: Impacts, Adaptation, and Vulnerability. Part A: Global and Sectoral Aspects. Contribution of Working Group II to the FifthAssessment Report of the Intergovernmental Panel on Climate Change. Cambridge University Press, Cambridge, United Kingdom and New York, NY, USA, pp. 1–32.

IPCC, 2014b. In: Core Writing Team, Pachauri, R.K., Meyer, L.A. (Eds.), Climate Change 2014: Synthesis Report. Contribution of Working Groups I, II and III to the Fifth Assessment Report of the Intergovernmental Panel on Climate Change. IPCC, Geneva, Switzerland, p. 151.

Kato, T., Ellis, J., 2016. Communicating Progress in National and Global Adaptation to Climate Change. COM/ENV/EPOC/IEA/SLT(2016)1. Organisation for Economic Co-operation and Development, Paris, France.

Leiter, T. & Olivier, J. (2017). Synergies in monitoring the implementation of the Paris Agreement, the SDGs and the Sendai Framework. Climate Change Policy Brief. Deutsche Gesellschaft für Internationale Zusammenarbeit (GIZ) GmbH. Eschborn, Germany.

Leiter, T., Kranefeld, R., Olivier, J., Brossmann, M., & Helms, J. (2017). Can climate vulnerability and risk be measured through global indices? Climate Change Policy Brief. Deutsche Gesellschaft für Internationale Zusammenarbeit (GIZ) GmbH. Eschborn, Germany, http://www.adaptationcommunity.net/monitoring-evaluation/policy-briefs/

Leiter, T., 2015. Linking monitoring and evaluation of adaptation to climate change across scales: avenues and practical approaches. New Direct. Eval. 2015 (147), 117–127. Available from: https://doi.org/10.1002/ev.20135.

Lesnikowski, A., Ford, J., Biesbroek, R., Berrang-Ford, L., Jody Heymann, S., 2016. National-level progress on adaptation. Nat. Clim. Chang. 6. Available from: https://doi.org/10.1038/NCLIMATE2863.

Magnan, A.K., Ribera, T., 2016. Global adaptation after Paris. Science 352 (6291), 1280–1282. Available from: https://doi.org/10.1126/science.aaf5002.

Magnan, A.K., Ribera, T., Treyer, S., 2015. National Adaptation Is Also a Global Concern. Available at: http://www.iddri.org/Publications/National-adaptation-is-also-a-global-concern.

Mimura, N., Pulwarty, R.S., Duc, D.M., Elshinnawy, I., Redsteer, M.H., Huang, H.Q., et al., 2014. Adaptation planning and implementation. In: Field, C.B., Barros, V.R., Dokken, D.J., Mach, K.J., Mastrandrea, M.D., Bilir, T.E., Chatterjee, M., Ebi, K.L., Estrada, Y.O., Genova, R.C., Girma, B., Kissel, E.S., Levy, A.N., MacCracken, S., Mastrandrea, P.R., White, L.L. (Eds.), Climate Change 2014: Impacts, Adaptation, and Vulnerability. Part A: Global and Sectoral Aspects. Contribution of Working Group II to the Fifth Assessment Report of the Intergovernmental Panel on Climate Change. Cambridge University Press, Cambridge, United Kingdom and New York, NY, USA, pp. 869–898.

Naswa, P. et al., 2015. Good practice in designing and implementing national monitoring systems for adaptation, UDP/CATIE/CTCN.

Ngwadla, X., El-Bakri, S., 2016. The Global Goal for Adaptation Under the Paris Agreement: Putting Ideas Into Action. CDKN, London, UK.

Noble, I.R., Huq, S., Anokhin, Y.A., Carmin, J., Goudou, D., Lansigan, F.P., et al., 2014. Adaptation needs and options. In: Field, C.B., Barros, V.R., Dokken, D.J., Mach, K.J., Mastrandrea, M.D., Bilir, T.E., Chatterjee, M., Ebi, K.L., Estrada, Y.O., Genova, R.C., Girma, B., Kissel, E.S., Levy, A.N., MacCracken, S., Mastrandrea, P.R., White, L.L. (Eds.), Climate Change 2014: Impacts, Adaptation, and Vulnerability. Part A: Global and Sectoral Aspects. Contribution of Working Group II to the Fifth Assessment Report of the Intergovernmental Panel on Climate Change. Cambridge University Press, Cambridge, United Kingdom and New York, NY, USA, pp. 833–868.

OECD, 2015. National Climate Change Adaptation: Emerging Practices in Monitoring and Evaluation. OECD Publishing, Paris.

Olhoff, A., Bee, S., Puig, D., (2015). The Adaptation Finance Gap Update. With Insights from the INDCs. UNEP and UNEP DTU Partnership, Copenhagen, Denmark.

Sanahuja, H.E., (2011). Tracking Progress for Effective Action: A Framework for Monitoring and Evaluating Adaptation to Climate Change, Climate-Eval Community of Practice, Global Environment Facility.

Singh, H., Harmeling, S., Chamling Rai, S., (2016). Global Goal on Adaptation: From Concept to Practice.

Spearman, M., McGray, H., (2011). Making adaptation count: concepts and options for monitoring and evaluation of climate change adaptation. Deutsche Gesellschaft für Internationale Zusammenarbeit (GIZ), Bundesministerium für wirtschaftliche Zusammenarbeit und Entwicklung (BMZ), and World Resources Institute (WRI).

Swart, R., Biesbroek, R., Lourenco, T.C., 2014. Science of adaptation to climate change and science for adaptation. Front. Environ. Sci. 2. Available from: https://doi.org/10.3389/fenvs.2014.00029.

UN Climate Resilience Initiative, (2017). Anticipate, Absorb, Reshape: Current Progress on Three Key Capacities for Climate Resilience. Available at: https://static1.squarespace.com/static/5651e0a2e4b0d031533efa3b/t/5911b65e725e256f43f30e18/1494333030141/A2R_infobrief_web_singlepages.pdf.

UNEP, 2014. The Adaptation Gap Report. United Nations Environment Programme, Nairobi.

UNEP, 2016. The Adaptation Finance Gap Report. United Nations Environment Programme (UNEP), Nairobi.

UNEP (2017). The Adaptation Gap Report 2017. United Nations Environment Programme (UNEP), Nairobi.

UNFCCC (2015). The Paris Agreement. FCCC/CP/2015/L.9/Rev.1. Available at: http://unfccc.int/resource/docs/2015/cop21/eng/10a01.pdf.

UNFCCC (2016a). Ad Hoc Working Group on the Paris Agreement, Second part of the first session Marrakech, 7–14 November 2016, Agenda item 4. Further guidance in relation to the adaptation communication, including, inter alia, as a component of nationally determined contributions, referred to in Article 7, paragraphs 10 and 11, of the Paris Agreement. Available at: http://unfccc.int/resource/docs/2016/apa/eng/inf02a01.pdf.

UNFCCC (2016b). Aggregate effect of the intended nationally determined contributions: an update. Synthesis report by the secretariat. Available at: http://unfccc.int/resource/docs/2016/cop22/eng/02.pdf.

Vallejo (2017), "Insights from national adaptation monitoring and evaluation systems", OECD/IEA Climate Change Expert Group Papers, No. 2017/03, www.oecd.org/environment/cc/Insights%20from%20national%20adaptation%20monitoring%20and%20evaluation%20systems.pdf.

Wiseman, V. (2016). The UNFCCC National Adaptation Planning Model: A Foundation for Fulfilling Post-2015 Commitments? Available at: http://napglobalnetwork.org/2016/04/the-unfccc-national-adaptation-planning-model-a-foundation-for-fulfilling-post-2015-commitments/.

World Bank (2010). Mainstreaming adaptation to climate change in agriculture and natural resources management projects. Guidance Note 8: Monitoring and Evaluation of Adaptation Activities. Available at: http://siteresources.worldbank.org/EXTTOOLKIT3/Resources/3646250-1250715327143/GN8.pdf.

World Health Organisation (WHO) (2015). Climate and Health Country Profiles - 2015. A Global Overview. WHO, Geneva. Available at: http://apps.who.int/iris/bitstream/10665/208855/1/WHO_FWC_PHE_EPE_15.01_eng.pdf?ua = 1.

FURTHER READING

Dupuis, J., Biesbroek, R., 2013. Comparing apples and oranges: the dependent variable problem in comparing and evaluating climate change adaptation policies. Glob. Environ. Chang.-Human Policy Dimens 23, 1476−1487. Available from: https://doi.org/10.1016/j.gloenvcha.2013.07.022.

Ford, J.D., et al., 2013. How to track adaptation to climate change: a typology of approaches for national-level application. Ecol. Soc. 18 (3). Available from: https://doi.org/10.5751/ES-05732-180340.

Preston, B.L., Mustelin, J., Maloney, M.C., 2014. Climate adaptation heuristics and the science/policy divide. Mitig. Adapt. Strateg. Glob. Chang. . Available from: https://doi.org/10.1007/s11027-013-9503-x.

Chapter 5

Evolution of Climate Change Adaptation Policy and Negotiation

Saleemul Huq[1], Yousuf Mahid[1] and Nadine Suliman[1,2]

[1]International Centre for Climate Change and Development, Dhaka, Bangladesh, [2]The University of Alberta, Alberta, Canada

Chapter Outline

5.1 INTRODUCTION

Climate change, one of the greatest challenges presently facing mankind, is pressing widespread and detrimental effects to both current and future generations. The Intergovernmental Panel on Climate Change (IPCC) has underscored the veracity of current changing weather patterns and the need for immediate and rigorous action. The predictions from scientific communities have endorsed that mitigation efforts alone, even at their most stringent, would be insufficient in responding to expected impacts of climate change over the next few decades, noting the variety of biophysical and socioeconomic repercussions (Helgeson and Ellis, 2015; The Committee on Approaches to Climate Change Adaptation, 2010). This indicates an increasing need for adaptation initiatives and policies in parallel with long-term mitigation action to ensure the security, safety, and sustainable development of all countries. As such, the state of adaptation in the global negotiation platform known as United Nations Framework Convention on Climate Change (UNFCCC) has evolved significantly in the recent years. This current notion portrays adaptation as one of the imperative pillars for upcoming climate change negotiations of the conventions.

5.2 OVERVIEW OF CLIMATE CHANGE ADAPTATION: AN EVOLVING CHALLENGE

The significance of climate change adaptation can be observed from the definition provided by various apex bodies and scientific or policy communities. The definitions may vary slightly but consist of a meaning in a common usage and include grander implications in the global setting of climate change negotiations. The definition of adaptation had been taken into consideration for the first time in the Third Assessment Report of IPCC, which defines it as "Adjustment in

Resilience. DOI: https://doi.org/10.1016/B978-0-12-811891-7.00005-0

natural or human systems in response to actual or expected climatic stimuli or their effects, which moderates harm or exploits beneficial opportunities" (IPCC, 2001). Another slightly different definition can be observed from the UNFCCC website; "Adaptation refers to adjustments in ecological, social, or economic systems in response to actual or expected climatic stimuli and their effects or impacts. It refers to changes in processes, practices, and structures to moderate potential damages or to benefit from opportunities associated with climate change" (UNFCCC, 2017a). The root cause in acquiring variations in the above-mentioned definitions is the fundamental difference in defining climate change conveyed by the UNFCCC and the IPCC (Levina and Tirpak, 2006). UNDP has defined adaptation as "a process by which strategies to moderate, cope with and take advantage of the consequences of climatic events are enhanced, developed, and implemented" (Mwandingi, 2006). Whereas, UK Climate Impact Programme emphasized adaptation as an outcome rather than only a process and offered additional interpretation (Levina and Tirpak, 2006). All of these seemingly small differences in the definitions present different expectations from different stakeholders and provide significant implications in the orientation of policy and financial matters.

Fundamentally and predominately established, adaptation is considered as local intervention where the response measures in the face of climate change impacts are tailored to local settings (Burton et al., 2006). Although, extracted and observed from the definitions and a significant number of literatures, the idea of climate change adaptation can be conceptualized as multidimensional, encompassing numerous activities in different sectors at various levels. Expert opinions suggest that in order to make adaptation efforts more robust and sustainable, local and sectoral level measures should be guided and supported by national strategies and policies. This can further be facilitated by global policy interference. At this stage, the debate comes into force whereas in many cases, these sectors do not have a direct mandate to respond to climate change itself. This in turn, generates questions revolving around the extent different aspects of adaptation should or can be addressed by the UNFCCC or by other global policy agendas and forums. This dispute limits the understanding of the whole scope of adaptation policy needs. New and emerging issues may come to the forefront which will indicate the requirements of introducing new technologies and reforming policies. This makes adaptation a dynamic and an evolving process. However, there is a prior need for any global agreement and policy orientation to have a clear delineation of the scope of the concept to deal with the evolving challenge. It is very much imperative to agree on such things that will define the scope of adaptation requirement and effective execution of what it encompasses. Following through, UNFCCC, within the process, has identified five general components of climate change adaptation activities as follows, to clearly exhibit this objective (UNFCCC, 2017b):

- Observation of climatic and nonclimatic variables;
- Assessment of climate impacts and vulnerability;
- Planning;
- Implementation; and
- Monitoring and evaluation of adaptation actions.

Besides the convention, in most cases, other global forums and agendas have considered these elements to articulate and convey adaptation policy intervention. The next sections of the chapter will present a brief overview of adaptation policy initiatives undertaken within the umbrella of UNFCCC and other global agendas and multilateral forums. In addition, the sections will highlight the current state of adaptation including major achievements that have been made so far and the gaps that remain in terms of policy interventions.

5.3 CURRENT POLICY AND LEGAL FRAMEWORK FOR CLIMATE CHANGE ADAPTATION UNDER THE UNFCCC

5.3.1 Adoption in the Convention

Since the official launch in 1990, international dialogue has been considering climate change adaptation as one of the key issues to be addressed inclusively (Levina, 2007). After the adoption of UNFCCC in 1992, objectives, principles, several commitments, and financial provisions regarding adaptation have been set within the different articles of the convention (UNFCCC, 1992).

The major objective of the convention, mentioned in Article 2, was to achieve a level of stabilized greenhouse gas concentration in the atmosphere within a fixed time frame so that the ecosystem can adapt naturally to the changing climate. In Article 3.3, one of the key principles was set to formulate such policies and measures to deal with the

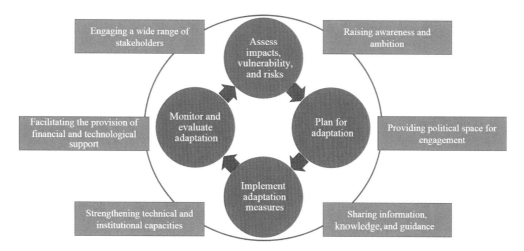

FIGURE 5.1   Role of convention to support adaptation. *UNFCCC, 2013. The State of Adaptation under the United Nations Framework Convention on Climate Change. Retrieved from: http://unfccc.int/files/adaptation/cancun_adaptation_framework/adaptation_committee/application/pdf/ac_2013_report_high_res.pdf.*

irreversible damage and serious threat that should cover the relevant issues related to adaptation and encompass all economic sectors. In addition, the following commitments on adaptation have been made in Article 4 of the convention:

Article 4.1 (b) highlights that in parallel to addressing anthropogenic emission for mitigating climate change impacts, all Parties shall formulate, implement, and regularly update measures to facilitate adequate adaptation.
Article 4.1 (e) calls for cooperation among the Parties to get prepared for adaptation in the face of climate change. Moreover, it calls for ensuring protection and rehabilitation of certain areas, in particular areas in Africa, which are more affected by climatic hazards like drought, desertification, etc. It emphasizes the development and elaboration of an integrated plan for the coastal zone management, water resources and agriculture sectors.
Article 4.1 (f) indicates the necessity of taking climate change adaptation into consideration in formulating social, economic, and environmental policies and actions. In addition, Article 4.1 (g) urges Parties to cooperate in order to conduct scientific, technical, technological, and socioeconomic research and to provide training, education, etc. to raise public awareness.

Moreover, the convention provides guidance on how the activities of adaptation would be supported and facilitated by developed country Parties for the vulnerable countries. In its Article 4.3, the convention directs specific financial obligations and requires that new and additional financial resources, including transfer of technology need to be provided by developed countries to cover the cost incurred by developing countries in implementing adaptation measures. Consecutively, to deal with the adverse effects of climate change, Article 4.4 of the convention obliges developed countries and other developing countries in the Annex-II to ensure assistance for developing countries in meeting the costs of adaptation (Levina, 2007; UNFCCC, 1992). Fig. 5.1 briefly depicts the role of the Convention in supporting adaptation processes.

5.4   CONFERENCE OF PARTIES (COP) AND MAJOR ADAPTATION MILESTONES UNDER THE CONVENTION

This section outlines the key COP decisions (Box 5.1 highlights the different COP that have been organized so far) taken to formulate and mobilize climate change adaptation policy and will briefly present the major thematic adaptation milestones achieved under the UNFCCC.

5.4.1   Impact Observation and Assessment of Risks and Vulnerability

As mentioned earlier, when the convention entered into force in 1994, the initial focus was on biophysical impacts through climate change mitigation; to reduce greenhouse gas emissions. With a view to construct a wide range of long-term planning scenarios, the Parties to the convention carried out global impact assessments, however, these impact assessment models were not sufficient enough in detailing out the national or regional or local level impacts. Following

BOX 5.1 Conference of Parties (COP) List

Meeting	Year	Location
COP 1	1995	Berlin
COP 2	1996	Geneva
COP 3	1997	Kyoto
COP 4	1998	Buenos Aires
COP 5	1999	Bonn
COP 6	2000	The Hague
COP 6	2001	Bonn
COP 7	2001	Marrakech
COP 8	2002	New Delhi
COP 9	2003	Milan
COP 10	2004	Buenos Aires
COP 11	2005	Montreal
COP 12	2006	Nairobi
COP 13	2007	Bali
COP 14	2008	Poznań
COP 15	2009	Copenhagen
COP 16	2010	Cancún
COP 17	2011	Durban
COP 18	2012	Doha
COP 19	2013	Warsaw
COP 20	2014	Lima
COP 21	2015	Paris
COP 22	2016	Marrakech
COP 23	2017	Bonn

through, the Parties looked forward to reporting their vulnerability and adaptation assessment in the National Communication. Going forward, the process of second generation assessments complemented the first generation scenario based assessment and included risk assessment and allowed more refined climate change scenarios for long-term socioeconomic and environmental planning.

At the first COP to the convention in 1995, with its Decision 11/CP.1, the Parties established guidance and defined a three-stage framework for actions on climate change adaptation. Stage-I was designed to conduct studies on possible impacts of climate change and to identify vulnerable countries and policy options for adaptation to deal with the impacts. This stage was envisaged to consider short-term adaptation activities. Stage-II and Stage-III were intended to perform medium- and long-term activities; where Stage-II would focus on the preparation for adaptation including capacity building activities and Stage-III would provide efforts on implementation and entail measures to facilitate adequate adaptation (UNFCCC, 1995; Levina, 2007; Burton et al., 2006).

In 1997, the third COP took place in Kyoto, Japan, where a framework of specific targets was defined and set for the developed countries to cut their greenhouse gas emissions. This is known as the Kyoto Protocol (Shah, 2002). Like the convention, the Kyoto Protocol was envisaged to assist the countries to undertake adaptation initiatives. Article 10 of the Kyoto Protocol, reaffirming the commitments made in the Article 4 of the Convention, urges Parties to facilitate adaptation through formulating and updating national and in some cases regional programme measures. This article also emphasizes adaptation technologies and methods for improving spatial planning to further assist climate change adaptation. In addition, Article 12.8 of the Kyoto Protocol has made a special provision for financing adaptation; "…Parties to this Protocol shall ensure that a share of the proceeds from certified project activities is used to cover administrative expenses as well as to assist developing country Parties that are particularly vulnerable to the adverse effects of climate change to meet the costs of adaptation" (UNFCCC, 1998). Moreover, in order to strengthen technical and institutional capacity within the convention, Parties in 1999 established the Consultative Group of Experts on National Communications from Parties not included in Annex I to the Convention (CGE), which over the years, has produced intense training modules to provide developing countries with hands-on training workshops at the regional level (UNFCCC, 2013).

5.4.2 Moving Towards Planning and Implementation Phase

With the publication of the Third Assessment Report of IPCC in 2001, ample evidence clearly indicated changing weather patterns and the need for adaptation actions along with mitigation. Since then, adaptation issues started receiving increased attention from the global community. The shift of thinking from "do we need to adapt?" to "how do we adapt?" (UNFCCC, 2013) became evident during that time. These circumstances provided impetus to the convention process and climate negotiation in addressing adaptation more seriously and effectively.

The Marrakech Accord, adopted by COP 7 in 2001, is reflected as the landmark for adaptation within the UNFCCC. For the first time, Parties recognized the intrinsic relationship between development and climate change issues (Helgeson and Ellis, 2015) and understood the high degree of vulnerability that developing countries were experiencing due to the climate variability. Following through, under the Decision 5/CP.7, Parties established a separate work programme for the Least Developed Countries (LDC), known as the LDC Work Programme, to address specific and immediate needs in the face of climate change (UNFCCC, 2001) and to increase adaptive capacity of the LDCs (Helgeson and Ellis, 2015). This work programme includes the development and implementation of the National Adaptation Programmes of Action (NAPA) as a mean for LDCs to exhibit the urgent and immediate adaptation needs on a priority basis, and the formation of the Least Developed Countries Expert Group (LEG). The Parties, under the Decision 5/CP.7, also established specific funds; Special Climate Change Fund (SCCF) to provide finance in regards to vulnerability and adaptation and Least Developed Countries Fund (LDCF) to support the LDC work programmes particularly to fund the preparation and implementation of NAPAs (Levina, 2007; UNFCCC, 2013). Furthermore, with the Decision 10/CP.7, an Adaptation Fund, under the Kyoto Protocol, was established to facilitate developing countries and to finance projects or programmes on climate change adaptation (Levina, 2007).

5.4.3 Exchange of Information and Lessons Learned

As the momentum was building up on the planning and implementation of adaptation, there emerged a need for collecting cross-country evidence including good practices; sharing of information and the lessons learned or experienced throughout the process. At COP 10 in 2004, the Parties concluded the negotiation emphasizing further implementation of Decision 5/CP.7 and with the Decision 1/CP.10 agreed to "The Buenos Aires programme of work on adaptation and response measures" (Levina, 2007; Helgeson and Ellis, 2015). The programme was requested to implement the actions required to deal with the adverse effects of climate change through developing methodologies; reporting in Parties' National Communication the vulnerability assessment and adaptation measures; collecting, sharing, and disseminating the information among the Parties etc. This programme further led to the COP 11 (2005) agreement, under the Decision 2/CP.11, on the adoption of a "5-year programme of work on impacts, vulnerability and adaptation to climate change." In 2006, at COP 12, the work programme got renamed as the "Nairobi work programme (NWP) on impacts, vulnerability and adaptation to climate change" (Levina, 2007). The overall objectives of the NWP have been to assist developing country Parties to better understand and measure vulnerability and adaptation with a view to inform them on practical adaptation actions to avoid the impact. In addition, during the period of 2006–2007, a series of workshops and events helped the developing countries get an opportunity to learn from the exchange of knowledge and good practices. Both the Parties, during that period, observed and pointed out the importance of a coordinated approach to address adaptation issues in the context of sustainable development (UNFCCC, 2013).

5.4.4 Scaling up Implementation Strategies and Coordinating the Evolving Adaptation Agenda

Published in 2007, the Fourth Assessment Report (AR4) of IPCC stressed a wide array of possible adaptation options. It also reaffirmed that more extensive adaptation than earlier was required to reduce the future vulnerability as a result of the greenhouse gas concentration in the atmosphere due to past emissions (IPCC, 2007). Against the backdrop of the fourth assessment report, the Parties adopted the Bali Action Plan at COP 13 in 2007, which established adaptation as one of the four pillars under the UNFCCC (Helgeson and Ellis, 2015). The Bali Road Map propelled a comprehensive process of effective and sustained implementation strategy and established an Ad Hoc Working Group (AWG) on Long-Term Cooperative Action. The implementation of the road map further offered lessons learned for future implementation through long-term inclusive actions (Global Greenhouse Warming, 2010). One of the major learnings was that, as a result of different circumstances and capacities, the nature of adaptation actions should be country driven, gender sensitive, and particularly participatory.

TABLE 5.1 Overview of Technical and Financial Support to Adaptation

	Facilitating the Provision of Financial and Technological Support	Strengthening Technical and Institutional Capacities	
Coordinating and/or Advisory Committees/ Bodies	Technology Executive Committee (TEC)	Standing Committee on Finance (SCF)	Adaptation Committee Least Developed Countries Expert Group (LEG) Consultative Group of Experts on National Communications from Parties not included in Annex I to the Convention (CGE)
Providers of Support	Climate Technology Centre and Network (CTCN)-UNEP	Green Climate Fund (GCF)–GCF Board Adaptation Fund (AF)–AF Board Least Developed Countries Fund (LDCF)-GEF Special Climate Change Fund (SCCF)-GEF	
Exchange of Information		Work Programme on Long-Term Finance	Durban Platform on Capacity-Building

Source: UNFCCC, 2013. The State of Adaptation under the United Nations Framework Convention on Climate Change. Retrieved from: http://unfccc.int/files/adaptation/cancun_adaptation_framework/adaptation_committee/application/pdf/ac_2013_report_high_res.pdf.

Following the next 3 years of negotiations, at COP 16 in 2010, Parties highlighted adaptation with the same level of priority as mitigation and adopted the Cancun Adaptation Framework (CAF) with an objective to enhance cooperative actions on adaptation. After gathering lessons learned following the NAPA process and recognizing the need for LDCs to develop medium- and long-term adaptation planning, the National Adaptation Plans (NAPs) preparation was formulated under the CAF. To provide further technical assistance and guidance to the LDC and to advise the COP on adaptation strategy, Parties formed an Adaptation Committee (AC) as a part of the CAF: to strengthen the existing institutional arrangements; to promote synergies between national and international institutions; to disseminate relevant information; and to coordinate the expanding agenda of adaptation under the convention. Additionally, Parties under the CAF adopted a Technology Mechanism consisting of a Technology Executive Committee (TEC) and a Climate Technology Centre and Network (CTCN) (UNFCCC, 2013).

In addition to technical support, a wide range of financial support was introduced and provided to the LDCs in order to implement the NAP and other adaptation actions. With the Decision 1/CP.16 of CAF, Parties established a Standing Committee on Finance (SCF) to assist the convention in regards to relevant issues of climate finance (UNFCCC, 2017c). Moreover, with a view to long-term, scaled-up, new and additional finance, Parties formed the Green Climate Fund (GCF) under the CAF (Decision 1/CP.16) as an operating entity of the financial mechanism of the convention (UNFCCC, 2016). Parties agreed that funding from GCF will be mobilized and channeled to adaptation actions through a separate thematic window. Table 5.1 provides a general overview of the technical and financial support to address the adaptation issue.

Besides adaptation initiatives, CAF introduced a new thematic work programme to address climate change induced loss and damage. Following through on COP 18 in 2012, where Parties acknowledged the need for comprehensive and strategic responses, the convention adopted an institutional arrangement known as Warsaw International Mechanism for Loss and Damage at COP 19 in 2013. The key goals of the initiative are to broaden global dialogues on the topic, to share knowledge and expertise, and to extend financial assistance to poorer countries vulnerable to climate change impacts. Within the loss and damage context, numerous adaptation approaches were rendered that can effectively reduce the risks of climatic impacts (UNFCCC, 2013).

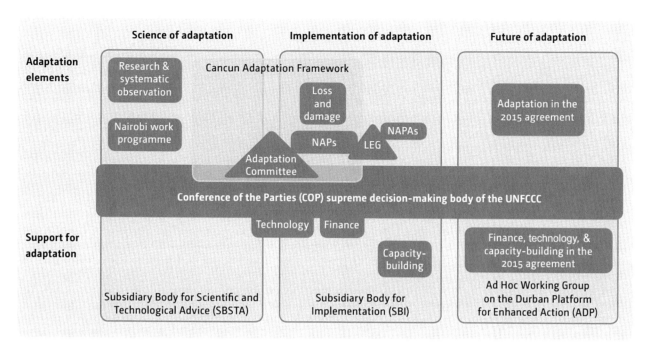

FIGURE 5.2   Overall institutional structure on adaptation under the convention. *UNFCCC, 2013. The State of Adaptation under the United Nations Framework Convention on Climate Change. Retrieved from: http://unfccc.int/files/adaptation/cancun_adaptation_framework/adaptation_committee/ application/pdf/ac_2013_report_high_res.pdf.*

5.4.5   The Paris Agreement and Adaptation

The Paris Agreement, bringing all the Parties under a shared cause, was adopted by the convention in 2015 at COP 21 with an aim to strengthen global response to tackle climate change impacts by agreeing to limit global temperature rise well below 2°C, pursuing an offer to set even below 1.5°C (UNFCCC, 2017d). The base of the agreement started building up during the establishment of the Durban Platform at COP 17 in 2011. The platform initiated the need for developing another legally binding instrument that can further mobilize climate change actions. In turn, adopted by 197 countries and ratified by 148 countries to date, this has become the legally binding global climate deal (DAG, 2016).

The Paris Agreement has predominately incorporated adaptation into the intention of the convention in order to effectively deal with the current scenario needs. Article 4.7 of the agreement highlighted mitigation cobenefits as part of adaptation interventions and economic diversification plans. In addition, a global goal has been established on climate change adaptation to ensure sustainable development through strengthening adaptive capacity and reducing the vulnerability of communities. Moreover, the agreement recognizes the "urgent and immediate needs" for the countries that are more vulnerable and conveys adaptation intervention as the "significant" current need to reduce the impacts associated with climate change (UNFCCC, 2015; Oettlé, 2016). Overall institutional structure on adaptation under the Convention is represented at a glance in Fig. 5.2.

5.5   CLIMATE CHANGE ADAPTATION INITIATIVES IN OTHER INTERNATIONAL FORUMS

This section briefly covers adaptation initiatives directly or indirectly undertaken over the years in global forums other than the UNFCCC. This will also present the overview of the different funding provisions available to assist those initiatives.

5.5.1   The Sendai Framework for Disaster Risk Reduction 2015–30, Hyogo Framework for Action 2005–15, and International Strategy for Disaster Risk Reduction

The expression "disaster risk reduction" functions as one of the main tools of adaptation to climate change. In formulating climate change strategies, adaptation acts as one of the subsets within disaster risk reduction. Similarly, disaster risk

BOX 5.2 The Relationship of Disaster Risk Reduction, Resilience and Climate Change Adaptation

Adaptation, which has multiple descriptions, definitions, and explanations, is the runner up in the course of climate change responses on a global scale as it is interdisciplinary and addresses the vulnerability of a system in the long term and on a global scale (Klein, 2014; Cannon and Müller-Mahn, 2010; Thomalla et al., 2006). Due to its diverse nature, adaptation can be linked to several climate change-related concepts/actions such as resilience as well as disaster-risk reduction. Adaptation mainly refers to any means of modifying/adjusting ecological, social, or economic systems in the face of climate-related stresses, potential environmental risks or disasters (Adger et al., 2009, 2005). Adaptive capacity of a system is a determining factor of its resilience and is commonly defined as the ability to "cope" with ill-effects of climatic changes and extreme weather events (or environmental disasters) (Adger et al., 2009, 2005; Gallopín, 2006). Resilience addresses the bigger picture and focuses on cross-scale dynamics involved in socioeconomic characteristics of world systems as well as climate change, poverty, inequality, etc. (Brown, 2016). Disaster-risk reduction relies on engineering and natural science approaches to risk management, with a focus on event, exposure, and technological solutions. It has mainly been a short-term approach but is increasingly long term with an emphasis on preparedness and awareness on a local–community scale (Thomalla et al., 2006).

Although it remains highly theoretical, disaster-risk reduction has been found to optimize adaptation action, which in turn has been seen to enhance the adaptive capacity of a system, and hence its resilience (Adger et al., 2009, 2005; Thomalla et al., 2006). As each of these networks/concepts are extremely complex, broad, and with separate aims and applications, research is yet to clearly and empirically report on the (potential) relationship between adaptation, resilience, and disaster-risk reduction (Thomalla et al., 2006). There has been little or no data reporting on the ability of adaptation to enhance a system's resilience and/or the impact of disaster-risk reduction on enhancing adaptation activities (Thomalla et al., 2006). Examining the theoretical basis for the links between all three concepts, it is logical that advanced technological solutions for monitoring/predicting climatic changes would enable a community/system to properly adapt to expected impacts (Thomalla et al., 2006). Furthermore, the better adapted a system is to potential climatic changes and their biophysical and socioeconomic impacts, the more resilient it is and the more able it is to regenerate itself while maintaining service provisions in a transformative and sustainable manner (Cannon and Müller-Mahn, 2010; Adger et al., 2009; Leach, 2008; Folke, 2006; Walker et al., 2004; Folke et al., 2002).

reduction strategies always depict sectoral measures and cooperate with other interests with a view to ensure improved adaptive capacity of those sectors. The relationship among disaster risk reduction, resilience, and adaptation has been illustrated briefly in Box 5.2.

The Sendai Framework for Disaster Risk Reduction highlighted the synergies between these two terms. In paragraph 47d, the framework embeds the implementation strategy and instructs that disaster risk reduction measures are to be incorporated into all development assistance programmes relevant to any sector including "poverty reduction, sustainable development, natural resource management, environment, urban development and adaptation to climate change" (Kelman, 2015). This statement helps move forward disaster risk reduction strategy incorporating different relevant sectors and accounting the veracity of the synergies. The Hyogo Framework for Action, with no exception, also represents a similar connotation. The mid-term review (2010–11) of the Hyogo Framework was directed towards a wide-range of dialogue to bring adaptation and disaster risk reduction onto one platform to reduce the impacts of extreme climatic events (UNISDR, n.d.). Furthermore, International Strategy for Disaster Reduction (ISDR), established in 2000 by the UN General Assembly, carries out such activities that are highly relevant to the works of UNFCCC and in some cases; both share the same agenda in managing disastrous climatic events (Levina, 2007).

5.5.2 Convention on Biological Diversity (CBD)

One of the major adaptation options would be the conservation and the sustainable use of biodiversity resources. Adopted in 1992, the Convention on Biological Diversity (CBD) with an objective to conserve biodiversity and to ensure equitable usage of the resources has successfully incorporated climate change adaptation issues within its work programme. Parties to the Convention have called for enhancing provision to integrate the issue furthermore. This clearly provides an opportunity to identify synergies between the UNFCCC and CBD to take mutually beneficial implementing activities regarding climate change adaptation. Although many initiatives have been taken by the Parties of the CBD to enhance synergies between these two Conventions, the synergies have not yet been able to see light due to the lack of understanding and coordination among national and international agencies and agreements (Levina, 2007; Helgeson and Ellis, 2015).

5.5.3   United Nations Convention to Combat Desertification

With an aim to combat desertification and to mitigate adverse effects in the drought prone areas/countries, the legally binding international agreement United Nations Convention to Combat Desertification (UNCCD) was established in 1994. Due to the dynamics of and nexus among land, biodiversity, and climate change, the UNCCD works closely with other two Rio Conventions, CBD and UNFCCC (UNCCD, 2017). Unlike the CBD, the UNCCD mentioned this collaborative approach in Article 8 of the convention in order to safeguard the sustainable use of natural resources and to derive maximum benefits when addressing climate change adaptation issues attributed through each agreement (Levina, 2007; Helgeson and Ellis, 2015).

5.5.4   European Climate Adaptation Platform

The partnership between the European Commission and the European Environment Agency, known as the European Climate Adaptation Platform (CLIMATE-ADAPT), has been developed to support particularly the European Countries in adapting to climate change (UNCCD Knowledge Hub, 2017). This platform helps Europe understand the adaptation initiatives taken by other countries to develop their own adaptation strategy. Also, this platform provides an opportunity for research communities and practitioners to share relevant adaptation information experienced across Europe (European Commission, n.d.).

Moreover, there are available funding provisions for climate change adaptation outside of UNFCCC as well. In order to facilitate adaptation activities which do not fall under the UNFCCC, the following funds have been set up, which can be framed as non-UNFCCC Adaptation Funds (Syrovátka, n.d.).

- MDG Achievement Fund (Environment and climate change thematic window)−UNDP Spain
- Cool Earth Partnership−Japan
- Global Climate Change Alliance−European Commission
- International Climate Initiative−Germany
- Climate Change and Development—Adapting by Reducing Vulnerability−UNEP-UNDP Denmark
- Africa Adaptation Programme−UNDP Japan
- Pilot Program for Climate Resilience−World Bank
- Global Facility for Disaster Reduction and Recovery−World Bank
- Agriculture Smallholder Adaptation (ASAP)

5.6   NATIONAL LEVEL PROGRESS ON ADAPTATION POLICY INITIATIVES

One of the main global initiatives to prioritize adaptation was the NAPA. The NAPA, as mentioned earlier, is an initiative led by the UNFCCC developed to urge and enable LDCs to assess and evaluate urgent climate change adaptation actions needed locally to reduce vulnerability and prevent high future costs of potential damage (UNFCCC, 2008). Each of the LDCs was assigned the task of developing its own NAPA and submitting it to the UNFCCC for publishing on a public database. Such an initiative was deemed beneficial in several aspects: the decentralization of developing contextual plans of action; the creation of a public platform for knowledge sharing and collaboration between stakeholders; and the identification of vulnerable sectors for efficient project development. More importantly, the NAPAs are envisaged to act as frameworks/lists of adaptation projects, a prerequisite for LDCs to be able to apply for the LDC Fund (McGray, 2014).

Most LDCs successfully submitted their NAPA to the UNFCCC secretariat, and analysts and researchers alike claimed they hold immense potential to spearheading and streamlining global adaptation efforts (UNFCCC, 2008). One of the main elements identified by most, if not all, NAPAs is capacity building and education. The wide focus by many LDCs on education as one of the top priority sectors for climate change adaptation and the inclusion of education and capacity building in projects was seen to resonate with adaptation goals and long term objectives.

Different from the NAPAs, another initiative led by Parties to the UNFCCC in 2011 (COP 17) in Durban, the NAP pertained to additional aspects of adaptation (McGray, 2014; Kissinger and Namgyel, 2013). Described as more inclusive, the NAP process, as compared to the NAPA, addressed medium- and long-term adaptation planning. While NAPA was mandatory for LDCs to be able to access the LDC fund, other developing countries were also invited to prepare and submit their NAP (Kissinger and Namgyel, 2013). The NAP process can be considered to complement NAPAs (for LDCs) as it further facilitates the inclusion of adaptation into existing frameworks of action. The objectives of NAP, as set during COP 17, are as follows (UNFCCC, 2011):

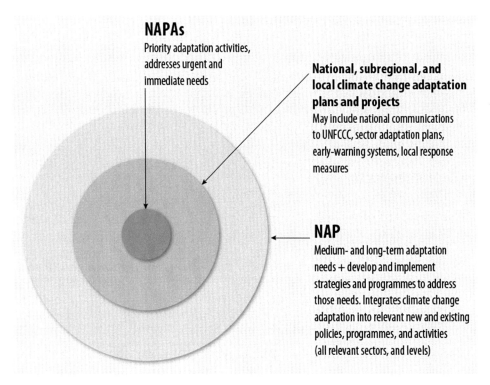

NAPAs
Priority adaptation activities, addresses urgent and immediate needs

National, subregional, and local climate change adaptation plans and projects
May include national communications to UNFCCC, sector adaptation plans, early-warning systems, local response measures

NAP
Medium- and long-term adaptation needs + develop and implement strategies and programmes to address those needs. Integrates climate change adaptation into relevant new and existing policies, programmes, and activities (all relevant sectors, and levels)

FIGURE 5.3 National adaptation action and the link between NAPA and NAP. *Kissinger, G., Namgyel, T., 2013. NAPAs and NAPs in Least Developed Countries. Retrieved from: https://ldcclimate. files.wordpress.com/2013/12/ ldcp13_napas-and-naps.pdf.*

1. "reduce vulnerability to the impacts of climate change, by building adaptive capacity and resilience"; and
2. "facilitate integration of climate change adaptation, in a coherent manner, into relevant new and existing policies, programmes and activities, in particular development planning processes and strategies, within all relevant sectors and at different levels, as appropriate"

As mentioned earlier, NAPA and NAP are two distinct initiatives in the adaptation response arena; however, they can be seen as complimentary, building off each other for efficient and effective action (see Fig. 5.3). Linking both initiatives allows for synergies and reduces duplication of efforts, enabling strengthened and sustainable adaptation on local and national scales.

In line with country-driven initiatives, we can find that the NAPAs and NAPs were just the starting point for national level adaptation action where many LDCs and developing countries conducted further adaptation action. Bangladesh as an example has pioneered in mainstreaming adaptation in national policies as the country identified specific key adaptation policies and further developed an inclusive climate change strategy and action plan (MoEF, 2015).

5.7   MAJOR GAPS AND THE WAY FORWARD

The Paris Agreement in Article 6 on Adaptation agreed to work towards a Global Goal on Adaptation (GGA). Also in Article 13 on measurement and transparency it included the need to include adaptation. Hence the current work on implementing Article 6 on Adaptation as well as the Adaptation parts of Article 13 are being actively discussed amongst both the research as well as policy-making communities; that needs to materialize. Article 11 of the Paris Agreement highlights the need to develop in-country systems for sustained capacity building of all citizen's to enable them to tackle climate change.

Despite the immense emphasis on adaptation at global, regional, national, and local levels, major gaps remain impeding the way for effective and sustainable implementation. Unlike mitigation, adaptation still has a long way to go in terms of its mainstreaming in policy, planning, and implementation. Researchers, scientists, and critics may claim that the reasons behind the status of adaptation in global action on climate change include its wide scope. However, the initiatives mentioned earlier and the conceptual work that has been done on adaptation so far has allowed it to gain momentum and build a strong foundation based on contextual and empirical evidence. In order to advance and progress adaptation action, gaps must be strategically identified and analyzed with a focus on solutions and future plans.

Harnessing the power of adaptation lies in the North and the South committing to global goals and initiatives as well as supporting work on the national and local levels.

Financial demands of adaptation along with fund allocation are major obstacles to adaptation activities, specifically in developing countries and LDCs (Thompson, 2011). The costly nature of adaptation to climate change, despite its significance to the economic and social stability of a country, impedes its prioritization in developing countries' strategies and plans of action. The initiatives, forums, and agreements mentioned earlier have been ratified by several developed nations and plans were made for adaptation funds to be granted, yet there hasn't been much follow through. This is often the case where developed nations have no incentive due to their individual economic and social priorities (Thompson, 2011). Commitments were made and continue to be made year after year, however, financial support given does not match any of the pledges and it is estimated that around only half of the declared amounts are actually granted.

To battle one side of this obstacle and prevent any delays in implementation, many LDCs have delved into investigating potential "no regret" options or adaptation solutions at low cost to support their medium- and long-term adaptation plans (Kissinger and Namgyel, 2013). Countries that are exploring "no regret" options include Nepal and Bangladesh. Nepal is focusing on urgent and immediate needs of vulnerable groups and supporting local level adaptation which can be community-based (Kissinger and Namgyel, 2013). Bangladesh has also invested greatly in community-based disaster response systems and managed to drastically reduce the number of casualties during natural disasters such as floods and cyclones; two phenomena which have been increasingly prevalent in the country. Some low cost, community-based adaptation methods include early warning systems which rely on existing infrastructure, such as mosques loudspeakers. This is not to claim that "no regret" options are to replace the necessary funds, yet they do present a short-term solution. Developed nations must step up and lead by example in terms of abiding by the commitments and plans in place to fund and support adaptation efforts.

Climate change impacts are not bound to political or geographical nations' borders; hence, responses need to be as such aligned. Regional collaboration on climate change action and, in particular, adaptation planning and implementation, is an additional weakness to effective adaptation action (Thompson, 2011). National adaptation research and plans, which are gaining most of the attention but still requires strengthening, is important yet is not sufficient on its own. Climate change discussions on a global level have often identified adaptive needs on a national level, such as the NAPAs and NAPs, but little has been conducted to identify regional cooperation needs. This includes budgeting for and allocating funds for regional cooperation on adaptation as well, which has not been done (Thompson, 2011). The adaptation unit of action needs to be redefined to include regional collective action in addition to the national and local which have been dominating all initiatives.

Focusing on national level adaptation efforts, another gap lies in the absence of effective institutionalization and awareness among ministerial decision-makers (Productivity Commission, 2012). Capacity development and institutional reform are required to avoid wasting time and duplicating efforts. Robust institutional arrangements backed up with capacity development may also allow for synergies in adaptation action as potential cross-over between different sectors may emerge; e.g., adaptation and sustainable development/poverty eradication (Ngwadla and El-Bakri, 2016).

An additional gap lies in the lack of fixed indicators to effective adaptation. Metrics to measure the success and/or progress of adaptive measures are yet to be developed and utilized to support the process (Productivity Commission, 2012). Placing flexible guidelines and frameworks which are sensitive to the uncertainty nature of adaptation needs and setting assessment methodologies will allow for cross-country comparisons and global and regional aggregation that are required to match the learning-by-doing attributes of adaptation action. It will also allow for improved budgetary allocation for adaptation measures as it will provide traceable and trackable empirical evidence and well documented experiences.

5.8 CONCLUSION

From the above discussion, it is evident that over the last few decades, numerous adaptation initiatives have taken place in both global and national platforms to address the vulnerability due to climate change. Mainly due to the lack of financial support and inadequate measures of tracking adaptation initiatives, policy- and decision-makers, along with practitioners, face difficulties in adopting adaptation measures on large scales. To ensure such policies are effective and are properly implemented, the promotion and facilitation of multistakeholder collaborations are much needed. Furthermore, new policies and mandates, where needed, will have to be adopted by the policy-makers to address and reduce the existing gaps. It is crucial to start the process where adaptation has already been mainstreamed in daily lives, be it small or large-scale initiatives either referred to as adaptation or otherwise. Many communities in the south are great harbors of adaptation knowledge, represented through various practices, activities, and traditions. By resorting to capacity development to enhance the quality and the applicability of such knowledge and its creators, response to climate change impacts would thrive beyond current day progress and scenarios post-1.5 degrees could be alleviated.

REFERENCES

Adger, W.N., Arnell, N.W., Tompkins, E.L., 2005. Successful adaptation to climate change across scales. Glob. Environ. Chang. 15 (2), 77−86.

Adger, W.N., Dessai, S., Goulden, M., Hulme, M., Lorenzoni, I., Nelson, D.R., et al., 2009. Are there social limits to adaptation to climate change? Clim. Chang. 93 (3−4), 335−354.

Brown, K., 2016. Resilience, Development and Global Change. Routledge, New York.

Burton, I., Diringer, E., Smith, J., 2006. *Adaptation to Climate Change: International Policy Options.* Retrieved from c2es.org: http://citeseerx.ist.psu.edu/viewdoc/download? http://dx.doi.org/10.1.1.651.2376&rep=rep1&type=pdf.

Cannon, T., Müller-Mahn, D., 2010. Vulnerability, resilience and development discourses in context of climate change. Nat. Hazards 55 (3), 621−635.

DAG, 2016. What Is the History of the Paris Agreement (2015)? Retrieved from: http://ask.un.org/faq/120272.

European Commission, n.d. European Climate Adaptation Platform. Retrieved from: https://ec.europa.eu/clima/sites/clima/files/docs/climate_adapt_-flyer_en.pdf.

Folke, C., 2006. Resilience: the emergence of a perspective for social−ecological systems analyses. Glob. Environ. Chang. 16 (3), 253−267.

Folke, C., Carpenter, S., Elmqvist, T., Gunderson, L., Holling, C.S., Walker, B., 2002. Resilience and sustainable development: building adaptive capacity in a world of transformations. AMBIO. 31 (5), 437−440.

Gallopín, G.C., 2006. Linkages between vulnerability, resilience, and adaptive capacity. Glob. Environ. Chang. 16 (3), 293−303.

Global Greenhouse Warming, 2010. Bali Action Plan. Retrieved from: http://www.global-greenhouse-warming.com/bali-action-plan.html.

Helgeson, J.F., Ellis, J., 2015. The Role of the 2015 Agreement in Enhancing Adaptation to Climate Change. Retrieved from researchgate.net: http://www.oecd-ilibrary.org/docserver/download/5jrxg3xb0h20-en.pdf?expires=1518349457&id=id&accname = guest&checksum=AC55B386 D85159B3D8976992C838FD5E.

IPCC, 2001. IPCC Third Assessment Report: Glossary of Terms. Retrieved from ipcc.ch: http://www.ipcc.ch/pdf/glossary/tar-ipcc-terms-en.pdf.

IPCC, 2007. IPCC Fourth Assessment Report on Climate Change 2007: Impacts, Adaptation and Vulnerability. Retrieved from: https://www.ipcc.ch/pdf/assessment-report/ar4/wg2/ar4_wg2_full_report.pdf.

Kelman, I., 2015. Climate Change and the Sendai Framework for Disaster Risk Reduction. Retrieved from: https://link.springer.com/article/10.1007/s13753-015-0046-5.

Kissinger, G., Namgyel, T., 2013. NAPAs and NAPs in Least Developed Countries. Retrieved from: https://ldcclimate.files.wordpress.com/2013/12/ldcp13_napas-and-naps.pdf.

Klein, N. 2014. This changes everything: Capitalism vs. the climate. London: Allen Lane.

Leach, M., 2008. Re-Framing Resilience: Trans-Disciplinarity, Reflexivity and Progressive Sustainability−A Symposium Report. http://opendocs.ids.ac.uk/opendocs/bitstream/handle/123456789/2315/Re-framing%20Resilience.pdf?sequence=1.

Levina, E., 2007. Adaptation to Climate Change: International Agreements for Local Needs. Retrieved from oecd.org: https://www.oecd.org/env/cc/39725521.pdf.

Levina, E., Tirpak, D., 2006. Adaptation to Climate Change: Key Terms. Retrieved from oecd.org: http://www.oecd.org/env/cc/36736773.pdf.

McGray, H., 2014. Clarifying the UNFCCC National Adaptation Plan Process. Retrieved from: http://www.wri.org/blog/2014/06/clarifying-unfccc-national-adaptation-plan-process.

MoEF, 2015. Bangladesh: Intended Nationally Determined Contributions (INDC). Retrieved from: http://www4.unfccc.int/ndcregistry/PublishedDocuments/Bangladesh%20First/INDC_2015_of_Bangladesh.pdf.

Mwandingi, M., 2006. UNDP-GEF Climate Change Adaptation. Retrieved from unfccc.int: https://unfccc.int/files/adaptation/adverse_effects_and_response_measures_art_48/application/pdf/200609_undp_support.pdf.

Ngwadla, X., El-Bakri, S., 2016. The Global Goal for Adaptation under the Paris Agreement: Putting Ideas into Action. Retrieved from: https://cdkn.org/wp-content/uploads/2016/11/Global-adaptation-goals-paper.pdf.

Oettlé, N., 2016. Adaptation and Loss & Damage in the Paris Agreement. Retrieved from: http://www.adaptationnetwork.org.za/2016/02/adaptation-and-loss-damage-in-the-paris-agreement/.

Productivity Commission, 2012. Barriers to Effective Climate Change Adaptation. Retrieved from: http://www.pc.gov.au/inquiries/completed/climate-change-adaptation/report/climate-change-adaptation.pdf.

Shah, A., 2002. COP3—Kyoto Protocol Climate Conference. Retrieved from: http://www.globalissues.org/article/183/cop3-kyoto-protocol-climate-conference.

Syrovátka, M., n.d. Financing Adaptation to Climate Change in Developing Countries. Retrieved from: http://www.development.upol.cz/uploads/dokumenty/Syrovatka_financing_adaptation_to_climate_change.pdf.

The Committee on Approaches to Climate Change Adaptation, 2010. Approaches to Climate Change Adaptation. Retrieved from env.go.jp: https://www.env.go.jp/en/earth/cc/adapt_guide/pdf/approaches_to_adaptation_en.pdf.

Thomalla, F., Downing, T., Spanger-Siegfried, E., Han, G., Rockström, J., 2006. Reducing hazard vulnerability: towards a common approach between disaster risk reduction and climate adaptation. Disasters 30 (1), 39–48. Available from: https://pdfs.semanticscholar.org/ca31/ff31276949630fc64ebb2c5fd780026bc8cc.pdf.

Thompson, B.H., 2011. The Need for (and Obstacles to) Regional Collective Action in Climate Adaptation. Retrieved from: https://www.law.northwestern.edu/research-faculty/searlecenter/workingpapers/documents/Thompson_Regional_Collective_Action.pdf.

UNCCD, 2017. About the Convention. Retrieved from: http://www2.unccd.int/convention/about-convention.

UNCCD Knowledge Hub, 2017. European Climate Adaptation Platform (CLIMATE-ADAPT). Retrieved from: http://knowledge.unccd.int/publications/european-climate-adaptation-platform-climate-adapt.

UNFCCC, 1992. Text of the United Nations Framework Convention on Climate Change. Retrieved from: http://unfccc.int/files/essential_background/convention/background/application/pdf/convention_text_with_annexes_english_for_posting.pdf.

UNFCCC, 1995. Report of the Conference of the Parties on Its First Session. Retrieved from: http://unfccc.int/resource/docs/cop1/07a01.pdf#page = 34.

UNFCCC, 1998. Kyoto Protocol to the United Nations Framework Convention on Climate Change. Retrieved from: http://unfccc.int/resource/docs/convkp/kpeng.pdf.

UNFCCC, 2001. Report of the Conference of the Parties on its Seventh Session. Retrieved from: http://unfccc.int/resource/docs/cop7/13a01.pdf#page = 32.

UNFCCC, 2008. National Adaptation Programmes of Action (NAPA). Retrieved from: http://unfccc.int/national_reports/napa/items/2719.php.

UNFCCC, 2011. Report of the Conference of the Parties on Its Seventeenth Session. Retrieved from: https://unfccc.int/resource/docs/2011/cop17/eng/09a01.pdf.

UNFCCC, 2013. The State of Adaptation under the United Nations Framework Convention on Climate Change. Retrieved from: http://unfccc.int/files/adaptation/cancun_adaptation_framework/adaptation_committee/application/pdf/ac_2013_report_high_res.pdf.

UNFCCC, 2015. Paris Agreement. Retrieved from: https://unfccc.int/files/essential_background/convention/application/pdf/english_paris_agreement.pdf.

UNFCCC, 2016. Green Climate Fund. Retrieved from: http://unfccc.int/cooperation_and_support/financial_mechanism/green_climate_fund/items/5869.php.

UNFCCC, 2017a. Focus: Adaptation. Retrieved from unfccc.int: http://unfccc.int/focus/adaptation/items/6999.php.

UNFCCC, 2017b. Elements of Adaptation. Retrieved from unfccc.int: http://unfccc.int/adaptation/items/7006.php#Observation.

UNFCCC, 2017c. Standing Committee on Finance. Retrieved from: http://unfccc.int/bodies/standing_committee/body/6973.php.

UNFCCC, 2017d. The Paris Agreement. Retrieved from: http://unfccc.int/paris_agreement/items/9485.php.

UNISDR, n.d. Climate Change Adaptation. Retrieved from: https://www.unisdr.org/we/advocate/climate-change.

Walker, B., Holling, C.S., Carpenter, S.R., Kinzig, A., 2004. Resilience, adaptability and transformability in social-ecological systems. Ecol. Soc. 9 (2), 5.

Adaptation Actions—Hazards, Ecosystems, Sectors

Flood Risk Management in the United Kingdom: Putting Climate Change Adaptation Into Practice in the Thames Estuary

Darren Lumbroso and David Ramsbottom

HR Wallingford, Wallingford, United Kingdom

Chapter Outline

6.1 INTRODUCTION

London and the Thames Estuary's floodplain in the southeast of England are susceptible to flooding from storm surges. Storm surges are generally a result of high winds coinciding with high tides, and low atmospheric pressure, pushing sea water towards the coast causing it to pile up there. The funnel shaped nature of the North Sea basin and the Thames Estuary increases its susceptibility to storm surges. The frequency of extreme sea levels and storm surge levels will increase in the future in the Thames Estuary, primarily as a result of increases in mean sea level and changes in local meteorology resulting from climate change (Lowe et al., 2001; Weisse et al., 2012).

The flooding resulting from the storm surge that struck the east coast of England on Saturday 31 January 1953 caused 307 deaths, damage to some 24,000 properties and approximately 65,000 ha of agricultural land inundated with saltwater, and the loss of major transportation links (Summers, 1978). The 1953 flood acted as a catalyst for implementing a forecasting system for coastal surges and the UK's largest ever improvements in coastal defences, which eventually resulted in the implementation in 1984 of the Thames Barrier, shown in Fig. 6.1 (Lumbroso and Vinet, 2011).

The Thames Barrier is a moveable barrier in the River Thames east of central London (see Fig. 6.1), that together with 300 km of flood defences and nine other major barriers, protect billions of pounds worth of assets and critical infrastructure in London and the Thames Estuary from storm surges (see Table 6.1) with an annual recurrence probability of up to 1 in 1,000 years. The Thames Barrier spans 520 m across the River Thames near Woolwich. It has 10 rotating steel gates. The flood gates are circular segments in cross-section, and operate by rotating, as shown in Fig. 6.2. They can be raised to allow "underspill" to allow operators to control upstream levels and a complete 180 degree rotation for maintenance, as shown in Fig. 6.1.

Owing to climate change, sea levels are rising. In addition, there is also a long-term lowering of ground levels in southeast England. Since the end of the last Ice Age 20,000 years ago, land and sea levels around the UK coastline have changed in response to the retreat of the ice sheets. As the ice melted, the reduction in load resulted in the British

Resilience. DOI: https://doi.org/10.1016/B978-0-12-811891-7.00006-2

FIGURE 6.1   Photograph of the Thames Barrier.

TABLE 6.1 People, Properties and Assets at Risk of Flooding in the Thames Estuary Floodplain

People and Properties	Transport	Critical Infrastructure
1.30 million residents (plus commuters, tourists and other visitors)	167 km of railways	16 hospitals
551,000 homes	300 km of roads	8 power stations
40,500 commercial and industrial properties	35 subway stations	1000 electricity substations
Property value £275 billion	51 railway stations	400 schools

Source: Environment Agency, 2009. TE2100 plan: Consultation document, April 2009; Environment Agency, 2016. TE2100 5 year review, Non-technical summary, July 2016.

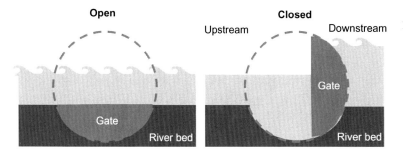

FIGURE 6.2   Diagram showing how the Thames Barrier gates operate.

landmass slowly tilting back up in the north and down in the south, a process called isostatic adjustment. This means that land levels in southeast England are going down by around 1.5 mm per year primary as a result of this effect (Environment Agency, 2009). The Thames Estuary's flood defence system was originally designed to provide a 1 in 1,000 year level of protection up to 2030 and took into account early estimates of sea level rise which have turned out to be conservative. These factors coupled with increasing development in the Thames Estuary's floodplain means that the risk of flooding is increasing and that flood defences, including the Thames Barrier, need to be upgraded, if the same level of protection is to be provided in the future.

The Thames Barrier and its associated flood defences are examples of "long-lived infrastructure." These are pieces of infrastructure that usually can be maintained for a significantly greater number of years (i.e., 50–100 years) than most other capital assets. When planning the upgrading of long-lived infrastructure, such as the flood defences in the Thames Estuary, it is cheaper and easier to account for long-term trends upfront in decisions today, rather than retrofitting at a later date (Fankhauser et al., 1999). The uncertainty in future climate change projections makes it impossible to directly use the output of a single climate model as an input for infrastructure design. As a result, instead of optimizing based on the climate conditions projected by models, the planning and implementation of long-lived infrastructure should be made more robust to possible future changes in climate conditions (Hallegatte, 2009).

In 2002, the Environment Agency, the organization responsible for flood risk management in England, established the Thames Estuary 2100 project (TE2100) with the objective of developing a strategic flood risk management plan for London and the Thames estuary through to the end of the century. Over a 6-year period, studies were undertaken to gain a thorough understanding of how flood risk is managed today and the options that could manage flooding through to 2100. TE2100 was the first major flood risk project in the United Kingdom to have put climate change adaptation at its core (Environment Agency, 2009).

The size (up to £9 billion) and irreversibility of the potential investments, the risks associated with their failure, and the long lifetimes and lead times of the infrastructure means that the investments are sensitive to climate change and hence the potential for maladaptation is significant (Reeder and Ranger, 2012). As part of TE2100 an approach to adaptation was developed that takes account of the impacts of climate change, lowering ground levels, and deterioration of the existing flood defence system (Lavery and Donovan, 2005).

The approach was based on a number of flood risk management thresholds including flood defence crest levels in different parts of the estuary and the frequency of flood barrier closures. Strategic options were developed that consisted of sequences of interventions to be implemented when each threshold is reached. The options were tested for a range of future scenarios to demonstrate that the TE2100 plan is robust to a wide range of future conditions. Indicators of change are monitored and used to update the dates when the thresholds will be reached and therefore the interventions will be needed. This chapter gives an overview of the approach taken.

6.2 FUTURE CHANGE AND ADAPTATION

6.2.1 Future Changes

The TE2100 project plan was designed to adapt to future changes which are outlined below.

6.2.1.1 Climate Change

There is uncertainty surrounding how much the sea level will have risen. The current rate in the Thames Estuary is of the order of 3 mm per year relative to the land level; however, the range of future climate-induced sea level rise remains highly uncertain with continued concern that large increases in the 21st century cannot be ruled out (Nicholls et al., 2011). The TE2100 Plan is based on a relative sea level rise estimate of 90 cm by 2100, but is adaptable to differing rates of sea level rise up to 2.7 m by the end of the 21st century (Environment Agency, 2016).

6.2.1.2 Socioeconomic Developments

In the Thames Estuary there is a large amount of existing development with more planned. Although there are various development plans for the Thames Estuary floodplain, the future economy will determine exactly the form these will take and this could vary considerably under the planning time frame of the TE2100 project.

6.2.1.3 Deterioration of Existing Flood Defence Assets

The majority of the current flood risk management infrastructure was constructed at least 30 years ago (in some cases longer ago), and is gradually deteriorating. It has been estimated that much this infrastructure will come to the end of its useful life between 2030 and 2060 (Environment Agency, 2009).

6.2.1.4 The Physical Environment, Including Estuary Morphology

The estuary has an essentially fairly stable history over the last century or so in terms of its physical development. The outer sandbanks in the estuary protect it from the worst effects of wave attack. However sea level rise could disturb this picture and it will be essential to monitor the state of the estuary into the future.

6.2.1.5   Public Awareness of Flooding

The awareness of flooding on the estuary is low, and the present high standard of protection means that there is little need for the public to be aware of the risk. However with an uncertain future, it may be desirable for this situation to change so that the public can be better prepared for the unexpected.

6.2.2   Adaptation

Within the TE2100 plan there were various forms of adaptation envisaged to cope with future changes up until the year 2100. These are outlined below.

6.2.2.1   Changes to the Timing of New Interventions

The TE2100 plan has a preferred option which will manage flood risk throughout the century and beyond given the envelope of change that is considered to be most likely based on current information. It consists of a sequence of interventions, each of which is implemented when a certain threshold is reached. The rates of change of key variables, such as sea level rise, are being monitored. If the rate of change stays within certain limits, the preferred options will be implemented, but the dates of implementation will change depending on the rate of change. In the UK large floods often act as a catalyst for policy changes (see Johnson et al., 2005), hence a large flood event may lead to implementation of some flood risk management measures at an earlier date than originally planned.

6.2.2.2   Ability to Change Between Options

If the rate of change of a key factor is forecast or observed to change significantly above the expected rate of change when the preferred option was selected, interventions will be needed much sooner than originally envisaged. Under these circumstances the preferred option would be reviewed to see if there is an alternative sequence of interventions that would be more efficient and cost effective. If, e.g., accelerated melting of the ice caps means that there may be an increase of more than 2 m in maximum sea levels in the next 100 years, it may be necessary change to the alternative sequence of interventions.

6.2.2.3   Adaptation of Engineering Responses

Engineering responses should be designed so that they can be adapted to changing circumstances, e.g., by providing foundations for flood defences that can deal with higher future loadings caused by increased flood water levels. The preferred options have been designed to cope with a range of future change. For example, when defences are replaced, the new defences are designed so that they can be raised in the future and designed in such a way that they can contribute to future options that are designed to cope with more extreme climate change scenarios.

6.2.2.4   Land Use Planning That Provides Flexibility in the Selection of Options

Each flood risk management option will require land for new defences, enlarged defences, new areas of habitat creation, and in some cases flood storage. The planning system needs to take cognizance of the land required for the preferred option and any likely alternative options so that it can be safeguarded. Although the exact timing of the implementation of interventions is not given in the TE2100 plan because of uncertainty over future change, the indicative timings provide a guide to when land will be required.

6.2.2.5   Adaptation to New Infrastructure

New infrastructure on the Thames Estuary could have a major impact on flood risk management. For example, the proposed new ports in the estuary would require free access for navigation, which affect options to build a new barrier downstream of the existing one. New transport links could provide the opportunity to combine a new crossing of the estuary with a new barrage. The plan should be flexible enough to accommodate major changes such as these.

6.3   THE OVERALL APPROACH TO FLOOD RISK MANAGEMENT UNDER THE TE2100 PLAN

As detailed above flood risk is increasing in the Thames Estuary. Fig. 6.3 shows the approach adopted to flood risk management taken by the TE2100 plan. If there are no interventions then flood risk in the Thames Estuary will increase as the climate changes and the sea level rises, this is shown by the dotted black line in Fig. 6.3. One possible way of

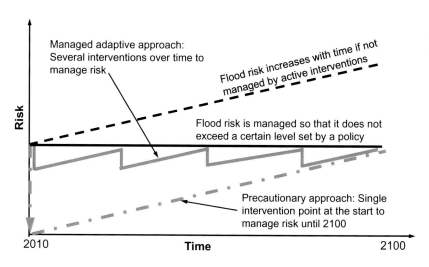

FIGURE 6.3 Managing flood risk through the century using the TE2100 managed adaptive approach.

decreasing flood risk is the "precautionary approach" shown by the dotted grey line in Fig. 6.3. This could comprise a single intervention point to manage flood risk until the end of the century. One example of such an approach would be the construction of a new barrier downstream of the existing Thames Barrier, which would be able to deal with a range of climate change scenarios. However, given the uncertainty in the predicted climate change scenarios such an approach is not flexible and it could result in an overdesigned solution which would be costly.

In the TE2100 plan a more flexible approach has been taken. Each area of the Thames Estuary floodplain has been assigned a flood risk policy, which sets the level of flood risk management activity or investment that can be justified in that area (Environment Agency, 2012). This is shown as the horizontal solid black line on Fig. 6.3. An option is made up of a combination (or portfolio) of different interventions which act together to achieve the recommended policy. This is the TE2100 managed adaptive approach, shown by the grey saw-tooth line in this Fig. 6.3 (Environment Agency, 2012). It is important to know when interventions will be required owing to the lead time to implement them, some interventions require a lead-in time of 20 years or more. The timing of interventions is related to a variety of thresholds for different indicators. This is discussed below.

A key output from the TE2100 plan was a set of options from which the preferred estuary wide option was selected together with associated local interventions. The preferred option is designed for an increase in peak surge tide level in excess of 4 m by a series of improvements to the flood defence system. The preferred option allows for a choice when the design surge tide level reaches the maximum allowable level at the Thames Barrier: improvement of the barrier or a new barrier downstream. Alternative options are needed in case a major change occurs and the preferred option is no longer the optimal solution for flood risk management on the estuary.

The method for implementing the TE2100 plan was as follows:

1. The main drivers of flood risk management used to develop the options were identified.
2. For each driver, indicators were identified which described the impact of the drivers and which could be monitored.
3. For each indicator, the thresholds where responses were needed to maintain the required level of flood risk were identified.
4. The preferred options include portfolios of responses for each threshold.
5. The lead time for planning and constructing each portfolio of responses was estimated to determine the length of time that decisions should be taken before the responses are needed. Fig. 6.4 shows the concept of lead times and decision points.
6. The TE2100 plan included assumed dates when the responses will be required, and assumed dates when decisions must be made.
7. The indicators are monitored. The monitoring results are used to update the estimated dates when portfolios of responses must be implemented, and the dates when decisions must be made.
8. The timing of a decision to implement an intervention is based on:
    a. The rate of change of the indicator (which is unlikely to be linear).
    b. The threshold value when an intervention is required.
    c. An estimate of how the indicator will continue to change, in order to estimate the date when it reaches the threshold value.
    d. The lead time for planning and constructing the intervention.

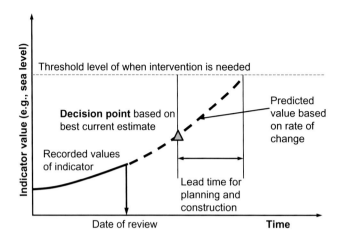

FIGURE 6.4   Thresholds, lead times, and decision points.

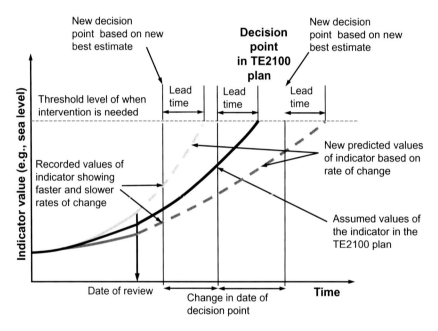

FIGURE 6.5   Impact of different rates of sea level change on decision points.

9. The process is illustrated in Fig. 6.5. Fig. 6.5 shows the effect of faster and slower rates of change compared with the TE2100 plan. Faster rates of change bring forward decisions, and slower rates of change delay decisions.

10. The TE2100 plan will be updated using the revised estimates of the dates when thresholds will be reached and decisions must be taken.

11. If significant changes occur in the expected dates when thresholds will be reached, the choice of options is reviewed. This is because an alternative option may be more effective for managing flood risk under the changed circumstances.

12. Alternative options were included in the TE2100 plan.

13. The preferred options and the alternative options will be reappraised, using the updated best estimates of future change. This could lead to a change in the selected option. For this reason, responses should be as adaptable as possible to avoid undertaking work that may not be needed in the future.

The procedure outlined above will take place over a number of years. The preferred options and the alternatives all involve a similar approach until a critical water level threshold is reached at the Thames Barrier. The critical drivers for this are mean sea level and peak surge tide level, and the current assumed date for major interventions is 2070 based on the climate change scenarios developed by the British Government in 2006.

6.4 BACKGROUND TO THE INDICATORS

There are 10 indicators, known as "indicators of change," that are monitored as part of the TE2100 plan. These are briefly outlined below.

- *Indicator 1—Sea level* The Environment Agency monitor changes in sea level in the estuary over time, which confirms that sea levels in the Thames Estuary are rising. Currently sea level rise is taking place within the limits of the TE2100 plan (Environment Agency, 2016).
- *Indicator 2—Peak surge level* which is the maximum water level which occurs as a result of tidal surges.
- *Indicator 3—Peak river flows* These are used for flood management in the areas of the estuary and tributaries which are influenced by river flows
- *Indicator 4—Condition of flood defence assets* There are approximately 3,800 separate flood defence assets in the estuary and the aim is to maintain at least 97% of these in "fair" or "good" condition (Environment Agency, 2016).
- *Indicator 5—Operation of the Thames Barrier and other barriers* The Environment Agency monitors how the operations of the Thames Barrier, and other major flood barriers, are changing over time, including how frequently they need to be operated and how reliable they are. The lifetime of the Thames Barrier needs to be maximized. Improvements in the accuracy of forecasts of barrier closures will reduce the number of closures and help to prolong the life of the Barrier (Environment Agency, 2016).
- *Indicator 6—Development* The numbers of people and properties in the floodplain have increased since the TE 2100 plan was completed; however, these increases are consistent with the plan's expectations.
- *Indicator 7—Erosion and deposition* The Environment Agency is interested in how and where geomorphological changes are occurring in the Thames Estuary as it can have a detrimental impact on the flood defences (e.g., undermining foundations of flood defences).
- *Indicator 8—Habitat* As part of the TE2100 plan the Environment Agency is committed to ensuring the amount of intertidal habitat does not decline in the Thames Estuary. Habitat replacement needs to take place before it is lost. The Environment Agency expects to create 1,200 ha of replacement habitat over the lifetime of the TE2100 plan (Environment Agency, 2016).
- *Indicator 9—Land use planning and development activities* The TE2100 Plan included objectives for supporting and informing—The land use planning process to promote appropriate, sustainable, and flood resilient development in the Thames Estuary. There are also objectives for protecting the social, cultural, and commercial value of the River Thames, its tidal tributaries and its floodplain. This indicator is used to help us understand how these objectives are being met by those who are helping to carry out the TE2100 plan's recommendations (Environment Agency, 2016)
- *Indicator 10—Public/institutional attitudes to flood risk* Since the TE2100 plan was implemented there has been no significant changes in the attitude to flood risk.

A review of the changes in the above indicators is carried out by the Environment Agency every 5 years.

6.5 THE PHILOSOPHY BEHIND THE DEVELOPMENT OF FLOOD RISK MANAGEMENT OPTIONS FOR THE THAMES ESTUARY

The options for the Thames Estuary were developed to address each of the key drivers. In some cases there was an overlap between the key drivers and indicators, e.g., the amount of habitat loss is dependent on sea level rise and the geomorphological changes (Ramsbottom and Reeder, 2008). Portfolios of responses were developed as part of the TE2100 plan—an example is given in Fig. 6.6. In this example the main driver for the choice of flood risk management interventions is the peak surge tide level.

One important threshold is the maximum annual number of closures of the Thames Barrier before its reliability is compromised (Ramsbottom and Reeder, 2008). Once the maximum number of closures has been reached the intervention is to raise upstream flood defences. The maximum number of closures per year is primarily influenced by mean sea level rise, which means raising upstream flood defences is also driven by increases in mean sea level. The development of interventions for increases in the maximum sea level is shown in Fig. 6.6.

The monitoring of indicators is required to assist with decision making. The approach to flood risk management in the TE2100 plan includes a method for developing portfolios of options, together with a regular process in which options and decisions are reviewed in order to take into account changing circumstances (Ramsbottom and Reeder, 2008).

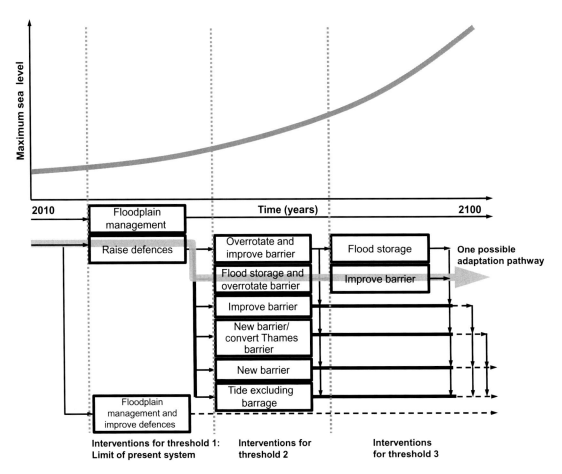

FIGURE 6.6 Responses to manage increases in maximum sea level.

The TE2100 plan takes a flexible approach to adaptation by combining three methods:

- *The implementation of "low-regret" measures in the near-term.* These are measures that reduce risk immediately and cost-efficiently under a wide range of future climate change scenarios. Low regret measures include raising existing flood defences.
- *The incorporation of "structural" flexibility,* such as designing in flexibility, so that infrastructure can be adjusted or enhanced in the future at minimal additional cost. For example, the current Thames Barrier can be overrotated to cope with greater than expected sea level rise. Safety margins can be incorporated in flood defence infrastructure to cope with greater than expected change; this approach is effective where the marginal cost is low (Ranger et al., 2013).
- *Pathway flexibility* The TE2100 plan adopted an "adaptation pathways" approach which means that flexibility is built into the long-term strategy (Ranger et al., 2013) and that the timing of new interventions and the interventions themselves can be changed over time depending on whether the thresholds levels of the various indicators have been exceeded. This approach helps the decision maker to identify the timing and sequencing of possible pathways of adaptation over time under different scenarios (Ranger et al., 2013). The grey arrow in Fig. 6.6 shows one of many potential adaptation pathways that could be taken in the future.

6.6 CONCLUSIONS

Climate change poses several new challenges for traditional flood risk management. In terms of the flood risk in the Thames Estuary these challenges are related to the nonstationarity of risk and the uncertainty in future flood risk as a result of the uncertainty in the future climate change scenarios. The TE2100 project was one of the first major infrastructure projects to explicitly recognize and address the challenge of the uncertainty in future climate risk throughout the planning process (Ranger et al., 2013).

The approach to flood risk management presented in the TE2100 plan includes a method of developing options together with a regular updating process in which options and decisions are reviewed taking account of changing circumstances. Interventions will be introduced as the need arises, taking account of the lead time needed for planning, design, and construction. A shift from "single solution" driven approaches to processes of continual management rests on understanding the complex interactions between socioeconomic, natural, and technical solutions. A portfolio of responses, including engineering options (e.g., flood defences, barriers), social (e.g., early warning systems, emergency preparedness), and land use planning solutions (e.g., building codes, flood proofing), are more likely to be effective and adopted than the simple strategies.

A preferred estuary-wide option has been identified by the appraisal process and is being implemented. This will set the direction for flood risk management on the estuary, but will be reviewed as required based on the results of monitoring and other significant changes in the estuary. The preferred option is designed for an increase in peak surge tide level in excess of 4 m by a series of improvements to the flood defence system. Alternative options are needed in case a major change occurs and the preferred option is no longer the optimal solution for flood risk management on the estuary. This could lead to significant changes in the interventions. The method of developing the TE2100 plan includes a large degree of flexibility, and it is likely that the flood risk management system for the Thames Estuary will evolve in response to changes in the drivers of flood risk management.

To conclude, there is scope for the methodology utilized in the TE2100 project to be used by planners in low income countries, where a lack of information often creates uncertainties. The adaptation pathways approach is relevant where it is known that climate change is likely to affect the decision and uncertainties mean that it is challenging to choose between different adaptation options (Reeder and Ranger, 2012). This will often be the case when dealing with long-term decisions, involving high sunk costs, but could also be relevant for shorter-term decisions that have far-reaching implications, such as sectoral planning, e.g., when considering how to best to invest in food supply chains or import facilities (Reeder and Ranger, 2012).

REFERENCES

Environment Agency, 2009. TE2100 Plan: Consultation Document, April 2009.

Environment Agency, 2012. Thames Estuary 2100: Managing Risks Through London and the Thames Estuary, TE2100 Plan.

Environment Agency, 2016. TE2100 5 Year Review, Non-Technical Summary, July 2016.

Fankhauser, S., Smith, J.B., Tol, R.S., 1999. Weathering climate change: some simple rules to guide adaptation decisions. Ecol. Econ. 30 (1), 67−78.

Hallegatte, S., 2009. Strategies to adapt to an uncertain climate change. Glob. Environ. Chang. 19 (2), 240−247.

Johnson, C.L., Tunstall, S.M., Penning-Rowsell, E.C., 2005. UK Floods as catalysts for policy change: historical lessons from England and Wales. Int. J. Water Resour. Dev. 21 (4), 561−575.

Lavery, S., Donovan, B., 2005. Flood risk management in the Thames Estuary looking ahead 100 years. Philos. Trans. Royal Soc. London A 363 (1831), 1455−1474.

Lowe, J.A., Gregory, J.M., Flather, R.A., 2001. Changes in the occurrence of storm surges around the United Kingdom under a future climate scenario using a dynamic storm surge model driven by the Hadley Centre climate models. Clim. Dynam. 18 (3), 179−188.

Lumbroso, D.M., Vinet, F., 2011. A comparison of the causes, effects and aftermaths of the coastal flooding of England in 1953 and France in 2010. Nat. Hazards Earth Syst. Sci. 11, 2321−2333.

Nicholls, R.J., Marinova, N., Lowe, J.A., Brown, S., Vellinga, P., De Gusmao, D., et al., 2011. Sea-level rise and its possible impacts given a 'beyond 4 C world' in the twenty-first century. Philos. Trans. Royal Soc. London A 369 (1934), 161−181.

Ramsbottom, D., Reeder, T., 2008. Adapting flood risk management for an uncertain future: flood management planning on the Thames estuary, 43rd Defra Flood and Coastal Management Conference, Manchester University, UK, 1 to 3 July 2008.

Ranger, N., Reeder, T., Lowe, J., 2013. Addressing 'deep' uncertainty over long-term climate in major infrastructure projects: four innovations of the Thames Estuary 2100 Project. EURO J. Decision Proces. 1 (3-4), 233−262.

Reeder, T., Ranger, N., 2012. How Do You Adapt in an Uncertain World? Lessons from the Thames Estuary 2100 Project. World Resources Report, Uncertainty series, Washington, DC.

Summers, D., 1978. The East Coast Floods, Published by David and Charles, 1978, ISBN 10: 0715374567.

Weisse, R., von Storch, H., Niemeyer, H.D., Knaack, H., 2012. Changing North Sea storm surge climate: an increasing hazard? Ocean Coastal Manage. 68, 58−68.

FURTHER READING

Lonsdale, K.G., Downing, T.E., Nicholls, R.J., Parker, D., Vafeidis, A.T., Dawson, R., et al., 2008. Plausible responses to the threat of rapid sea-level rise in the Thames Estuary. Clim. Chang. 91 (1-2), 145−169.

Chapter 7

The Science of Adaptation to Extreme Heat

Ethan D. Coffel[1], Alex de Sherbinin[1], Radley M. Horton[1,5],
Kathryn Lane[2], Stefan Kienberger[3] and Olga Wilhelmi[4]

[1]*Columbia University, New York, NY, United States,* [2]*New York City Department of Health & Mental Hygiene, New York, NY, United States,*
[3]*University of Salzburg, Salzburg, Austria,* [4]*National Center for Atmospheric Research, Boulder, CO, United States,* [5]*Lamont Doherty Earth Observatory, Palisades, NY, United States*

Chapter Outline

7.1 INTRODUCTION

In recent years, research has focused on extreme heat as a serious risk to human health (Petkova et al., 2013; Hayhoe et al., 2010; Petkova et al., 2014; Kalkstein and Greene, 1997; Glaser et al., 2016), agriculture (Schlenker and Roberts, 2009; Lobell and Gourdji, 2012; Ray et al., 2015; Battisti and Naylor, 2009), infrastructure (Coffel and Horton, 2015; Williams and Joshi, 2013; Bhaduri, 2012; Coffel et al., 2017; Williams, 2017), and economic performance (Lobell and Gourdji, 2012; Burke et al., 2015; Kjellstrom et al., 2009). Recent heat waves, which have been responsible for tens of thousands of additional deaths and damage to infrastructure and agriculture (Dole et al., 2011), have been linked to climate change (Stott et al., 2004; Black et al., 2004; Christidis et al., 2015; Meehl, 2004; Barriopedro et al., 2011), and the IPCC has concluded that the frequency, severity, and duration of extreme heat events is very likely to increase worldwide in the future (IPCC & Press, C. U. et al., 2013). The substantial impacts that result from present-day heat waves serves as a warning that future extreme temperatures are likely to cause extensive harm to people around the world, in rural and urban areas, and in low- and high-income countries (Herold et al., 2017). Adaptation to reduce these impacts is essential, and scientific research plays a key role in determining which strategies are most effective in varying climatic and societal conditions, as well the best methods of targeting adaptation to the most vulnerable populations. In this chapter we first address projections of future heat, the urban heat island (UHI) effect, and urban microclimates, and then we turn to population vulnerability, the health impacts of heat stress, and the interaction with air quality and infrastructure. Subsequent sections address adaptation strategies informed by science and human behavior, which is a critical element in responses to extreme heat. A concluding section addresses the importance of science to heat-related adaptation initiatives (Fig. 7.1).

7.2 HEAT PROJECTIONS

Climate change resulting from greenhouse gas emissions is expected to result in a mean global temperature increase of 1.5−4°C by the end of the 21st century (IPCC & Press, C. U. et al., 2013). Warming will likely continue into the 22nd century depending on how rapidly net emissions of greenhouse gases and other radiatively important agents like

Resilience. DOI: https://doi.org/10.1016/B978-0-12-811891-7.00007-4

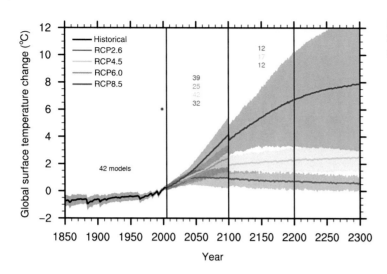

FIGURE 7.1 Global mean temperature projections through 2300. *From IPCC AR5, Ch. 12. http://www.ipcc.ch/report/ar5/wg1/.*

aerosols are reduced, as well as the success of potential negative emissions technology. Temperature changes over land will be larger than the global mean (Karl et al., 2015), with more warming expected over high latitude land masses than tropical ones. There is also evidence that the most extreme temperatures may increase more than the mean in some regions (Kodra and Ganguly, 2014; Horton et al., 2015, 2016).

Climate change has already resulted in 0.8−1°C of warming as compared with preindustrial times (Walsh et al., 2014). This upward shift in the temperature distribution has been manifest as increases in the frequency, duration, and intensity of heat waves (Easterling et al., 2000; Donat et al., 2013; Alexander et al., 2006), as well as a rise in the global land area coverage of above average temperatures (Hansen et al., 2012). Extreme heat events have intensified due to changes in the mean temperature alone (Ballester et al., 2010); there is some evidence, however, that temperature variability may also increase in the future due to a variety of physical processes (Coumou et al., 2014; Cohen et al., 2014; Schär et al., 2004), although this has not yet been observed on a global scale (Huntingford et al., 2013) and may be accounted for by varied rates of mean temperature change (Argüeso et al., 2016).

Climate change will intensify the hottest heat events, but an equally serious effect will be to drastically increase the frequency of temperatures that are currently considered severe (Mora et al., 2017; Forzieri et al., 2017). In the tropics, where temperature variability is lowest, by mid- to late-century a large portion of the year may be spent above the current average annual maximum temperature (Horton et al., 2016), resulting in significant increases in exposure to extreme heat events such as the heat wave that struck India in 2015 (Zommers et al., 2016). The mid-latitudes too will see significant changes; recent research projects that the number of days exceeding the annual maximum temperature could increase by a factor of 10−20 in cities across the United States (Horton et al., 2015) by mid-century (2050−70). As an example, this means that Baltimore, MD, could potentially experience 15−20 days above the current average once-per-year hottest temperature of 37°C. These changes in the frequency of extreme events will occur rapidly in the coming decades, regardless of emissions reductions.

While many impacts of heat depend on temperature alone, human health and well-being also depend on humidity. The human body is highly efficient at controlling its core temperature through evaporative cooling as long as relative humidity is low. However, as relative humidity rises, thermal regulation becomes difficult and the risk of physiological stress is amplified. A variety of metrics are used to estimate the combined impacts of temperature and humidity. The most commonly reported heat stress indicator is the heat index, or the "feels like" temperature, which is widely presented in weather reports in the United States, for example. There are many variations of the heat index algorithm (Anderson et al., 2013), but all are based on empirical formulae calibrated to relate temperature and humidity conditions to physiological impacts (Rothfusz and Headquarters, 1990). This task requires that the index include assumptions about body type, clothing, wind, and activity level—varying these parameters can have a substantial impact on potential heat risk, making the heat index a problematic general heat stress index.

The wet-bulb globe temperature (WBGT) is the most widely used heat stress index in workplace safety, athletics, and the military (Budd, 2008). It is a weighted average of the dry bulb, wet bulb, and globe (also known as the mean radiant temperature) temperatures, and can be measured directly or estimated using an empirical model. There is extensive calibration data relating WBGT thresholds to levels of heat risk; however, as with the heat index, differences in clothing, activity level, and other individual factors can substantially degrade its ability to predict heat illness. Research suggests that most heat stress indices perform similarly in predicting heat-related mortality, likely due to their strong correlation with temperature (Barnett et al., 2010).

Other, less widely used indices exist, such as the apparent temperature (Steadman, 1984), HUMIDEX (Masterton and Richardson, 1979), the environmental stress index (Moran et al., 2001), and the human thermal comfort index (Harlan et al., 2006). A more physically based measure is the wet-bulb temperature, which represents the lowest temperature that can be achieved through evaporative cooling alone. While there is little empirical evidence relating wet bulb temperature to heat stress impacts, the direct relationship between wet-bulb temperature and evaporative cooling makes it relevant to human health since the body sheds heat through the evaporation of sweat. Recent research has posited that when the wet-bulb temperature exceeds the skin temperature, the body will no longer lose heat through evaporative cooling, and heat illness will occur with prolonged exposure. This wet-bulb temperature threshold, about 35°C, may pose a fundamental limit to heat adaptation (Sherwood and Huber, 2010), although in practice those exerting themselves outdoors, exposed to sunlight, and those more vulnerable to heat would be unable to tolerate the heat at much lower temperature and humidity combinations.

Climate change has been linked to a rise in the WBGT since the 1970s (Knutson and Ploshay, 2016), and there is evidence that wet-bulb temperatures could regularly approach the 35°C limit by late century in certain densely populated regions (Pal and Eltahir, 2015; Im et al., 2017). Thus far, the literature on heat stress and heat-induced mortality has focused on the effects of high air temperatures (Anderson and Bell, 2011), which in many locations are strongly correlated with high wet-bulb temperatures and other measures of heat stress. However, the possibility of reaching wet-bulb temperatures of 35°C indicates an urgent need for research into the effects of extreme heat and humidity on human health. Each area should assess the relationship between heat and mortality as differences in climate, population vulnerability, infrastructure, and built environment, could affect the threshold at which heat is most dangerous. In addition, livestock and other large animals may be severely affected by rising temperatures, potentially with impacts on agriculture and food supply.

7.2.1 Microclimates and the Urban Heat Island

Temperatures can vary across small spatial scales, resulting in different levels of heat exposure between neighborhoods within a city or, to a lesser extent, between localized regions in rural areas. These microclimates have a variety of causes, including land cover (Manteghi et al., 2015), vegetation, material use, local air quality (Karner et al., 2010), and localized weather associated with topography and coastal effects. In addition, development patterns and socioeconomic conditions can result in large differences in heat stress vulnerability between or within neighborhoods (Weber et al., 2015). In many high-income cities, dense development, lack of vegetation, and scarce open space are linked to higher temperatures. These areas also are often occupied by people with fewer economic and social resources to cope with heat (Harlan et al., 2006). By focusing on both climate and social microenvironments, heat adaptation strategies can be targeted to the most at-risk populations.

The UHI effect is a long studied phenomena (Oke, 1973) which can amplify temperatures in metropolitan areas by 1−4°C during the day and up to 10°C at night. The UHI is observed in long term temperature records and is large enough to necessitate a correction factor in calculations of global mean temperature (McKitrick and Michaels, 2007). There are five key physical causes of the UHI, the most important of which is the loss of convective efficiency between the surface and the lower atmosphere (Zhao et al., 2014) in urban areas due to the reduced surface friction over smooth, paved city landscapes. Other drivers of the UHI are reduced evaporative cooling in cities due to lower levels of vegetation cover, enhanced energy absorption due to reduced albedo, increased energy storage by artificial materials, and direct waste heat release from buildings, industry, and vehicles.

Approximately 50% of the world's population lives in urban areas (Sidiqui et al., 2016); given the strong warming effect of the UHI—especially with regard to nighttime temperatures—mitigating urban warming could substantially reduce global heat exposure (Weber et al., 2015). Effective mitigation of the UHI will require localized adaptation, as the relative importance of the physical drivers listed above depends on the background climate and the characteristics of the city, and each physical driver can be combated through different adaptation strategies. Most research on the UHI to date has focused on high-income cities; more work is needed to quantify the effect and causes of the UHI in low- and middle-income countries, as well as the influence of local air pollution on urban temperatures (Cao et al., 2016). A first-ever global estimate of the UHI across all major cities using land surface temperature data found that the biggest factor in determining the degree of UHI is the size of city, and that many tropical cities experience more than a 3°C nighttime temperature differential between urban and surrounding rural areas (Center for International Earth Science Information Network CIESIN, 2013).

7.2.2   Population Vulnerability

Vulnerability to heat stress and mortality during heat waves varies widely among populations (Hondula et al., 2012; Medina-Ramón et al., 2006). The elderly (Ellis and Nelson, 1978) and those with preexisting health conditions—especially limited mobility, obesity, and serious mental illness (Vandentorren et al., 2006)—are more susceptible to heat illness and death. However, the young (including infants (Berko et al., 2014)), too, are vulnerable to heat. Early heat stress research focused on military training and athletics (Budd, 2008; Kerr et al., 2013), and those performing rigorous physical labor or wearing confining clothes can experience heat illness, no matter their age (Jay and Kenny, 2010). Some classes of medications, such as antipsychotics, can impair thermoregulation, potentially increasing risk during hot weather (Cuddy, 2004). Research also suggests that heat illness can have profound and long-lasting health impacts, contributing to a higher risk of future illness and early mortality (Wallace et al., 2007).

Experience from recent heat waves has shown that social isolation is a major risk factor for heat illness; in the 1995 Chicago heat wave, many deaths occurred in single-person households in neighborhoods without convenient public transportation or strong community structures (Klinenberg, 2001); these regions also have decreased community cohesiveness, with higher crime rates, declining population levels, more empty housing stock, and reduced levels of business and other street activities (Klinenberg, 2002). Many studies in the United States have also found that African-Americans are at higher risk of heat-health outcomes (Gronlund, 2014), likely reflecting lower access to air-conditioning and greater baseline health challenges due to historic marginalization.

A consensus is emerging that large-scale heat-related mortality events are primarily social disasters, which can be ameliorated through behavioral adaptation. Evidence of reduced heat risk can be found in France: during the 2003 European heat wave, an estimated 15,000 people died of heat-related illness (Kovats and Hajat, 2008). After that event, heat adaptation measures were put into place at the national and local levels including opening cooling centers—free public spaces with air-conditioning and water—and promoting public heat-safety awareness; in 2006, another heat wave occurred, and statistical comparisons to the 2003 event suggest that adaptation measures reduced heat-related mortality by approximately two-thirds (Fouillet et al., 2008).

Exposure to high temperatures during a heat wave can vary across a population due to socioeconomic factors (Kuras et al., 2015). A variety of mapping studies have helped to identify those neighborhoods most likely to be vulnerable to extreme heat, which can help in targeting adaptation activities (Weber et al., 2015; Sidiqui et al., 2016; Wolf and McGregor, 2013). Air-conditioning is the most effective way to prevent heat-related illness and death, and lack of air-conditioning, which is closely tied to income, is a major vulnerability factor (Lane et al., 2014). Cities in the United States with higher air-conditioning prevalence have been found to have lower rates of heat-related mortality (Medina-Ramon and Schwartz, 2007), and the long-term decline in heat-related mortality may largely be due to the increasing prevalence of air-conditioning (Petkova et al., 2014; Bobb et al., 2014). However, as long as electricity is supplied by fossil fuel sources, increased air-conditioner use will worsen the underlying problem of greenhouse gas emissions.

Housing and neighborhood characteristics also play a role in risk; during the 2003 heat wave in France, those living on the top floor of a building or those in buildings with little nearby vegetation were more likely to die, likely due to exposure to localized higher temperatures (Vandentorren et al., 2006). People in more vulnerable housing are often those with the least social resources, concentrating risk in marginalized groups of society (Klinenberg, 2001). In addition, areas with higher poverty levels may be home to more people with multiple chronic health conditions and fewer resources to prevent and treat illness (Krieger et al., 2005). These neighborhoods may also have more physical vulnerabilities. In New York City, e.g., neighborhoods with less green space, higher surface temperatures, and more people needing financial assistance were associated with more excess deaths during and after heat waves (Madrigano et al., 2015).

In many urban heat events, existing resources such as cooling centers have not been utilized by vulnerable people for a variety of reasons including inability to get to the center, concerns about theft, personal safety, and the well-being of pets (Hall et al., 2004), as well as concern about the stigma of accepting public assistance (Sampson et al., 2013) and preference for staying home (Lane et al., 2014). Research and experience show that awareness of a heat warning does not necessarily translate into an individual taking protective action (Bassil and Cole, 2010). Close attention to public perception of heat adaptation strategies and awareness of behavioral tendencies, along with effective risk communication, is essential to effective urban adaptation. Simply providing the opportunity to seek shelter from heat isn't enough; people must be aware of the danger of heat, willing to admit their personal risk, and be able and motivated to seek help when they need it.

Less research has been conducted on vulnerability in lower- and middle-income countries, but it is likely that similar factors apply: the elderly, those with preexisting health conditions, and those without access to air-conditioning or safe water are at high risk during heat waves. Other factors likely are important as well, especially the relationship between heat, drought, and water/food insecurity in low-income regions. In addition, in many low-income rural areas, a high proportion of the population works outdoors, making occupational exposure to heat stress a major concern (Srinivasan et al., 2016).

7.2.3 Health Impacts of Heat Stress

7.2.3.1 Physiology

Humans must maintain an extremely narrow core temperature range to live, but the body is capable of tolerating temporary high levels of heat stress (Périard et al., 2015). Heat illness (hyperthermia) results from the body's failure to maintain its normal core temperature, and exposure to extreme heat can result in a range of health impacts. Heat illness can range from fatigue, syncope, and cramps to heat exhaustion, which is generally reversible given prompt medical treatment, and finally to heat stroke, a medical emergency which can cause long-term organ damage and rapid death through multiorgan failure (Hess et al., 2014).

Acclimatization to a wide variety of climates is possible through technology and behavioral adaptation, and to some extent, physiological (Nielsen et al., 1993) changes. Physiological adjustment to heat, however, happens slowly over the course of repeated exposures, as when one relocates to a warmer climate; physiological change is unlikely to protect against infrequent extreme heat events (Chalmers et al., 2014). The degree of acclimatization varies in different regions and for different climatic variables. Research has suggested that in the United States acclimatization to cold is more spatially uniform than for heat, as suggested by the relatively similar levels of temperature-induced mortality in cold weather across the country. Heat has been found to produce more varied mortality responses, with some regions—like the hot southwest—seeing smaller changes in mortality during heat waves than cooler regions under similar departures from the local mean temperature (e.g., a hot day reaching the 90th percentile relative to the local temperature distribution) (Medina-Ramon and Schwartz, 2007). This difference may be due in part to better air-conditioning coverage in hotter regions, as opposed to the Northeast United States or Europe, where coverage varies (Anderson and Bell, 2011). In addition, the risk of widespread heat illness may be greatest when temperatures rise far above the local climatologically normal range, which is most likely to occur in the mid-latitudes where temperature variability is greatest.

Emergency room visits and hospital admissions for certain cardiovascular, respiratory, renal, and mental health (Peng et al., 2017; Hansen et al., 2008) conditions have been observed to spike during heat waves (Fletcher et al., 2012; Lin et al., 2009). Similarly, on days with extremely high temperatures, overall mortality from natural causes is generally greater than what would be expected during summer. For example, a 2010 heat wave in Ahmedabad, Gujarat, India, a city that routinely deals with hot temperatures, resulted in 1344 excess deaths, or a 43% increase in all-cause mortality (Azhar et al., 2014). In 2015 a heat wave in India resulted in over 2300 deaths (NRDC, 2016), and the severe June 2015 heat wave in Pakistan resulted in a 17-fold increase in heat-related mortality in Karachi; this is likely an underestimate due to the challenges of tracking and recording deaths in an underresourced area (Ghumman and Horney, 2016).

The World Health Organization (WHO) estimates that only four countries in Africa—Algeria, Mauritius, Seychelles, and South Africa—have high quality daily death data (WHO, 2017). In countries lacking enough death data for analysis, heat-related mortality and morbidity may go unrecorded. Excess mortality associated with heat also occurs during hot, but not extreme, summer conditions (Gasparrini et al., 2015). Because there are more of these days in any given summer, substantial mortality may occur outside of extreme events. Rising average temperatures due to climate change may lead to increases in this non-extreme excess mortality as well.

7.2.3.2 Historical Trends and Future Projections of Heat Wave Mortality

Many studies have quantified the excess deaths attributable to heat across the world and have projected how these mortality rates may change in the future (Petkova et al., 2013, 2014; Hayhoe et al., 2010; Kalkstein and Greene, 1997; Anderson and Bell, 2011; Samet et al., 2000; Staddon et al., 2014; Chen et al., 2017). Heat-related mortality in the United States is found to have declined throughout the 21st century (Petkova et al., 2014), likely due to the spread of air-conditioning throughout the country and generally improving health (Bobb et al., 2014); however, some studies indicate that the decline has stopped (Sheridan et al., 2009). Almost all research concludes that heat-related mortality is likely to substantially increase in the future (Kovats and Hajat, 2008; Peng et al., 2011), due to both higher temperatures and much higher levels of exposure (Jones et al., 2015), and that this increase will not be offset by declining cold-related mortality (Staddon et al., 2014).

Most projections of future mortality rely on statistical models fit to historical daily mortality data. It is difficult to estimate the impacts of adaptation on future mortality, and there is some evidence that aggressive strategies to protect the most vulnerable populations can have a large impact (Fouillet et al., 2008). However, there is also evidence that heat stress may begin to approach fundamental limits for prolonged human tolerance (Sherwood and Huber, 2010; Pal and Eltahir, 2015); it is unknown how mortality will respond to such thresholds, but it is reasonable to expect substantial increases without aggressive adaptation measures. More research into the relationship between extremely high wet-bulb temperatures and mortality is urgently needed, as some of the world's most densely populated regions in India, Africa, and the Middle East could experience wet-bulb temperatures approaching the theoretical limit for human tolerance—35°C—by mid- to late-century (Im et al., 2017). More research is also needed on the relationship between heat and other stressors outside the scope of this chapter such as drought, food and water insecurity, political instability, and conflict.

The UHI results in urban populations suffering greater exposure to extreme heat, but research does not necessarily find that urban areas experience the highest rates of heat-related mortality (Chen et al., 2017). While temperatures are higher in cities, urban residents may also have more access to air-conditioning, water, and social support, especially in low-income countries with substantial income divides between rural and urban areas. More research is needed on mortality in rural areas, especially in low-income regions.

7.2.4 Urban Heat, Air Quality, and Infrastructure

7.2.4.1 Air Quality

Air pollution is estimated to result in the premature deaths of approximately three million people each year, mostly in low- and middle-income countries, in both urban and rural areas (Lelieveld et al., 2015). The main emission sources that contribute to air pollution are electricity generation, heavy industry, vehicle engines, both natural and human biomass burning, and agriculture. Like heat, air pollution is a physiological stressor which can both cause acute illness—especially respiratory and cardiovascular conditions—as well as exacerbate existing health problems. In addition, heat stress and poor air quality are often coincident, amplifying their impact; a substantial proportion of the deaths attributed to the 2003 European heat wave may have been caused by or related to air pollution (Fischer et al., 2004). Future particulate emissions are highly uncertain, but under a business as usual scenario premature air quality-related mortality could double by mid-century (Lelieveld et al., 2015) due to a rise in both population and emissions, especially in Asia and Africa.

High air temperatures increase local air pollution, even without additional emissions (Fiore et al., 2012). Background ground-level ozone (O_3) pollution in particular is correlated with temperature. High temperatures accelerate the photochemical processes that generate O_3, and they may also increase natural biomass burning (for example, wildfires) which generates O_3 precursors (Jacobson and Streets, 2009). However, precipitation and humidity tend to reduce pollution through wet deposition and convective atmospheric venting. Most models project increases in warm-season background O_3 levels and more intense O_3 pollution during heat waves (Rieder et al., 2013), a phenomena known as the "ozone climate penalty" (Rasmussen et al., 2013). The effect of climate change on particulate matter is more uncertain (Tai et al., 2010), but some studies suggest that if global aridity increases, dust and wildfire emissions could contribute to at least localized increases in particulate and aerosol pollution (Fiore et al., 2012).

There is also evidence that climate change may increase the frequency of meteorological conditions favorable to poor air quality in some regions (Horton et al., 2012, 2014). Air pollution episodes are associated with calm winds, low humidity, and temperature inversions, where the atmospheric lapse rate becomes negative near the surface, causing temperature to increase with height and acting as a block on vertical convective air movement. In some regions, notably China, these conditions are expected to become more likely in the future (Horton et al., 2012). Precipitation helps to clear the air of particulate matter (Fiore et al., 2012); while more of the globe is projected to see precipitation increase than decrease under climate change due to increasing atmospheric water vapor (Willett et al., 2007), even regions with more overall precipitation may see longer dry spells with a larger fraction of precipitation falling in a smaller number of events (Rahmstorf et al., 2012; O'Gorman and Schneider, 2009), allowing pollution to accumulate. It is also possible that pollen production could increase in some regions due to a lengthened growing season, potentially worsening allergies and asthma for some populations (Beggs and Bambrick, 2006).

7.2.4.2 *Infrastructure*

High temperatures have strongly negative effects on urban infrastructure, affecting electricity generation (Sathaye et al., 2013; Franco and Sanstad, 2008) and transportation (Schweikert et al., 2014; Meyer et al., 2010), both of which can contribute to human health impacts by reducing the availability of air-conditioning, medical treatment, and in low-income countries, food and safe water. In severe heat waves, the loss of electricity due to excessive demand or heat-induced equipment failure can result in the potentially dangerous loss of air-conditioning and refrigeration (Auffhammer et al., 2017). For vulnerable populations, air-conditioning can be a life-saving necessity during the most extreme heat, and thus maintaining electricity supply and ensuring equitable access to air conditioning and electricity can be critical. Electricity demand is expected to rapidly rise due to development and population increase, especially in Asia and Africa (Isaac and van Vuuren, 2009). Many of these regions are also prone to extreme heat, and since peak electricity load is strongly correlated with temperature (largely due to air-conditioning) (Sathaye et al., 2013; Deschênes and Greenstone, 2011), it will be important to consider resilience to increasing temperatures in the design of electricity transmission grids and generation equipment.

Transportation infrastructure, including roads, railroads, and the aviation system (Coffel and Horton, 2015; Williams and Joshi, 2013; Bhaduri, 2012; Coffel et al., 2017), can experience performance loss or failure in hot conditions (Smoyer-Tomic et al., 2003). In addition, the large spatial extent of heat waves means that there is the potential for significant disruption to transportation networks during severe heat events. Since even localized failures can result in cascading delays throughout a transportation network, it will be important to consider changing extreme temperature events in the design of future infrastructure.

7.3 ADAPTATION STRATEGIES

A variety of adaptation strategies exist to mitigate the impacts of extreme heat and the UHI. Some of these practices are standard in high-income cities around the world, such as air-conditioning access, public cooling centers, and weather forecasts warning of the dangers of heat waves. Other techniques require architectural or urban design changes and are not yet widely used. This section presents the evidence supporting some of the most prominent methods to both cool urban areas and reduce the human impacts of heat.

7.3.1 Urban Design

7.3.1.1 *Green Spaces*

Green spaces, from large parks to trees and grass, can reduce local temperatures by $1-6°C$ during the day, in both the sun and shade, through albedo change, reduced surface heat storage, and increased evapotranspiration (Zommers et al., 2016; Bowler et al., 2010; Oliveira et al., 2011; Georgi and Zafiriadis, 2006; Solecki et al., 2005; Chen et al., 2012; Susca et al., 2011). Trees, plants, and green spaces can also reduce local air pollution, improve human well-being in dense urban areas (Givoni, 1991; Bertram and Rehdanz, 2015), and increase property values (Bolitzer and Netusil, 2000).

The amount of cooling that a green space provides depends on its design, the characteristics of the surrounding urban area, and the local climate. Case studies of specific green spaces have found some provide the strongest cooling effect during the day (Oliveira et al., 2011), while others are most effective at night (Zoulia et al., 2008) especially in regions with high building height-to-street-width ratios. Research in an arid city in China found that small but densely-spaced green patches were more efficient at reducing surface temperature than sparse but large spaces; this may be especially relevant in water-constrained regions (Maimaitiyiming et al., 2014). While green spaces are generally cooler than their surroundings, they have been found to have little or no impact on ambient temperatures outside of their boundaries (Bowler et al., 2010; Oliveira et al., 2011; Chen et al., 2012). The effectiveness of a green space may be enhanced in dry climates, as low relative humidity allows for evaporative cooling of water transpired by plants; studies in humid regions have generally shown less of a green-space cooling effect (Hamada and Ohta, 2010).

The type of greenery has an impact on the effectiveness of a green space at reducing urban temperatures. Parks with trees may have a stronger cooling effect than those without (Shashua-Bar et al., 2010). Even single trees can reduce energy costs by providing shade on buildings; a study in Sacramento, CA in the United States found that each residential tree providing shade to a house reduces air-conditioning energy use by 7% while increasing winter heating energy use by only 2% (Simpson and McPherson, 1998). However, the impact of street trees on ambient temperatures is uncertain; some studies have found reduced temperatures near trees, while others have found little effect (Oliveira et al., 2011).

Current research suggests that green spaces and urban parks are essential for well-being in dense cities and can reduce air pollution and provide limited cooling, but they are unlikely to substantially reduce the average temperature across large urban areas.

7.3.1.2 Cool Roofs

Modifying roof color can raise the albedo and thus lower the energy absorption of a large building, providing a cost effective method of reducing cooling energy use in warm climates; recent research has suggested that cool roofs, whether white (in color) or green (e.g., vegetation-covered), be adopted in warm climates (Susca et al., 2011; Gaffin et al., 2009; Rossi et al., 2014). A large fraction—perhaps up to 10%—of urban air-conditioning energy use goes towards cooling the heat added by the UHI (Akbari et al., 2001), and air-conditioning units also produce substantial waste heat. A dark colored roof, the standard in the United States, absorbs up to 80% of incoming solar radiation, whereas a white roof reflects 50%−80% (Akbari et al., 2001), potentially reducing building energy consumption by up to 20% (Akbari et al., 2009; Touchaei et al., 2016). Modeling studies, backed by empirical tests, suggest that white and green roofs applied across a city could reduce urban temperatures by 1−2°C, substantially countering the UHI (Santamouris, 2014; Li et al., 2014; Zhang et al., 2016; Synnefa et al., 2008). Green roofs reduce ambient temperature mainly through evapotranspiration, while white and reflective roofs cool by raising the mean albedo of an urban area. The effect of both is greatest during the day, with limited night-time cooling (Li et al., 2014). When applied globally, one study suggests white and green roofs could provide the cooling equivalent of removing 1−4 years of current human CO_2 emissions from the atmosphere (approximately 0.05°C in global mean cooling) (Akbari et al., 2012). In metropolitan regions, these temperature reductions account for a substantial fraction of the current UHI and could offset a portion of the heat effects of future climate change for large populations (Georgescu et al., 2014).

Green roofs are more expensive than white roofs and are less effective at raising albedo; however, they can reduce storm water runoff (Sproul et al., 2014) and can provide localized evaporative cooling (Santamouris, 2014). Green roofs may also improve ecological connectivity, and can provide public green space if so configured (Braaker et al., 2014; Berardi et al., 2014). Their cost-effectiveness depends on the climate; energy savings may be outweighed by irrigation costs in arid regions, and by installation costs in temperate zones (Ascione et al., 2013). Green roofs may also necessitate structural modifications in some buildings, while white roofs may require frequent cleaning and repainting to maintain strong albedo benefits.

Overall, research suggests that cool roofs are a promising approach to reducing urban energy use and ambient temperature, and most research suggests that white roofs are the most effective and cost-efficient option.

7.3.1.3 Building Design

The geometry and material used in urban structures affect their thermal properties. Many common building materials, especially those with a low albedo like concrete and asphalt, store heat and contribute to UHI.

The geometry of urban streets can affect the strength of the UHI; urban canyons, or streets with tall buildings on each side, have reduced ventilation and generally higher temperatures (Block et al., 2012; Oke, 1981) than more open areas. The building height-to-street-width (H/W) ratio is correlated with the strength of the UHI; when this ratio is higher, meaning tall buildings surrounding narrow streets, temperature tends to drop more slowly at night. This effect is likely due to a decrease in radiative cooling, as more outgoing long-wave radiation is absorbed by surrounding buildings rather than escaping into the atmosphere (Oke, 1981).

New materials with retroreflective properties offer a method of reducing radiation absorption by closely-spaced buildings. Retroreflective materials reflect incoming radiation in the same direction from which it came; this means that incoming solar radiation can be reflected back to the atmosphere instead of reflecting off a surface and then being absorbed by the ground or a nearby building (Rossi et al., 2014).

Green facades—usually implemented with climbing plants and vines on the side of a building—can be an effective way of reducing local near-surface temperatures by 10°C or more (Block et al., 2012). In general, green facades are generally not as effective as cool roofs at reducing ambient temperatures, as the sides of buildings receive less direct sunlight and thus changing their albedo has less thermal effect. However, as a part of a comprehensive urban-greening strategy, green facades can play a significant role in reducing thermal absorption by man-made structures.

7.3.1.4 Urban Water Bodies

Urban water bodies, whether nearby oceans, lakes, rivers, or even small ponds and creeks, can provide a cooling effect. The air over a body of water can be 2−6°C cooler than over the surrounding city landscape (Manteghi et al., 2015), and regions downwind of the water body will experience some cooling, with many studies suggesting 1−2°C as a common temperature reduction (Chen et al., 2009). The spatial extent of the cooling effect depends heavily on the size of the water body, the prevailing winds, and the local humidity conditions; as evaporative cooling is the dominant mechanism for temperature reduction, water bodies will have less effect in humid conditions (Theeuwes et al., 2013). Moving or spraying water has a stronger cooling effect than stationary water bodies, and even small fountains can substantially cool nearby areas (Nishimura et al., 1998). Water provides increased latent and decreased sensible heat fluxes, which are likely to reduce air temperature during the daytime, but less so at night—and it should be noted that when air temperature drops below water temperature, bodies of water serve as strong warming influences, in some cases canceling out their daytime cooling effect and making them a net contributor to mean UHI (Steeneveld et al., 2014). In addition, increased evaporation can raise local relative humidity, and depending on the balance between temperature reduction and humidity increase, measures of heat stress may rise or fall near bodies of water (Saaroni and Ziv, 2003).

7.3.2 Behavior

7.3.2.1 Early Warning Systems

In recent years as the public health impacts of major heat waves have become more widely understood, countries have begun implementing heat wave early warning systems (HEWS) in an attempt to reduce heat-related mortality and prepare for climate change (Lowe et al., 2011; Ebi and Schmier, 2005). Early warning systems generally consist of a trigger, usually some combination of predicted extreme heat and humidity lasting for a specified duration, and a response plan once the trigger occurs. For early warning systems to be effective, they must be accompanied by active and targeted warnings for high-risk individuals (Kovats and Kristie, 2006), as well as advice on what protective measures people should take.

As climate change continues to intensify heat waves, early warning systems may need to be updated (Hess and Ebi, 2016) to reflect the changing characteristics of heat waves or changing public responses to extreme temperatures. Temperature−health relationships must also be monitored and HEWS can also be improved by monitoring their performance in past heat waves and adapting their trigger thresholds accordingly (Åström et al., 2015).

7.3.2.2 Personalized Heat Monitoring

Increased recognition of the localized nature of heat stress and the difficulty of predicting individual heat risk based on environmental conditions has spurred research into personalized heat exposure monitoring (Jay and Kenny, 2010; Bernard and Kenney, 1994; Reid et al., 2012). Tests of such devices have found substantial variation in exposure between individuals in the same neighborhood due to differences in activity level, clothing, and access to air-conditioning (Kuras et al., 2015; Bernhard et al., 2015). Most devices have been used only to measure heat exposure, but some have attempted to predict heat illness. These types of monitors usually take the form of wearable body temperature or heart rate sensors which can alert the user when their physiological parameters indicate unsafe heat exposure. More research is needed to identify and calibrate these physiological predictors of heat illness, as well as to test the efficacy of this adaptation approach in improving overall health outcomes during extreme heat events. As with many adaptation strategies, adopting the monitoring device is only the first step; users need to be trained in the proper response to a heat stress alert, and they must have the ability to seek air-conditioning, water, and rest.

Most research thus far has focused on high-income countries, and has generally found personal heat exposure to be strongly correlated with but generally less than outside air temperature, likely due to air-conditioning. Implementing personalized monitoring in low-income regions will likely prove more challenging, especially in rural areas where electricity access may be limited and equipment delivery and maintenance costs more expensive.

7.3.2.3 Encouraging Protective Action

Recent heat waves have shown that making cooling centers, water, and other services available to vulnerable populations isn't enough; there are often a variety of reasons that people choose not to take protective action without active encouragement. The use of air-conditioning is considered to be the most effective protection against heat-related illness, but it is not equitably distributed. Many at-risk individuals either do not have access or are reluctant to use it due to cost or personal preference (Lane et al., 2014). Many at-risk elderly people report checking on their friends or family during heat waves, a practice that may provide some protection by encouraging individuals to take protective actions (Nitschke et al., 2013). In urban areas, most heat-related deaths occur at home (Klinenberg, 2002; Lane et al., 2014);

more research is needed on the most effective methods of encouraging vulnerable populations to seek water and cooler spaces during heat waves. Several jurisdictions have programs to check on vulnerable residents; however, more research to evaluate and optimize these efforts is needed.

Optimizing interior spaces for hot weather can have significant impacts on heat exposure (Raja et al., 2001). Shutters and blinds can be closed on windows during the day. If air-conditioning is available, it can be used to cool a small area of the indoor space, and set to a low-cool setting to protect health while reducing energy use and costs. The use of fans to remove hot air during the day and to inject cool air during the night, as well as using shades to block sunlight during the hottest hours of the day, can reduce indoor temperatures in the absence of air-conditioning. However, during periods of extreme heat, there is no evidence that fans help prevent illness and death among vulnerable individuals (Gupta et al., 2012). Informing the public about strategies to mitigate indoor temperatures may be an effective way to reduce heat risk in some communities. In addition, frequent cool (but not cold) showers and baths may help to provide relief (Hajat et al., 2010).

7.4 CONCLUSIONS

Heat is a serious health risk in the current climate and will likely impose a greater burden on both urban and rural populations as heat waves intensify in the future. Experience in recent heat waves has shown that targeted intervention strategies, especially centered around alerting at-risk individuals to the risk of heat stress and providing them with the resources to protect themselves, can be highly effective at reducing mortality. Many of the heat wave adaptation strategies presented here have thus far been tested primarily in high-income countries; more research is needed on the most effective and efficient ways of implementing these and other evidence-based adaptation methods in low- and middle-income regions. In addition, most adaptation strategies thus far have focused on urban areas; while urban regions have a higher heat burden due to the UHI, rural regions face extreme heat as well, and often have less access to air-conditioning, safe water, and social support. More research is needed into rural adaptation, especially for agricultural workers in low-income countries, who may face an increasingly high heat risk in the future.

It is essential that adaptation measures be scientifically tested before being implemented on a large scale, especially where resources are limited. Some adaptations are much more effective than others, and many effective methods are not expensive. More research is also needed on implementation, barriers to adaptation, and the potential for cobenefits and costs of heat-related adaptation. Increasingly severe heat is poised to be one of the most directly observable impacts of climate change, as well as one with significant health impacts. Governments, both national and local, can reduce the burden of heat on their populations by implementing well-designed, equitable, and tested adaptation strategies.

REFERENCES

Akbari, H., Pomerantz, M., Taha, H., 2001. Cool surfaces and shade trees to reduce energy use and improve air quality in urban areas. Sol. Energy 70, 295−310.

Akbari, H., Menon, S., Rosenfeld, A., 2009. Global cooling: increasing world-wide urban albedos to offset CO2. Clim. Chang. 94, 275−286.

Akbari, H., Damon Matthews, H., Seto, D., 2012. The long-term effect of increasing the albedo of urban areas. Environ. Res. Lett. 7, 24004.

Alexander, L.V., et al., 2006. Global observed changes in daily climate extremes of temperature and precipitation. J. Geophys. Res. Atmos. 111, D05109.

Anderson, G., Bell, M.L., 2011. Heat waves in the United States: mortality risk during heat waves and effect modification by heat wave characteristics in 43 U.S. communities. Environ. Health Perspect. 119, 210−218.

Anderson, G., Bell, M.L., Peng, R.D., 2013. Methods to calculate the heat index as an exposure metric in environmental health research. Environ. Health Perspect. 121, 1111−1119.

Argüeso, D., Di Luca, A., Perkins-kirkpatrick, S.E., Evans, J.P., 2016. Seasonal mean temperature changes control future heat waves. Geophys. Res. Lett. . Available from: https://doi.org/10.1002/2016GL069408.

Ascione, F., Bianco, N., de' Rossi, F., Turni, G., Vanoli, G.P., 2013. Green roofs in European climates. Are effective solutions for the energy savings in air-conditioning? Appl. Energy 104, 845−859.

Åström, C., Ebi, L.K., Langner, J., Forsberg, B., 2015. Developing a heatwave early warning system for Sweden: evaluating sensitivity of different epidemiological modelling approaches to forecast temperatures. Int. J. Environ. Res. Public Health 12.

Auffhammer, M., Baylis, P., Hausman, C.H., 2017. Climate change is projected to have severe impacts on the frequency and intensity of peak electricity demand across the United States. Proc. Natl. Acad. Sci. U S A 114, 1886−1891.

Azhar, G.S., et al., 2014. Heat-related mortality in India: excess all-cause mortality associated with the 2010 Ahmedabad heat wave. PLoS One 9.

Ballester, J., Rodó, X., Giorgi, F., 2010. Future changes in Central Europe heat waves expected to mostly follow summer mean warming. Clim. Dyn. 35, 1191–1205.

Barnett, A.G., Tong, S., Clements, A.C.A., 2010. What measure of temperature is the best predictor of mortality? Environ. Res. 110, 604–611.

Barriopedro, D., Fischer, E.M., Luterbacher, J., Trigo, R.M., Garcia-Herrera, R., 2011. The hot summer of 2010: redrawing the temperature record map of Europe. Science (80-.). 332, 220–224.

Bassil, K.L., Cole, D.C., 2010. Effectiveness of public health interventions in reducing morbidity and mortality during heat episodes: a structured review. Int. J. Environ. Res. Public Health 7, 991–1001.

Battisti, D.S., Naylor, R.L., 2009. Historical warnings of future food insecurity with unprecedented seasonal heat. Science 323, 240–244.

Beggs, P.J., Bambrick, H.J., 2006. Is the global rise of asthma an early impact of anthropogenic climate change? CiÃ\textordfemeninencia & SaÃ\textordmasculinede Coletiva 11, 745–752.

Berardi, U., GhaffarianHoseini, A., GhaffarianHoseini, A., 2014. State-of-the-art analysis of the environmental benefits of green roofs. Appl. Energy 115, 411–428.

Berko, J., Ingram, D.D., Saha, S., Parker, J.D., 2014. Deaths attributed to heat, cold, and other weather events in the United States, 2006-2010. Natl. Health Stat. Rep. 1–15.

Bernard, T.E., Kenney, W.L., 1994. Rationale for a personal monitor for heat strain. Am. Ind. Hyg. Assoc. J. 55, 505–514.

Bernhard, M.C., et al., 2015. Measuring personal heat exposure in an urban and rural environment. Environ. Res. 137, 410–418.

Bertram, C., Rehdanz, K., 2015. The role of urban green space for human well-being. Ecol. Econ. 120, 139–152.

Bhaduri, A., 2012. Climate change. Econ. Labour Relations Rev. 23, 3–12.

Black, E., Blackburn, M., Harrison, R.G., Hoskins, B.J., Methven, J., 2004. Factors contributing to the summer 2003 European heatwave. Weather 59, 217–223.

Block, A.H., Livesley, S.J., Williams, N.S.G., 2012. Responding to the urban heat island: a review of the potential of green infrastructure. Vic. Cent. 1–62.

Bobb, J.F., Peng, R.D., Bell, M.L., Dominici, F., 2014. Heat-related mortality and adaptation to heat in the United States. Environ. Health. Perspect. 122, 811–816.

Bolitzer, B., Netusil, N.R., 2000. The impact of open spaces on property values in Portland, Oregon. J. Environ. Manage. 59, 185–193.

Bowler, D.E., Buyung-Ali, L., Knight, T.M., Pullin, A.S., 2010. Urban greening to cool towns and cities: a systematic review of the empirical evidence. Landsc. Urban. Plan. 97, 147–155.

Braaker, S., Ghazoul, J., Obrist, M.K., Moretti, M., 2014. Habitat connectivity shapes urban arthropod communities: the key role of green roofs. Ecology 95, 1010–1021.

Budd, G.M., 2008. Wet-bulb globe temperature (WBGT)-its history and its limitations. J. Sci. Med. Sport. 11, 20–32.

Burke, M., Hsiang, S.M., Miguel, E., 2015. Global non-linear effect of temperature on economic production. Nature 527, 235–239.

Cao, C., et al., 2016. Urban heat islands in China enhanced by haze pollution. Nat. Commun. 7, 12509.

Center for International Earth Science Information Network (CIESIN), C. U. Global Urban Heat Island (UHI) Data Set, 2013.

Chalmers, S., Esterman, A., Eston, R., Bowering, K.J., Norton, K., 2014. Short-term heat acclimation training improves physical performance: a systematic review, and exploration of physiological adaptations and application for team sports. Sport Med. 44, 971–988.

Chen, K., et al., 2017. Impact of climate change on heat-related mortality in Jiangsu Province, China. Environ. Pollut. 224, 317–325.

Chen, X., et al., 2012. Study on the cooling effects of urban parks on surrounding environments using Landsat TM data: a case study in Guangzhou, southern China. Int. J. Remote. Sens. 33, 5889–5914.

Chen, Z., et al., 2009. Field measurements on microclimate in residential community in Guangzhou, China. Front. Archit. Civ. Eng. China 3, 462.

Christidis, N., Jones, G.S., Stott, P.A., 2015. Dramatically increasing chance of extremely hot summers since the 2003 European heatwave. Nat. Clim. Chang. 5, 3–7.

Coffel, E., Horton, R., 2015. Climate change and the impact of extreme temperatures on aviation. Weather. Clim. Soc. 7, 94–102.

Coffel, E.D., Thompson, T.R., Horton, R.M., 2017. The impacts of rising temperatures on aircraft takeoff performance. Clim. Chang. 1–8. Available from: https://doi.org/10.1007/s10584-017-2018-9.

Cohen, J., et al., 2014. Recent Arctic amplification and extreme mid-latitude weather. Nat. Geosci. 7, 627–637.

Coumou, D., Petoukhov, V., Rahmstorf, S., Petri, S., Schellnhuber, H.J., 2014. Quasi-resonant circulation regimes and hemispheric synchronization of extreme weather in boreal summer. Proc. Natl. Acad. Sci. 111, 12331–12336.

Cuddy, M.L.S., 2004. The effects of drugs on thermoregulation. AACN. Clin. Issues 15, 238–253.

Deschênes, O., Greenstone, M., 2011. Climate change, mortality, and adaptation: evidence from annual fluctuations in weather in the US. Am. Econ. J. Appl. Econ. 3, 152–185.

Dole, R., et al., 2011. Was there a basis for anticipating the 2010 Russian heat wave? Geophys. Res. Lett. 38, n/a--n/a.

Donat, M.G., et al., 2013. Updated analyses of temperature and precipitation extreme indices since the beginning of the twentieth century: the HadEX2 dataset. J. Geophys. Res. Atmos. 118, 2098–2118.

Easterling, D.R., et al., 2000. Climate extremes: observations, modeling, and impacts. Science. (80-.). 289, 2068–2074.

Ebi, K.L., Schmier, J.K., 2005. A stitch in time: improving public health early warning systems for extreme weather events. Epidemiol. Rev. 27, 115–121.

Ellis, F.P., Nelson, F., 1978. Mortality in the elderly in a heat wave in New York City, August 1975. Environ. Res. 15, 504–512.

Fiore, A.M., et al., 2012. Global air quality and climate. Chem. Soc. Rev. 41, 6663.

Fischer, P.H., Brunekreef, B., Lebret, E., 2004. Air pollution related deaths during the 2003 heat wave in the Netherlands. Atmos. Environ. 38, 1083−1085.

Fletcher, B.A., Lin, S., Fitzgerald, E.F., Hwang, S.A., 2012. Association of summer temperatures with hospital admissions for renal diseases in New York state: a case-crossover study. Am. J. Epidemiol. 175, 907−916.

Forzieri, G., Cescatti, A., e Silva, F.B., Feyen, L., 2017. Increasing risk over time of weather-related hazards to the European population: a data-driven prognostic study. Lancet Planet. Heal 1, e200−e208.

Fouillet, A., et al., 2008. Has the impact of heat waves on mortality changed in France since the European heat wave of summer 2003? A study of the 2006 heat wave. Int. J. Epidemiol. 37, 309−317.

Franco, G., Sanstad, A.H., 2008. Climate change and electricity demand in California. Clim. Chang. 87, 139−151.

Gaffin, S.R., Khanbilvardi, R., Rosenzweig, C., 2009. Development of a green roof environmental monitoring and meteorological Network in New York City. Sensors 9, 2647−2660.

Gasparrini, A., et al., 2015. Mortality risk attributable to high and low ambient temperature: a multicountry observational study. Lancet 386, 369−375.

Georgescu, M., Morefield, P.E., Bierwagen, B.G., Weaver, C.P., 2014. Urban adaptation can roll back warming of emerging megapolitan regions. Proc. Natl. Acad. Sci. U S A 111, 2909−2914.

Georgi, N.J., Zafiriadis, K., 2006. The impact of park trees on microclimate in urban areas. Urban Ecosyst. 9, 195−209.

Ghumman, U., Horney, J., 2016. Characterizing the impact of extreme heat on mortality, Karachi, Pakistan, June 2015. Prehosp. Disaster. Med. 31, 263−266.

Givoni, B., 1991. Impact of planted areas on urban environmental quality: a review. Atmos. Environ. Part B, Urban Atmos. 25, 289−299.

Glaser, J., et al., 2016. Climate change and the emergent epidemic of CKD from heat stress in rural communities: the case for heat stress nephropathy. Clin. J. Am. Soc. Nephrol. 11, 1472−1483.

Gronlund, C.J., 2014. Racial and socioeconomic disparities in heat-related health effects and their mechanisms: a review. Curr. Epidemiol. Rep. 1, 165−173.

Gupta, S., et al., 2012. Electric fans for reducing adverse health impacts in heatwaves. Cochrane Libr. Available from: https://doi.org/10.1002/14651858.CD009888.pub2.

Hajat, S., O'Connor, M., Kosatsky, T., 2010. Health effects of hot weather: from awareness of risk factors to effective health protection. Lancet 375, 856−863.

Hall, M.J., et al., 2004. Psychological impact of the animal-human bond in disaster preparedness and response. J. Psychiatr. Pract. 10, 368−374.

Hamada, S., Ohta, T., 2010. Seasonal variations in the cooling effect of urban green areas on surrounding urban areas. Urban For. Urban Green. 9, 15−24.

Hansen, A., et al., 2008. The effect of heat waves on mental health in a temperate Australian City. Environ. Health. Perspect. 116, 1369−1375.

Hansen, J., Sato, M., Ruedy, R., 2012. Perception of climate change. Proc. Natl. Acad. Sci. 109, E2415−E2423.

Harlan, S.L., Brazel, A.J., Prashad, L., Stefanov, W.L., Larsen, L., 2006. Neighborhood microclimates and vulnerability to heat stress. Soc. Sci. Med. 63, 2847−2863.

Hayhoe, K., Sheridan, S., Kalkstein, L., Greene, S., 2010. Climate change, heat waves, and mortality projections for Chicago. J. Great. Lakes. Res. 36, 65−73.

Herold, N., Alexander, L., Green, D., Donat, M., 2017. Greater increases in temperature extremes in low versus high income countries. Environ. Res. Lett. 12, 34007.

Hess, J.J., Ebi, K.L., 2016. Iterative management of heat early warning systems in a changing climate. Ann. N. Y. Acad. Sci. 21−30. Available from: https://doi.org/10.1111/nyas.13258.

Hess, J.J., Saha, S., Luber, G., 2014. Summertime acute heat illness in U. S. emergency departments from 2006 through 2010: analysis of a nationally representative sample. Environ. Health. Perspect. 122, 1209−1215.

Hondula, D.M., et al., 2012. Fine-scale spatial variability of heat-related mortality in Philadelphia County, USA, from 1983-2008: a case-series analysis. Environ. Heal. A Glob. Access Sci. Source 11, 1−11.

Horton, D.E., Harshvardhan, Diffenbaugh, N.S., 2012. Response of air stagnation frequency to anthropogenically enhanced radiative forcing. Environ. Res. Lett. 7, 44034.

Horton, D.E., Skinner, C.B., Singh, D., Diffenbaugh, N.S., 2014. Occurrence and persistence of future atmospheric stagnation events. Nat. Clim. Chang. 4, 698−703.

Horton, R.M., Coffel, E.D., Winter, J.M., Bader, D.A., 2015. Projected changes in extreme temperature events based on the NARCCAP model suite. Geophys. Res. Lett. 42, 7722−7731.

Horton, R.M., Mankin, J.S., Lesk, C., Coffel, E., Raymond, C., 2016. A review of recent advances in research on extreme heat events. Curr. Clim. Chang. Rep. 2, 242−259.

Huntingford, C., Jones, P.D., Livina, V.N., Lenton, T.M., Cox, P.M., 2013. No increase in global temperature variability despite changing regional patterns. Nature 500, 327−330.

Im, E.-S., Pal, J.S., Eltahir, E.A.B., 2017. Deadlyheat waves projected in the densely populated agricultural regions of South Asia. Sci. Adv. 3, 1−8.

IPCC, 2013. Climate Change 2013: The Physical Science Basis. In: Stocker, T.F., Qin, D., Plattner, G.-K., Tignor, M., Allen, S.K., Boschung, J., Nauels, A., Xia, Y., Bex, V., Midgley, P.M. (Eds.), Contribution of Working Group I to the Fifth Assessment Report of the Intergovernmental Panel on Climate Change. Cambridge University Press, Cambridge, United Kingdom/New York, NY, USA, p. 1535. Available from: https://doi.org/10.1017/CBO9781107415324.

Isaac, M., van Vuuren, D.P., 2009. Modeling global residential sector energy demand for heating and air conditioning in the context of climate change. Energy Policy 37, 507–521.

Jacobson, M.Z., Streets, D.G., 2009. Influence of future anthropogenic emissions on climate, natural emissions, and air quality. J. Geophys. Res. Atmos. 114, 1–21.

Jay, O., Kenny, G.P., 2010. Heat exposure in the Canadian workplace. Am. J. Ind. Med. 53, 842–853.

Jones, B., et al., 2015. Future population exposure to US heat extremes. Nat. Clim. Chang. 5, 652–655.

Kalkstein, L.S., Greene, J.S., 1997. An evaluation of climate/mortality relationships in large U.S. cities and the possible impacts of a climate change. Environ. Health. Perspect. 105, 84–93.

Karl, T.R., et al., 2015. Possible artifacts of data biases in the recent global surface warming hiatus. Science. (80-.). 348, 1469–1472.

Karner, A.A., Eisinger, D.S., Niemeier, D.A., 2010. Near-roadway air quality: synthesizing the findings from real-world data. Environ. Sci. Technol. 44, 5334–5344.

Kerr, Z.Y., Casa, D.J., Marshall, S.W., Comstock, R.D., 2013. Epidemiology of exertional heat illness among U.S. High School Athletes. Am. J. Prev. Med. 44, 8–14.

Kjellstrom, T., Kovats, R.S., Lloyd, S.J., Holt, T., Tol, R.S.J., 2009. The direct impact of climate change on regional labor productivity. Arch. Environ. Occup. Health 64, 217–227.

Klinenberg, E., 2001. Dying alone. Ethnography 2, 501–531.

Klinenberg, E., 2002. Heat Wave: A Social Autopsy of Disaster in Chicago. University of Chicago Press, Chicago.

Knutson, T.R., Ploshay, J.J., 2016. Detection of anthropogenic influence on a summertime heat stress index. Clim. Chang. 138, 25–39.

Kodra, E., Ganguly, A.R., 2014. Asymmetry of projected increases in extreme temperature distributions. Sci. Rep. 4, 5884.

Kovats, R.S., Hajat, S., 2008. Heat stress and public health: a critical review. Annu. Rev. Public. Health 29, 41–55.

Kovats, R.S., Kristie, L.E., 2006. Heatwaves and public health in Europe. Eur. J. Public. Health 16, 592–599.

Krieger, N., Chen, J.T., Waterman, P.D., Rehkopf, D.H., Subramanian, S.V., 2005. Painting a truer picture of US socioeconomic and racial/ethnic health inequalities: the public health disparities geocoding project. Am. J. Public. Health 95, 312–323.

Kuras, E.R., Hondula, D.M., Brown-Saracino, J., 2015. Heterogeneity in individually experienced temperatures (IETs) within an urban neighborhood: insights from a new approach to measuring heat exposure. Int. J. Biometeorol. 59, 1363–1372.

Lane, K., et al., 2014. Extreme heat awareness and protective behaviors in New York City. J. Urban Heal. 91, 403–414.

Lelieveld, J., Evans, J.S., Fnais, M., Giannadaki, D., Pozzer, A., 2015. The contribution of outdoor air pollution sources to premature mortality on a global scale. Nature 525, 367–371.

Li, D., Bou-Zeid, E., Oppenheimer, M., 2014. The effectiveness of cool and green roofs as urban heat island mitigation strategies. Environ. Res. Lett. 9, 55002.

Lin, S., et al., 2009. Extreme high temperatures and hospital admissions for respiratory and cardiovascular diseases. Epidemiology 20, 738–746.

Lobell, D.B., Gourdji, S.M., 2012. The influence of climate change on global crop productivity. Plant. Physiol. 160, 1686–1697.

Lowe, D., Ebi, K.L., Forsberg, B., 2011. Heatwave early warning systems and adaptation advice to reduce human health consequences of heatwaves. Int. J. Environ. Res. Public. Health 8, 4623–4648.

Madrigano, J., Ito, K., Johnson, S., Kinney, P.L., Matte, T., 2015. A case-only study of vulnerability to heat wave – related mortality. Environ. Health. Perspect. 672, 672–678.

Maimaitiyiming, M., et al., 2014. Effects of green space spatial pattern on land surface temperature: implications for sustainable urban planning and climate change adaptation. ISPRS J. Photogramm. Remote Sens. 89, 59–66.

Manteghi, G., Bin Limit, H., Remaz, D., 2015. Water bodies an urban microclimate: a review. Mod. Appl. Sci. 9, 1–12.

Masterton, J.M., Richardson, F.A., 1979. Humidex: a method of quantifying human discomfort due to excessive heat and humidity. Environ. Canada, Atmos. Environ. Serv. .

McKitrick, R.R., Michaels, P.J., 2007. Quantifying the influence of anthropogenic surface processes and inhomogeneities on gridded global climate data. J. Geophys. Res. Atmos. 112, 1–14.

Medina-Ramon, M., Schwartz, J., 2007. Temperature, temperature extremes, and mortality: a study of acclimatisation and effect modification in 50 US cities. Occup. Environ. Med. 64, 827–833.

Medina-Ramón, M., Zanobetti, A., Cavanagh, D.P., Schwartz, J., 2006. Extreme temperatures and mortality: assessing effect modification by personal characteristics and specific cause of death in a multi-city case-only analysis. Environ. Health. Perspect. 114, 1331–1336.

Meehl, G.A., 2004. More intense, more frequent, and longer lasting heat waves in the 21st century. Science (80-) 305, 994–997.

Meyer, M.D., Amekudzi, A., Patrick, J., 2010. Transportation asset management systems and climate change. Transp. Res. Rec. J. Transp. Res. Board, No. 2160, Transp. Res. Board Natl. Acad. Washington, D.C. 2160, 12–20.

Mora, C., et al., 2017. Global risk of deadly heat. Nat. Clim. Chang. 7, 501–506.

Moran, D.S., et al., 2001. An environmental stress index (ESI) as a substitute for the wet bulb globe temperature (WBGT). J. Therm. Biol. 26, 427–431.

Nielsen, B., et al., 1993. Human circulatory and thermoregulatory adaptations with heat acclimation and exercise in a hot, dry environment. J. Physiol. 460, 467–485.

Nishimura, N., Nomura, T., Iyota, H., Kimoto, S., 1998. Novel water facilities for creation of comfortable urban micrometeorology. Sol. Energy 64, 197–207.

Nitschke, M., et al., 2013. Risk factors, health effects and behaviour in older people during extreme heat: a survey in South Australia. Int. J. Environ. Res. Public. Health 10.

NRDC, 2016. Expanding Heat Resilient Cities Across India.

O'Gorman, P.A., Schneider, T., 2009. The physical basis for increases in precipitation extremes in simulations of 21st-century climate change. Proc. Natl. Acad. Sci. 106, 14773–14777.

Oke, T.R., 1973. City size and the urban heat island. Atmos. Environ. 7, 769–779.

Oke, T.R., 1981. Canyon geometry and the nocturnal urban heat island: comparison of scale model and field observations. J. Climatol. 1, 237–254.

Oliveira, S., Andrade, H., Vaz, T., 2011. The cooling effect of green spaces as a contribution to the mitigation of urban heat: a case study in Lisbon. Build. Environ. 46, 2186–2194.

Pal, J.S., Eltahir, E.A.B., 2015. Future temperature in southwest Asia projected to exceed a threshold for human adaptability. Nat. Clim. Chang. 18203, 1–4.

Peng, R.D., et al., 2011. Toward a quantitative estimate of future heat wave mortality under global climate change. Environ. Health. Perspect. 119, 701–706.

Peng, Z., Wang, Q., Kan, H., Chen, R., Wang, W., 2017. Effects of ambient temperature on daily hospital admissions for mental disorders in Shanghai, China: a time-series analysis. Sci. Total. Environ. 590–591, 281–286.

Périard, J.D., Racinais, S., Sawka, M.N., 2015. Adaptations and mechanisms of human heat acclimation: applications for competitive athletes and sports. Scand. J. Med. Sci. Sport. 25, 20–38.

Petkova, E.P., Horton, R.M., Bader, D.A., Kinney, P.L., 2013. Projected heat-related mortality in the U.S. Urban Northeast. Int. J. Environ. Res. Public. Health 10, 6734–6747.

Petkova, E.P., Gasparrini, A., Kinney, P.L., 2014. Heat and mortality in New York City since the beginning of the 20th century. Epidemiology 25, 1.

Rahmstorf, S., Coumou, D., Correction for Cao, X., et al., 2012. Differential regulation of the activity of deleted in liver cancer 1 (DLC1) by tensins controls cell migration and transformation. Proc. Natl. Acad. Sci. 109, 4708.

Raja, I.A., Nicol, J.F., McCartney, K.J., Humphreys, M.A., 2001. Thermal comfort: use of controls in naturally ventilated buildings. Energy Build. 33, 235–244.

Rasmussen, D.J., Hu, J., Mahmud, A., Kleeman, M.J., 2013. The ozone–climate penalty: past, present, and future. Environ. Sci. Technol. 47, 14258–14266.

Ray, D.K., Gerber, J.S., MacDonald, G.K., West, P.C., 2015. Climate variation explains a third of global crop yield variability. Nat. Commun. 6, 5989.

Reid, C.E., et al., 2012. Evaluation of a heat vulnerability index on abnormally hot days: an environmental public health tacking study. Environ. Health. Perspect. 120, 715–720.

Rieder, H.E., Fiore, A.M., Polvani, L.M., Lamarque, J.-F., Fang, Y., 2013. Changes in the frequency and return level of high ozone pollution events over the eastern United States following emission controls. Environ. Res. Lett. 8, 14012.

Rossi, F., Pisello, A.L., Nicolini, A., Filipponi, M., Palombo, M., 2014. Analysis of retro-reflective surfaces for urban heat island mitigation: a new analytical model. Appl. Energy 114, 621–631.

Rothfusz, L.P., Headquarters, N.S.R., 1990. The heat index equation (or, more than you ever wanted to know about heat index). Fort Worth, Texas: National Oceanic and Atmospheric Administration, National Weather Service, Office of Meteorology.

Saaroni, H., Ziv, B., 2003. The impact of a small lake on heat stress in a Mediterranean urban park: the case of Tel Aviv, Israel. Int. J. Biometeorol. 47, 156–165.

Samet, J.M., Dominici, F., Zeger, S.L., Schwartz, J., Dockery, D.W., 2000. The national morbidity, mortality, and air pollution study. Part I: methods and methodologic issues. Res. Rep. Health. Eff. Inst. 5-14-84. doi:PubMed ID: 11354823.

Sampson, N.R., et al., 2013. Staying cool in a changing climate: reaching vulnerable populations during heat events. Glob. Environ. Chang. 23, 475–484.

Santamouris, M., 2014. Cooling the cities - a review of reflective and green roof mitigation technologies to fight heat island and improve comfort in urban environments. Sol. Energy 103, 682–703.

Sathaye, J.A., et al., 2013. Estimating impacts of warming temperatures on California's electricity system. Glob. Environ. Chang 23, 499–511.

Schär, C., et al., 2004. The role of increasing temperature variability in European summer heatwaves. Nature 427, 332–336.

Schlenker, W., Roberts, M.J., 2009. Nonlinear temperature effects indicate severe damages to U.S. crop yields under climate change. Proc. Natl. Acad. Sci. 106, 15594–15598.

Schweikert, A., Chinowsky, P., Kwiatkowski, K., Espinet, X., 2014. The infrastructure planning support system: analyzing the impact of climate change on road infrastructure and development. Transp. Policy 35, 146–153.

Shashua-Bar, L., Potchter, O., Bitan, A., Boltansky, D., Yaakov, Y., 2010. Microclimate modelling of street tree species effects within the varied urban morphology in the Mediterranean city of Tel Aviv, Israel. Int. J. Climatol. 30, 44–57.

Sheridan, S.C., Kalkstein, A.J., Kalkstein, L.S., 2009. Trends in heat-related mortality in the United States, 1975-2004. Nat. Hazards 50, 145–160.

Sherwood, S.C., Huber, M., 2010. An adaptability limit to climate change due to heat stress. Proc. Natl. Acad. Sci. 107, 9552–9555.

Sidiqui, P., Huete, A., Devadas, R., 2016. Spatio-temporal mapping and monitoring of Urban Heat Island patterns over Sydney, Australia using MODIS and Landsat-8. Earth Obs. Remote Sens. Appl. (EORSA), 2016 4th Int. Work. 217–221. Available from: https://doi.org/10.1109/EORSA.2016.7552800.

Simpson, J.R., McPherson, E.G., 1998. Simulation of tree shade impacts on residential energy use for space conditioning in Sacramento. Atmos. Environ. 32, 69–74.

Smoyer-Tomic, K.E., Kuhn, R., Hudson, A., 2003. Heat wave hazards: an overview of heat wave impacts in Canada. Nat. Hazards 28, 465–486.

Solecki, W.D., et al., 2005. Mitigation of the heat island effect in urban New Jersey. Environ. Hazards 6, 39−49.

Sproul, J., Wan, M.P., Mandel, B.H., Rosenfeld, A.H., 2014. Economic comparison of white, green, and black flat roofs in the United States. Energy Build. 71, 20−27.

Srinivasan, K., Maruthy, K., Venugopal, V., Ramaswamy, P., 2016. Research in occupational heat stress in India: challenges and opportunities. Indian J. Occup. Env. Med. 20, 73−78.

Staddon, P.L., Montgomery, H.E., Depledge, M.H., 2014. Climate warming will not decrease winter mortality. Nat. Clim. Chang. 4, 190−194.

Steadman, R.G., 1984. A universal scale of apparent temperature. J. Clim. Appl. Meteorol. 23, 1674−1687.

Steeneveld, G.J., Koopmans, S., Heusinkveld, B.G., Theeuwes, N.E., 2014. Refreshing the role of open water surfaces on mitigating the maximum urban heat island effect. Landsc. Urban. Plan. 121, 92−96.

Stott, P.A., Stone, D.A., Allen, M.R., 2004. Human contribution to the European heatwave of 2003. Nature 432, 610−614.

Susca, T., Gaffin, S.R., Dell'Osso, G.R., 2011. Positive effects of vegetation: urban heat island and green roofs. Environ. Pollut. 159, 2119−2126.

Synnefa, A., Dandou, A., Santamouris, M., Tombrou, M., Soulakellis, N., 2008. On the use of cool materials as a heat island mitigation strategy. J. Appl. Meteorol. Climatol. 47, 2846−2856.

Tai, A.P.K., Mickley, L.J., Jacob, D.J., 2010. Correlations between fine particulate matter (PM2.5) and meteorological variables in the United States: implications for the sensitivity of PM2.5 to climate change. Atmos. Environ. 44, 3976−3984.

Theeuwes, N.E., Solcerová, A., Steeneveld, G.J., 2013. Modeling the influence of open water surfaces on the summertime temperature and thermal comfort in the city. J. Geophys. Res. Atmos. 118, 8881−8896.

Touchaei, A.G., Hosseini, M., Akbari, H., 2016. Energy savings potentials of commercial buildings by urban heat island reduction strategies in Montreal (Canada). Energy Build. 110, 41−48.

Vandentorren, S., et al., 2006. August 2003 heat wave in France: risk factors for death of elderly people living at home. Eur. J. Public. Health 16, 583−591.

Wallace, R.F., Kriebel, D., Punnett, L., Wegman, D.H., Amoroso, P.J., 2007. Prior heat illness hospitalization and risk of early death. Environ. Res. 104, 290−295.

Walsh, J., et al., 2014. Chapter 2: our changing climate. Third US Natl. Clim. Assess. .

Weber, S., Sadoff, N., Zell, E., de Sherbinin, A., 2015. Policy-relevant indicators for mapping the vulnerability of urban populations to extreme heat events: a case study of Philadelphia. Appl. Geogr. 63, 231−243.

WHO, 2017. Civil Registration and Vital Statistics Systems. Available at: http://www.aho.afro.who.int/profiles_information/index.php/AFRO: Civil_registration_and_vital_statistics_systems.

Willett, K.M., Gillett, N.P., Jones, P.D., Thorne, P.W., 2007. Attribution of observed surface humidity changes to human influence. Nature 449, 710−712.

Williams, P.D., 2017. Increased light, moderate, and severe clear-air turbulence in response to climate change. Adv. Atmos. Sci. 34, 576−586.

Williams, P.D., Joshi, M.M., 2013. Intensification of winter transatlantic aviation turbulence in response to climate change. Nat. Clim. Chang. 3, 644−648.

Wolf, T., McGregor, G., 2013. The development of a heat wave vulnerability index for London, United Kingdom. Weather Clim. Extrem. 1, 59−68.

Zhang, J., Zhang, K., Liu, J., Ban-Weiss, G., 2016. Revisiting the climate impacts of cool roofs around the globe using an Earth system model. Environ. Res. Lett. 11, 84014.

Zhao, L., Lee, X., Smith, R.B., Oleson, K., 2014. Strong contributions of local background climate to urban heat islands. Nature 511, 216−219.

Zommers, Z., et al., 2016. Loss and Damage: The Role of Ecosystem Services. United Nations Environment Programme, Nairobi, Kenya.

Zoulia, I., Santamouris, M., Dimoudi, A., 2008. Monitoring the effect of urban green areas on the heat island in Athens. Environ. Monit. Assess. 156, 275.

Chapter 8

Measuring Drought Resilience Through Community Capitals

Andries J. Jordaan, Dusan M. Sakulski, Curtis Mashimbye and Fumiso Mayumbe

University of the Free State, Bloemfontein, South Africa

Chapter Outline

8.1 INTRODUCTION

Dry periods and droughts remain the major meteorological factor with devastating impacts on the livelihoods of most rural people in Africa. The agricultural sector specifically incurs millions of dollars in losses every year. Economic growth in most developing countries is severely hampered with every disastrous drought, even given the low contribution of agriculture to GDP in an industrialized economy.

Most people in agriculture acknowledge climatic extremes and the fact that South Africa (SA) will experience future dry periods, as a given. It is just a matter of when and how severe. The challenge though, is to prevent dry periods from developing into disaster droughts. Important, however, is the vulnerability and the resilience of the agricultural sector and of individual farmers as key factors in drought prevention and mitigation. One cannot assess drought risk by assessing precipitation, evaporation, and transpiration alone. These are variables used for drought hazard assessment and not total drought risk (Wilhite, 2000a,b; Jordaan, 2011; Jordaan et al., 2017). Hazard assessment is only one component of the risk assessment equation and that is clearly illustrated in this chapter.

Vulnerability and resilience are key factors to any disaster risk assessment and should always be assessed in relation to a specific hazard (Wisner et al., 2004)—drought in this case. Scientists have already acknowledged the integration of social, environmental (i.e., ecological), and economic factors in watershed management already since the 1980s. Any drought strategy should support efforts to increase resilience against droughts amongst all role players in agriculture.

This chapter explains an alternative methodology applied for a drought risk assessment in the Eastern Cape in South Africa. The community capitals framework proposed by Flora et al. (2007) was integrated with the drought hazard assessment in order to determine drought risk.

8.2 DESCRIPTION OF STUDY AREA

The Eastern Cape (EC) is one of the nine Provinces in South Africa and borders KwaZulu-Natal (KZN), Free State (FS), and Lesotho to the north, and the Northern Cape (NC) and Western Cape (WC) to the west. The Indian Ocean forms the southern and eastern borders of the Eastern Cape (Fig. 8.1).

The Eastern Cape was one of the regions most suitable to compare drought vulnerability, adaptation, coping, and resilience of commercial and communal subsistence farmers because of the historical demarcation of communal areas. Large areas in the Eastern Cape are still managed by tribal authorities with mainly common property right systems.

Resilience. DOI: https://doi.org/10.1016/B978-0-12-811891-7.00008-6

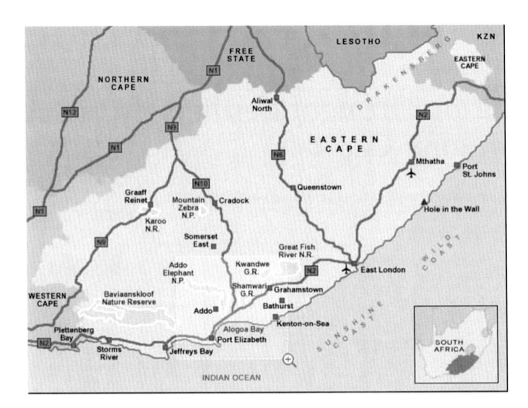

FIGURE 8.1 Eastern Cape Province.

These areas are entwined with well-planned commercial farms with well-defined individual or private property right systems. The Eastern Cape also covers different rainfall zones with annual precipitation of 1000 mm in the eastern coastal zones to less than 350 mm per annum in the western part of the province. Three districts with the largest diversity were selected as study areas namely Cacadu, OR Tambo, and Joe Gqabi.

The province is blessed with much fertile land and agriculture is the most important activity outside the metropolitan areas. Land use and vegetation cover are amongst the most important indicators for environmental vulnerability to drought. The total area of the Eastern Cape is 17.1 million ha of which 86.8%, or 14.8 million ha consists of farm land. Around 6.9%, or 1.2 million ha, is arable with 13.6 million ha available for grazing. About 3.7%, or 623,400 ha, consists of conservation areas with 140,000 ha under forestry and 1.49 million ha, or 8.7% of the land in the province, is used for other purposes. Eighty-four percent or 14.2 million ha of the province is utilized as natural grazing with 537,000 ha utilized for dry land cropping for commercial purposes and 642,000 ha of dry land cropping on communal land. Irrigation land adds up to 166,000 ha—mostly for dairy prodcution along the coastal zone and the Xhariep-Fish River irrigation scheme and fruit production in the Langkloof valley.

Agriculture in the Eastern Cape is classified as either commercial and developing commercial or as subsistence and small-scale. The number of commercial farmers are 4006 with 310,400 subsistence communal farmers; i.e., 24% of the 1.29 million subsistence farmers in South Africa. Two thirds of the population in the Eastern Cape lives in rural areas and agriculture is an important factor in the development of peoples' livelihoods in the province. Six hundred and forty-three thousand households, or 37.3% of total households in the Eastern Cape, are involved in agricultural activities. Of these 48.5% are involved in livestock production, 54.3% in poultry production and 60.5% in grains and food crops—mostly for subsistence—and 34.2% in fruit and vegetables (AgriSETA, 2010).

The most common land use activities amongst subsistence farmers are livestock rearing, some cultivation in high potential areas and exploitation of natural resources for fire wood, building materials and medicinal plants. The vast majority of livelihoods living on the developing land derive their income from on-farm as well as off-farm sources. Off-farm sources include wages, remittances from migrants and commuters, and state welfare grants, which in most areas is the main source of income (Andrew., 2003; STATSSA, 2011). The potential off-farm income is an important indicator for resilience against drought and contributes much toward the coping capacity of rural farmers against drought (Jordaan, 2011). Andrew (2003) found that small stock and its fibers are commonly traded on the market for cash income while cattle are mainly used for daily subsistence (from milk) and for cultural purposes. Cattle remain a cultural asset and rural peoples' wealth is calculated according to the number of cattle they own (Andrew., 2003; Jordaan et al., 2017).

The dualistic nature of farming systems in most cases creates the impression that small-scale subsistence and communal farming is wasteful, destructive, and economically unproductive. While there is much room for improvement, one cannot ignore the importance of the potential production derived from the 310,400 subsistence farmers in the Eastern Cape. Jordaan (2012), e.g., found that some of these so-called subsistence farmers are in fact commercial farmers in spite of the fact that they form part of the communal farming system. Jordaan (2012) analyzed the results of the Elundini livestock improvement program in the Mount Fletcher region and from a sample of 1200 farmers found the mean number of ewes owned per farmer was 60 (median = 40), with 10% of the farmers farming with more than 300 ewes, which put these farmers in the commercial domain.

Commercial agriculture consists mainly of livestock, with some pockets of crop production and horticulture in high potential areas. Livestock farming in the Eastern Cape consists mainly of sheep and more particularly wool sheep farming, angora goats, beef and dairy cattle, and game. Crop production contributes little, at less than 1% of total production in South Africa (DAFF, 2012). Livestock production, on the other hand, is an important sector for the Eastern Cape in relation to the rest of SA with 23% of cattle numbers, 30% of sheep numbers, 37% of goat numbers, and 6% of pig numbers in South Africa. The Eastern Cape also supplies about 25% of the milk in SA (ECDC, 2009).

The dependence on agriculture exacerbates the importance of drought and drought risk in the Eastern Cape. The drought risk methodology applied in the province is discussed in the following sections.

8.3 DROUGHT HAZARD RISK ASSESSMENT

Drought risk assessment originates from the hazard, which is drought caused by too little precipitation and too much evaporation (meteorological drought) (Wilhelmi and Wilhite, 2002). Hazard assessment is one of the variables in the drought risk assessment. Drought risk is calculated as a weighted sum of factors (capitals) including severity, intensity, and duration of a dry period. Drought hazard is the result of water shortage in support of normal biological production, and/or the lack of drinking water (Wilhite, 2000a,b). One therefore expects that most indicators for the hazard would be weather related, although scientists have also developed indicators based on the biological indicators affected by drought (Fouche et al., 1985, 1992; Du Pisani, 1998).

- *Meteorological drought:* Meteorological indicators such as percentage of rainfall and the Standard Precipitation Index (SPI) or the Standard Precipitation Evaporation Index (SPEI).

- *Agricultural drought:* Remote sensing satellite indices such as the Vegetation Condition Index (VCI) and the percentage of Average Seasonal Greenness (PASG). Measurement of agricultural droughts is also done through secondary indicators such as actual veld condition, grazing reserves, drinking water, and animal and crop condition.
- *Hydrological drought:* Actual measurements of dam levels, stream flow, and groundwater levels are used to measure hydrological drought.
- *Socioeconomic drought* is only measureable through secondary indicators such as impacts on individual farmers and the regional and larger economy.

The results for the hazard assessment are illustrated spatially in Fig. 8.2. The 12-month SPEI value was used as the preferred index for hazard assessment. The most influential factors contributing to the hazard rating in this study were exceedance probability, intensity, and duration. SPEI <-1.5 is an indication of severe drought whereas SPEI <-2 indicates extreme droughts. The assumption of drought based on a specific SPEI value, or any other meteorological indicator for that matter, is challenged since vulnerabilities and drought impacts differ from region to region, from system to system, and from community to community. Communal farmers farming on degraded land with no resources, e.g., are much more vulnerable to dry periods; they already experience *"man-made-droughts"* at SPEI <-1.2.

Calculating SPEI and drought severity was done according to the methodology proposed by McKee et al. (1993, 1995). Through these calculations the sum of severity could be calculated and used as an indicator for the drought hazard. The sum of severity and the exceedance probability are correlated to determine the fit for the severity sum to be used as a drought indicator. Exceedance probability on its own only reflects the probability for a dry period of SPEI <-1.5 and does not reflect severity, whereas the sum of severity combines probability, duration, and intensity. Severity sum was calculated for all quaternary catchments. The sum of severity was indexed on a scale of 1−5 and the drought hazard profile was developed for the three districts, Cacadu, Joe Gqabi, and OR Tambo. See Fig. 8.2.

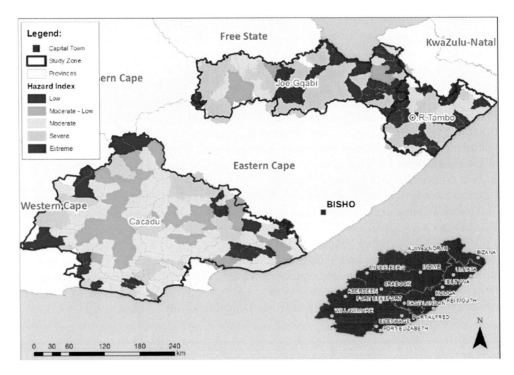

FIGURE 8.2 Drought hazard severity per catchment.

It is interesting to note that the regions with the highest hazard severity are in fact the higher rainfall zones located in the east. High hazard risk zones, however, are also located in arid regions. It is important to note that this map only shows drought hazard based on historical meteorological data. Drought resilience, adaptation, or vulnerability is not reflected in these results.

8.4 VULNERABILITY AND COPING CAPACITY INDICATORS

Although the lack of normal rainfall, is regarded as the main indicator for drought, the *impact* of drought becomes the decisive factor when analyzing drought risks. This section deals with the vulnerabilities, coping mechanisms, or resilience to drought.

Vulnerability, coping capacity and resilience are usually grouped as social, environment and economic, but this research proposes a more detailed classification according to the community capitals framework (Flora et al., 2007). The community capitals provided a more detailed framework and was used to calculate vulnerability and coping capacity, meaning that indicators were grouped under (1) human capital, (2) social capital, (3) cultural capital, (4) financial capital, (5) infrastructure capital, (6) environmental capital, and (7) political capital.

Vulnerability was calculated as a weighted average of (1) human, (2) social, (3) cultural, (4) financial, (5) infrastructure, (6) environmental, and (7) political capitals. Coping capacity was also calculated as a weighted average of (1) human, (2) social, (3) cultural, (4) financial, (5) infrastructure, (6) environmental, and (7) political capitals. The weighting of indicators are discussed in the next section.

8.4.1 Weighting of Indicators

In most cases the different indicators are not weighted owing to lack of data, hence they are considered as independent and equally important variables. That means that the effects of a certain combination of indicator values are not tested. Weighting of indicators is important since composite indicators in most cases should bear a higher weight than individual indicators. Dwyer et al. (2004) reported that weight indicator values are in most cases determined according to subjective perceptions of the importance of some indicators. Davidson (1997) comes to the conclusion that "no amount of clever mathematical manipulation will uncover the correct weights for social vulnerability indicators, because no single correct set of weights exists a priori." Some weighting techniques are derived from participatory methods such as analytical hierarchy processes and budget locations; other methods include statistical models; a combination of statistical models and expert judgements; others from correlation analyses and problem tree analyses. Weighting can be very subjective in the absence of adequate data and proper modeling, but previous studies have found that weights based on experience of the researcher as well as inputs from experts in most cases were better than applying no weights at all (Dwyer et al., 2004; Damm, 2010; Jordaan, 2011). In the context of this research weights were allocated after consultations and inputs from experts.

As a confirmation of Dwyer et al. (2004) and Jordaan's (2011) findings, the researchers' experience and first-hand knowledge of the topic attributed to the assurance of allocating weights. The allocation of weights reflected the relative importance of each indicator, and that was discussed with, and tested with, more than 50 commercial farmers, 260 communal farmers, and several other experts and scientists. Allocation of weights to the same indicators was repeated after 3 months and each new allocation was compared with the previous allocation; this method provided consistency and accuracy. The method of repeated weighting ratified the correctness of the allocation and prevented impulsive decisions (Jordaan, 2011).

As indicated already, the final allocation of weights to the seven capitals were as follows:

- Human capital   0.12
- Social capital   0.04
- Cultural capital   0.10
- Financial capital   0.27
- Infrastructure capital   0.08
- Environmental capital   0.35
- Political capital   0.04

Individual indicators were also weighted according the same methodology. It was interesting to note that the natural scientists interviewed allocated the highest weighting to natural resources, stating that "…you cannot make a success on a bad farm especially during dry periods." Commercial farmers, on the other hand, allocated a higher rating to human capital, stating that "a good manager through perseverance, initiative and hard work can make a success on any farm as long as you farm according the potential of the farm and climatic conditions." Communal farmers as a group, however, rate financial capital as the most important, stating "without money you cannot survive and do anything." The highest rating of 0.35 was eventually allocated to environmental capital followed by financial capital. The mentioned comments were specifically made in relation to drought resilience.

8.4.2 Community Capitals Analysis

The calculation of vulnerability and coping capacity was based on data generated as follows:

- Where available, the use of meta data and actual data from previous reports;
- GIS data from previous studies;
- Expert inputs from scientists, extension officers, and farmers;
- Individual observations and inputs from the research team; and
- Combining census data with individual observations.

The large number of indicators used limited potential errors with individual indicators. Calculating human, social, cultural, and political capital is mostly qualitative.

8.4.2.1 *Human Capital (Weight; 0,12)*[1]

Human capital vulnerability was calculated as a function of (1) education level (0,5), (2) age (0,2), and (3) health status (0,3), while human capital resilience was calculated as a function of (1) perseverance (0,2), (2) farming experience (0,3), (3) exposure to mentorship (0,1), and (4) management skills (0,4).

---

1. Weightings for indices are indicated as numbers in brackets.

TABLE 8.1 Mean Vulnerability Indices per Capital per District

District	Human	Social	Culture	Financial	Environm.	Political
Cacadu	2,03	1,95	1,05	1,83	1,70	2,68
Joe Gqabi	2,48	2,84	1,83	2,16	2,69	3,42
OR Tambo	3,98	3,68	4,42	4,61	3,83	4,67

8.4.2.2   Social Capital (Weight; 0,04)

Social capital vulnerability was calculated as a function of (1) formal networks (0,3), (2) informal support structures (0,4), and (3) safety and security status (0,3), while social capital resilience was calculated as a function of (1) formal farming institutions (0,2), (2) informal farming support structures (0,3), (3) private extension support (0,3), and (4) access to information (0,2).

8.4.2.3   Cultural Capital (Weight; 0,10)

Cultural capital vulnerability was calculated as a function of (1) dependency syndrome (0,6), (2) gender equality and beliefs (0,1), and (3) cultural beliefs and practices (0,3). Social capital resilience was calculated as a function of (1) innovative planning and approaches (0,3), (2) work ethics (0,4), and (3) life experience (0,3).

8.4.2.4   Financial Capital (Weight; 0,27)

Social capital vulnerability was calculated as a function of (1) price sensitivity of products (0,3), (2) market access (0,4), and (3) unemployment rate in region (0,3). Social capital resilience on the other hand was calculated as a function of (1) alternative on-farm income sources (0,3), (2) alternative off-farm income sources (0,1), (3) available of financial safety nets (0,3), and (4) availability of fodder banks (0,3).

8.4.2.5   Infrastructure Capital (Weight; 0,08)

Infrastructure capital vulnerability was calculated as a function of (1) irrigation infrastructure (0,3), (2) communication network access (0,1), and (3) farm planning (0,6). Infrastructure capital resilience was calculated as a function of (1) irrigation (0,3), (2) fencing (0,3), (3) water reticulation (0,3), and (4) access roads (0,1).

8.4.2.6   Environmental Capital (Weight; 0,35)

Environmental or ecological capital vulnerability was calculated as a function of (1) land degradation[2] (0,6), (2) land use (0,2), (3) mountainous index (0,1), and (4) predator threats (0,1), while social environmental resilience was calculated as a function of (1) soil quality (0,3), (2) groundwater supply (0,2), and (3) surface water supply (0,5).

8.4.2.7   Political Capital (Weight; 0,04)

Political capital vulnerability was calculated as a function of (1) government drought relief plans (0,2), (2) government efficiency (0,2), and (3) land ownership policy (0,6). Political capital resilience was calculated as a function of (1) political support for drought affected sector (0,4), and (2) government extension service (0,6).

8.5   RESULTS OF THE VULNERABILITY AND COPING CAPACITY ASSESSMENT

A summary of the results for vulnerability for each of the capitals per district is shown in Table 8.1 and illustrated in Fig. 8.2. According to these results, OR Tambo district was extremely vulnerable with an average vulnerability index of more than 4.[3] Cacadu as a district was the least vulnerable to drought even though it was the most arid district. This is again proof that aridity and drought are two separate concepts and one cannot simply classify high drought vulnerability to arid regions.

---

2. Land degradation is a composite indicator that includes overgrazing and encroachment of unwanted species.
3. One is low vulnerability and five is high vulnerability.

Fig. 8.3 is a radar graph illustrating results for vulnerability to drought according to the CCF7 analysis.

Vulnerability and coping capacity are key elements in drought risk in that these represent man-made factors that one can address in order to reduce drought risk. Drought hazard based on precipitation and evapotranspiration is a given within which farmers have to plan. Drought risk was the highest in OR Tambo district, which might be a surprise to most as this is the district with the highest average annual precipitation. High vulnerability and low coping capacity combined with hazard risk was responsible for the high drought risk in the OR Tambo district. Drought vulnerability was the highest in OR Tambo as illustrated in Fig. 8.4.

In spite of the relatively high precipitation in the east, vulnerability to drought was the highest due mainly to high land degradation, which in turn was partly the result of the communal land use system. Other factors that also contributed to high vulnerability were the land ownership system with open access to land, and social factors such as low levels of education, a dependency syndrome, and cultural beliefs. The value of this assessment, however, is not only in

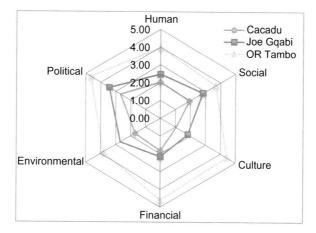

FIGURE 8.3   Vulnerability indices per capital per district.

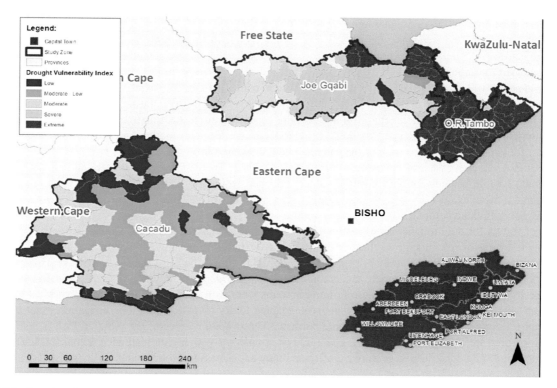

FIGURE 8.4   Drought vulnerability.

TABLE 8.2 Mean Coping Capacity Indices per Capital per District

District	Human	Social	Culture	Finance	Infrastr	Environ.	Political
Cacadu	2,74	3,63	4,22	3,21	3,83	3,00	1,41
Joe Gqabi	2,55	2,59	3,65	2,47	3,31	3,04	1,80
OR Tambo	1,85	1,79	2,79	1,05	1,94	4,30	2,40

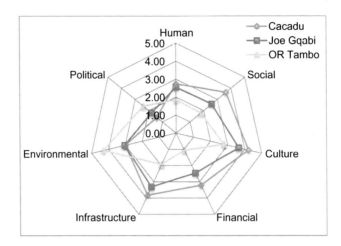

FIGURE 8.5   Coping capacity indices per capital per district.

the spatial illustration of drought risk, but rather in the identification of vulnerability and resilience factors. Coping capacity results for each of the capitals were also calculated for each district. It was clear from the results shown in Table 8.2 and illustrated in Fig. 8.5 that coping capacity was the highest in Cacadu and the lowest in OR Tambo.[4] The relatively high index for the environmental capital was mainly the result of the high quality soil and high rainfall. The fact that the land was degraded and overgrazed was overshadowed by the natural potential of the region.

The results for the capitals for each district are illustrated in the radar graph in Fig. 8.4.

Coping capacity was also relatively low in OR Tambo, due mainly to human, social, financial, and political factors. Soil quality and the availability of surface water due to the relatively high precipitation, on the other hand, increased the values for coping capacity. Catchments with high coping capacity were those with access to irrigation, since farmers then had alternative income sources and they could provide their own feed and fodder during dry spells.

The map for coping capacity for each of the catchments is illustrated in Fig. 8.6.

The net index scores for each of the capitals were also calculated. The negative values were an indication of vulnerable catchments while catchments with positive values indicated resilience. The higher the values the more resilient those catchments were against dry periods and drought. The method of calculating the net index score for each of the capitals has not been done before.

The final results for the drought risk assessment are illustrated in Fig. 8.6. The OR Tambo district was the district with the highest drought risk in spite of the fact that it is the district with the highest annual precipitation. Communal farmers in the district experienced normal dry periods as droughts simply because they did not have capacity to withstand dry periods, and they were extremely vulnerable to any exogenous shock. Cacadu district seems to be more resilient against dry periods and droughts because farming systems were well adapted to arid conditions. Most commercial farmers farm with wool sheep and goats, which are well adapted to dry climatic conditions. Irrigation water also played an important role in making farming systems more resilient. The irrigation areas of the Xhariep-Fish irrigation system and the coastal area are clearly illustrated on Fig. 8.7. Other catchments with higher resilience within Cacadu were areas where farmers applied a conservative grazing capacity and they had access to groundwater for irrigation.

---

4. Low coping capacity is one and high coping capacity is five.

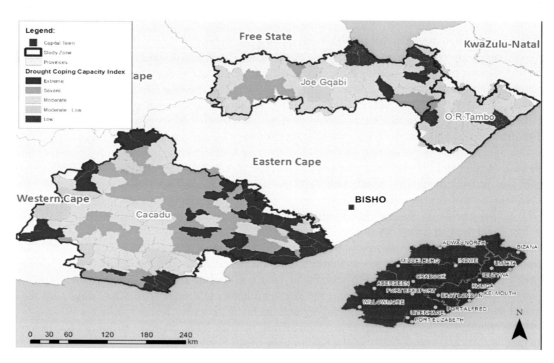

FIGURE 8.6   Drought coping capacity.

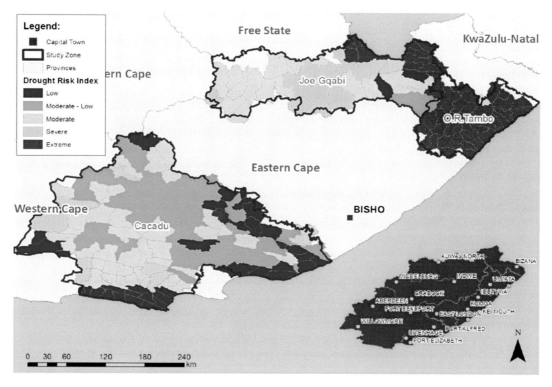

FIGURE 8.7   Drought risk.

8.6   CONCLUSION

The results for drought risk assessment clearly highlighted the importance of vulnerability and coping capacity as essential elements in drought risk. The importance of a drought risk assessment is not so much in the result illustrated on maps, but rather in the identification of indicators that resulted in drought vulnerability or contribute to drought resilience. It is important for extension services and development agencies to identify and understand these indicators and address the gaps through extension programmes and development plans.

Aridity and drought are two different concepts and the reason for misinterpretation of droughts. Not all dry periods are droughts and the impact of dry periods is different between different sectors depending on its vulnerability and resilience against dry periods, which was calculated by means of the seven capitals. The use of the prescribed indicators without the consideration of the sector specific characteristics is foolish. The difference between the communal farming sector and the commercial farmers in South Africa in terms of drought vulnerability and resilience is significant. Communal farmers and the smallholder farming sector are extremely vulnerable to drought because of (1) overgrazing, (2) land degradation, (3) poor infrastructure on their land, (4) a lack of grazing management systems, (5) poor quality animals, (6) lack of reserves, (7) imperfect markets, (8) lack of knowledge, and (9) cultural beliefs and other factors clearly highlighted in the community capitals based risk assessment (Jordaan, 2011; Jordaan et al., 2017). Communal farmers experience normal dry periods as droughts and they report significant drought losses every one in 3 years. A normal dry period could be disastrous for them while, on the other hand, most commercial farmers are able to manage the 1 in 3 and the 1 in 5 year dry periods.

The community capitals drought risk assessment methodology is also important in that it inform policy on how government should support land reform farmers and ensure food security under more intense and frequent droughts as suggested by most climate change models. The communal land use system and the high dependency on government support amongst communal farmers were amongst the key contributors to drought risk. The *"tragedy of the commons"* and government dependency are at the root of drought vulnerability. These two factors resulted in the lack of mitigation and preparedness planning amongst communal and land reform farmers. They continue to apply poor agricultural practices in anticipation that *"government will assist when drought comes."*

The categorization of indicators according the community capitals provides a detailed framework for drought risk assessment, which allows farmers, government, and development organizations to identify and prioritize key indicators. This research also contradicts the notion that climate change is to blame for recent droughts.

REFERENCES

AgriSETA, 2010. Sector Analysis: Agriculture. AgriSETA, Pretoria, South Africa.

Andrew, M.A., 2003. Evaluating Land and Agrarian Reform in South Africa. University of the WesternCape, Programme for Land and Agrarian Studies (PLAAS). School of Government, Bellville, South Africa.

DAFF, 2012. Abstract of Agricultural Statistics. Department of Agriculture, Forestry and Fisheries, Directorate Statistics and Economic Analysis, Pretoria, South Africa.

Damm, M., 2010. Mapping Social-Ecological Vulnerability to Flooding: A Sub-National Approach for Germany. UNU-EHS, Bonn, Germany.

Davidson, R., 1997. An urban earthquake disaster risk index. The John A. Blume Earthquake Engineering Center, Department of Civil Engineering, Stanford University, Report No. 121, Stanford.

Du Pisani, L.F., 1998. Assessing rangeland Drought in South Africa. Agri. Syst. 57 (3), 367−380.

Dwyer, A., Zoppou, C., Nielsen, O., Roberts, S., 2004. Quantifying Social Vulnerability: A Methodology for Identifying Those at Risk to Natural Hazards. Geoscience, Australia, Department of Industry, Tourism and Resources, Canberra, Australia.

ECDC, 2009. Economic Development Planning. Eastern Cape Development Cooperation. www.ecdc.co.za/about-the-eastern-cape/economy/ (accessed 11.02.14).

Flora, C., Flora, J., Fey, S., Emery, M., 2007. Community capitals framework. Biosecurity Bilingual Monograph, Learning Communities: International Journal of Learning in Social Contexts (Australia).

Fouche, H.J., 1992. Simulation of the Production Potential of Veld and the Quatification of Drought in the Central Free State. PhD Thesis, University of the Free State, Bloemfontein, RSA.

Fouche, H.J., de Jager, J.M., Opperman, D.P.J., 1985. A mathematical model for assessing the influence of stocking rate on the incidence of droughts and for estimating the optimal stocking rates. J. Grassland Soc. Southern Africa 2 (3), 4−6.

Jordaan, A.J., 2011. Drought Risk Reduction in the Northern Cape. PhD Thesis. University of the Free State, Bloemfontein, RSA.

Jordaan, A.J., 2012. An analysis of the Mngcunube "hands-on" mentorship program for small-scale stock farmers in the Eastern Cape, South Africa. South Afr. J. Agri. Ext. 40 (1), 48−57.

Jordaan, A.J., Sakulski, D.M., Muyambu, F., Mashimbye, C., 2017. Drought risk assessment. In: Jordaan, A.J. (Ed) Vulnerability, Adaptation to and Coping with Drought: The Case of Commercial and Subsistence Rain Fed Farming in the Eastern Cape; Vol II, WRC Report Nr. TT 716/2/17. ISBN: 978-1-4312-0885-2. Pretoria. pp. 5.2−5.33.

McKee, T.B., Doesken, N.J., Kleist, J., 1993. Drought monitoring with multiple time-scales. In: 9th Conference on Applied Climatology (pp. 233−236). Dallas, Texas, USA.

McKee, T.B., Doeskin N.J., Kleist, J., 1995. Drought monitoring with multiple time scales. In: Proceedings of the 9th Conference on Applied Climatology, January 15-20, 1995. American Meteorological Society (pp. 233−236), Boston, MA, USA.

STATSSA, 2011. Statistics South Africa, 2011. Government Printers, Pretoria.

Wilhelmi, O.V., Wilhite, D.A., 2002. Assessing vulnerability to agricultural drought: a Nebraska case study. Nat. Hazards 25, 37—58.

Wilhite, D.A., 2000a. Drought planning and risk assessment: status and future directions. Ann. Arid Zone 39, 211—230.

Wilhite, D.A., 2000b. In: Wilhite, D.A. (Ed.), Drought Preparedness in the Sub-Sahara Context, vol. I & II. Routledge, London, UK (2000).

Wisner, B., Blaikie, P., Cannon, T., Davis, I., 2004. At Risk: Natural Hazards, People's Vulnerability and Disasters, second ed Routledge, London.

FURTHER READING

Wilhite, D.A., Easterling, W.E., Wood, D.A. (Eds.), 1987. Planning for Drought. Toward a Reduction of Societal Vulnerability. Westview Press, London, UK.

Wilhite, D.A., Svoboda, M.D., Hayes, M.J., 2007. Understanding the complex impacts of drought: a key to enhancing drought mitigation and preparedness. Water Resourc. Manage. 21, 763—774.

Chapter 9

Community-Based Adaptation: Alaska Native Communities Design a Relocation Process to Protect Their Human Rights

Robin Bronen

Alaska Institute for Justice, Anchorage, AK, United States

Chapter Outline

9.1 INTRODUCTION

This chapter focuses on the climate change threats to the lives, livelihoods, homes, health, and basic subsistence of human populations who reside in coastal and riverine communities all over the world. Extreme weather events combined with accelerating slow onset biophysical change are increasing the vulnerability of the places where people live and maintain livelihoods. In addition, the cumulative impact of frequent extreme weather events is challenging the capacity of governance institutions at local, regional, and national levels to prepare for and respond to these events (IPCC, 2012; Bronen, 2011). Fiscally, governments are spending large amounts of money on disaster preparation and response, insurance payouts, and rebuilding damaged or destroyed infrastructure in order to protect people and the infrastructure on which they depend (Nichols and Bruch, 2008).

Slow ongoing environmental change, such as sea level rise and erosion, is also causing land to permanently disappear. Sea level rise is accelerating and is expected to worsen over the next century due to increased rates of ice sheet mass loss from Antarctica and Greenland (Nicholls and Cazenave, 2010; Joughin et al., 2014). Sea level rise contributes to flooding, storm surges, erosion, and salinization of land and water (IPCC, 2012). Locations currently experiencing coastal erosion and inundation will continue to do so in the future due to increasing sea levels (IPCC, 2012).

These climate-induced environmental changes will cause people to be displaced. Sea level rise is causing severe fiscal impacts (Grannis, 2011). A valuable government tax base and significant private investment will no longer exist as coastal land becomes increasingly marginal for human habitation. Economic activities, including livelihoods, will be disrupted and people will lose their homes and connection with heritage and cultural and spiritual ties to the land (Grannis, 2011). Sea walls and storm surge barriers may not be able to provide protection (Bronen, 2011).

Currently, no governance framework exists in the United States, or in any other country, to evaluate climate change impacts and determine when people can no longer be protected in place and need to relocate. Recognizing this institutional gap and the complex challenges of climate-induced population displacement, the Bicameral Task Force on

Resilience. DOI: https://doi.org/10.1016/B978-0-12-811891-7.00009-8

Climate Change recommended in their December 2013 report "Implementing the President's Action Plan: US Department of the Interior":

> that the Administration devote special attention to the problems of communities that decide they have little choice but to relocate in the face of the impacts of climate change. Because the relocation of entire communities due to climate change is such an unprecedented need, there is no institutional framework within the U.S. to relocate communities, and agencies lack technical, organizational, and financial means to do so.
>
> United States Congress Bicameral Task Force on Climate Change (2013)

President Obama's Task Force on Climate Preparedness and Resilience (Task Force) echoed this recommendation in November 2014 and affirmed that the federal government should take a lead role to establish a relocation institutional framework to respond to the complex challenges of climate-related population displacement (White House, 2014).

This lack of a governance framework hampers the ability of local, regional, and national government agencies to respond. If climate-induced environmental change renders the places where people live uninhabitable and causes land to disappear, new governance institutions need to be designed to determine whether people can be protected in place or require relocation (Bronen, 2011; Bronen and Stuart Chapin, 2013).

Relocation is "a process whereby a community's housing, assets, and public infrastructure are rebuilt in another location" (Abhas, 2010). In addition to moving infrastructure, relocation can also include the social processes of rebuilding livelihoods and social networks. Preventive relocations, which occur prior to an extreme weather event which causes mass population displacement, can be a critical disaster risk reduction tool that can save lives and offer long-term protection. International law requires nation state governments to protect vulnerable populations from climate-induced environmental change which threatens the civil, economic, social, and cultural rights fundamental to the inherent dignity of individuals as well as collective society. The duty to protect is inherent in the concept of sovereignty and implies that the nation state government has the primary responsibility for the protection of populations within its jurisdiction (CICISS, 2001, 2.15). In 2008, the European Court on Human Rights found in the *Budayeva* decision that the failure of state authorities to implement preventive land use planning and emergency evacuation policies in the hazardous area of a town in the Caucasus mountains caused the death of eight people (Kaelin, 2011). The Court ordered the State to pay compensation to the family members of the survivors (Kaelin, 2011).

The government responsibility to protect people through the implementation of preventive relocations may require the relocation of people against their will (Ferris, 2012). However, government-mandated relocations have been uniformly disastrous for the people relocated. Development projects, particularly dams, have displaced approximately 280−300 million people in the previous 20 years and 15 million people are displaced annually (Ferris, 2012). Governments have also forcibly relocated people for geopolitical motives (Tester and Kulchyski, 1994). During World War II, the US government forcibly relocated hundreds of Alaska Natives living in far western Alaska, on the Aleutian Islands, to theoretically protect them from the Japanese (Mobley, 2012). These government-mandated relocations weaken social, cultural, and political institutions and networks, disrupt subsistence and economic systems, and impact the cultural identity and traditional kinship ties within a community (Abhas, 2010). In Alaska, the forcible relocation of the Aleuts caused the death of approximately 10% of the relocated population (Mobley, 2012).

Currently, according to a World Bank study, most environment-related relocations mandated by government agencies occur in the aftermath of an extreme weather or geological event or when it is imminently going to occur (Correa, 2011). Government decisions to create no-build zones in the aftermath of an extreme weather event can create a de facto relocation process in order to prevent future vulnerability (Thomas, 2014). However, extreme weather events, which cause mass population displacement, may not be an appropriate indicator to use to evaluate whether people should be relocated. Without the scientific evidence to prove that future extreme weather events will cause the same location to be vulnerable and threaten the lives and livelihoods of residents, ad hoc no-build zones could be considered forcible evictions and violate people's human rights (Thomas, 2014). In addition, in the aftermath of an extreme weather event, most people want to return home and will, unless the land on which they lived no longer exists (Raleigh and Jordan, 2010).

To address both the severe consequences of government-mandated relocations and the lack of a uniform methodology to assess climate change risk in relation to the ability of people to remain where they currently live, this chapter posits the design of an adaptive governance relocation framework and specifically discusses one component of this governance framework—the design of a social-ecological monitoring and assessment tool. An adaptive governance framework means that governance institutions can dynamically respond to climate-induced environmental changes and shift their efforts from protection in place to managed retreat and community relocation (Bronen and Stuart Chapin, 2013; Bronen, 2011).

The governance framework needs to be designed to protect the well-being of relocated populations. In order to avoid or minimize the harmful effects of government-mandated relocations, the incorporation of an institutional mechanism which will engage community residents in a collaborative decision-making process with government representatives to determine whether and when to relocate may be a critical component of the relocation governance framework. Community-based integrated ecological and social assessments can facilitate this coproduction of knowledge for community residents, as well as government agencies (Kofinas, 2009). The assessment process could evaluate future vulnerability and risk, including an assessment of mitigation strategies to protect people in place (Correa, 2011).

These assessments may also foster transparency, encourage civic empowerment, create multilevel and interdisciplinary collaborative governance systems, and ensure that adaptation measures are locally designed and implemented. All of which may increase the likelihood that relocation will be less harmful to those relocated.

9.2 METHODS

This participatory action research focuses on the governance mechanisms designed to respond to extreme weather events and support hazard mitigation. The Alaska Institute for Justice (AIJ) invited 27 Alaska Native communities imminently threatened by flooding and erosion to participate in the project to design and implement a governance framework for community-led relocation. Fifteen agreed to participate and are pursuing a range of adaptation responses, from implementing flood and erosion control to enable protection in place, to facilitating community relocation. AIJ did not inquire into the reasons the remaining communities chose not to participate. All 15 communities are geographically remote, with only planes providing year-round access. Thirteen have no roads to or from their villages. Each village is a federally-recognized indigenous tribe, and subsistence hunting and gathering are central to their culture and survival (GAO, 2009). Village life revolves around these activities, with the resources obtained from the natural environment forming the basis for community cohesion, social identity, livelihoods, and cultural events. The villages have small cash economies, and only limited work opportunities. Food bought in stores is expensive because of the high cost of its transport to isolated areas, making subsistence activities vital to the communities' food security.

The first step to design the monitoring and assessment tool required an extensive literature review, such as hazard mitigation plans and erosion studies, to understand and synthesize the documentation for each community related to environmental change and storm events. We validated this information with each community and then asked each community to identify the most important environmental factors to monitor. We next identified the state and federal government and nonprofit agencies which had the authority and expertise to monitor these environmental changes. We are currently creating collaboration mechanisms between the Alaska Native communities and these governmental and nongovernmental agencies to design and implement the environmental component of the monitoring and assessment tool.

9.2.1 Governance Institutions to Protect in Place

US federal and state government agencies have a variety of tools they can use to facilitate protection in place and managed retreat from vulnerable coastal and riverine areas, but have very limited tools to facilitate a community-wide relocation. Significant limitations prevent governments from responding effectively and dynamically to climate-induced environmental changes. This section outlines the current governance mechanisms to control flooding and erosion in order to protect people where they live. It is important to understand these mechanisms, which have created the expectation that people can be protected in place, in order to determine the appropriate governance tools which could be designed to determine whether and when a preventive relocation needs to occur.

Traditionally, governments and property owners have protected coastal and river development from flooding and erosion with engineered structures, such as sea walls and storm surge barriers (Lewis, 2012). Although these solutions have been problematic because of their expense to build and maintain, the increase of flooding and erosion on neighboring properties, and the encouragement to develop vulnerable areas, they have been the primary response to erosion and flooding (Grannis, 2011; Lewis, 2012). In addition, governments can implement building codes to maximize the capacity of infrastructure to withstand flooding and provide an additional mechanism to protect in place.

Postdisaster relief also emphasizes rebuilding houses, reconstructing infrastructure, and rehabilitating livelihoods in the places where people lived prior to the disaster to attempt to restore these places to a predisaster reality (Bronen, 2011). Following a disaster, the common response of people is to relocate close to where they have been displaced, with approximately 30% of the affected population moving and with over 90% of these people returning at some later stage (Raleigh and Jordan, 2010). As a consequence, in the aftermath of a disaster, decision makers face significant difficulty implementing measures that restrict the ability of people to return to the places where they lived. For example, in Florida, after Category 3 Hurricane Dennis damaged and destroyed coastal property, landowners demanded the right to return to the places where they had lived. In response, local governments constructed 15 ft sea walls along 26 miles of the Florida coast (Grannis, 2011). In Alabama, hurricanes have destroyed dwellings and infrastructure on Dauphin Island 10 times in the past 40 years, most recently in 2005 by Hurricane Katrina. Despite the destruction and vulnerability caused by these repeated extreme weather events, since 1979, Dauphin Island has received $80 million in federal funding—more than $60,000 per resident—plus an additional $72 million in federal flood insurance payouts to rebuild the community (Siders, 2013). Dauphin Island residents have contributed to these costs by paying $9.3 million in insurance premiums (Siders, 2013).

Managed retreat is another government mechanism which can protect people from extreme weather events, erosion, and flooding. Land use restrictions, such as setbacks, and acquisition programs are two of the tools government agencies can use. Setbacks are building restrictions that establish a distance from a boundary line where landowners are prohibited from building structures. In waterfront areas, the boundary line is often the tide line. Erosion-based setbacks require the local jurisdiction to have scientific data of historic erosion rates, and in the context of a rapidly changing environment caused by sea level rise and thawing permafrost, prospective erosion rates (Grannis, 2011).

Acquisition programs, where property owners voluntarily decide to be bought out by the government, offer another mechanism to orchestrate a managed retreat. With this program, state or local government offers to buy, with public funds, property vulnerable to hazards (Lewis, 2012). The acquisition programs normally occur on a household level where individual property owners make the decision to participate in the program. Implementation of these buy-out programs is not uniform, and depends on several factors including socioeconomic background of the homeowner. The US Federal Emergency Management Agency (FEMA) discourages using this program to relocate communities in their entirety, although a few communities have chosen to do so (Bronen and Stuart Chapin, 2013).

Government's ability to proactively use managed retreat to protect people will depend on people understanding they live in vulnerable locations. Acquisition programs are voluntary and government decisions that affect property rights have legal implications. Property owners can use the law to support their desire to remain protected in place and prevent the government from implementing land use and building restrictions (Siders, 2013). Landowners who are prevented from returning home or whose land values are diminished by the US government can challenge new land use regulations by arguing that the government has "taken" their property, in violation of the Fifth Amendment of the Constitution, without just compensation (Siders, 2013). As a consequence, these legal standards can delay the ability of government actors to proactively move people from vulnerable coastal and riverine areas.

While government agencies may be able to use land use restrictions and acquisition programs as a way of managing coastal retreat, these tools will be insufficient to respond to the complexity of mass population displacement and the need to rebuild livelihoods, homes, and community infrastructure in a new location. Climate-induced environmental change presents a completely different scenario than that contemplated by current laws designed to respond to extreme weather events, erosion, and flooding. In the context of a rapidly changing environment, the ongoing effort to protect people in place may lead to recovering in ways that recreate or even increase existing vulnerabilities, and preclude longer term planning and policy changes for enhancing resilience and adaptation (Grannis, 2011; Lewis, 2012).

Slow ongoing environmental change, such as sea level rise and its consequential environmental impacts such as erosion, will cause large areas of land to permanently disappear. The current governance tools to protect people and infrastructure may only offer short-term protection because of increased rates of sea level rise, and the consequent increased intensity of storm surges, flooding, and erosion. Decision makers will need to balance many trade-offs when deciding whether to protect people in situ, such as the degree of threat to people and property, cost to build the erosion or flood control infrastructure, value of the threatened property or infrastructure, long-term costs to maintain, environmental impacts, and the physical conditions of the land (Grannis, 2011). Similarly, setbacks also will be less effective over the long term as sea level rise inundates broad areas of low-lying land (Grannis, 2011). For this reason, it is critical to understand the rates of environmental change in order to plan for a preventive relocation prior to the land's disappearance.

9.2.2 Alaska

In Alaska, the combination of decreased arctic sea ice extent, thawing permafrost and repeated extreme weather events is causing several Alaska Native communities to choose relocation as the best long-term adaptation strategy. The institutional mechanisms, discussed above, to control erosion and flooding have not been able to provide long-term protection to some communities, despite spending millions of dollars.

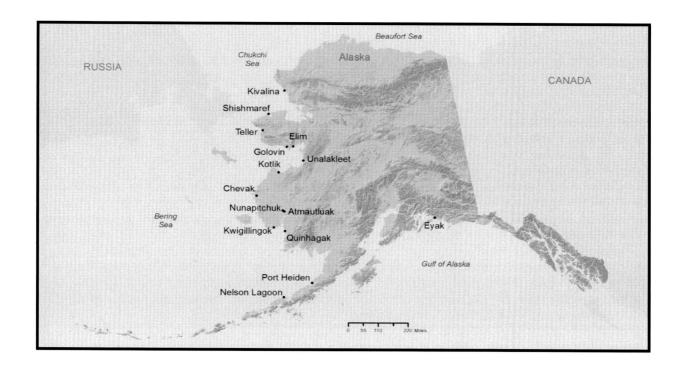

9.2.3 Shishmaref

Shishmaref is an Inupiat village of around 600 people located on Sarichef Island in the Chukchi Sea and 50 km (30 miles) south of the Arctic Circle. Sarichef is a barrier island (USACE, 2006). Between 1973 and 2013, 10 flooding events were recorded in Shishmaref, seven of them declared state emergencies and three federal emergencies.

Erosion and littoral drift are causing Sarichef's footprint to move. Since 1969, Shishmaref has lost an estimated 60 m (200 ft) of land (AECOM, 2016). Between 1973 and 2015, 11 erosion-related events occurred in the village, four of them declared state disasters and two federal disasters. Erosion has undermined buildings and infrastructure, causing several structures to collapse into the sea. Protection measures have been ineffective in anything but the short term. Numerous control and facility relocation projects have been undertaken in an attempt to protect the community in place and provide more time to relocate it. Between 1973 and 2009, the state, federal, and tribal governments invested about $16 million in shoreline protection (GAO, 2009) A rock-wall barrier was constructed along significant portions of Shishmaref's sea front in 2009 and 2010, but around a third of the village, including the airport, homes, and community infrastructure, remain exposed. The 2009 USACE report stated that severe damage was expected by 2019.

In Shishmaref, local and tribal governments documented the environmental changes occurring within the community and used this information to decide that relocation is the only long-term sustainable adaptation strategy (Bronen and Stuart Chapin, 2013). Shishmaref has been planning for relocation since 1976, and residents voted to do so in 2002. Two years later the Shishmaref Erosion and Relocation Coalition chose Tin Creek, around 20 km (12 miles) from the village's current location, as its preferred relocation site. The community reiterated its decision to relocate and Tin Creek as its preferred site in 2007 (BEESC, 2010). The most recent vote to relocate took place in August 2016 to decide on their chosen relocation site since receiving the most recent geotechnical report regarding relocation site vulnerability to future climate hazards, such as thawing permafrost (AECOM, 2016; Alaska Dispatch News, 2016). The governance challenges to relocate Shishmaref have been enormous. Shishmaref still has not relocated.

Equally challenging has been the process to determine when protection in place is no longer possible and community relocation is required. Erosion is the principal threat to the habitability of Alaska Native villages (USACE, 2006, 2009). With diminishing Arctic sea ice extent to protect coastal communities, waves and storm surges are accelerating erosion (GAO, 2003, 2009).

9.2.4 Quinhagak

Quinhagak is surrounded by shallow lakes and wetlands, with elevation ranging from less than 6 ft to approximately 22 ft above sea level. The community is the home to approximately 700 primarily Yup'ik Eskimo residents (POWTEC,

2012; City of Quinhagak Hazard Mitigation Planning Team, 2012). Erosion, river flooding, coastal storm surges, and thawing permafrost threaten residential dwellings, critical community infrastructure, and livelihoods. Flood hazards are high because the developed areas of Quinhagak are adjacent to the floodplain of the Kanektok River. Because of its close proximity to the Bering Sea, Quinhagak is also exposed to storm surges (POWTEC, 2012; City of Quinhagak Hazard Mitigation Planning Team, 2012).

The critical community infrastructure affected by these hazards includes the only functional dock in the community, the health care clinic, and the sewage lagoon (POWTEC, 2012; City of Quinhagak Hazard Mitigation Planning Team, 2012). A peninsula of land protects the City Dock from severe tidal currents. Accelerating rates of erosion are narrowing this land barrier, which will leave the dock unprotected. Should this land barrier completely erode, the City Dock will be no longer be sheltered from open coastal waters (POWTEC, 2012).

In addition, vessels have great difficulty navigating the channels leading to the dock because of silt and large tidal action (POWTEC, 2012). Fuel barges have been stuck, and are often damaged (POWTEC, 2012). The inaccessibility of the dock by the barges sometimes requires that fuel be flown into the community at a huge expense to the community. Coastal Villages Seafood moved the fish processing plant, which employed approximately 100 local people and produced millions of pounds of fish, from Quinhagak because boats were unable to access the plant (POWTEC, 2012).

In addition, coastal erosion also affects the new sewage lagoon, which has no barrier between it and advancing coastal erosion (Pleasant, 2013). The community has a housing crisis, with one-third of the homes in the community unfit for human habitation due to significant subsidence and the infiltration of mold and rot. Erosion threatens residences and fish camps (POWTEC, 2012).

The Alaska Legislature established the Alaska Climate Change Mitigation Program (ACCMP) in 2008 with funding to address the immediate planning needs of communities imminently threatened by climate-induced environmental change, such as erosion (Bronen and Stuart Chapin, 2013). Through this program, Quinhagak received funding to complete a Hazard Impact Assessment (HIA). In addition, the City of Quinhagak completed a Hazard Mitigation Plan in January 2012, a month prior to the completion of the HIA (City of Quinhagak Hazard Mitigation Planning Team, 2012).

The HIA recommends monitoring rates of environmental change, such as sea level rise and erosion, in order to reduce the cost of repairing and replacing infrastructure and to address the critical need for data to better predict rates of climate-induced environmental change (POWTEC, 2012). However, no mechanism was left in place to facilitate this monitoring and no financial or technical assistance resources were provided to assist community residents with this critical process. Difficult decisions need to be made regarding whether protection in place is a viable long-term strategy to protect community residents from the threats caused by climate change. The combination of antiquated and damaged infrastructure needing replacement or repair coupled with the unknown projected erosion and flooding risk elucidates the need to implement and design a community-based social-ecological monitoring and assessment tool. The HIA process demonstrates the complexity of the issues facing communities threatened by climate-induced environmental change.

Based on the ACCMP legislation, Quinhagak is eligible for additional adaptation funding so it is possible that these funds could be used for monitoring and assessment. However, with limited funding and many critical adaptation needs, it is uncertain whether this funding will be used for monitoring and assessment. In addition, the state of Alaska is currently in a fiscal crisis and no funding has been allocated to continue this work.

9.2.5 Relocation Institutional Framework

An adaptive governance relocation framework would incorporate all of the institutional mechanisms to protect people in the places where they live, such as engineered sea walls and land use policies, and also create new mechanisms to implement a relocation process so that national, state, local, and tribal governments can dynamically shift their efforts from protection in place to managed retreat and community relocation (Bronen and Stuart Chapin, 2013; Bronen, 2011).

Human rights principles, based on the fundamental freedoms inherent in human dignity, need to provide the foundation upon which this adaptive governance framework is implemented (Moyn, 2010; Bronen, 2011; Tanner et al., 2015). The right to self-determination is the most important human rights principle to guide climate change adaptation. In the context of climate-induced environmental change that threatens the habitability of the places where people live, self-determination means that people have the right to make decisions regarding adaptation strategies (Bronen, 2011, 2015). The right to self-determination also means that people have the right to make fundamental decisions about when, how, where, and if relocation occurs to protect them from climate-induced environmental threats. In order to operationalize this right, people need the capacity to assess and document the environmental changes and sociological impacts and vulnerabilities caused by climate change (May and Plummer, 2011). In this way, they can determine whether the risks can be mitigated at the original location where they live. Community-based integrated assessments can foster empowerment, promote human rights protections, and encourage transparent decision-making processes—all elements of good governance (Alfredsson, 2013). However, the ability of this community-based process to foster human rights will depend on the capacity of governance institutions to collaborate, be transparent in decision-making, and be inclusive of all sectors of society.

Implementing an adaptive governance relocation framework also requires multilevel and diverse governmental and nongovernmental actors to engage in a collaborative process of knowledge production and problem solving (Kofinas, 2009). Adaptive capacity in social systems refers to the ability of institutions to balance power among interest groups and engage in an iterative learning process that can generate knowledge and be flexible in solving problems. Networks of multiple and diverse organizations are critical to building adaptive capacity (Armitage and Plummer, 2010; Berkes, 2009).

To integrate the concept of collaboration into conventional risk management, those most directly affected by the hazard must actively participate in the gathering of data during the risk assessment process (May and Plummer, 2011). Adaptation responses require community-based information of environmental changes. Local populations document their experiences with the changing climate, particularly extreme weather events, in many different ways. This self-generated knowledge can uncover existing capacity and shortcomings within the community (IPCC, 2012). A community-based integrated social-ecological monitoring and assessment process, designed with local populations and technical experts, can be an essential component of this adaptive governance framework because of the potential to build adaptive capacity and collaboration amongst diverse institutions.

Objective assessment of a hazard, the social perception of that hazard and the ability to anticipate the sociological impacts of ongoing environmental changes are critical to the development of sustainable adaptation strategies (Correa, 2011). Understanding the rate of climate-induced environmental change is essential in order for individuals, communities and governments to adapt. People need to believe that they cannot be protected in situ and that relocation is not implemented for discriminatory reasons or as a pretext to allow the land on which they live to be used for a different purpose, such as commercial development. If people perceive the threat to their lives and livelihoods of staying where they are to be high, they may be more likely to consider relocation (Ferris, 2012). They also need to have resources to recreate lives and livelihoods in a new location. The political decision-making process also becomes more viable if people believe that relocation is the best adaptation option to provide long-term protection (Correa, 2011).

Local level environmental change is also essential in order to identify the relocation indicators which can assess when protection in place is no longer provides a community with long-term sustainable adaptation to climate hazards. Global, regional, and national climate change assessments have generally aggregated information above the level of resolution required for effective community policy (IPCC, 2012). Local landscape change can influence microclimate conditions and outweigh the influence of larger geospatial analyses of long-term climate change predictions (Sallenger et al., 2012). For example, at local and regional levels, sea level rise varies and may exceed averaged global projections, depending on a variety of reasons, including the topography and geologic factors (Sallenger et al., 2012).

In addition, exposure and vulnerability to climate change are dynamic, varying across temporal and spatial scales, and depend on economic, social, geographic, demographic, cultural, institutional, governance, and environmental factors (IPCC, 2012). Consistent monitoring of environmental change and the impact of these changes on individuals, households, and the larger community offers the opportunity to capture the dynamic nature of a community's vulnerability and resilience to the changes. Decision-makers at the local level need to understand how their particular locality will be affected by global and regional projections of climate-induced environmental change and have the governance tools to help them effectively identify and evaluate which are the best policy options to adapt to their local context.

Designing and implementing adaptation strategies also require the involvement of multilevel institutions. Community-based integrated social ecological assessments can facilitate communication between community residents and local, state, regional, and national actors who can bring broader knowledge to local scenarios in order to better understand local dynamics. They can also bring technical expertise that may not exist at the local level to better assess and implement adaptation strategies. Local knowledge can provide not only a long-term historical perspective but an understanding of the connections between people and the environment, while western scientific approaches can generate projections of future change in the context of a broader global scientific data analysis and modeling (Kannen and Forbes, 2011). Government agencies that may not have access to local information to understand local scenarios need this information in order to integrate this information into regional or national models of climate change scenarios (Lewis, 2012). In this way, both residents and government agencies together can anticipate vulnerability in order to implement a dynamic and locally informed institutional response.

Finally, these assessments can also be the tool to determine whether and when relocation needs to occur. Social ecological indicators can be used to assess vulnerability and guide the transition from protection in place to community relocation. Unlike government-mandated relocation programs where the government makes the decisions regarding the timing of the relocation, climate-induced relocations require a dynamic process closely connected with changes in the environment that affect the well-being of community residents.

9.2.6   Components of an Integrated Social-Ecological Assessment and Monitoring Tool

An integrated social-ecological assessment and monitoring tool has the potential to incorporate the components of a HIA, and also include a health and livelihoods component to the assessment. The aim of our work with 15 Alaska Native Tribes is to design and implement a relocation process that affirms the collective rights of communities to self-determination and ensure that their social and cultural rights are protected before, during and after their move. The design of this community-based integrated social-ecological monitoring and assessment tool is based on the monitoring and assessment done by Shishmaref, which began documenting the impact of accelerating rates of erosion, caused by the combination of decreased arctic sea ice, thawing permafrost, and repeated extreme weather events, on the health of community residents and on community infrastructure in the 1980s (GAO, 2009, 2003). This monitoring and assessment, although intermittent due to limited funding, garnered government technical assistance and also provided the basis upon which community residents decided that the relocation of their entire community was the only long-term adaptation strategy which could protect community residents (Bronen and Stuart Chapin, 2013).

A first step to the design of the monitoring and assessment tool has been to focus on environmental monitoring, specifically monitoring rates of erosion and thawing permafrost as well as flooding and storm surges. The State of Alaska, Department of Natural Resources, Division of Geological and Geophysical Surveys (DGGS) Coastal Hazards Program is currently working with Alaska Native communities to customize tools to monitor erosion, map shoreline change, and document storm surge impacts. Permafrost monitoring will also be integrated with erosion monitoring to assist communities to have a more comprehensive understanding of the environmental change occurring because warmer air and ocean waters.

Monitoring health impacts of climate-induced environmental change will also be included in the assessment tool and is a critical component. Alaska residents describe a variety of climate change impacts on health, including morbidity and mortality caused by unpredictable and extreme weather, mental health, changes to lifestyle, and damage to water and sanitation infrastructure (Brubaker et al., 2011). Similar to the monitoring of environmental change, preventing negative health outcomes requires a local scale understanding of the type, timing, and rate of change, as well as the direct and indirect health effects (Brubaker et al., 2011). Integrated health assessments can systematically identify and quantify the many pathways through which climate change can affect health in different social and ecological contexts. The World Health Organization suggests that a natural point of entry for Health Impact Assessments is during the planning process for climate-induced relocations because the assessment can put the key health issues in front of the policymakers who directly influence the implementation of plans (Winkler et al., 2013).

Finally, the integrated assessment will incorporate a component that focuses on the environmental impacts on livelihoods, which would include the availability of subsistence foods. The integration of climate change impacts on infrastructure, health, and livelihoods has the potential to facilitate community-based adaptation, which dynamically addresses ongoing environmental change.

9.3   CONCLUSION

The combination of extreme weather events and slow ongoing environmental change will challenge the capacity of people and the governance institutions charged with protecting them. Land will permanently disappear because of sea level rise and the consequential impacts of erosion. Preventive relocations provide an institutional mechanism to proactively protect people before the land on which they live and maintain livelihoods is no longer habitable or ceases to exist. However, no governance framework in the United States currently addresses the essential issues of deciding when a preventive relocation should occur and who should make the decision that relocation is warranted. Government-mandated relocations have impoverished relocated populations and caused the rupturing of kinship ties and social networks. New governance institutions need to be designed and implemented so that the adverse impacts of relocation are minimized or avoided. Being able to monitor gradual and continuous environmental processes and also capture unexpected ecological feedback loops that may drastically impact the ability of communities to remain protected in place will enhance the resilience of relocated populations.

Community-based integrated socio-ecological assessments, which create multilevel multidisciplinary knowledge production with local communities leading the data gathering effort, can be a critical component of this new governance framework. The work being done by Alaska Native communities to design and implement these community-based assessments can provide a model for other coastal communities faced with increasing risk caused by climate-induced environmental change.

REFERENCES

Abhas, K.J., 2010. Safer Homes, Stronger Communities a Handbook for Reconstructing after Natural Disasters Global Facility for Disaster Reduction and Recovery. World Bank, Washington, DC.

Alfredsson, G., 2013. Good governance in the arctic. In: Loucheva, N. (Ed.), Polar Textbook II. Nordic Council of Ministries, Denmark, pp. 187−198.

Armitage, D., Plummer, R., 2010. Adapting and transforming: governance for navigating change. In: Armitage, D., Plummer, R. (Eds.), Adaptive Capacity and Environmental Governance. Springer, Germany.

Berkes, F., 2009. Evolution of co-management: role of knowledge generation, bridging organizations and social learning. J. Environ. Manage. 90, 1692−1702.

Bronen, R., 2011. Climate-induced community relocations: creating an adaptive governance framework based in human rights doctrine. N Y U Rev. Law Soc. Chang. 35 (2), 101−148.

Bronen, R., 2015. Climate-induced community relocations: Using integrated social-ecological assessments to foster adaptation and resilience. Ecol. Soc. 20 (3), 36.

Bronen, R., Stuart Chapin, F., III, 2013. Adaptive governance and institutional strategies for climate-induced community relocations in Alaska. In: Proceedings of the National Academy of Sciences, Washington, DC.

Brubaker, M.Y., Bell, J.N., Berner, J.E., Warren, J.A., 2011. Climate change health assessment: a novel approach for Alaska Native communities. Int. J. Circumpolar Health 70 (3), 266−273.

CICISS, 2001. (Canadian International Commission on Intervention and State Sovereignty) Responsibility to Protect. International Development Research Centre, Canada.

City of Quinhagak Hazard Mitigation Planning Team, 2012. City of Quinhagak Hazard Mitigation Plan. City of Quinhagak, Alaska.

Correa, E., 2011. Resettlement as a disaster risk reduction measure: case studies. In: Correa, E. (Ed.), Preventive Resettlement of Populations at Risk of Disaster: Experiences from Latin America. World Bank, Washington, DC.

Ferris, E., 2012. Protection and Planned Relocations in the Context of Climate Change. UN High Commission of Refugees, Division of International Protection, Geneva, Switzerland.

GAO, 2003. Alaska Native Villages: Most Are Affected by Flooding and Erosion, but Few Qualify for Federal Assistance. Government Accountability Office, Washington, DC.

GAO, 2009. Alaska Native Villages: Limited Progress Has Been Made on Relocating Villages Threatened by Flooding and Erosion. Government Accountability Office, Washington, DC.

Grannis, J., 2011. Adaptation Tool Kit: Sea-Level Rise and Coastal Land Use. Georgetown Climate Center, Washington, DC.

IPCC, 2012. Summary for policymakers. In: Field, C.B., Barros, V., Stocker, T.F., Qin, D., Dokken, D.J., Ebi, K.L., et al., Managing the Risks of Extreme Events and Disasters to Advance Climate Change Adaptation: A Special Report of Working Groups I and II of the Intergovernmental Panel on Climate Change. Cambridge University Press, Cambridge, UK, and New York, NY, USA, pp. 3−21.

Joughin, I., Smith, B.E., Medley, B., 2014. Marine ice sheet collapse potentially under way for the Thwaites Glacier Basin, West Antarctica. Science 344, 735−738.

Kaelin, W., 2011. A Human Rights-Based Approach to Building Resilience to Natural Disasters. Brookings, Washington, DC.

Kannen, A., Forbes, D.L., 2011. Contributing Authors: R. Cormier, J. Salamon Integrated assessment and response to Arctic coastal change. In: D.L. Forbes (ed.), State of the Arctic Coast 2010 − Scientific Review and Outlook. International Arctic Science Committee, Land-Ocean Interactions in the Coastal Zone, Arctic Monitoring and Assessment Programme, International Permafrost Association. Helmholtz-Zentrum, Geesthacht, Germany.

Kofinas, G.P., 2009. Adaptive co-management in social-ecological governance. In: Chapin III, F.S., Kofinas, G.P., Folke, C. (Eds.), Principles of Ecosystem Stewardship: Resilience-Based Natural Resource Management in a Changing World. Springer, New York, NY.

Lewis, D.A., 2012. The relocation of development from coastal hazards through publicly funded acquisition programs: examples and lessons from the Gulf Coast. Sea Grant Law Policy J. 5 (1), 98−139.

May, B., Plummer, R., 2011. Accommodating the challenges of climate change adaptation and governance in conventional risk management: adaptive collaborative risk management (ACRM). Ecol. Soc. 16 (1), 47.

Mobley, C., 2012. World War II Aleut Relocation Camps in Southeast Alaska. National Park Service, Anchorage, Alaska.

Moyn, S., 2010. The Last Utopia, Human Rights in History. President and Fellows of Harvard College, Cambridge, MA.

Nichols, S.S., Bruch, C., 2008. New frameworks for managing dynamic coasts: legal and policy tools for adapting U.S. coastal zone management to climate change. Sea Grant Law Policy J. 1 (1), 19−41.

Nicholls, R.J., Cazenave, A., 2010. Sea-level rise and its impact on coastal zones. Science 328, 1517.

Pleasant, J., 2013. Increased coastal erosion due to storm activity. Alaska Native Tribal Health Consortium. Local Environment Observer October 15, 2013 [online] http://www.anthctoday.org/.

POWTEC, L.L.C., 2012. with Tetra Tech Qinhagak Hazard Impact Assessment. POWTEC, Bremerton, Washington, DC.

Raleigh, C., Jordan, L., 2010. Climate change and migration: emerging patterns in the developing world. In: Mearns, R., Norton, A. (Eds.), Social Dimensions of Climate Change Equity and Vulnerability in a Warming World. The International Bank for Reconstruction and Development World Bank, Washington, DC, pp. 103—131.

Sallenger Jr, A.H., Doran, K.S., Howd, P.A., 2012. Hotspot of accelerated sea-level rise on the Atlantic coast of North America. Nat. Clim. Chang. 2, 884—888.

Siders, A., 2013. Managed Coastal Retreat: A Legal Handbook on Shifting Development Away from Vulnerable Areas. Columbia Center for Climate Change Law, Columbia Law School, New York, NY.

Tanner, T., Lewis, D., Wrathall, D., Bronen, R., Huq, S., Lawless, C., et al., 2015. Livelihood resilience: Preparing for sustainable transformations in the face of climate change. Nat. Clim. Chang. 5, 23—26.

Tester, F.J., Kulchyski, P., 1994. Tammarnit (Mistakes): Inuit Relocation in the Eastern Arctic, 1939—63. UBC Press, Vancouver.

Thomas, A., 2014. Philippines: Typhoon Survivors Face Obstacles to Recovery. Refugees International, Washington, DC.

USACE, 2006. Alaska Village Erosion Technical Assistance Program: An Examination of Erosion Issues in the Communities of Bethel, Dillingham, Kaktovik, Kivalina, Newtok, Shishmaref, and Unalakleet. U.S. Army Corps of Engineers, Alaska, Anchorage.

USACE, 2009. Study Findings and Technical Report: Alaska Baseline Erosion Assessment. US Army Corps of Engineers, Alaska District.

United States Congress Bicameral Task Force on Climate Change, 2013. Implementing the President's Action Plan: U.S. Department of the Interior. United States Congress, Washington, DC.

White House, 2014. President's State, Local and Tribal Leader's Task Force on Climate Preparedness and Resilience, Recommendations to the President. White House, Washington, DC.

Winkler, M.S., Krieger, G.R., Divall, M.J., Cissé, G., Wielga, M., Singer, B.H., et al., 2013. Untapped potential of health impact assessment. Bull. World Health Org. 91 (4), 298—305.

FURTHER READING

Alaska Division of Community and Regional Affairs, 2013. Understanding and Evaluating Erosion Problems. Alaska Department of Commerce, Community and Economic Development, Anchorage, Alaska.

Burkett, V.R., Davidson, M.A. (Eds.), 2012. Coastal Impacts, Adaptation and Vulnerability: A Technical Input to the 2012 National Climate Assessment. Cooperative Report to the 2013 National Climate Assessment, Washington, DC.

Smith, A.O., 2009. Introduction. In: Smith, A.O. (Ed.), Development & Dispossession: The Crisis of Forced Displacement and Resettlement. School for Advanced Research Press, Santa Fe, NM.

Chapter 10

California: It's Complicated: Drought, Drinking Water, and Drylands

Gillisann Harootunian[1,2]

[1]California State University, Fresno, CA, United States, [2]University of Maryland, University College, Adelphi, MD, United States

Chapter Outline

10.1 INTRODUCTION: THE DROUGHT, DRYLANDS, AND ECOSYSTEM SERVICES

Geoscientists used paleoclimate reconstructions of drought and precipitation for central and southern California to define the recent Drought as the most severe in 1200 years (Griffin and Anchukaitis, 2014). A study by Adams et al. (2015) expanded on the conclusions of Griffin and Anchukaitis. Adams et al. reconstructed 2000-year-long streamflow records for the southwestern Sierra Nevada of California from moisture-sensitive tree ring chronologies, and then simulated annual level fluctuations of the major regional lake—Tulare Lake—to show that 2015 was the driest water year in the last 2015 years and that the 2012−15 drought was the driest 4-year period in the record. Multidrive cores into the Tulare Lake Bed also indicate rising sea level temperatures will lead to a decrease in precipitation for the central and southern Sierra Nevada (Blunt and Negrini, 2015).

The story of how and why agriculture came to dominate the Central Valley of California provides much needed context. In *Cadillac Desert: The American West and Its Disappearing Water*, Mark Reisner documented how nation-building—at time zealous—drove the settlement of the west coast of the US. Agriculture was the overwhelming economic base of early California. Its profits, combined with the drive for nation-building, led to the corruption of national politics in many ways. One example is the Homestead Act, which was designed to promote family farms through granting 160 acres of land to settlers, but was undermined to enable corporate agriculture. In general, water development was more aggressive than sustainable (Reisner, 1986). Mark Arax wrote an in-depth examination of the crown of corporate agriculture in the San Joaquin Valley in *The King of California: J.G. Boswell and the Making of a Secret American Empire*. The title *King of California* points toward conflict between the powerful influence of corporate agriculture and the democratic intent of legislation such as the Homestead Act. California's economy today is diversified. It is the most populous state in the United States by a large margin, with over 38 million residents or 11.9% of the US population (U.S. Census, American Community Survey, 2015). Yet, the legacy of the state's early years persists. Continual learning and research is needed for course corrections that manage water resources to provide sustainable services for all.

California has been the national leader in agriculture for over 50 years. US Department of Agriculture statistics for 2014−15 show California accounts for close to 13% of national farm commodity value (U.S. Department of Agriculture, 2016). The Central Valley is immense, approximately 47,000 km^2 (18,000 square miles). The Central Valley has about 75% of the irrigated land in California, which is 17% of irrigated land in the United States. The

Resilience. DOI: https://doi.org/10.1016/B978-0-12-811891-7.00010-4

Central Valley has approximately 35,000 irrigated agricultural operations on over 2.8 million hectares (7 million acres) (California State Water Resources Control Board, 2016–2017). It produces 25% of US food, including 40% of its fruits, nuts, and table foods. The 250-plus crops grown in the Central Valley are valued at $17 billion annually (USGS, California Water Science Center). For 2016 alone, the estimated total economic impacts of the Drought on California's statewide agriculture was $603 million, with $597 million of that being in the Central Valley (Medellín-Azuara, August 15, 2016). The Central Valley is so large that it has two major valleys within it: the Sacramento Valley is the northern half, and San Joaquin Valley is the southern—and agriculturally intensive—half (Fig. 10.1).

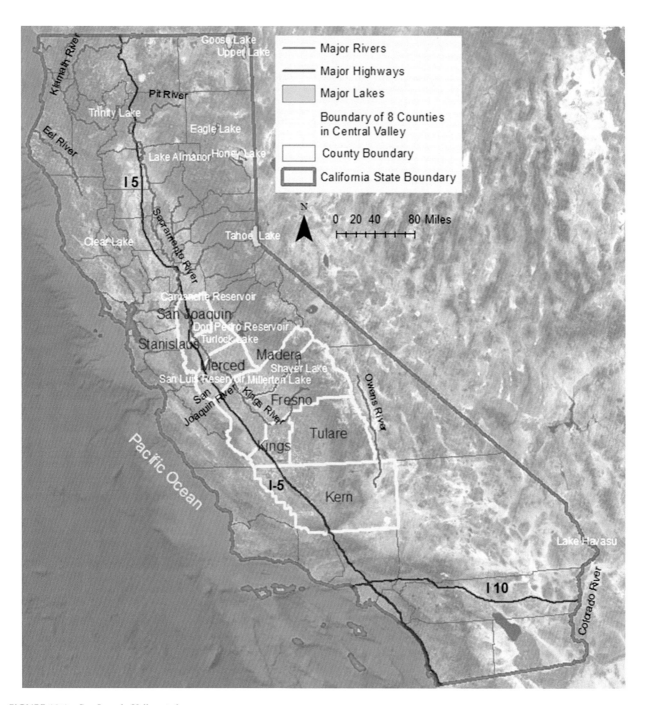

FIGURE 10.1 San Joaquin Valley study area.
Yellow border indicates the eight counties that are the heart of agriculture in California. *Dr. Fayzul Pasha and Dr. Dilruba Yeasmin (California State University, Fresno).*

The consequences of establishing agribusiness of such magnitude in the arid Central Valley have been significant: it has altered and had an impact on the rivers of California more than any other industry (Mount, 1995). Agribusiness demands up to 85% of the state's developed water sources (California State Lands Commission, 1994). The result is that the Central Valley—the heart of agriculture in the state—is in a "largely permanent structural drought" (Lund, December 28, 2016).

Not only the Central Valley but much of California are drylands. This has placed the state in a largely permanent position of having to invest billions of dollars in water infrastructure. Drylands are "deluge versus drought" climates because they depend on "a remarkably limited number of storms over a short period of time" (in California, the winter season) (Mount, 1995, p. 146). Mountain ranges enclose the Central Valley. These mountain ranges create both the rain shadows making it drylands and the rivers creating floods during the winter rainstorms or during the rapid snowmelt in late winter and early spring. The Sierra Nevada range towers up to 4,420 m (14,500 ft) and creates California's largest rivers. The Central Valley has 6,216 km (3,863 miles) of the 12,553 km (7,800 miles) of rivers in the state, or about 49% total (California State Lands Commission, 1994). Sierra Nevadan rivers have the capacity for savage flooding. To illustrate, California Governor Jerry Brown had declared a drought State of Emergency on January 17, 2014 (California Governor's Drought Task Force, 2015). When a series of strong rainstorms finally arrived in January 2017, Governor Brown declared a state of emergency from flooding (California, Office of the California Office of the Governor, 2017). The drylands "deluge versus drought" climate creates the need for a water supply that can control floods in the wet season (winter and spring) and can deliver a reliable supply in the dry season (summer).

The Department of Water Resources choose to focus on the Central Valley as a model when it drafted a statewide *System Reoperation Study* (February 2014):

> this region has the highest integration of water supply and flood management facilities. Additionally, the greatest potential for ecosystem restoration through infrastructure reoperation is found in the Central Valley because the existing infrastructure has had a profound effect on aquatic ecosystems. (p. 1).

The US Army Corps of Engineers' *Central Valley Integrated Floor Management System: Watershed Plan* (November 2015) clarifies that "profound effect on aquatic ecosystems":

> Today less than 4 percent of the historical riparian forests that lined the valley streams remain, with a significant portion of this forest growing on, or close to, levees of the [State Plan of Flood Control] …. Major dams on the system regulate flows and hydrologic processes that also reduce natural geomorphic processes and limit the amount of materials in the channel for gravel and silt deposition. Consequently, valuable habitats continue to decline with no opportunity for the local or regional system to recover. (pp. 1−21).

Finally, water rights are a key feature of drylands. Settlements in the ancient Negev desert appear to have developed a system of water rights to provide enough security to farmers to risk investing in long-term agriculture (Evenari et al., 1961). In California, complex water rights determine who gets water—and how much and when. Water rights determine allocations not only among the three main types of users (agricultural, municipal and industrial, environmental) but also among the same type of users (e.g., different kinds of agricultural operations). Water rights have long been a source of conflict in California, summed up in the popular epithet for the West: "Whiskey is for drinking, and water is for fighting over."

When drylands become even drier, during a drought, conflicts sharpen. Security for one service (agricultural irrigation) can impinge on another (drinking water). In the San Joaquin Valley, the Drought resulted in less surface water, which in turn resulted in agribusinesses with expensive and deep wells pumping more groundwater. As water tables dropped, the shallower and cheaper wells of a number of smaller communities, especially economically disadvantaged ones, and of their residents were affected. These wells draw from the upper layers of an aquifer, and the wells increasingly ran dry or became contaminated by an increasing concentration of pollutants, such as nitrates from fertilizers or chemicals from pesticides such as simazine, 1,2-dibromo-3-chloropropane ("DBCP"), and Trichloropropane ("1,2,3-TCP") (Burow et al., 2008; Keller and Wegley, 2015; September 2015). Moreover, the depletion of groundwater impacts has postdrought impacts—some aquifers in the San Joaquin Valley have become so depleted that recovery will take decades, and a few aquifers may never recover. The ecosystem's sustainability and ability to deliver services like drinking water and agricultural irrigation water is being compromised for the long term.

10.2   SCIENTIFIC EVIDENCE: CLIMATE CHANGE AND THE US WEST

A range of scientific research has suggested a link between anthropogenic warming and an increase in the occurrence, strength, and length of droughts in California and the US West (U.S. Global Change Research Program, 2017; IPCC,

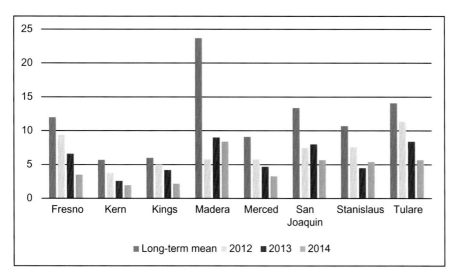

FIGURE 10.2 Historical rainfall in the eight counties (San Joaquin Valley).
The 2012–14 3-year period was "by far the hottest and driest" in the state's full instrumental record from 1895 (Mann and Gleick, March 31, 2015, p. 3858). *Dr. Fayzul Pasha and Dr. Dilruba Yeasmin (California State University, Fresno).*

2014; Ingram and Malamud-Roam, 2013; Yoon et al., 2015; Griffin and Anchukaitis, 2014; Diffenbaugh et al., 2015; Cook et al., 2014, 2015). Extreme weather events such as floods and the current drought surpass the natural climate variability of the region (Ingram and Malamud-Roam, 2013; Yoon et al., 2015). The need for adaptations is pressing. For "Droughts, Floods, and Hydrology" in the West, the US Global Change Research Program presents this key finding "… [A]ssuming no change to current water-resource management, chronic, long duration hydrological drought is increasingly possible by the end of this century. (*Very high confidence*)" (August 2017, p. 281) (Fig. 10.2).

Snow measurements during the Drought offer a glimpse of a future without adaptations. The State of California uses the Sierra Nevada snow pack as a major reservoir. During the Drought, California's Department of Water Resources found no snow on the Phillips Station measuring plot at 2,072 m (6,798 ft) of altitude in the Sierra Nevada (April 1, 2015). The historical average depth of snow on that date for Phillips Station had been 1.69 m (5.5 ft). The Department of Water Resources (2015) made note of the warming trend that has made California's winter of 2014–15 the warmest in its recorded history. The Department also noted that the impact of "cumulative conditions—multiple dry years, low snowpack, warm temperatures—" have created the conditions for a new term: Snow Drought (September 2016).

The risk of increasing severity and length of droughts in California is predicted to escalate and is increasingly linked to anthropogenic warming. Annual rainfall shortages are over twice as likely to lead to drought if the year is relatively warm, and anthropogenic warming has increased the likelihood of the dry warm years that create drought. Climate model simulations demonstrate that California's warming increases when both human and natural forcings are included but does not increase when only natural forcings are included. The conclusion is that human forcing has caused the observed increase in probability of dry warm years. Continued global warming presents the risk of a future regime where almost every single annual rainfall deficit will coincide with increased temperatures, i.e., a scenario with nearly 100% risk for dry warm years (Diffenbaugh et al., 2015).

In general, the risk of megadroughts is high for the US Southwest in the latter half of the 21st century. Increased greenhouse gas concentrations are driving cross-model drying trends, and the resulting increases in evapotranspiration will likely counterbalance any increases in rainfall (Cook et al., 2014). When scientific uncertainties about anthropogenic influence on future climates are analyzed within drought variability and millennial-long historical and paleoclimate records, findings demonstrate that an increase in evapotranspiration is one of the dominant drivers of global drought trends. Under a business as usual emissions scenario, the risk of multidecadal droughts between 2050 and 2099 in the US Southwest is more than 80%. Both high and moderate future emissions scenarios demonstrated megadrought potentials (Cook et al., 2014, 2015).

10.3 TAKING EVERY DROP: THE MINING OF WATER SOURCES AND THE DRINKING WATER CRISIS

California has a "water-rich" north and a "water-poor" south. Seventy-five percent of California's runoff is located in the northern part of the state while 80% of the demand is located in the southern part. Transferring water from the northern to the southern part of the state led to "the development of the largest water engineering program in the world" (Mount, 1995, p. 313):

> to capture, control, and redistribute more than 60 percent of the water that runs off of the surface of California, the state's engineers, our own highly evolved breed of beavers, have built more than 1,400 dams and thousands of miles of levees, canals, and aqueducts that channelize surface water through the state, shepherding it from the water-rich to the water-poor, protecting us from flooding, and supplying us with electrical power. The domestication of California's rivers, which has fueled this country's largest state economy, has left few rivers in their natural state. (Mount, 1995, p. 6).

Even those numbers do not capture the full scale. The Central Valley Project manages just over 11.1 million megaliters of water (9 million acre feet), and of that delivers about 8.6 million megaliters annually (7 million acre feet). Approximately 6.167 million of those megaliters (5 million acre feet) go to agriculture to irrigate about 1.2 million hectares (3 million acres) (U.S. Department of Interior, 2016). The State Water Project, the largest state-built and operated water and power system in the United States, has total reservoir storage over 7.1 million megaliters (5.8 million acre feet) and an average annual water delivery over 2.9 million megaliters (2.4 million acre feet) (1990–2009). Its record annual delivery was over 4.5 million megaliters (3.7 million acre feet) (2006) (California, Department of Water Resources, 2016).

In California, the goal of more than one surface water project is to take every drop. Well known examples include these:

- US Bureau of Reclamation's Friant Dam that dried up the San Joaquin River for 80 km (50 miles);
- US Army Corps of Engineers Pine Flat Dam that dried up Tulare Lake (Mount, 1995, p. 129);
- Los Angeles aqueduct that dried up the Owens Lake (until massive pollution from dust forced L.A. to leave enough water to create mud flats); and
- O'Shaughnessy Dam that flooded the Hetch Hetchy valley in Yosemite National Park to capture Tuolmne River water for San Francisco (and a few smaller municipalities).

Legal battles drag on about every drop of water in California. No less than decades of litigation are required to attempt to redress an action. For example, a US Congressional Act (2006) was needed to settle an 18-year lawsuit filed by the National Resources Defense Council to restore the 80 km stretch of the San Joaquin River that was dried up by the Friant Dam. Known as the San Joaquin River Restoration Settlement Act, the Congressional Act ended over 50 years of litigation over the Friant Dam. Constant attempts are being made to overturn that Act. US Congressional District CA-21 is comprised of parts of the four counties in the San Joaquin Valley that are an "agricultural powerhouse" (Valadao, David, Congressman, 2017a). Its Congressman introduced H.R. 2898 (2015) to repeal the San Joaquin River Restoration Program, an effort maintained and stepped up by H.R. 23 (2017b).

Even taking every drop from those surface water projects, however, does not supply enough water to meet demand. A second ecosystem—groundwater—provides about 40% of California's water supply in an average year, 45% in dry years, and up to 60% in multiyear drought years (California Department of Water Resources, April 30, 2014).

Scientists estimate that by 2015 the cumulated groundwater overdraft under the Drought was approximately 18.5 million megaliters (15 million acre feet) (Howitt et al., 2015, p. 13). Estimates for the Drought in 2016 included likely surface water supply losses of well over 3 million megaliters (2.6 million acre feet), with increased groundwater pumping replacing about 2.34 million megaliters (1.9 million acre feet). The vast majority—about 1.8 million megaliters (1.46 million acre feet)—of that pumping would occur in the San Joaquin Valley's Tulare Lake Basin (Medellin-Azuara, August 15, 2016) (Fig. 10.3).

Pumping of the groundwater at such rates is mining, i.e., extracting the resource at larger than natural recharge rates. Groundwater mining has made the San Joaquin Valley the most severe example of land subsidence in the United States. As mining depletes groundwater, rocks and soil compact and fall in on themselves, providing less solid support to the ground and to infrastructure above. The subsoil can compact to the extent that the storage capacity of an aquifer is permanently damaged (USGS, Water Science School). New technologies allowed researchers from NASA's Jet Propulsion Lab to map subsidence to the resolution of centimeters during the Drought. Two particularly acute subsidence bowls are evident, centered in the town of Corcoran (lower left) and south of El Nido (upper center) (indicated by red) (NASA Jet Propulsion Laboratory, 2015). The California Department of Water completed its own land survey along the State Water Project's California Aqueduct to discover about 130 km (70 miles) of the aqueduct's path had subsided 0.38 m (1.25 ft) in 2 years (Farr, Jones, and Liu, 2015) (Fig. 10.4).

FIGURE 10.3 Dr. Poland who investigated land subsidence in the San Joaquin Valley poses for a photograph showing the subsidence from 1925 to 1977 (USGS, December 9, 2015). *Photo Credit: http://water.usgs.gov/ogw/pubs/fs00165/ - Land Subsidence in the United States, USGS — USGS Fact Sheet-165—00.*

As water tables plummeted in the San Joaquin Valley, the wells of smaller communities, especially economically disadvantaged ones, and of their residents were affected. These shallow and cheaper wells draw from the upper layers of an aquifer. The wells dried up or became contaminated as pollutants became more concentrated. Decades of agricultural fertilizers and animal waste in the region has led to nitrate infiltration of the aquifer that will persist, worsen, and spread in future decades (Lund et al., 2012). Decades of pesticides has also polluted the aquifers with chemicals such as simazine, 1,2-dibromo-3-chloropropane ("DBCP"), and Trichloropropane ("1,2,3-TCP") (Burow et al., 2008; Keller and Wegley, 2015; September 2015). Seven economically disadvantaged communities that present one of the worst drinking water crisis are clustered in the San Joaquin Valley, specifically in heavily agricultural Tulare County. These seven communities are further clustered within one of Tulare County's Irrigation Districts—the Alta Irrigation District. The Alta Irrigation District has aptly described the drinking water crisis: it has become a "never-ending problem" (Keller and Wegley, 2015, pp. 1—1).

The seven communities formally fall under the state designation of Disadvantaged Communities. That designation is applied to communities which are economically disadvantaged, i.e., their average median household income is at least less than 80% of the statewide median household income. Moreover, some of these Disadvantaged Communities are unincorporated: they are not connected, or historically were never able to connect, to municipal water and sanitation services (Flegel et al., 2013). Neither the residents nor the Disadvantaged Communities have the means to extend their wells or to install new, deeper wells. The lack of drinking water in these Disadvantaged Communities eventually prompted a tour by a United Nations Special Rapporteur for the Human Right to Safe Drinking Water and Sanitation (United Nations General Assembly, 2011). In turn, this prompted legislative action: California became the first state in the United States to enact a Human Right to Water Bill (Assembly Bill 685) in 2012. By 2017, a range of adaptations had been proposed (Fig. 10.5).

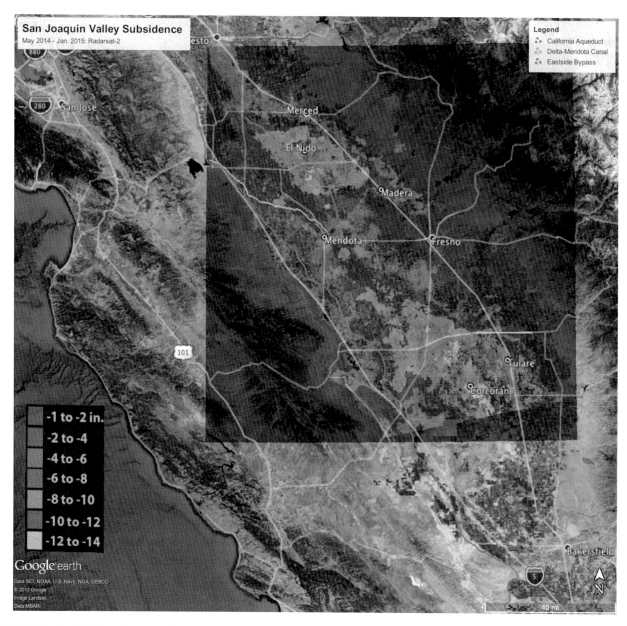

FIGURE 10.4   NASA's Jet Propulsion Laboratory map showing total subsidence in California's San Joaquin Valley for the period May 3, 2014 to January 22, 2015. *NASA's Jet Propulsion Laboratory, based on Canada's Radarsat-2 satellite (Canadian Space Agency/NASA/JPL-Caltech).*

10.4   LEGISLATIVE RESPONSE: SURFACE WATER AND GROUNDWATER

California law, though highly detailed, must meet state-wide objectives. The California Environmental Quality Act (popularly known as "CEQA") must be broad enough to apply to all projects statewide, from constructing a dam in the San Joaquin Valley to expanding a shopping mall in Los Angeles. The CEQA standard is a blanket one, and its aim is to avoid negative impacts. If negative impacts are significant and unavoidable, CEQA then aims to mitigate the impacts, if feasible. In sum, CEQA aims to prevent further environmental damage. CEQA was groundbreaking decades ago when it passed in 1970. CEQA—and its national counterpart ("NEPA")—were the first laws to retard the ongoing rampant destruction of the environment.

Since CEQA aims to halt damage to the environment, legislators and policy-makers have begun to deploy advisory actions and fiscal incentives as tools to encourage restoring damaged ecosystems. As examples, California Proposition

50 (2002), Proposition 84 (2006), and Proposition 1 (2014) tie eligibility for state funding for water infrastructure to two coequal goals: ecosystem restoration and participation of Disadvantaged Communities.

One result illustrates how a law such as CEQA that was an important adaptation in its time can later be deployed as a maladaptation. Today, CEQA can be used as a "loop hole" to avoid having to restore ecosystems. For example, some major dams in the Central Valley have been built and are owned by the Irrigation Districts, whose clients are largely agribusiness. When the Irrigation Districts need to repair an old or complete a new water project, they can use funding other than the funds provided by the newer Proposition 50, Proposition 84, or Proposition 1 that tie eligibility to ecosystem restoration and participation of Disadvantaged Communities. When they use other funding, the Irrigation Districts only have to meet the older CEQA requirement of avoiding or mitigating negative impacts; they do not meet the newer Proposition requirements of providing benefits to the environment (Fig. 10.6).

Groundwater pumping was never regulated, and its mining continued until the Drought raised the threat of long-term, irreversible damage. That spurred the passage in 2014 of the Sustainable Groundwater Management Act (SGMA, or "Sigma"). Sigma aims first to halt the mining of groundwater and then to reverse the overdraft in the long term. The date set is 2040 for the state's 200 or so newly formed Groundwater Sustainability Agencies to achieve their goals. Sigma was significant because it broke through a 160-year labyrinth of litigation over water rights. While all parties benefit from sustainable groundwater management, Sigma in particular makes agriculture sustainable in the long term. Sigma provides a needed regulation to support agribusiness in the San Joaquin Valley, particularly smaller farms with lesser resources.

Box: National-Level Mal/Adaptations

The federal government has a high level of involvement in water development in California. Two federal agencies have built major water projects in California. The US Bureau of Reclamation builds water projects for irrigation development and the US Army Corps of Engineers builds water projects for flood control. The concern is whether having the federal government finance massive water infrastructure can distort state-level perceptions and priorities.

For example, the federal government has its own environmental protection act: the National Environmental Policy Act ("NEPA"). A NEPA Environmental Impact Statement may serve to fulfill the requirements of California's Environmental Quality Act ("CEQA"). Currently, a NEPA Environmental Impact Statement is fulfilling California's CEQA requirements for a proposed Temperance Flats dam/reservoir on the San Joaquin River. One reason the NEPA Environmental Impact Statement is being used to meet the CEQA requirements is that drafting an Environmental Impact Statement is costly. California's Department of Water Resources has long been known for its "meager yearly budget" (Mount, 1995, p. 96) that has rendered it "unable to provide CEQA review" for Temperance Flats (US DOI, August 2014, ES-2).

How might Californians perceive and prioritize Temperance Flats if their own state taxes had to finance the Environmental Impact Statement (as well as all subsequent costs)? To what extent might the federal government's largesse dampen Californians' commitment to smaller local and regional projects? Ones that might develop water supplies at a more modest cost to Californians' wallets and their environment? The power of the federal government to build massive water infrastructure and provide the water at very low rates enabled the 20th century population explosion in Californian drylands. Yet, to what extent is federal largesse now propelling a 21st century California along a "single track" massive water infrastructure that is increasingly less than ideal for drylands (Bhattacharya et al., 2004; Agarwal, 2001)? Has it become a maladaptation?[1]

These questions indicate the need for a thorough review of federal policies and programs to ascertain what impact these might be having on adaptation for climate change in California. As another example, federal insurance policies set up decades ago may now be functioning as entrenched systemic barriers to adaptation efforts (Frisvold, 2015; Christian-Smith et al., 2014). Federal crop insurance and agricultural disaster assistance could be inhibiting incentives to change by providing generous payments to farmers despite the reluctance of some to adapt. Revenue losses for crops were significant during the Drought, with $246 million estimated for 2016. Of that, $222.8 million will occur in the Tulare Lake Basin (Medellín-Azuara, August 15, 2016). Yet, during this time, gross revenue increased for many of San Joaquin Valley's biggest farmers. One reason for the revenue increase is crop substitution: farmers switch to growing crops such as almonds, which are water-intensive but high-profit (Frisvold, 2015; Christian-Smith et al., 2014). In the end, continual learning, research, and outreach are needed to keep adaptations from becoming entrenched maladaptations.

---

1. Bureau of Reclamation Environmental Impact Statements can influence public perception in other ways, too: "Again and again in the documents that supported yet another dam, the economic value of the existing river was shown as nonexistent, while the value of the fishing and boating to take place in the new lake was shown as very high" (California State Lands Commission, 1994). The Bureau of Reclamation's draft EIS for Temperance Flats does exactly that: it showcases one new boat-in campground and one new boat ramp for the new lake as environmental and recreational benefits—and leaves the river a blank (U.S. Department of Interior, August 2014).

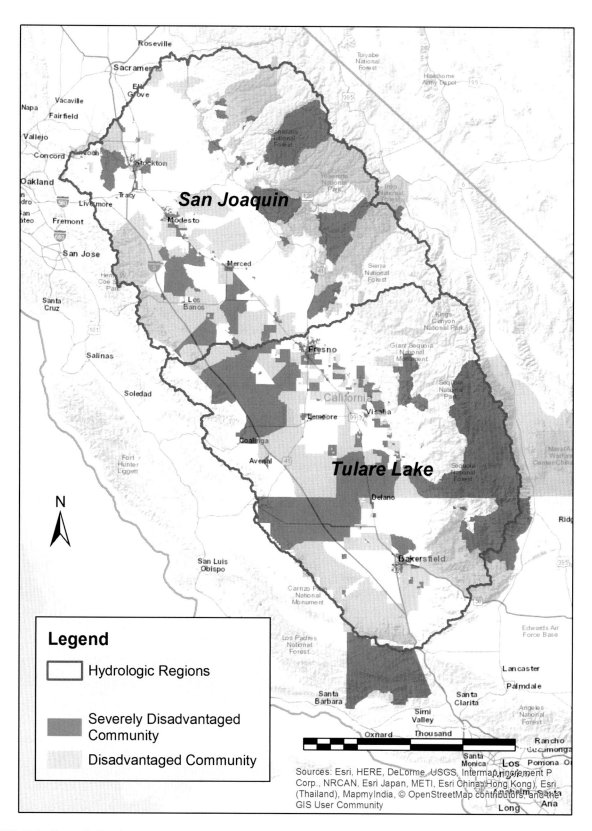

FIGURE 10.5   Economically Disadvantaged Communities ("DAC"s) mapped by hydrologic regions within the San Joaquin Valley. *David Drexler (California State University, Fresno), based on California, Department of Water Resources data (2015).*

FIGURE 10.6   Taking every drop. The dried up San Joaquin River: Diversions from California's vast systems of dams and canals can drain rivers. The San Joaquin River, pictured here, ran dry for miles for decades until the San Joaquin River Restoration Settlement Act was reached. *Photo Credit: Deanna Lynn Wulff.*

10.5   LEGISLATIVE RESPONSE: DRINKING WATER CRISIS

The Drought also triggered new drinking water legislation. Tulare County in the San Joaquin Valley found itself the subject of intense media coverage and a visit from the UN Special Rapporteur for the Human Right to Safe Drinking Water and Sanitation (February 22—March 4, 2011) (United Nations General Assembly, 2011). The seven Disadvantaged Communities experiencing a drinking water crisis are clustered within the boundaries of Tulare County's Alta Irrigation District.[2] The Alta Irrigation District's "Depth to Groundwater" maps indicate that from fall 2011 to fall 2015, depth to groundwater plummeted from 6 m (20 ft) to 18—21 m (60—70 ft) in the area of those seven communities (clustered along the southeastern edge of the Alta Irrigation District). The Alta Irrigation District's southwestern portion has more concentrated agribusiness. There, depth to groundwater plummeted from about 21 m (70 ft) to nearly 40 m (130 ft) (website, accessed December 5 and 6, 2016) (Fig. 10.7).

Tulare County's Office of Emergency Services began providing potable water to the affected residents, on a costly emergency basis. In June 2015, California Senate Bill 88 finally empowered the State Water Resources Control Board to require certain water systems to consolidate with or extend their public water systems to Disadvantaged Communities. Senate Bill 88 ("SB 88") is effective for two types of Disadvantaged Communities. First, SB 88 will impact "fringe" Disadvantaged Communities which are located on the margins of municipalities. Second, SB 88 will impact "island" Disadvantaged Communities which are located within the municipalities that grew up around them.

A third and final type of Disadvantaged Community cannot be reached by municipal systems, however. Called "legacy" Disadvantaged Communities, these sprang up in the early to mid-20th century with the advent of industrialized agriculture. The legacy Disadvantaged Communities are located close to the agribusiness which employ the farm

---

2. These communities are Cutler, Orosi, East Orosi, Sultana, Monson, Yettem, and Seville.

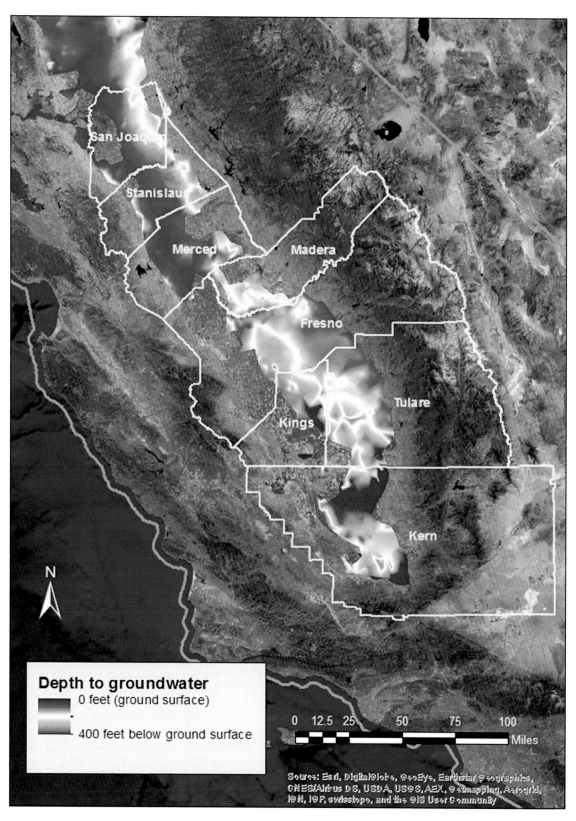

Depth to groundwater

0 feet (ground surface)

400 feet below ground surface

0   12.5 25   50   75   100
Miles

FIGURE 10.7   Depth to groundwater, spring 2014, in the eight counties of the San Joaquin Valley (outlined in *yellow*). *David Drexler (California State University, Fresno), based on California Department of Water Resources data (2014). Maps. Appendix 1 in Zelezny.*

laborers. The legacy Disadvantaged Communities are isolated, small, and poor (over 80% have fewer than 500 residents, and many have fewer than 100 residents) (Flegel et al., 2013).[3] Legacy Disadvantaged Communities rely on pumping of groundwater to meet domestic water needs. They comprise about 30% of the state's Disadvantaged Communities. Because no municipalities are near enough to extend their public water systems, how can these legacy communities end their reliance on wells?

The proposed adaptation is to create a "Drinking Water Authority" through a joint powers agreement between the Alta Irrigation District and the authorities of the Disadvantaged Communities (e.g., Public Utility Districts). The joint powers agreement would legally enable the multiple stakeholders to build a "North Tulare County Regional Surface Water Treatment Plant." The joint powers agreement is necessary because the Alta Irrigation District's clients own the relevant water rights. Moreover, the Alta Irrigation District has a policy not "to detach upon development." Whereas many other Irrigation Districts detach upon municipal development, Alta Irrigation District's decision was to protect its "surface water rights and groundwater authority, easements, and facilities along with utilizing all available water resources for the better" of Alta Irrigation District (Keller and Wegley, 2015; September 2015).

The proposal for a joint Drinking Water Authority to build a regional surface water treatment plant comes at a time when the Drought has dramatically increased the authority of State Water Resources Control Board to administer surface and groundwater resources (Pearah, 2016). Each year, California's agriculture irrigates close to 3.9 million hectares (9.6 million acres). That requires close to 42 million megaliters (34 million acre feet) out of the 53 million-plus megaliters (43 million acre feet) of California's developed water resources (California Department of Water Resources, "Agricultural Water Use," 2016). Therefore, agricultural water use presents the State Water Resources Control Board with large opportunities for immediate impact, especially in the agriculturally-intensive San Joaquin Valley. The State Water Resources Control Board can wield cease and desist orders or implement a reward system of "larger apportionment of available water rights" to those who adapt (Pearah, 2016).

10.6 FISCAL RESPONSE: DRINKING WATER CRISIS

A State of California fiscal response is also necessary because the seven Disadvantaged Communities cannot afford to construct or maintain the North Tulare County Regional Water Treatment Plant. Construction costs are estimated to be $27.3 million ($15.5 million for the plant and $11.8 million for the conveyance pipes). The Alta Irrigation District claimed it would pursue $20 million in Proposition 1 (2014) public monies to help construct the plant, and obtain the remaining $7 million in low interest loans. The Alta Irrigation District is allowed to pursue the maximum amount of Prop 1 funding available for a project—$20 million—because the project provides regional benefits shared among multiple users, at least one of which is a Disadvantaged Community. Moreover, Proposition 1's nonstate cost share requirement of not less than 50% (in this case, $10 million) will be reduced or waived because the project directly benefits a Disadvantaged Community. An open question remains however about postconstruction costs, particularly the ongoing monthly fee for surface water delivery (Keller and Wegley, 2015; September 2015; Kapheim, October 19, 2016; Kapheim, November 21, 2016).

The ongoing monthly fee is estimated to be $33.89 per household (Kapheim, October 19, 2016; Keller and Wegley, 2015; September 2015). An outstanding question is how much or all of the $33.89 ongoing monthly fee should be borne by agribusiness?

The US Census indicated there are 3,610 households in the seven communities, so the $33.89 fee per month per household would be $122,343 per month and $1,468,115 per year. The Alta Irrigation District has 44,920 cropped hectares (111,000 cropped acres) (Alta Irrigation District, January 4, 2016). This means the annual cost of the fee per hectare would be $32.68 ($1,468,115/44,920) (or, $13.23 per acre: $1,468,115/111,000). Furthermore, the *2015 Tulare County Annual Crop and Livestock Report* shows that 146,923 "bearing" hectares (363,064 "bearing" acres) resulted in

---

3. An on-site tour of these Disadvantaged Communities was conducted on November 27, 2016. The first visited, Sultana, exists only as a Census-Designated Place with a population of 775 (US Census 2010). Sultana seems dropped in the middle of farm fields. Immediately upon entering from the west is a trailer park. Nearby is another low income area, a conglomerate of mobile homes, shacks, and side-by-sides. Of note is that directly across the street (Road 105) was a new housing development, albeit starter homes. Sultana turned out to be not atypical of the other towns. On the immediate east outskirts of Yettem, e.g., are trailers. Some aged middle class homes are present, along with a few much newer larger homes.

a Total Crop Value for crops of $5,262,976,800.[4] That makes the average annual Total Crop Value $35,821 per hectare (or, $14,496 per acre).

The question remains of how much of the fee should be borne by agribusiness? The State Water Resources Control Board commissioned the University of California, Davis to conduct a study of nitrate contamination in the Tulare Lake Basin and Salinas Valley areas. The UC, Davis study found that agriculture was responsible for 96% of the nitrate pollution in groundwater (Lund et al., 2012). The contamination spans decades and has to be measured in gigagrams. Previous studies discuss agribusiness' "profligate overuse" of fertilizers, up to 1.089 trillion tonnes (1.2 trillion tons) annually (Mount, 1995, p. 259; California State Lands Commission, 1994; California State Water Resources Control Board, 1992). The UC, Davis study recommended the State implement a nitrate fertilizer use fee to compensate the affected communities for costs of mitigation (Lund et al., 2012).

A well heard response from agribusiness is they are being "fee-ed to death." This response echoes that of many Californians. In 1978, Californians approved Proposition 13. "Prop 13" divorced assessed property values for taxation from actual market value. One consequence was the perversion of housing and land use markets. In turn, this impacted California's ability to use tax and fiscal tools to incentivize climate change adaptations in those markets. To meet chronic fiscal shortfalls, California developed a labyrinth system of impact fees and use taxes. For example, Californians now routinely pay fees for public services such as trash pick-up, and property developers pay "development-impact fees" to offset their impact on public infrastructure and services. The City of Fresno has multiple water fees, which it is trying to reform into one uniform "water capacity fee" (Sheehan, December 8, 2016). While many Californians complain of being "fee-ed to death," the worst impacts of impact fees and use taxes are felt in Disadvantaged Communities whose residents pay out of more limited income.

The State Water Resources Control Board has reiterated that factors such as climate change are bringing longer and deeper droughts, and increasing concentrations of pollutants are leading to higher costs for drinking water. Disadvantaged Communities face multiple financial barriers to meeting these costs, especially: (1) the lack economies of scale in small water systems; (2) costs being distributed over a small low-income rate base; and (3) the lack of a good rate base effectively barring access to capital (California State Water Resources Control Board, 2017).

One final and key question relevant to the debate remains about who should pay the ongoing monthly fee: Who really stands to benefit from the new Tulare County Regional Surface Water Treatment Plant? The obvious and immediate beneficiaries are residents in the seven Disadvantaged Communities who will have clean drinking water. Yet, US Census (2010) demographics show the seven communities have these rates for renter-occupied housing units: Cutler — 56%; East Orosi — 59%; Monson — 45%; Orosi — 44%; Seville — 49%; Sultana — 66%; Yettem — 75%. The owners of the properties (not the renters) will find themselves the recipients of benefits in the long term because the owners will find themselves in possession of California land with state-of-the-art water infrastructure. California land with a reliable water supply has high value, a value that will climb with "the continuing degradation of the groundwater supply and/or the implementation of more restrictive water quality standards" (Keller and Wegley, 2015; September 2015). Moreover, due to Prop 13, the taxes on that property can only increase in very small increments until and unless the property is sold.

The property value will increase even more in the face of anticipated population growth. The California Department of Finance projects that from 2010 to 2030, the state's household population will grow from 36,412,423 to 43,136,741. In Tulare County, the household population will grow from 437,407 to 572,515. That 31% increase from 2010 to 2030 will jump to a 76% increase by 2060.

The state's "Safe Drinking Water" program that financed the feasibility study for "North Tulare County Regional Study Water Treatment Plant" caps growth at 10% of the communities' current Average Daily Demand (for water). However, the feasibility study proposes "a large capital investment" to acquire a site that would allow doubling the plant's capacity by "constructing an identical facility, if needed, in the future." The study also proposes that the canal's turnout structure and raw water pipeline be sized to accommodate doubling capacity. The ostensible reasons cited for building a twin plant are the projected growth in the seven communities and in the region as well as increased requests for service from declining groundwater quality and more restrictive water quality standards and regulations (Keller and Wegley, 2015, pp. 1−26, 2−7, 2−14A and 3−5−3−10).

---

4. Tulare County has a lucrative milk industry (24.6%). The milk industry's value ($1,718,001,000) was subtracted from Tulare County's total crop and livestock reported value ($6,980,977,800) to arrive at a value of $5,262,976,800 for crops only. The reason for calculating a value for crops only is the Alta Irrigation District website provides a self-description of 44,920 "cropped" hectares. Tulare County's average value per hectare ($35,821) was multiplied by the Alta Irrigation District's 44,920 cropped hectares to arrive at the total value of $1,609,079,230 for the Alta Irrigation District's cropped land.

At no point is climate change mentioned in the feasibility study. The feasibility study's predicted scenario is increased population growth coupled with declining ground water quality and increasing water quality regulations. The cumulative impact of those three factors will be a challenge in itself, and the extent that climate change will amplify those impacts merits discussion. In addition, an amplified impact will amp up the value of that California land with a new state-of-the-art water supply.

10.7 CONCLUSION

Scientists from NASA, Cornell, and Columbia (Cook et al., 2015) predicted climate change will bring megafloods and megadroughts to California, exacerbating the "deluge or drought" climate of drylands. These mega-events will represent a fundamental and unprecedented climate shift from the past millennium. They are outside of the experience of both the ecosystems and human systems in the region. Adaptations will be needed to require change at all levels. The adaptations will need to restore and to manage the ecosystems to provide sustainable services for critical uses—environmental, municipal and industrial, and agricultural—and for the range of stakeholders within those three categories.

This case study illustrates that adaptations will find themselves debated within a larger state of play involving many actors and many factors. Those forces can conflict with one another, influencing both outcomes and whether today's adaptations become tomorrow's maladaptations and whether today's adaptations move forward swiftly or at a sclerotic pace. That regional drinking water treatment plant, sustainably financed, would change the state of play in its region. Many course shifts will be needed. The solution for a "never-ending problem" is never-ending research and learning.

REFERENCES

Adams, K.D., Negrini, R.M., Cook, E.R., Rajagopal, S., 2015. Annually resolved late Holocene paleohydrology of the southern Sierra Nevada and Tulare Lake, California. Water Resour. Res. 51. Available from: https://doi.org/10.1002/2015WR017850.

Agarwal, A., 2001. Drought? Try capturing the rain. In: Briefing paper for members of the Parliament and state legislatures: An occasional paper from the Centre for Science and Environment. Retrieved from: http://cseindia.org/challenge_balance/readings/Drought%20briefing-Aggarwal.pdf.

Alta Irrigation District, http://www.altaid.org/about-alta-id-mainmenu-95 (accessed 4.01.16.).

Alta Irrigation District, Depth to Groundwater Maps: Fall 2011-Fall 2015. Retrieved from: http://www.altaid.org/ground-water-mainmenu-96/ground-water-maps.

Bhattacharya, K., Azizi, P.M., Shobair, S.S., Mohsini, M.Y., 2004. Drought impacts and potential for their mitigation in southern and western Afghanistan. Colombo, Sri Lanka: International Water Management Institute (IWMI) v, 19p. (IWMI Working Paper 91/Drought Series: Paper 5). https://doi.org/10.3910/2009.266.

Blunt, A.B., Negrini, R.M., 2015. Lake levels for the past 19,000 years from the TL05-4 cores, Tulare Lake, California, USA: Geophysical and geochemical proxies. Quaternary International 387, 122−130. Available from: http://dx.doi.org/10.1016/j.quaint.2015.07.001.

Burow, K.R., Jennifer, L.S., Dubrovsky, N.M., 2008. Regional nitrate and pesticide trends in groundwater in the eastern San Joaquin Valley, California. J. Environ. Quality. 37 (2008). Available from: https://doi.org/10.2134/jeq2007.0061S-249−S-263.

California, Department of Water Resources, 2016. Agricultural Water Use. Retrieved from: http://www.water.ca.gov/wateruseefficiency/agricultural/.

California, Department of Water Resources, 2016. California State Water Project at a Glance. Retrieved from: http://www.water.ca.gov/recreation/brochures/pdf/swp_glance.pdf.

California, Department of Water Resources, September 2016. Drought and Water Year: 2016. Retrieved from: http://water.ca.gov/waterconditions/docs/a3065_Drought_8page_v8_FINALsm.pdf.

California, Department of Water Resources, April 30, 2014. Report to the Governor's Drought Task Force - Groundwater Basins with Potential Water Shortages and Gaps in Groundwater Monitoring. http://www.water.ca.gov/waterconditions/docs/Drought_Response-Groundwater_Basins_April30_Final_BC.pdf.

California, Department of Water Resources, 2015. Sierra Nevada Snowpack Is Virtually Gone; Water Content Now Is Only 5 Percent of Historic Average, Lowest Since 1950. 1 April 2015. Retrieved from: http://www.water.ca.gov/news/newsreleases/2015/040115snowsurvey.pdf.

California, Department of Water Resources, February 2014. System Reoperation Study: Strategy Formulation and Refinement: Draft: Phase 2 Report. Retrieved from: http://www.water.ca.gov/system_reop/docs/System%20Reop%20Phase%202%20Report%20-%20February%202014%20Draft.pdf

California Governor's Drought Task Force, 2015. Retrieved from: http://drought.ca.gov/.

California Office of the Governor, January 23, 2017. Governor Brown declares state of emergency in counties across California following severe winter storms. Retrieved from: https://www.gov.ca.gov/news.php?id=19668.

California State Lands Commission, 1994, 2nd ed. California's Rivers: A Public Trust Report. Sacramento.

California State Water Resources Control Board, February 2017. Affordable safe drinking water initiative. Retrieved from: http://www.waterboards.ca.gov/water_issues/programs/conservation_portal/assistance/docs/notice_asdw.pdf.

California State Water Resources Control Board, 2016−2017. Central Valley Water Program Board Fact Sheet: Irrigated Lands Regulatory Program. Retrieved from: http://www.waterboards.ca.gov/centralvalley/board_info/exec_officer_reports/fy1617_program_updates/fy1617_ilrp_factsheet.pdf.

California State Water Resources Control Board, 1992. Water Quality Assessment. California Environmental Protection Agency, Sacramento.

Christian-Smith, J., Levy, M.C., Gleick, P.H., 2014. Maladaptation to drought: a case report from California, USA. Sustain. Sci. 10 (2015), 491−501. Available from: https://doi.org/10.1007/s11625-014-0269-1. Retrieved from: http://droughtmonitor.unl.edu/SupplementalInfo/Forecasts.aspx.

Cook, B.I., Autl, T.R., Smerdon, J.E., 2015. Unprecedented 21st century drought risk in the American Southwest and Central Plains. Sci. Adv. 1 (1), 1−7. Retrieved from: http://advances.sciencemag.org/content/1/1/e1400082.full.

Cook, B.I., Smerdon, J.E., Seager, R., Coats, S., 2014. Global warming and 21st century drying. Clim. Dynam. 43, 2607−2627. Retrieved from: http://www.ldeo.columbia.edu/res/div/ocp/pub/seager/CookBetal21st2014.pdf.

Diffenbaugh, N.S., Swain, D.L., Touma, D., 2015. Anthropogenic warming has increased drought risk in California. Proc. Natl Acad. Sci. 112 (13), 3931−3936. Retrieved from: http://www.pnas.org/content/112/13/3931.abstract?sid = 9cfa7dea-8123-4091-bf8d-a541772dea30.

Evenari, M., Shanan, L., Tadmor, N., Aharoni, Y., 1961. Ancient agriculture in the Negev. Science 133 (3457), 979−996.

Farr, T.G., Jones, C., Liu, Z., August 2015. Progress report: subsidence in the Central Valley, California. Retrieved from: http://water.ca.gov/groundwater/docs/NASA_REPORT.pdf.

Flegel, C., Rice, S., Mann, J., Tran, J., 2013. California unincorporated: mapping disadvantaged communities in the San Joaquin Valley. Policy Link, in partnership with California Rural Legal Assistance, Inc., and California Rural Legal Assistance Foundation. Retrieved from: http://www.policylink.org/find-resources/library/california-unincorporated-mapping-disadvantaged-communities-in-the-san-joaquin-valley.

Frisvold, G.B., 2015. Water, Agriculture, and Drought in the West Under Changing Climate and Policy Regimes. National Resources Journal Retrieved from: http://digitalrepository.unm.edu/nrj/vol55/iss2/5/.

Griffin, D., Anchukaitis, K.J., 2014. How unusual is the 2012−2014 California drought? Geophys. Res. Lett. 41 (24), 9017−9023. Available from: https://doi.org/10.1002/2014GL062433. Retrieved from: http://onlinelibrary.wiley.com/doi/10.1002/2014GL062433/epdf.

Howitt, R., MacEwan, D., Medellin-Azuara, J., Lund, J., Sumner, D., Economic analysis of the 2015 California drought on agriculture: a report for the California Department of Food and Agriculture. University of California, Davis: Center for Watershed Sciences. Retrieved from: https://watershed.ucdavis.edu/files/biblio/Economic_Analysis_2015_California_Drought__Main_Report.pdf.

Ingram, B.L., Malamud-Roam, F., 2013. The West Without Water: What Past Floods, Droughts, and Other Climatic Clues Tell Us about Tomorrow. University of California Press, Berkeley.

IPCC, 2014. Climate change 2014: impacts, adaptation, and vulnerability. Part A: global and sectoral aspects. In: Field, C.B., Barros, V.R., Dokken, D.J., Mach, K.J., Mastrandrea, M.D., Bilir, T.E., White, L.L. (Eds.), Contribution of Working Group II to the Fifth Assessment Report of the Intergovernmental Panel on Climate Change (Vol. 1). Cambridge, United Loss and Damage: The Role of Ecosystem Services Reference 63. Cambridge University Press, Kingdom and New York, NY, USA, Retrieved from: https://www.ipcc.ch/pdf/assessment-report/ar5/wg2/WGIIAR5-FrontMatterA_FINAL.pdf.

Kapheim, C., November 21, 2016. Personal Interview. Alta Irrigation District Headquarters, Dinuba, CA.

Kapheim, C., October 19, 2016. Presentation. Kings Basin Water Authority Advisory Committee Meeting. Fresno County Farm Bureau, Fresno, CA.

Keller, D.R., Wegley, J.H., February 2015. North Tulare County regional surface water treatment plant study. Project financed by the State of California Safe Drinking Water State Revolving Fund.

Keller, D.R., Wegley, J.H., September 2015. North Tulare County regional surface water treatment plant study [addendum]. Project financed by the State of California Safe Drinking Water State Revolving Fund.

Lund, J.R., Harder, T., et al., January 2012. Addressing nitrate in California's drinking water: With a focus on Tulare Lake Basin and Salinas Valley Groundwater. Report prepared for the State Water Resources Control Board report to the Legislature. University of California, Davis: Center for Watershed Sciences. http://groundwaternitrate.ucdavis.edu/files/138956.pdf.

Lund, J., December 28, 2016. Out with the Old Drought and in with the New. University of California, Davis: Center for Watershed Sciences: California WaterBlog. Retrieved from: https://californiawaterblog.com/2016/12/28/out-with-the-old-drought-and-in-with-the-new/.

Mann, M.E., Gleick, P.H., 2015. Climate change and California drought in the 21st century. Proc. Natl. Acad. Sci. U.S.A. 112 (13), 3858−3859. Available from: https://doi.org/10-1073/pnas.1503667112.

Medellin-Azuara, J., et al., August 15, 2016. Economic analysis of the 2016 California drought on agriculture: A report for the California Department of Food and Agriculture. University of California, Davis: Center for Watershed Sciences. Retrieved from: https://watershed.ucdavis.edu/droughtimpacts.

Mount, J.F., 1995. California Rivers and Streams: The Conflict between Fluvial Process and Land Use. Illustrations by Janice C. Fong. University of California Press, Berkeley.

NASA Jet Propulsion Laboratory, August 19, 2015. California Drought Causing Valley Land to Sink. Retrieved from: http://www.jpl.nasa.gov/news/news.php?feature = 4693.

Pearah, P.J., 2016. Summer 2016 Keeping the Desert at Bay: Adapting California Water Management to Climate Change, Vol. 22. Envt'l L. & Pol'y, Hastings W.-Nw. J, pp. 137−172.

Reisner, M., 1986. Cadillac Desert: The American West and Its Disappearing Water. Viking Penguin, New York.

Sheehan, T. December 8, 2016. Price of new homes in Fresno goes up by thousands; streets, parks are the reason. *Fresno Bee*. Retrieved from: http://www.fresnobee.com/news/local/article119792478.html.

United Nations General Assembly, 2011. Report of the special rapporteur on the human right to safe drinking water and sanitation, Catarina de Albuquerque. GE.11-15379, August 2, 2011. Retrieved from: http://www2.ohchr.org/english/bodies/hrcouncil/docs/18session/A-HRC-18-33-Add4_en.pdf.

U.S. Army Corps of Engineers, Sacramento District, November 2015. Central Valley Integrated Flood Management Study, California: Draft Watershed Plan. Retrieved from: http://www.spk.usace.army.mil/Portals/12/documents/civil_works/CVIFMS/CVIFMS_Draft_Watershed_Plan_Public_Release_DEC2015.pdf

U.S. Census, American Community Survey, 2015. Retrieved from: https://factfinder.census.gov/faces/nav/jsf/pages/community_facts.xhtml.

U.S. Department of Agriculture: National Agricultural Statistics Service. California Agricultural Statistics Review: 2014-2015. 2016. Retrieved from: https://www.nass.usda.gov/Statistics_by_State/California/Publications/California_Ag_Statistics/Reports/2014cas-all.pdf.

U.S. Department of Interior: Bureau of Reclamation: Mid-Pacific Region, August 2014. Upper San Joaquin River Basin Storage Investigation: Draft Environmental Impact Statement. Retrieved from: https://www.usbr.gov/mp/sccao/storage/.

U.S. Department of Interior: Bureau of Reclamation, 2016. About the Central Valley Project. Retrieved from: https://www.usbr.gov/mp/cvp/about-cvp.html.

U.S. Global Change Research Program, August 2017. Climate Science Special Report (CSSR): Third-Order Draft (TOD). Retrieved from: https://www.nytimes.com/interactive/2017/08/07/climate/document-Draft-of-the-Climate-Science-Special-Report.html?rref = collection%2Ftimestopic%2FNational%20Academy%20of%20Sciences&action = click&contentCollection = timestopics&region = stream&module = stream_unit&version = latest&contentPlacement = 2&pgtype = collection.

USGS Water Science School. Groundwater Depletion. Retrieved from: http://water.usgs.gov/edu/gwdepletion.html.

Valadao, David, Congressman, 2017a. California's 21st District. Retrieved from: https://valadao.house.gov/district/.

Valadao, David, Congressman, 2017b. Rules Committee Print 115—Text of H.R. 23: Gaining Responsibility on Water Act of 2017. Retrieved from: https://valadao.house.gov/issuesandlegislation/sponsoredlegislation.htm.

Yoon, J.-H., et al., Extreme Fire Season in California: A Glimpse into the Future? Explaining Extreme Events of 2014 from a Climate Perspective. American Meteorological Society: Special Supplement to the Bulletin 96:12. December 2015. S5-S14. Retrieved from: https://www.ncdc.noaa.gov/news/explaining-extreme-events-2014.

FURTHER READING

Adler, R.W., 2010. Climate change and the hegemony of state water law. Stanford Environ. Law J. 29, 1.

Arax, Mark, Wartzman, Rick, 2003. The King of California: J.G. Boswell and the Making of a Secret American Empire. Public Affairs, New York.

California, Department of Finance, Demographics Research Unit. Total Population Projections for California and Counties: July 1, 2015 to 2060 in 5-year Increments. Retrieved from: http://www.dof.ca.gov/Forecasting/Demographics/Projections/.

California, Department of Finance, Demographics Research Unit. Projected Households, Household Population, Group Quarters and Persons per Household for the Counties and State of California. Retrieved from: http://www.dof.ca.gov/Forecasting/Demographics/Projections/.

County of Tulare, August 2016. 2015 Tulare County Annual Crop and Livestock Report.

County of Fresno, 2016. 2015 County of Fresno Annual Crop and Livestock Report.

Igler, D., 2001. Industrial Cowboys: Miller & Lux and the Transformation of the Far West, 1850-1920. University of California Press, Berkeley.

Kates, R.W., Travis, W.R., Wilbanks, T.J., 2012. Transformational adaptation when incremental adaptations to climate change are insufficient. Proceedings of the National Academy of Sciences in North America 109 (19), 7156−7161. Permalink: http://www.pnas.org/content/109/19/7156.

Lund, J., 2017. California's Wettest Drought-2017. University of California, Davis: Center for Watershed Sciences, California WaterBlog, February 5, 2017. Retrieved from: https://californiawaterblog.com/2017/02/05/californias-wettest-drought-2017/.

Mann, M.L., et al., 2016. Incorporating anthropogenic influences into fire probability models: Effects of human activity and climate change on fire activity in California. PLoS One. Available from: https://doi.org/10.1371/journal.pone.0153589.

Null, S.E., Prudencio, L., 2016. Climate change effects on water allocations with season dependent water rights. Sci. Total Environ. 571, 943−954. Available from: https://doi.org/10.1016/j.scitotenv.2016.07.081.

Rheinheimer, D.E., Null, S.E., Viers, J.H., 2016. Climate-adaptive water year typing for instream flow requirements in California's Sierra Nevada. J. Water Resour. Plan. Manage. 142 (11), 04016049.

Rose, C.M., 1990. Energy and efficiency in the realignment of common-law water rights. J. Legal Stud. 19 (2), 261−296.

Schwarz, A.M., 2015. California Central Valley Water rights in a changing climate. San Francisco Estuary Watershed Sci. 13 (2), Permalink: http://escholarship.org/uc/item/25c7w914.

Scott, A., Coustalin, G., 1995. The evolution of water rights. Fall Nat. Resour. J. 35, 821−979.

Sheehan T., Economic Analysis of the 2015 Drought for California Agriculture, 17 August 2015. Retrieved from: https://watershed.ucdavis.edu/files/biblio/Final_Drought%20Report_08182015_Full_Report_WithAppendices.pdf.

Tewari, D.D., 2005. An evolutionary history of water rights in South Africa. Water World, IWHA 157−182.

Treadwell, E., 2005. The Cattle King: A Dramatized Biography. Great West Books, Lafayette, CA.

United Nations General Assembly, August 2, 2011. Report of the special rapporteur on the human right to safe drinking water and sanitation, Catarina de Albuquerque. GE.11-15379. Retrieved from: http://www2.ohchr.org/english/bodies/hrcouncil/docs/18session/A-HRC-18-33-Add4_en.pdf.

University of California, Davis: Center for Watershed Sciences, California Water Blog, Accessed at https://californiawaterblog.com.

U.S. Drought Monitor, 2015. Produced jointly by the National Oceanic and Atmospheric Administration, the U.S. Department of Agriculture, and the National Drought Mitigation Center at the University of Nebraska-Lincoln. Retrieved from: http://droughtmonitor.unl.edu/.

U.S. Geological Service, California Water Science Center. "California's Central Valley: Regional Characteristics." Retrieved from: http://ca.water.usgs.gov/projects/central-valley/about-central-valley.html.

Valadao, David, Congressmen, 2015. Western Water and American Food Security Act of 2015. Retrieved from: https://www.congress.gov/bill/114th-congress/house-bill/2898/text.

Vranesh, George, 1986. The historic relationship of water quantity and water quality. Nat. Resour. Environ. 1 (4), 3−5. 46-48.

Zelezny, L., Fu, X., Harootunian, G., Drexler, D., Avalos, A., Chowdhury, N., et al., 2015. Impact of the Drought in the San Joaquin Valley of California Retrieved from: http://www.fresnostate.edu/academics/drought/.

Chapter 11

Advancing Coastal Climate Resilience: Inclusive Data and Decision-Making for Small Island Communities

Roger-Mark De Souza[1] and Judi Clarke[2]

[1]Sister Cities International, Washington, DC, United States, [2]J F Clarke Consulting Inc., Bridgetown, Barbados

Chapter Outline

11.1 INTRODUCTION AND CONTEXT

In the Caribbean, climate change threatens coastal ecosystems, community livelihoods, and national economies. However, these threats could be mitigated by engaging community members, local leaders, and private sector actors in an inclusive adaptation planning process for coastal resource management and resilience.

The coastal tourism industry, which is critical to most Caribbean islands' economic development, provides livelihoods for local communities but also competes with them for resources. Therefore, these two constituencies must collaborate to ensure that coastal resources are protected and livelihoods are made resilient (Amelung et al, 2007). The challenge remains, however, that Caribbean coastal communities often lack access to climate data and environmental information to support their own decision-making; and they are routinely excluded from discussions to advocate their needs, as well as decisions about tourism, infrastructure, and development—all of which affect them (Becken et al, 2013).

This chapter suggests an approach to advance the scientific basis for assisting coastal communities in integrating climate-related information into their planning and management decisions. This is done through the development of a *Toolkit* that provides a detailed template for conducting a participatory, interactive, and place-based adaptation planning process in small coastal communities that are primarily dependent on tourism.

Through sustained collaboration and partnership, this approach could increase the integration of local knowledge into adaptation planning and coastal development decision-making. It carries the added advantage of providing an inclusive methodology for evaluating the costs/benefits of adaptation strategies and a pathway for incorporating these strategies into national/regional decision-making. By informing coastal planning processes, such an approach could ultimately help increase community resilience to climate change impacts.

In small island coastal communities in the Caribbean—particularly those dependent on tourism—climate change threatens coastal ecosystems, community livelihoods, and local/national economies. However, the severity of these threats could be mitigated by engaging community members, local leaders, and private sector tourism actors in an inclusive adaptation planning process for coastal resource management. The tourism industry, which is critical to most Caribbean islands' economic development, and highly dependent on natural resources, could therefore play a vital role in this process.

Resilience. DOI: https://doi.org/10.1016/B978-0-12-811891-7.00011-6

Tourism provides local livelihoods but simultaneously competes for resources with local communities; it is both a source of environmental degradation and can be a powerful ally in efforts to address climate impacts.

But in many local communities in the Caribbean and other small islands, climate change isn't part of the vernacular: it is not a topic of conversation or a pressing concern. There are several reasons for this:

1. Communities often lack access to climate data and other environmental information, and they may not see the connections between land use, resource scarcity, and environmental change.
2. As policymakers in coastal regions develop plans for adaptation at the national level, local communities are often left out of the process, as the community members and leaders lack the tools needed to assess the costs and benefits of adaptation strategies and to advocate for their needs.
3. Communities are often not engaged in decision-making processes around tourism, infrastructure, and economic development projects that use coastal resources—all of which will be affected by climate change and will impact adaptation plans.

11.2 AN APPROACH TO COASTAL COMMUNITY ENGAGEMENT TO BUILD THE EVIDENCE BASE

One successful approach based on participatory rural appraisal (PRA) methodologies involves directly engaging communities in the collection of data and practical usable information. The approach is premised on conducting a participatory, interactive, and place-based adaptation planning process in small coastal communities dependent on tourism and the process also brings together key sectors that may not have collaborated previously.

Such an approach allows for the inclusion of tourism actors and can be refined and reviewed by experts and partners from other tourism-dependent coastal communities to ensure its transferability. In principle this approach, if successfully implemented, will help often-ignored coastal communities to identify and assess coastal risks and vulnerabilities to climate change; and to identify means to address intersecting stressors (e.g., climate change and tourism development) to enhance coastal resilience and livelihoods.

The approach's methodology draws primarily on the extensive livelihoods and climate change vulnerability and impact analysis work conducted by the project partners and collected in the CARIBSAVE Climate Change Risk Atlas (2008−2012) and CCCCC (2009). This multidisciplinary approach does the following:

- Draws on existing information and reinforces ongoing/recent research and planning processes;
- Engages local policymakers and communities to fill gaps in existing research by capturing local knowledge;
- Provides guidance for working with the local community to analyze its own adaptive capacity and climate change vulnerability, and identify its own priorities;
- Combines community knowledge and scientific data for a greater understanding of local impacts of climate change; and
- Combines information gained at different levels and from different sources to draw plausible conclusions about the adaptive capacity and vulnerability in the target community hence suitable adaptation solutions.

The CARIBSAVE Climate Change Risk Atlas was implemented in 15 Caribbean countries. Each resulting document (or Risk profile) presented a practical, evidence-based comprehensive report of the threats climate change poses to the 15 countries and what they can do about them. Through the lens of the tourism sector, and with focal areas such as community livelihoods, gender, poverty and development, agriculture and food security, energy, water quality and availability, and comprehensive disaster management, the Risk Atlas provides communities with response strategies that are specific to individual countries and key sectors within each country, to build capacity, reduce vulnerability, and enhance resilience. In particular, communities were engaged through participatory processes such as interviews, focus groups and vulnerability mapping exercises. Creating maps allowed community members to visualise, not only areas that were already known to be vulnerable, but also those at risk from future climatic events.

We can benefit from and build upon the experience of Climate Change Risk Atlas in three ways:

First, the Climate Change Risk Atlas was created by project partners through in-depth consultation with regional, public, private, and community stakeholders across the Caribbean to ensure that the work would be useable by all stakeholders from every sector (CARIBSAVE, 2012). The extensive work conducted with communities was fundamental to bridging the gap between science and local experiences of climate-related impacts and provides an important basis to apply the lessons of that national-level project at a very local scale.

Second, this approach could build on the lessons for reaching out and engaging broader Caribbean initiatives to expand the interest in, and potential use of, the project's outputs. The work coming out of the Risk Atlas has been served as the evidence-base and rationale for a number of successful projects across the Caribbean, including:

- The Climate Change, Coastal Community Enterprises—Adaptation, Resilience and Knowledge (CCCCE-ARK) Project in the Bahamas, Barbados, Belize, and Jamaica.
- The Partnership for Canada-Caribbean Community Climate Change Adaptation Project (ParCA) conducting community-based vulnerability assessments in Tobago and Canada (and Prince Edward Island).
- The Caribbean Fish Sanctuaries Partnership Initiative (C-Fish) collaborating with regional, national, and local stakeholders to build resilience to climate change and improve the sustainability of livelihoods in coastal communities in Jamaica, Saint Lucia, St. Vincent and the Grenadines, and Grenada.
- The Global Islands Vulnerability Research for Adaptation Policy and Development (GIVRAPD) Project—a technical study designed to integrate scientific and local knowledge from comparative learning sites in the Caribbean (Jamaica and Saint Lucia) and in the Indian Ocean (Mauritius and the Seychelles) by using a community-based vulnerability assessment framework.

Third, the Risk Atlas offers an understanding of climate change impacts on tourism-related coastal development, adaptive capacity at the community level, the complex factors influencing livelihood choices, and the differential climate change vulnerability challenges of men and women.

This approach also draws on a number of other methodologies and tools, namely:

- Caribbean Risk Management Guidelines developed for the Caribbean Community (CARICOM);
- The Vulnerability and Capacity Assessment methodology developed by the Caribbean Community Climate Change Centre;
- The methodology developed by IISD and the World Bank in exploring the social dimensions of climate change;
- The Department for International Development's approach in linking the sustainable livelihoods approach with reducing disaster vulnerability;
- The Climate Vulnerability and Capacity Analysis (CVCA) model; and
- Methodology used by the FAO in its work on smallholder livelihoods in sub-Saharan Africa.

11.3 KEY STEPS

11.3.1 Contextualize the Approach

It will be important to contextualize this approach taking into account information and data at both the national and local scales including (but not limited to):

- Land use change (past, current baseline, and future trends) related to major economic activities (i.e., tourism, agriculture, aquaculture);
- Ecosystem services as related to the main economic activities in the coastal community of interest;
- Exposure to climatic events based on current climate variability and future changes (extremes in temperature and rainfall; hurricanes); and slow onset (sea-level rise) events;
- Sensitivity to current variability and projected changes in tourism, water resources, and coastal areas;
- Natural, social, economic, political, and human capital related to adaptive capacity.

11.3.2 Survey/Interviews

Online surveys and phone interviews with stakeholders in the project location would be used to capture knowledge/data resources and gaps, existing climate adaptation activities and future needs, and identify barriers to decision-making (see Kuriakose et al., 2009). The survey will be administered at the beginning (baseline), middle (mid-term evaluation), and conclusion of the process, to inform the next steps; to assess changes as a result of the process; and provide results and feedback for future iterations of the process. Stakeholders will be identified using the following considerations:

- Key persons and entities that will be involved and impacted by the project;
- Information the stakeholders can contribute and/or want to receive through this work; and
- Their role in governance at the national and/or local levels and leadership in the sustainability and application of information from the exercise.

11.3.3 Participatory Mapping

A participatory mapping exercise will be conducted with select stakeholders from the project site. One exercise we would suggest is focused on "Participatory GIS" (PGIS), wherein participants will map local data and knowledge about coastal resource use, tourism sector development, climate changes and other key variables.

PGIS is the practice of gathering data through traditional methods such as interviews, questions, and focus groups and by using paper maps to record spatial details (See Morrow, 1999). This information is then digitized so that it can be analyzed and interrogated using computer GIS software, and results can be communicated using computer-drawn maps.

The practice is multidisciplinary and relies on the integration of "expert" knowledge with socially and gender-differentiated local knowledge. It builds on high levels of stakeholder participation in the processes of spatial learning, decision-making, and action and brings together diverse groups of stakeholders from community-based and non-governmental organizations, technical agencies, and policymakers to exchange ideas, perspectives, and information on a more even playing field. This process thus strengthens and builds new relationships to support practical and inclusive decision-making and implementation of adaptation measures.

This participatory GIS step will also be used to assess the impact of community livelihood activities, tourism development, and climate trends on local coastal resources now and at select points in the future. It will capture the perceptions of stakeholders on the different groups and assets that exist and operate within the pilot study area, identify their location, and then apply information on climate trends and projections to paint a picture of current and potential future vulnerability or resilience to climatic changes. Specifically, the mapping exercise will attempt to qualify the following in spatial terms:

- The general population in the community;
- "High-impact" areas based on biophysical impacts (sea-level rise, storm surge, flooding, mangrove or coral declines, vulnerability to hurricane winds);
- "High-vulnerability" areas in respect to the priority sectors—tourism, coastal resources, community livelihoods and development (critical infrastructure, institutions, homes, natural resources);
- Applicable criteria for the "most vulnerable" groupings (of persons and assets/capital);
- Other areas where each of the most vulnerable groups or physical/infrastructural assets has significant numerical presence;
- Intersection or overlap of most vulnerable groupings and high-risk areas for biophysical impacts; and
- Any actions or infrastructure that add to the area's capacity to adapt to biophysical impacts.

The participants in the exercise will be encouraged to lead the map development and detailing process, so that a sense of trust and ownership is built into the process. It is also critical at this juncture that external facilitators play only a guiding or mediating role, to ensure that participants understand what they are to map. This helps to ensure that external biases do not become embedded in the process and outputs. Differentiation can also be built into the process where two or more subgroups create separate maps, to discern and visualize how perspectives on personal, cultural, or sectoral priorities may converge or differ.

Once maps are completed, the project team will then work to translate the drawn map outputs to geo-referenced and projected spatial layers on a GIS platform (ideally ESRI's ArcGIS) to establish new and credible spatial data that can be integrated not only for the pilot project's assessment and reporting purposes, but for any similar work that can be done in the future.

11.3.4 Risk Assessment and Adaptation Analysis

It will be important to convene local community members and leaders, coastal resource and climate scientists, and representatives from the tourism sector to review the results of the survey and participatory mapping exercise, along with the latest projections of climate impacts for the pilot site/country.

Together, the participants will engage in a risk assessment exercise and develop a matrix of adaptation strategies coming out of the processes from steps 1 to 3 above. These strategies will be designed to increase resilience to the climate challenges now and in the future. The adaptation strategies produced by this process are intended to contribute to development objectives (including poverty reduction) at all levels within the country. More specifically, in the project area the strategies will:

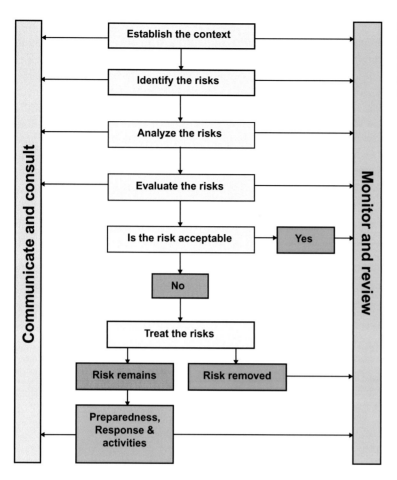

1. Promote climate-resilient livelihood strategies and capacity building for adaptive capacity and action planning among tourism enterprises;
2. Reduce disaster risks and reduce the effects of climate-related impacts on vulnerable households and tourism-related livelihoods;
3. Develop the capacity of local civil society and governmental institutions to provide better support to communities, households, and individuals in their adaptation efforts; and
4. Address the underlying causes of vulnerability, such as poor governance, lack of control over resources, or limited access to basic services.

Identifying the main climate risks, impacts, and adaptation options requires a socially inclusive, broad-based consultative process within the community. This will ensure that any proposed adaptation actions will build on local experiences and reflect the views and needs of vulnerable groups.

This could be done through workshops or focus group discussions. The key steps in the process are based on the Risk Management Guidelines for Climate Change Adaptation Decision-Making, which were developed under the "Adaptation to Climate Change in the Caribbean" project which was overseen by the World Bank, with support provided by CARICOM and funded by CIDA (now Global Affairs Canada) (see Fig. 11.1). Once completed, this activity leads the community to commence the adaptation planning process that will serve to validate and provide fine resolution risk and adaptation intervention inputs at the local level.

11.4 WHERE COULD SUCH AN APPROACH BE TESTED?

Ideally, such an approach would lend itself to locations where negative climate change impacts are already affecting local livelihoods and tourism activities and where there is a mix of tourism sector actors who are engaged in community activities. One such location is Punta Cana in the Dominican Republic.

Tourism has become the most important economic driver in the Dominican Republic, representing the leading source of foreign exchange and an important contributor to the national economy. In 2011, the tourism industry directly contributed more than USD 100 billion (4.7%) to the Dominican GDP and is indirectly responsible for more than USD 325 billion (15.1%). One out of every 20 Dominicans works in tourism, accounting directly or indirectly for over half a million jobs nationwide.

Punta Cana is the most popular tourist destination in the Dominican Republic, and one of the most important in the entire Caribbean, with more than 36,000 hotel rooms spread over 30 miles of coastline and more than 2 million visitors arriving annually, many of them from the United States. The rapid tourism and real estate development in the Punta Cana region over the last 20 years has attracted large numbers of Dominicans and Haitians from all over Hispaniola in search of jobs and economic opportunities. Unfortunately, the arrival of workers has far outpaced the creation of infrastructure, housing, and basic services needed to support them, putting enormous pressure on coastal resources.

Potential stakeholders to be engaged include:

Tourism actors
- Altragracia Hotel Association (the regional association in Punta Cana) http://www.puntacanabavarohotels.com/index.php
- The Punta Cana Ecological Foundation which is a not for profit providing technical assistance, training and support to the local community
- ASONAHORES (the national hotel and restaurant association of the Dominican Republic) http://www.asonahores.com/
- National Network of Pro-Environment Businesses (ECORED) http://www.ecored.org.do/

Government/Community Leaders
- Mayors' offices of Veron, Punta Cana, and Bavaro Municipalities
- Dominican Ministry of Climate Change
- Local neighborhood associations

Climate and Coastal Resource Management Scientists
- The Caribbean Community Climate Change Centre
- The University of the West Indies
- Dr. Jocelyn Widmer, University of Florida (currently collaborating with Grupo Puntacana Foundation on a community mapping project in Punta Cana).

11.5 CONCLUSION

We believe that this approach provides valuable insight and a unique opportunity to help Caribbean coastal communities prepare for and recover from extreme weather events, climate hazards, and changing climate conditions. By filling key information gaps and providing additional scientific and participatory analyses at the community level, this approach will help Non-governmental Organizations (NGOs), local government, and tourism companies develop integrated coastal resilience strategies. The work will increase understanding of coastal risks and impacts and improve existing planning tools. The effort will also include significant outreach to ensure that the results inform emerging National Adaptation Plans (NAPs), particularly related to coastal resilience, being developed by ministries and government offices across the region under the UN Framework Convention on Climate Change's Cancun Adaptation Framework.

Specific benefits of the process outlined in this chapter include:

- Better understanding of scientific gaps that could inform community decision-making and engagement;
- Improved incorporation of scientific information with tourism development activities together with community engagement;
- Improved understanding by the scientific community of community needs and tourism sector development, which will allow for better scientific study of ways to build coastal resilience; and
- Consideration of these results as Caribbean governments develop NAPs that incorporate coastal resilience programs.

Small Island Developing States are the most vulnerable in the world to climate change. While the initial pilot workshop would be in the Dominican Republic, the final toolkit will be designed to be used in most tourism-dependent coastal communities facing similar climate challenges, including US islands and territories (Hawai'i, Guam, Puerto Rico, and US Virgin Islands) and other coastal areas. Lessons learned from the pilot, from after-workshop surveys, and from consultations with other peer experts will be incorporated into the final toolkit design, which will include region-specific guidance to support its implementation in other coastal areas.

The key steps in the process are based on the CARICOM *Risk Management Guidelines for Climate Change Adaptation Decision-Making*[1] outlined in Fig. 11.1.

ACKNOWLEDGMENT

A special thank you to the following collaborators who helped refine the approach identified in this chapter: Maxine Burkett (Jamaica/University of Hawai'i); Nicholas Fields (graduate student, Barbados/Yale University), Jake Kheel (Grupo PuntaCana Foundation, Dominican Republic), Meghan Parker (Wilson Center, Washington, DC), and Kalim Shah (Trinidad and Tobago/Indiana University).

REFERENCES

Amelung, B., Nicholls, S., Viner, D., 2007. Implications of global climate change for tourism flows and seasonality. J. Travel Res 45 (3), 285–296.

Becken, S., Mahon, R., Rennie, H., Shakeela, A., 2013. The tourism disaster vulnerability framework: an application to tourism in small island destinations. Nat. Hazards 71 (1), 955–972.

Caribbean Community Climate Change Centre (CCCCC), 2009. Vulnerability and Capacity Assessment: A Guidance Manual for Conducting and Mainstreaming Vulnerability and Capacity Assessment in the Caribbean Region. Caribbean Community Climate Change Centre (CCCCC).

CARIBSAVE, 2012. Briefing Note: CARIBSAVE Climate Change Risk Atlas (CCCRA).

Kuriakose, A.T., Bizikova, L., Bachofen, C.A., 2009. Assessing vulnerability and adaptive capacity to climate risks: methods for investigation at local and national level. Social Development Working Papers.

Morrow, B.H., 1999. Identifying and mapping community vulnerability. Disasters 23 (1), 1–18.

FURTHER READING

Adger, W.N., Dessai, S., Goulden, M., Hulme, M., Lorenzoni, I., Nelson, D.R., et al., 2009. Are there social limits to adaptation to climate change? Clim. Change 93 (3–4), 335–354.

Allison, E., Perry, A., Badjeck, M., Neil Adger, W., Brown, K., Conway, D., et al., 2009. Vulnerability of national economies to the impacts of climate change on fisheries. Fish Fisher. 10 (2), 173–196.

Association of Caribbean States. Land Cover Maps: Mexican Institute of Statistics and Geography.

Becken, S., Hay, J., 2007. Tourism and Climate Change. Channel View Publications, Clevedon, UK.

Bujosa, A., Riera, A., Torres, C., 2015. Valuing tourism demand attributes to guide climate change adaptation measures efficiently: the case of the Spanish domestic travel market. Tourism Manage. 47, 233–239.

Calgaro, E., Lloyd, K., Dominey-Howes, D., 2013. From vulnerability to transformation: a framework for assessing the vulnerability and resilience of tourism destinations. J. Sustain. Tourism 22 (3), 341–360.

Cannon, T., Twigg, J., Rowell, J., 2003. Social Vulnerability, Sustainable Livelihoods, and Disasters. Community Risk Assessment and Action Planning Project.

CARE International, 2009. Climate Vulnerability and Capacity Analysis Handbook.

CARICOM, 2003. Caribbean Risk Management Guidelines for Climate Change Adaptation Decision Making. Adapting to Climate Change in the Caribbean (ACCC) Project.

Cigliano, J.A., Meyer, R., et al., 2015. Making marine and coastal citizen science matter. Ocean Coastal Manage. 115, 77–87.

Cinner, J., McClanahan, T., Graham, N., Daw, T., Maina, J., Stead, S., et al., 2012. Vulnerability of coastal communities to key impacts of climate change on coral reef fisheries. Global Environ. Change 22 (1), 12–20.

Connors, J.P., Lei, S., Kelly, M., 2011. Citizen Science in the Age of Neogeography: Utilizing Volunteered Geographic Information for Environmental Monitoring. Ann. Assoc. Am. Geogr. 1267–1289.

Conrad, C.C., Hilchey, K.G., 2010. A review of citizen science and community-based environmental monitoring: issues and opportunities. Environ. Monit. Assess. 176 (1), 273–291.

Devictor, V., Whittaker, R.J., Beltrame, C., 2010. Beyond scarcity: citizen science programmes as useful tools for conservation biogeography. Diver. Distrib. 16 (3), 354–362.

Dickinson, J.L., Shirk, J., et al., 2012. The current state of citizen science as a tool for ecological research and public engagement. Front. Ecol. Environ. 10 (6), 291–297.

Diedrich, A., 2007. The impacts of tourism on coral reef conservation awareness and support in coastal communities in Belize. Coral Reefs 26 (4), 985–996.

---

1. Developed under the "Adapting to Climate Change in the Caribbean" project (2001–2004) funded by the Canadian International Development Agency (CIDA) — now known as Global Affairs Canada.

Duarte, C., Losada, I., Hendriks, I., Mazarrasa, I., Marbà, N., 2013. The role of coastal plant communities for climate change mitigation and adaptation. Nat. Clim. Change 3 (11), 961–968.

Ellul, C., Gupta, S., Haklay, M., et al., 2013. A Platform for Location-Based App Development for Citizen Science and Community Mapping. Prog. Location-Based Serv. 71–90.

Few, R., Brown, K., Tompkins, E.L., 2007. Public participation and climate change adaptation: avoiding the illusion of inclusion. Clim. Policy 7 (1), 46–59.

Flores, R., Alberto, R., Taddia, A.P., et al., 2016. Climate change projections in Latin America and the caribbean: review of existing regional climate models' outputs. Inter-Am. Dev. Bank.

Intergovernmental Panel on Climate Change, 2001. Climate Change 2001: Impacts, Adaptation, and Vulnerability.

Kaján, E., Saarinen, J., 2013. Tourism, climate change and adaptation: a review. Curr. Issues Tourism 16 (2), 167–195.

Kelly, P.M., Adger, W.N., 2000. Theory and practice in assessing vulnerability to climate change and Facilitating adaptation. Clim. Change 47 (4), 325–352.

Levine, A.S., Feinholz, C.L., 2014. Participatory GIS to inform coral reef ecosystem management: mapping human coastal and ocean uses in Hawai'i. Appl. Geogr. 59, 60–69.

Matso, K.E., Becker, M.L., 2014. What can funders do to better link science with decisions? Case studies of coastal communities and climate change. Environ. Manage. 54 (6), 1356–1371.

Moreno, A., Becken, S., 2009. A climate change vulnerability assessment methodology for coastal tourism. J. Sustain. Tourism 17 (4), 473–488.

Mycoo, M., 2013. Sustainable tourism, climate change and sea-level rise adaptation policies in Barbados. Nat. Resour. Forum 38 (1), 47–57.

Newman, G., Wiggins, A., et al., 2012. The future of citizen science: emerging technologies and shifting paradigms. Front. Ecol. Environ. 10 (6), 298–304.

O'Brien, K., Leichenko, R., Kelkar, U., et al., 2004. Mapping vulnerability to multiple stressors: climate change and globalization in India. Global Environ. Change 14 (4), 303–313.

Papakonstantinou, A., 2016. Coastline zones identification and 3D coastal mapping using UAV spatial data. ISPRS Int. J. Geo-inf. 5 (6), 75.

Pelling, M., Uitto, J.I., 2001. Small island developing states: natural disaster vulnerability and global change. Global Environ. Change Part B: Environ. Hazards 3 (2), 49–62.

Preston, B.L., Yuen, E.J., Westaway, R.M., 2011. Putting vulnerability to climate change on the map: a review of approaches, benefits, and risks. Sustain. Sci. 6 (2), 177–202.

Roy, H.E., Pockock, M.J.O., et al., 2012. Understanding citizen science and environmental monitoring: final report on behalf of UK Environmental Observation Framework. Centre for Ecology & Hydrology, p. 173.

Ruckelshaus, M., Doney, S., Galindo, H., Barry, J., Chan, F., Duffy, J., et al., 2013. Securing ocean benefits for society in the face of climate change. Marine Policy 40, 154–159.

Scott, D., 2011. Why sustainable tourism must address climate change. J. Sustain. Tourism 19 (1), 17–34.

Scott, D., Simpson, M., Sim, R., 2012. The vulnerability of Caribbean coastal tourism to scenarios of climate change related sea-level rise. J. Sustain. Tourism 20 (6), 883–898.

Shakeela, A., Becken, S., 2014. Understanding tourism leaders' perceptions of risks from climate change: an assessment of policy-making processes in the Maldives using the social amplification of risk framework (SARF). J. Sustain. Tourism 23 (1), 65–84.

Sheppard, S.R.J., Shaw, A., et al., 2011. Future visioning of local climate change: a framework for community engagement and planning with scenarios and visualization. Futures 43 (4).

Silvertown, J., 2009. A new dawn for citizen science. Trends Ecol. Evolut. 24 (9), 467–471.

Smit, B., Pilifosova, O., 2003. Adaptation to climate change in the context of sustainable development and equity. Sustain. Dev. 8 (9), 9.

Soden, R., Palen, L., 2014. From Crowdsourced Mapping to Community Mapping: The Post-earthquake Work of OpenStreetMap Haiti. In: Proceedings of the 11th International Conference on the Design of Cooperative Systems, pp. 311–326.

Tamondong, A., 2016. A framework for capacity building in mapping coastal resources using remote sensing in the Philippines. In: International archives of the photogrammetry, remote sensing and spatial information sciences (XLI-B6), pp. 149–153.

United Nations Convention on Biological Diversity. National Reports.

United Nations Framework Convention on Climate Change. Non-Annex I national Communications.

United Nations Office for Disaster Risk Reduction. National Policy, Plans & Statements.

Van Aalst, M.K., Cannon, T., Burton, I., 2008. Community level adaptation to climate change: the potential role of participatory community risk assessment. Global Environ. Change 18 (1), 165–179.

Weaver, D., 2011. Can sustainable tourism survive climate change? J. Sustain. Tourism 19 (1), 5–15.

Widmer, J., 2016. Veron Urban Survey. *University of Florida*, Gainesville FL.

Yusuf, A.A., Francisco, H., 2009. Climate change vulnerability mapping for Southeast Asia. In: Economy and Environment Program for Southeast Asia (EEPSEA), Singapore, pp. 10–15.

Chapter 12

Building Urban Resilience to Address Urbanization and Climate Change

Julie Greenwalt[1], Nina Raasakka[2] and Keith Alverson[3]

[1]Cities Alliance, Brussels, Belgium, [2]UN Environment, Bangkok, Thailand, [3]International Environmental Technology Center, UN Environment, Osaka, Japan

Chapter Outline

12.1 URBANIZATION AND CLIMATE CHANGE: DEFINING THE 21ST CENTURY

Climate change and urbanization are increasingly defining our planet yet they are both characterized by inequitable impacts and inadequate responses interspersed with innovation and new visions for the future. Despite different drivers, these two trends are shaping the state of the planet with the most impact in the rapidly urbanizing developing countries in Africa, Asia and the Pacific, and Latin America and the Caribbean.

The 21st century has been defined by a historic shift in urbanization rates so that now over half of the world's population lives in cities and this is expected to increase across all regions in the coming decades. The number of megacities of over 10 million inhabitants, currently at 27, is expected to grow. For instance, in the Asia and the Pacific region, urbanization has been described as a "megatrend," and it is projected that by 2030, 28 of the 41 megacities will be located in the region (UNEP, 2016a,b, 150). Although these megacities often dominate headlines, research, and regional economies, they actually house a minority of urban inhabitants; as almost half of the urban population live in cities of fewer than 500,000 inhabitants. The most growth in cities of all sizes is projected for Africa and Asia and the Pacific.

Urbanization has enhanced economic development and improved the lives of many people in the past few decades, however basic living conditions still elude millions of urban residents around the world. Despite progress in improving slums and decreases in the percentage of people living in slums (from 39% to 30%), there has been an increase in absolute numbers with over 800 million people living in slums in 2015 (UN, 2015).

Cities' share of global carbon dioxide emissions are estimated at 75%, however studies about the collective contribution of all urban areas to global greenhouse emissions (GHG) are limited as emissions vary by country and even between cities within the same country, although generally cities in developed countries have higher emissions (Seto et al., 2014). Cities in developing countries are generally more vulnerable to climate change, and thus have greater adaptation needs, in contrast to their relatively low GHG emissions, as compared to cities in the developed world. In 2014, over 50% of cities with at least 300,000 inhabitants were at high risk of exposure to at least one of six types of disasters (cyclones, floods, droughts, earthquakes, landslides, and volcano eruptions) (UN DESA, 2016). This high level of exposure combined with high levels of vulnerability in expanding informal settlements in urban and peri-urban areas has negative implications that may counteract any sustainable development.

Resilience. DOI: https://doi.org/10.1016/B978-0-12-811891-7.00012-8

Understanding how urbanization and climate trends will interact and how that impacts on potentially vulnerable groups, political equity, and urban governance will be central to sustainable urban development in the coming decades (Martine and Schensul, 2013).

Ecosystems, both at local and peri-urban scale, play a significant role in sustaining some of the diverse and intertwined components of natural features that make urban landscapes safe and liveable environments. Local ecosystems provide a variety of services that directly benefit urban dwellers, such as water drainage (through permeable areas that soak up precipitation) and provision of shade and absorption of heat by foliage (to counter the urban heat island effect for more detail on urban heat island effect please see Chapter 7: The Science of Adaptation to Extreme Heat). Peri-urban ecosystems provide provisioning and regulating services more indirectly, such as watersheds that maintain water quality, as well as wetlands, mangroves, or forests that can help mitigate the impacts of storm surges in coastal areas or prevent soil erosion on banks and riverbeds due to the complex root systems of the vegetation (Guadagno et al., 2013). Urban land expansion rates have been higher than urban population growth in the past few decades and this trend is expected to continue (Seto et al., 2011). The inefficient expansion of urban areas causes increased conversion of agricultural lands, forests, and other ecosystems (among other impacts) and increased pressures on environmental quality, which is exacerbated by significant climate extremes projected across all regions. Fig. 12.1 represents how urban population growth and urban land expansion is increasingly encroaching on biodiversity hotspots around the world. This is due partly to the high endemism of species concentrated in, or close to, the urban areas inhabited by humans along coastlines and islands, presenting a threat to biodiversity and ecosystem productivity as well as increasing the need for adaptation and mitigation in and around those cities (Seto et al., 2012).

In high-income countries, improving urban spaces is critical for reducing GHG emissions and there is great potential for multiple benefits from upgrading urban areas—increased mobility, more reliable energy sources, reducing vulnerabilities to disaster while transitioning to a low-carbon economy powered by cities.

In developing countries, where the majority of urbanization is set to occur, socioeconomic potential is critically interlinked with urban development and city expansion. With urbanization and limited housing options, many people are living in informal settlements in vulnerable areas, raising questions about how to prevent climate hazards and antiurbanization policies from combining to put some of the most vulnerable urban dwellers in Africa and Asia at risk (Martine and Schensul, 2013).

There is a positive correlation between human development, measured by the Human Development Index, and the percentage of urban population in the country (UN DESA, 2006). Given that much of the urban expansion projected over the coming decades is yet to occur in Africa and Asia, there is significant potential for socioeconomic development as well as for infrastructure construction and adopting planning solutions that support a climate-smart approach instead of getting locked into infrastructure that promotes high-carbon lifestyles. Further, the potential benefits for a healthier environment from reducing pollution and improving water and food security can make the case for the return on investments in sustainable urban development from Latin America to sub-Saharan Africa. However there is also a danger that cities are not able to address underlying causes of inequality, integrate climate action and environmental consideration into planning, such that negative trends are reinforced.

12.2 URBAN TROUBLES TRIFECTA: ENVIRONMENTAL DEGRADATION, CLIMATE CHANGE, AND VULNERABLE POPULATIONS

Air, land, and water pollution have been major issues affecting the liveability of cities worldwide for many decades, whether it is the solid waste accumulation on riverbanks and coastlines; red alert air pollution days; or malodorous, dirty, nonpotable water in rivers and lakes from the dumping of chemicals and waste. The source of degradation spans across sectors and includes industrial pollution from factories, lack of residential sanitation, poor waste management, and coal burning stoves. Many cities in developed countries from Singapore to London have worked to address pollution and environmental degradation, from the implementation of national policies such as clean air acts and vehicle emission regulations to city-led clean-ups and recycling systems, however there are still many neighborhoods and cities in developed countries that are not free from environmental crises and polluted environment, e.g., the lead water contamination in Flint, Michigan. Furthermore, the ability of ecosystems to provide regulating services to combat this pollution is lessened as a result of multiple, interconnected factors impacting the environment in and around cities, compounding the problem.

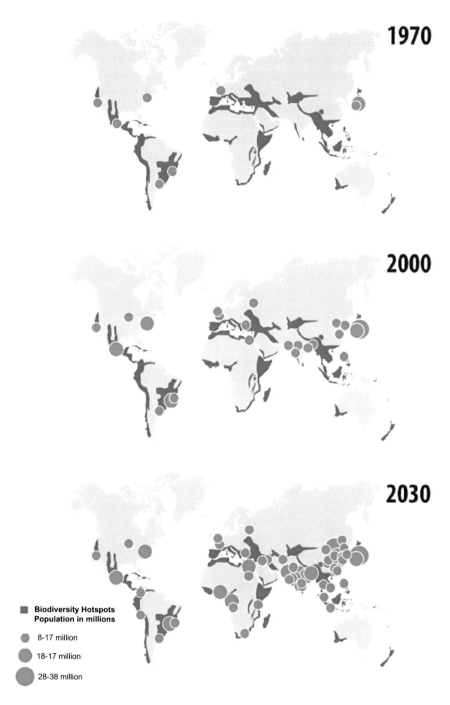

FIGURE 12.1   Urban expansion (*green* (dark gray in print versions) areas indicate urban areas with large populations) in biodiversity hotspots (in *blue* (light gray in print versions) color) (McPhearson et al., 2017).

The Fifth Assessment of the IPCC confirms that there are many climate change-related risks for urban areas and they are on the rise (Revi et al., 2014). The risks include: rising sea levels, storm surges, heat stress, extreme precipitation, inland and coast flooding, landslides, drought, increased aridity, water scarcity, and air pollution. These risks are heightened for those living in informal settlements, lacking housing, services, and infrastructure. Climate change effects will exacerbate the existing stressors on important urban and peri-urban ecosystems and the valuable services they provide that benefit city-dwellers both directly (e.g., cool spaces provided by shade in an urban forest areas or park) and indirectly (e.g., watersheds for clean water provision to cities).

Currently half of Asia's urban population is located in low elevation coastal zones, and places like Bangladesh, with megacity Dhaka, may lose up to 17.5% of their urban areas, leaving millions homeless with sea level rise of just one meter (McGranahan et al., 2007). In Africa, there are multiple large, medium, and small cities—most of which are growing—and economically important to the region, e.g., Lagos, Accra, Dar-es-salaam in low elevation coastal zones around the entire continent. While the impact of sea level rise could be felt citywide, in many cities, such as Monrovia, the informal settlements are in areas extremely vulnerable to sea level rise so those communities will see the greatest impacts. Unless the urban land issues that exclude low-income households from formal housing are addressed, it is difficult to see how equitable adaptation efforts in coastal cities will happen (Martine and Schensul, 2013).

Beyond sea level rise, water scarcity and quality are major issues in urban areas, especially in Asia, Africa, and Small Island Developing States worldwide. African cities are heavily reliant on surface water with one in five urban residents primarily dependent on surface water supply; they may face water security risks due to high sediment or nutrient (TNC, 2016). In addition, 46% of the urban population lives in dryland cities. In Asia, approximately 36% of urban dwellers (540 million people) live in dryland cities and it is projected that 94 million people will be living in cities with perennial water shortages in 2050 (UNEP, 2016a,b, 152). This problem may be more prevalent in growing secondary cities than primary cities. A recent study of 34 Afghan cities illustrated that the five major cities were located in areas with plentiful freshwater sources but the secondary cities had less plentiful and reliable water sources (Government of the Islamic Republic of Afghanistan, 2015).

Water shortages are not just an issue in developing countries with recent water crises in California and southern England impacting on residents and the agricultural industry. As with sea level rise, it is important to keep the perspective that water disproportionately affects those living in informal settlements and with limited finances to adjust to the higher costs for water (Balk et al., 2009).

Climate change will also impact the built environment and large-scale infrastructure including sanitation, transport, and telecommunications as well as straining existing health and emergency services. A major area where climate change will have an impact is on the energy provision to cities. Changes in demand and supply are likely to occur as a result; for instance, air-conditioning demand is expected to increase in line with increasing number of extremely hot days (see Chapter 7: The Science of Adaptation to Extreme Heat). In addition to this, large-scale electricity production that supplies cities with their power supply such as coal or hydropower may be impacted by changes in streamflow affecting the cooling water available for coal-fired power plants or water flow for hydropower plants. In Albania, 90% of domestic energy is generated through hydropower from the River Drin. Already, low rainfall in the years 2002−08, coupled with existing inefficiencies in domestic energy supply, was linked with lower domestic power production. One study estimates that a 20% decrease in runoff driven by climate change could lead to 15% less electricity generation from large hydropower plants by 2050 in Albania, with countrywide effects on various sectors (Ebinger et al., 2009). Adequate planning for climate-responsive building codes that incorporate risk and adaptation measures appropriately, e.g., active ventilation systems are needed in order to foresee possible restrictions in energy supply.

Despite this forecast, most adaptation efforts to date have focused on rural areas. In part this was based on a false perception that people are more vulnerable there, because incomes are lower. But vulnerability is not solely about income and it is clear that land tenure, housing, basic services, and connectivity issues which marginalized urban residents face are being exacerbated by climate change. The focus on rural adaptation is also because there is a perception that projects funded from the outside are easier to implement in rural areas, where stakeholders are limited in number, and their patterns of vulnerability are more straightforward and linear. Farmers need water and soil productivity. Fishermen need to live near the coast. Relatively simplistic maps can reasonably approximate vulnerability and impact patterns, and interventions can be designed based on these. In cities adaptation is a far more complex problem.

The environmental degradation and climate change projections for cities are compounded by the vulnerability of urban poor marginalized, living in informal settlements, and working in the informal economy. This becomes most apparent when natural disasters hit. The Labutta administrative township area, located in the southern tip of Myanmar's Irrawaddy region and home to around 315,000 people, with a population of around 46,000 living in Labutta town itself, was severely damaged by Cyclone Nargis in 2008. The township is still recovering from the cyclone, which resulted in 130,000 deaths and widespread damage. Labutta remains highly vulnerable to the impacts of climate change, due to interconnecting socioeconomic, ecological, and infrastructure conditions as demonstrated by a recent vulnerability assessment (Fee et al., in press). For instance, the highly productive deltaic mangrove ecosystem in Labutta is degrading rapidly due to land conversion for agro-industry and domestic fuelwood. The rate of deforestation is so fast that with current trends, mangrove cover will be entirely lost by 2019, further increasing sensitivity of the township's population to climate change and events similar to the deadly Cyclone Nargis.

Women are generally more vulnerable to changes in the climate than men, including in the urban context. This is due to their greater sensitivity to the primary and secondary impacts from natural disasters combined with a lesser access to critical resources such as education, finance, and as a consequence, land, as a result of higher poverty rate and social norms. During natural disasters, women's and children's fatalities are higher than men's owing to a number of factors (UN-Habitat, 2015) and it is expected that this would also apply to urban context if gender mainstreaming measures are not integrated into urban resilience planning.

12.3 TOWARDS URBAN RESILIENCE AND GLOBAL SUSTAINABLE DEVELOPMENT

12.3.1 Multilateral Agreements on Cities, Climate Change, and Sustainable Development

Critical global milestones were reached in the period 2015−16 that have set the scene for sustainable development and action on climate change. These international agreements are defining a vision of inclusive, low-carbon green cities. Some cities have also developed strategies and plans for resilient development, low-carbon development, and/or climate change adaptation but much work still needs to be done. Networks of cities such as C40—Cities Climate Leadership Group, ICLEI—Local Governments for Sustainability, and 100 Resilient Cities are building capacity and commitment at the local level while providing a platform for sharing knowledge and experience. In addition, commitments such as the Global Covenant for Climate and Energy are engaged directly with cities and raising the profile of cities in relation to global agreements. These networks as well as others working in the field such as Cities Alliance, UN Environment, UN-Habitat, and World Resources Institute (WRI) have contributed to improved adaptation in cities through technical support to the development of adaptation and resilience plans at the city-level, e.g., the Resilient City Strategies of 100RC; tools such as the Urban Community Resilience Assessment by WRI and the Climate Risk Adaptation Framework and Taxonomy (CRAFT) by C40; and attempts to connect local initiatives with financing opportunities as with the Transformative Actions Program by ICLEI. These efforts have increased political will, recognition of the need for urban adaptation and resilience, and support exchange of knowledge and experiences across a wide array of cities. Conversely, adaptation in cities has been stifled by the difficulty of measuring adaptation and resilience in the urban context, lack of local capacity, lack of integrated or mainstream approaches which address adaptation, mitigation, DRR, and the broader basic functioning of a city. These problems are compounded by the significant gap in finance between mitigation and adaptation, which climate funds such as the Least Developed Countries Fund, Special Climate Change Fund, and the Green Climate Fund (GCF) aspire to closing, but this remains a work in progress.

12.3.1.1 Agenda 2030

Cities play an important role in the United Nations 2030 Agenda for Sustainable Development, the comprehensive global vision for attaining sustainable development, adopted by UN member states in 2015. A dedicated goal on cities has been included as one of the 17 Sustainable Development Goals (SDGs): *Make cities and human settlements inclusive, safe, resilient and sustainable* (Goal 11). Despite a dedicated goal cities will play a critical role in the implementation of all 17 of the SDGs, e.g., environmental sustainability in cities will directly contribute to the achievement of Goal 2 (zero hunger), Goal 3 (good health and well-being), Goal 6 (clean water and sanitation), Goal 7 (affordable and clean energy), Goal 12 (responsible consumption and production), and Goal 13 (climate action) as shown by Fig. 12.2.

12.3.1.2 Paris Agreement on Climate Change

The Paris Agreement, adopted at the 21st Conference of Parties (COP21) of the United Nations Framework Convention on Climate Change (UNFCCC) in December 2015, pledges to keep global temperatures from rising more than 2°C by 2100 with an ideal target of keeping temperature rise below 1.5°C. The Nationally-Determined Contributions (NDCs) provide an indication of how cities need to be involved with both mitigation efforts to reduce emissions and promote adaptation initiatives, especially in developing countries. A rapid review of NDCs demonstrated that 110 countries had urban content in their NDCs with either explicit recognition of cities or with regards to urban-oriented sectors like transportation and energy (UN-Habitat, 2016). The role of cities and other subnational authorities has been increasingly

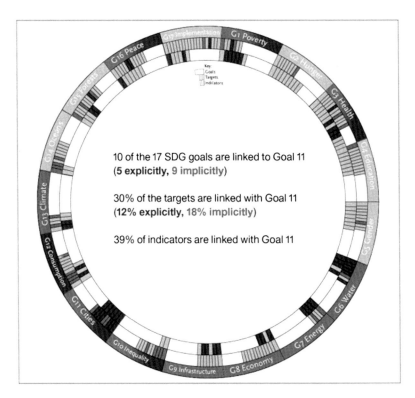

FIGURE 12.2   The urban dimension of Agenda 2030 illustrated by the numerous explicit and implicit linkages between Goal 11 and other goals, targets, and indicators of the SDGs. *Cities Alliance, 2015. Sustainable Development Goals and Habitat III: Opportunities for a successful New Urban Agenda. Brussels, Belgium (Cities Alliance, 2015).*

acknowledged in the UNFCCC, which encourages efforts to scale up actions to reduce emissions and/or to build resilience and decrease vulnerability to the adverse effects of climate change. The Lima Paris Action Agenda included a focus on Cities and Subnationals and now the Marrakech Partnership for Global Climate Action adopted by UNFCCC Member States at the 22nd Conference of the Parties (COP22) in 2016 includes a section on Human Settlements. On June 1, 2017, the United States of America (USA) announced its intention to withdraw from the Paris Agreement which many worry will reduce the effectiveness of the agreement and limit the availability of climate finance for developing countries via the GCF and others. However, the decision at the national level in the United States also demonstrated the leadership role that cities and subnationals can play, with many cities and states coming forward to pledge their continued commitment to the Paris Agreement with mayors from Pittsburgh to Salt Lake City showcasing efforts already underway and commitments for the future—especially on energy and transportation.

12.3.1.3 New Urban Agenda

The New Urban Agenda, adopted at the Habitat III conference in Quito, Ecuador in October 2016, sets out a roadmap for sustainable and inclusive development in cities with a large recognition of the issues of climate change, pollution, and resilience. With the New Urban Agenda, UN member states committed to promoting cleaner cities, especially reducing air pollution, and to increase the use of renewable energy, which will eventually make cities more resilient in terms of social aspects. It also contained strong wording on improving urban planning, infrastructure, and local responses to minimize the impact of disasters as well as the importance of establishing partnerships with businesses and civil society for innovative green and sustainable solutions. Specifically on climate change, the New Urban Agenda was especially focused on the involvement of local government and other local partners, such as civil society to take climate action in cities.

12.3.1.4 Sendai Framework

A fourth critical global agreement is the Sendai Framework for Disaster Risk Reduction 2015—30, which was adopted in March 2015 at the Third United Nations World Conference on Disaster Risk Reduction. The goal of the framework is to substantially reduce disaster risk and losses in lives, livelihoods, and health in communities and countries with specific recognition of the need to address land use and urban planning, building codes, and environmental and resource management. The Sendai Framework includes seven targets and the UNISDR Making Cities Resilient campaign has emphasized what cities can do to reduce disaster risk and meet these seven targets.

12.3.2 Urban Resilience: Local Action, Cooperation and Understanding

While these global agreements are critical for setting standards and communicating political commitment, they do not reach maximum utility without the requisite capacity, governance, knowledge, and finance at the local level.

12.3.2.1 Urban Form for the Future

Although further urbanization presents challenges in cities with environmental degradation, it also presents opportunities to improve resource efficiency and mitigate climate change. Compact, mixed-use urban development can reduce energy consumption, transport emissions, and carbon dioxide emissions. (Seto et al., 2011).

While cities in developed countries have many assets in terms of technological innovation and limited land conversion due to urban expansion, the Fifth Assessment report of the IPCC from Working Group II highlighted that there are great opportunities for mitigation and adaptation in cities that are still growing because decisions on institutions and infrastructure are yet to be taken rather than already being set in stone (Revi et al., 2014). This opportunity is especially relevant in the rapidly expanding small and medium-sized cities, however capacity and finance may constrain the extent to which these secondary cities can capitalize on this opportunity.

The urban expansion in developing countries provides an opportunity to ensure that these cities are not locked in to inefficient urban forms with high emissions and increased degradation and conversion and degradation of ecosystems. However, the optimum urban form for sustainability depends on which issues are most important for those cities as denser areas are better for mitigating climate change but can result in negative externalities such as air pollution affecting more urban residents (Lohrey and Creutzig, 2016). The urban form of a city is shaped by many factors, including government regulation, housing needs of the population, and economic development. Similarly, the impact of cities on ecosystems is influenced by varying factors including consumption and production patterns of urban residents, transportation infrastructure, and the geographical constitution of the urban and peri-urban areas. Conceptualizing these changes as a process and utilizing a process-based framework for analyzing urbanization and land use change is useful for identifying areas for intervention to improve sustainability (Seto et al., 2012b).

Community and Climate Change Adaptation in Cities

There is a rich history of engaging community groups in environmental efforts such as biodiversity conservation in rural areas and in the past decade, Community-Based Adaptation (CBA) has been recognized as an important methodology for sustainable climate change adaptation at the local level. Efforts for community-led adaptation or incorporating local knowledge or local community structures have largely been limited to rural areas. Conversely in cities, there have been growing efforts of community engagement and the formation of community groups—in developing and developed countries alike—to address a wide array of issues community parks, urban agriculture, microsavings and loans, physical mapping and environmental campaigns. While climate change is increasingly on the radar of citizens in rural and urban areas, there have been limited attempts to draw together the lessons from CBA application in rural areas with the dynamic community efforts in cities.

Research on mangroves has demonstrated that when communities are empowered and have legitimate rights to the land, there are ecological benefits to the mangroves (Brooks, 2017). Similarly, in urban areas, efforts to organize community members around one project can have additional, unintended benefits. For example, a UN-Habitat project in Afghanistan to create Community Development Councils in the five major cities to spend community funds also helped to increase a sense of community and also create a space for women to engage with each other. In Port Harcourt in Nigeria, the city and surrounding areas struggle from severe environmental degradation as a result of the oil industry and lack of basic urban services. The Human City Project in Port Harcourt is a community-driven media, architecture, urban planning, and human rights initiative. The initiative provides skills training on media and advocacy, while increasing local

(Continued)

(Continued)

knowledge through community-led mapping—bringing people together to empower them to plan and build their neighbourhood in sustainable ways (www.cmapping.net). Initiatives such as this in urban areas have not typically been labeled CBA but they are vital to reducing vulnerability, building resilience, and adapting to climate change at the neighbourhood level (Figs. 12.3—12.4).

FIGURE 12.3 Flooding in Port Harcourt, Nigeria due to heavy rains and improper drainage. Photo credit: Michael Uwemedimo/CMAP.

FIGURE 12.4 Community-led mapping of informal settlements in Port Harcourt, Nigeria. Photo credit: Michael Uwemedimo/ CMAP.

12.3.2.2 *Urban Ecosystem Services and Ecosystem Based Adaptation Application*

As urbanization continues, especially in many developing countries and emerging economies in the world, there is an opportunity to shape urban form both at the core and in peri-urban areas with a recognition of how urban areas impact on ecosystems. Urban areas, like ecosystem services, transcend both habitat boundaries and governance and administrative boundaries resulting in challenges for determining both causality and the best management modalities. However, in order to promote interventions which are beneficial for people and the environment, it is important to understand how urban development impacts on ecosystem functioning and how in turn, ecosystems can continue providing services that benefit urban areas and increase the well-being of their inhabitants, for instance through reducing sensitivity to climate change impacts and sequestering carbon.

Climate change is already having negative effects on the provision of certain ecosystem services; while increased consumption, land expansion and conversion in and around urban areas is destroying valuable habitats. For instance, in Gorakhpur in India, the loss of water bodies and permeable areas that absorb monsoon rains and construction on floodplain zones, combined with increasingly intense precipitation periods and insufficient maintenance of waterways, have resulted in a change to the nature of the flooding and waterlogging hazards for the city (Mitra et al., 2015).

Ecosystem dynamics in urban landscapes are poorly understood, especially when it comes to designing, creating, and restoring ecological processes, functions, and services in urban areas (Elmqvist et al., 2015). The link between enhanced ecosystem services, sustainable development, and urban resilience is not well known, partly owing to a lacking evidence base and little research. Improved integration of ecosystem services into urban planning to enhance complementary land use and green infrastructure provides great opportunities for strengthening ecosystem functions and restoring degraded ecosystems, but this integration has been limited to date.

Ecosystem-based adaptation (EbA) is the use of biodiversity and ecosystem services as part of an overall adaptation strategy to help people, including those living in cities, to adapt to the adverse effects of climate change (CBD, 2010). EbA has gained attention at the international level as a cost-effective adaptation measure and promising approach also in the urban context, whereby it refers to harnessing the diverse ecosystem services which exist in urban and peri-urban systems to help city dwellers adapt to the adverse impacts of climate change. Indeed, the use of ecosystems for urban climate change adaptation and disaster risk reduction is a rapidly growing research area, and one with huge potential and huge challenges (Brink et al., 2016).

Effectiveness of Nature-Based Solutions and EbA for Adaptation in Cities and Urban Areas

The use of nature and ecosystems to help urban communities increase their wellbeing while at the same time adapt to climate change and address environmental degradation has become increasingly recognized in the last decade as a vital area for further research and action. Numerous complementary and mutually reinforcing concepts are in use, ranging from "nature-based solutions" (NBS) to "urban green infrastructure" (UGI) in addition to EBA. "NBS" remains an umbrella term that describes solutions inspired and supported by nature to provide economic, social, and environmental benefits, EbA refers specifically to adaptation.

These approaches complement "engineering" or "grey" solutions such as dikes, levees, seawalls for coastal protection; drainage systems for floodwater managements and air-conditioning or cooling centers for heat waves have generally been more used in helping cities to adapt to climate change as they provide a rapid, replicable, and mostly durable protection which is suitable to the urban planning context. However, these solutions are also hugely resource-intensive and risk locking in infrastructure that may not be sufficiently flexible to cope with the climate extremes emerging in the coming decades if these bypass the most recent projections under which they were planned. For instance, coastal protection structures failed when hurricanes Katrina (2005) and Sandy (2012) hit New Orleans and New York City, overtopping the levees and sea walls that were engineered to protect inhabitants.

The existing long-term evidence base for EbA in both urban and rural areas is still scarce (Depietri and McPhearson, 2017), although results from smaller scale pilot studies (Rao et al., 2012; Badola and Hussein, 2005) indicate that green interventions may bring considerable benefits particularly in terms of coastal protection, with a large potential for cobenefits.

While both type of technologies have benefits to building cities' resilience, there is also merit in combining them to create a hybrid "green—gray" approach. In the example of Lami Town located on Fiji, a comparative analysis of EbA and engineering options including a mixed approach combining both revealed that the greatest benefit-to-cost ratio was associated with EbA options, with a benefit of $19.50 for every dollar spent, with an assumed damage avoidance of 10%—25%. Use of a mixed approach with an emphasis on ecosystem adaptation options could increase the assumed damage avoidance up to 25% with a benefit-to-cost ratio of $15 (Rao et al., 2012). According to Depietri and McPhearson (2017) a mix of gray and green infrastructure options merits substantial consideration to improve the management of urban water, heat and other climate-driven risks by urban planners, who should review the use of hybrid approaches as a more flexible and sustainable approach to adaptation to climate change.

(Continued)

(Continued)

The majority of academic studies to assess the effectiveness of NBS and EBA solutions originate from urban areas in European countries where urban planning is already quite advanced (Geneletti and Zardo, 2016; Wamsler, 2015; Frantzeskaki et al., 2017). There is a critical need now to further assess how EBA and gray—green infrastructure solutions could be applied and scaled up in megacities in developing countries, where governance frameworks are weaker and where the rapid rate of urban expansion does often not allow for comprehensive urban planning in ways that can integrate long-term resilience and sustainability considerations.

Urban EbA solutions offer a strong case for building resilience and improving the liveability of cities in a cost-effective manner, however, their time frame for success differs from traditional infrastructure solutions. For instance, reforestation, afforestation, and forest conservation contribute to maintaining healthy watersheds, which provide water management benefits such as riverine flood control and regulating water supply and quality through their root systems; however these benefits can take more time to emerge as compared the installation of designated infrastructure, see example, of Thames flood barriers in Chapter 6, Flood Risk Management in the United Kingdom: Putting Climate Change Adaptation Into Practice in the Thames Estuary, of this volume. Additionally, the disconnect between short-term actions and decision-making cycles and long-term goals can present a barrier where nature-based solutions for building resilience in urban areas are concerned (Kabisch et al., 2016). Adopting a long-term view would allow the recognition of additional cobenefits that accrue and appreciate such as carbon sequestration, biodiversity benefits, improved air quality, and recreational opportunities, among others. Unlike grey infrastructure, ecosystems or other types of "green infrastructure" can even appreciate over time, resulting in potential cost savings but only when measured over a sufficiently long timescale (UNEP, 2014).

Although a number of studies have been carried out on urban EbA, there is still plenty of scope for the expansion of the evidence base on urban EbA interventions' effectiveness for building resilience (Brink et al., 2016; Kabisch et al., 2016). For instance, valuation of ecosystem services and their comparison with "hard" infrastructure-focused adaptation interventions can demonstrate the return on benefits such as cultural acceptance and quality of life, which may turn out to generate highly positive valuation for ecosystem-services.

While ecosystem-based approaches to adaptation are gradually starting to be utilized in urban areas, engaging the local urban community has largely been missing. Urban residents need to be engaged in a participatory process to identify the biggest risks (e.g., flooding) and their greatest needs (e.g., sanitation) to ensure that EbA solutions can serve the requirements of urban communities. Urban EbA also presents an opportunity for developing new or improved livelihoods based on rehabilitated ecosystems for urban residents, especially the urban poor. Brink et al. (2016) and Geneletti and Zardo (2016) raise a related point that future research should concentrate on power relations and potential increase in inequalities arising from EbA implementation in urban areas, as this often results in use of large areas of land that may cause conflicts with other land uses especially where land is scarce. This reinforces the notion that for EbA interventions to be most effective for cities, the involvement of local communities as well as relevant local and regional authorities—including those involved in energy, transport, and housing—is crucial to ensure that interventions are sustainable and meet the needs of the population. There are diverse entry points for mainstreaming EbA; however, cities that have integrated general environmental and climate change considerations appear to have progressed better in adaptation mainstreaming according to Wamsler (2015), hence raising awareness on environmental conservation and climate change issues among communities can also advance the EbA agenda. Science and technology from both the urban planning and environment fields, such as GIS for spatial planning and vulnerability assessments to develop city-specific options, could be better integrated to find site-specific solutions. Recognizing the above, projects piloting EbA interventions in urban settings that involve training policy-makers and sharing knowledge regionally and globally have been recently approved by global funding bodies such as the Global Environment Facility.

Waste Management for Urban Resilience

When disaster strikes a city, the first difficulty that many residents and humanitarian responders face is clearing debris. Such disaster waste impedes the efficiency and effectiveness with which early response and recovery can be carried out. Furthermore, in the frenzy of immediate post disaster activities, waste may be collected and dumped indiscriminately. Such actions ignore potentially positive use—such as rubble (concrete, steel, timber) as a resource for reconstruction—as well as potential health

(Continued)

hazards associated with failure to separate hazardous wastes (asbestos, pesticides, oils, and solvents) or provide long-term well-managed repositories. Finally, improper treatment of waste in post disaster situations, including healthcare waste, often leads to infection and disease outbreaks, such as cholera. In addition to impeding post disaster actions, waste can exacerbate the impacts of the disasters themselves. For example drainage systems are more easily clogged if waste is plentiful, exacerbating flooding, and masses of floating debris can cause more structural damage than the water itself. Extreme events, whether "natural" or "man made," often make it difficult to properly manage normal waste streams. The possibility of collapse of urban municipal solid waste services, including possible loss of experienced waste managers and facilities can lead to lack of collection services and uncontrolled dumping of waste. Many urban sewage systems are designed to overflow in the event of a 1 in 100-year flooding event, thereby minimizing damage to the sewage system itself, while tolerating infrequent discharge directly into the environment. Unfortunately, with climate change causing such events to occur more frequently such safety overflows may no longer be infrequent or environmentally sound. Moreover, responsibilities for emergency response and waste management are inevitably in different departments or agencies, leading to administrative obstacles for disaster waste management. Preventive waste management is thus an important component of disaster risk reduction. In some cases, large open dumpsites can even themselves be the source of disasters for nearby residents, when intense rainfall causes deadly artificial landslides of waste.

In addition to assisting with preparation for and response to disasters, waste management is a key component of everyday resilience building in the urban environment. High population density in urban areas necessitates sound waste management for economic success, for human health and well-being, and for healthy urban and peri-urban ecosystems. As a result, sustainable access to the financial, social, and environmental capital that underpins resilience requires sound management of waste. Solutions include reuse, reduction, and recycling of waste, and integrated national and local waste management strategies with disaster risk reduction and response plans. In a resilient city waste is minimized, and where and when it must exist is used as a resource—be it from recycling, waste to energy electricity and heat generation, post disaster rebuilding materials, or industrial symbiosis (Fig. 12.5).

FIGURE 12.5   Waste meets water in Monrovia, Liberia; photo credit: Julie Greenwalt/Cities Alliance.

12.3.2.3   Improving Governance and Service Delivery to Build Resilience

The approach for building city resilience to climate change and disasters is not radically different from the solutions needed to address urban issues which have been plaguing cities for decades, such as better infrastructure, improved service delivery, affordable housing, proper solid waste management, etc. Addressing many of the current social, economic, and environmental problems in cities combined with education about the changing environment and its risks, would increase resilience in local communities. An integrated approach to governance, infrastructure, housing, economic growth, and urban planning would improve livelihoods and result in sustainable cities. For example, developing well-located quality affordable housing could steer new urban development away from locations likely to be prone to future climate hazards and address the housing issues that many cities struggle with.

Improving governance of cities to address issues of sprawl is fundamental to urban resilience and to sustainable natural resource management and maintaining an ecological balance on the planet. Building resilience for cities cannot be restricted to administrative boundaries and therefore need intermunicipal cooperation as the urban landscape extends beyond the city boundaries and peri-urban areas are critical for ecosystem services, adaptation, and managing sustainable urban land expansion.

Cities are central to a sustainable solution not only for countries but also globally. Building human and institutional capacity on adaptation and mitigation at the local level is crucial. Cooperation between national governments, local governments, and the scientific community is critical for improvements in infrastructure, urban planning, economic development, conservation of ecosystems, and addressing climate change. This cooperation can also be utilized to strengthen local government and community adaptation capacity. Long-term monitoring and research of EbA intervention results and their comparison with and combination with alternative adaptation approaches (namely hard infrastructure) is critical in order to build an evidence base to chart the costs, benefits, and cobenefits of ecosystem-based solutions in the long run in order to further determine their suitability as local measures and to provide a stronger rationale for decisions involving EbA (Geneletti and Zardo, 2016).

To truly achieve resilient development pathways, in cities as well as countries, requires two paradigm shifts. We need to move beyond using GDP—a crude measure of national annual income—as an indicator of our development success, and focus more on sustaining national wealth, including social, environmental, and economic capital assets. GDP income that is generated solely, or mostly, by decreasing wealth, is not sustainable. Furthermore, development and adaptation actions need to move beyond projects as a means of implementation to a more holistic definition of success in development and to transformational, system-wide change that becomes the norm.

Given the plethora of international agreements, national policies, and in some cases city plans, it is important to align policies and incentives for sustainable development, climate change adaptation, mitigation, DRR, and other high-priority urban agendas within cities to streamline effective development that addresses the diversity of goals without excluding different communities.

Solutions vary by country, city, and local community as a dryland secondary city in Ethiopia like Dire Dawa faces different challenges than a coastal megacity in Asia such as Dhaka or a city working on urban renewal as with Detroit, USA, both in terms of urbanization and pressures as well as climate change impacts. However, sustainable development in cities across the globe will be intrinsically linked to adaptation to climate change that is equitable and effective.

REFERENCES

Badola, R., Hussein, S.A., 2005. Valuing ecosystem functions: an empirical study on the storm protection function of Bhitarkanika mangrove ecosystem, India. Environ. Conserv. 32, 85–92.

Balk, D., et al., 2009. Mapping urban settlements and the risks of climate change in Africa, Asia and South America. In: Guzmán, J.M., et al., (Eds.), Population Dynamics and Climate Change. UNFPA and IIED, New York and London, pp. 80–103.

Brink, E., Aalders, T., Adam, D., Feller, R., Henselek, Y., Hoffman, A., et al., 2016. Cascades of green: a review of ecosystem-based adaptation in urban areas. Global Environ. Change 36, 111–123.

Brooks, S. 2017. The human element of mangrove management − CIFOR forests news. CIFOR Forests News. Available from: https://forestsnews.cifor.org/47992/the-human-element-of-mangrove-management?fnl = cn&utm_source = CIFOR + Website&utm_medium = Slide + show + bar&utm_campaign = Forests + News.

Cities Alliance, 2015. Sustainable Development Goals and Habitat III: Opportunities for a successful New Urban Agenda. Brussels, Belgium.

Convention on Biological Diversity, 2010. CBD COP 10 Decision X/33.

Depietri, Y., McPhearson, T., 2017. Integrating the Grey, Green, and Blue in Cities: Nature-Based Solutions for Climate Change Adaptation and Risk Reduction inKabisch, N. et al. (eds.), Nature-based Solutions to Climate Change Adaptation in Urban Areas, Theory and Practice of Urban Sustainability Transitions, pp. 91–109.

Ebinger, J., et al., 2009. An Assessment of Climate Change Vulnerability, Risk, and Adaptation in Albania's Power Sector. World Bank.

Elmqvist, T., Setälä, H., Handel, S.N., van der Ploeg, S., Aronson, J., Blignaut, J.N., et al., 2015. Benefits of restoring ecosystem services in urban areas. Curr. Opin. Environ. Sustain. 14, 101–108.

Fee, L., Gibert, M., Bartlett, R., Capizzi, P., Horton, R., Lesk, C., in press. Climate Change Vulnerability Assessment of Pakokku Township (Magway, Dry Zone Area, Myanmar) 2016–2050, MCCA and UN-Habitat Myanmar.

Frantzeskaki, N., Borgström, S., Gorissen, L., Egermann, M., Ehnert, F., 2017. Nature-Based Solutions Accelerating Urban Sustainability Transitions in Cities: Lessons from Dresden, Genk and Stockholm Cities. In: Kabisch, N. et al. (eds.), Nature-based Solutions to Climate Change Adaptation in Urban Areas, Theory and Practice of Urban Sustainability Transitions, pp. 65−91.

Geneletti, D., Zardo, L., 2016. Ecosystem-based adaptation in cities: an analysis of European urban climate adaptation plans. Elsevier, Land Use Policy (50), 38−47. Available from: https://doi.org/10.1016/j.landusepol.2015.09.0030264-8377.

Guadagno, L., Depietri, Y., Paleo, U.F., 2013. Urban disaster risk reduction and ecosystem services. In: Renaud, F.G., Sudmeier-Rieux, K., Estrella, M. (Eds.), The Role of Ecosystems in Disaster Risk Reduction. United Nations University Press, Tokyo, pp. 389−416. In: Renaud, F.G., Sudmeier-Rieux, K., Estrella, M. (Eds.), The Role of Ecosystems in Disaster Risk Reduction. United Nations University Press, Tokyo, pp. 389−416.

Government of the Islamic Republic of Afghanistan, 2015. State of Afghan Cities 2015. GoIRA, Kabul.

Kabisch, N., Frantzeskaki, N., Pauleit, S., Naumann, S., Davis, M., et al., 2016. Nature-based solutions to climate change mitigation and adaptation in urban areas −perspectives on indicators, knowledge. gaps, barriers and opportunities for action. Ecol. Soc. 21 (2), 39.

Lohrey, S., Creutzig, F., 2016. A 'sustainability window' of urban form. Transport. Res. Part D: Transport Environ. 45, 96−111.

Martine, George, Schensul, Daniel (Eds.), 2013. The Demography of Adaptation to Climate Change. UNFPA, IIED and El Colegio de México, New York, London and Mexico City.

McGranahan, G., Balk, D., Anderson, B., 2007. The rising tide: assessing the risks of climate change and human settlements in low elevation coastal zones. Environ. Urban. 19 (1), 17−37.

McPhearson, T., Karki, M., Herzog, C., Santiago Fink, H., Abbadie, L., Kremer, P., et al., 2017. Urban ecosystems and biodiversity. In: Rosenzweig, C., Solecki, W., Romero-Lankao, P., Mehrotra, S., Dhakal, S., Ali Ibrahim, S. (Eds.), Climate Change and Cities: Second Assessment Report of the Urban Climate Change Research Network. Cambridge University Press, New York.

Mitra, A., Wajih, S., Singh, B., 2015. Wheezing ecosystems, livelihood services and climate change resilience in Uttar Pradesh. Asian Cities Climate Resilience. Working Paper Series 18:2015.

Revi, A., Satterthwaite, D.E., Aragón-Durand, F., Corfee-Morlot, J., Kiunsi, R.B.R., Pelling, M., et al., 2014. Urban areas. In: Field, C.B., Barros, V. R., Dokken, D.J., Mach, K.J., Mastrandrea, M.D., Bilir, T.E., et al., Climate Change 2014: Impacts, Adaptation, and Vulnerability. Part A: Global and Sectoral Aspects. Contribution of Working Group II to the FifthAssessment Report of the Intergovernmental Panel on Climate Change. Cambridge University Press, Cambridge, United Kingdom and New York, NY, pp. 535−612.

Rao, N.S., Carruthers, T.J.B., Anderson, P., Sivo, L., Saxby, T., Durbin, T., et al., 2012. A comparative analysis of ecosystem−based adaptation and engineering options for Lami Town, Fiji. A synthesis report by the Secretariat of the Pacific Regional Environment Programme.

Seto, K.C., Fragkias, M., Güneralp, B., Reilly, M.K., 2011. A Meta-analysis of global urban land expansion. PLoS One 6 (8), e23777. Available from: https://doi.org/10.1371/journal.pone.0023777.

Seto, K.C., Güneralp, B., Hutyra, L.R., 2012. Global forecasts of urban expansion to 2030 and direct impacts on biodiversity and carbon pools. Proc. Natl. Acad. Sci. 109 (40), 16083−16088.

Seto, K.C., Reenberg, A., Boone, C.G., Fragkias, M., Haase, D., Langanke, T., et al., 2012. Urban land teleconnections and sustainability. Proc. Natl. Acad. Sci. 109 (20), 7687−7692.

Seto, K.C., Dhakal, S., Bigio, A., Blanco, H., Delgado, G.C., Dewar, D., et al., 2014. Human settlements, infrastructure and spatial planning. In: Edenhofer, O., Pichs-Madruga, R., Sokona, Y., Farahani, E., Kadner, S., Seyboth, K., et al., Climate Change 2014: Mitigation of Climate Change. Contribution of Working Group III to the Fifth Assessment Report of the Intergovernmental Panel on Climate Change. Cambridge University Press, Cambridge, United Kingdom and New York, NY.

TNC, 2016. Sub-Saharan Africa's Urban Water Blueprint: Securing Water Through Water Funds and Other Investments in Ecological Infrastructure. The Nature Conservancy, Nairobi, Kenya.

United Nations, 2015. The Millennium Development Goals Report.

United Nations, Department of Economic and Social Affairs, Population Division, 2006. World Urbanization Prospects: The 2005 Revision. Working Paper No. ESA/P/WP/200.

United Nations Department of Economic and Social Affairs, 2016. The World's Cities in 2016: Data Booklet. New York.

UNEP, 2014. Green Infrastructure Guide for Water Management: Ecosystem-based management approaches for water-related infrastructure projects. United Nations Environment Programme, Nairobi, Kenya.

UNEP, 2016a. GEO-6 Regional Assessment for Africa. United Nations Environment Programme, Nairobi, Kenya.

UNEP, 2016b. GEO-6 Regional Assessment for Asia and the Pacific. United Nations Environment Programme, Nairobi, Kenya.

UN-Habitat, 2016. Sustainable urbanization in the Paris Agreement. UN-Habitat, Nairobi, Kenya.

UN-Habitat, 2015. Urbanization and Climate Change in Small Island Developing States. UN-Habitat, Nairobi, Kenya.

Wamsler, C., 2015. Mainstreaming ecosystem-based adaptation: transformation toward sustainability in urban governance and planning. Ecol. Soc. 20 (2), 30. Available from: http://dx.doi.org/10.5751/ES-07489-200230.

FURTHER READING

Andersson, E., Barthel, S., Borgström, S., Colding, J., Elmqvist, T., Folke, C., et al., 2014. Reconnecting cities to the biosphere: stewardship of green infrastructure and urban ecosystem services. Ambio 43 (4), 445−453.

City of London Corporation, 2010. Rising to the Challenge − The City of London Climate Change Adaptation Strategy. City of London Corporation, London. Available from: https://www.cityoflondon.gov.uk/services/environment-and-planning/sustainability/Documents/climate-change-adaptation-strategy-2010-update.pdf.

IUCN, 2017. Ecosystem Based Adaptation and Mitigation. Available from: https://www.iucn.org/commissions/commission-ecosystem-management/our-work/cems-thematic-groups/ecosystem-based-adaptation-and-mitigation (accessed 2 April).

Kremer, P., Andersson, E., McPhearson, T., Elmqvist, T., 2015. Advancing the frontier of urban ecosystem services research. Ecosys. Serv. (12), 149−151.

McDonald, R.I., Marcotullio, P.J., Güneralp, B., 2013. Urbanization and global trends in biodiversity and ecosystem services. In: Elmqvist, T., Fragkias, M., Goodness, J., Güneralp, B., Marcotullio, P.J., McDonald, R.I., et al.,Urbanization, Biodiversity and Ecosystem Services: Challenges and Opportunities. Springer Netherlands, Dordrecht, The Netherlands, pp. 31−52. Available from: http://dx.doi.org/10.1007/978-94-007-7088-1_3.

McMichael, A.J., Lindgren, E., 2011. Climate change: present and future risks to health, and necessary responses. J. Inter. Med. 270 (5), 401−413.

McPhearson, T., et al., 2015. Resilience of and through urban ecosystem services. Ecosys. Serv. 12, 152−156.

Nursey-Bray, M., 2014. Gender, Governance and Climate Change Adaptation. Handbook of Climate Change Adaptation. Springer, Berlin, p. 2014.

Pickett, S.T.A., et al., 2008. Beyond urban legends: an emerging framework of urban ecology, as illustrated by the Baltimore Ecosystem Study. BioScience 58 (2), 139−150.

TEEB- The Economics of Ecosystems and Biodiversity for Local and Regional Policy Makers, 2010.

TEEB − The Economics of Ecosystems and Biodiversity, 2011. TEEB Manual for Cities: Ecosystem Services in Urban Management.

UN-Habitat, 2008. State of the World's Cities 2008−2009: Harmonious Cities. Earthscan. Available from: https://sustainabledevelopment.un.org/content/ documents/11192562_alt-1.pdf.

Uzhydromet, 2016. Third National Communication of the Republic of Uzbekistan under the UN Framework Convention on Climate Change. Centre of Hydrometeorological Service (Uzhydromet) at the Cabinet of Ministers of the Republic of Uzbekistan, Tashkent, 2016.

Zhou, W., Cadenasso, M.L., Schwarz, K., Pickett, S.T., 2014. Quantifying spatial heterogeneity in urban landscapes: integrating visual interpretation and object-based classification. Remote Sensing 6 (4), 3369−3386.

Chapter 13

Climate-Smart Agriculture in Southeast Asia: Lessons from Community-Based Adaptation Programs in the Philippines and Timor-Leste

Alvin Chandra and Karen E. McNamara
The University of Queensland, Brisbane, QLD, Australia

Chapter Outline

13.1 INTRODUCTION

In Southeast Asia, food security and agriculture depend on smallholder and peasant farming. This chapter investigates how rural smallholder farmers in this region have implemented strategies that integrate adaptation and mitigation to pursue climate-smart agriculture (CSA). While smallholder agriculture is vital to food security, poverty alleviation and employment, small-scale farmers are widely recognized as one of the most climate-vulnerable groups (Bhatta and Aggarwal, 2016). Some of the region's poorest households live on small-scale farms to supplement their nutrition and income. The increasingly extreme weather events in Southeast Asia highlight the vulnerability of agriculture and smallholder farmers to the impacts of climate change and extreme weather events. In order to grow food for their livelihoods, rural communities must cope with the impacts of climate change. However, under future climate conditions, the capacity of smallholder farmers to access resources to adapt to the impacts of climate change is and will continue to be challenging.

The impacts of climate change are now being experienced by small-scale food producers. These impacts include shifts in agroecological zones, unpredictable rainfall, dried land leading to rainwater run-off, flash flooding washing away crops and soil, heat waves destroying crops, dramatic changes in planting seasons, and changes to seasonal employment contracting opportunities (IPCC, 2007, 2014). The available evidence suggests that climate change will place increasing pressure on Southeast Asia's ability to feed itself. There is an essential and timely need to mobilize measures to support small-scale food producers to adapt and employ low cost technology/best practices to reduce GHG emissions, and increase their resilience to climatic shocks. These measures can ensure that vulnerable smallholder

Resilience. DOI: https://doi.org/10.1016/B978-0-12-811891-7.00013-X

communities are not pushed further into poverty and can proactively play a role in feeding the region under increasingly difficult agricultural production conditions.

The close linkage between adaptation, mitigation and food security (primarily because of the need for both adaptation and mitigation if the world's people are to have their food needs met) has led to calls for addressing adaptation and mitigation in an integrated way. The CSA has emerged as an approach to sustainably increase agricultural productivity, mitigate increasing greenhouse gas (GHG) emissions, enhance resilience, and support national food security and development goals (see Box 13.1 for review of CSA). Since CSA's emergence in 2010, researchers and policy makers have used it to strengthen the link between adaptation, mitigation and food security, while simultaneously aiming to influence climate policy and finance (Lipper et al., 2014). In recent years, the concept has developed, partnerships have flourished and international research has expanded, as is evident through the policies and research leadership of the United Nations Food and Agriculture Organisation, the Global Alliance for CSA, and the Consultative Group on International Agricultural Research (CGIAR).

The Association of Southeast Asian Nations Climate Resilience Network (ASEAN-CRN) has emerged as a regional platform for sharing information, experiences, and expertise about CSA. The platform aids in mobilizing support towards a regional governance structure for the integration of climate change concerns in the agriculture sector. Both international and regional efforts in Southeast Asia aim to better link policy makers to scientific

BOX 13.1 Summary of CSA Literature, Practice and Characteristics

CSA is a new approach to bridge the growing divide between the two discourses and foster long-term climate-resilient development in the agriculture sector. CSA is defined by FAO (2010, 5) as "agriculture that sustainably increases productivity, resilience (adaptation), reduces/removes GHGs (mitigation), and enhances achievement of national food security and development goals." Adaptation, mitigation and food security are the three key pillars of CSA (Lipper et al., 2014). In contrast, the World Bank's definition of "climate-smart" references agricultural transition to low-carbon growth, in addition to enhancing development and reduces vulnerability (WB, 2009). In contrast, FAO coined the CSA definition within three pillars—adaptation, mitigation, and food security (FAO, 2010). Both the FAO and the World Bank definitions view CSA as integral to increasing the level of food production that may involve several agricultural technologies or practices. The World Bank extends the definition of CSA as a "governance framework" where the right tools, technologies (energy efficiency, carbon capture and storage, and next-generation renewables), and institutions are integral to climate-smart approaches (WB, 2009). The FAO articulation of CSA identified that innovative financial mechanism as a prerequisite for operationalizing CSA.

CSA represents a combination of practices that have historically been used in the environmental ecology, conservation, climate change, and agriculture fields. CSA practices described in the literature include diverse on-farm practices such as agronomy, agroforestry, livestock, forestry, land use, pastoral and grazing, water and soil management, and bioenergy (Bryan et al., 2013; Thorn et al., 2016). Similarly, the literature states that smart practices can provide policy directions for mainstreaming of climate change, health and nutritional benefits, finance and infrastructure development (Harvey et al., 2014). In theory, effective CSA practices and technologies should address three core components: sustainably increasing productivity, supporting farmers' adaptation to climate change, and reducing levels of GHGs (Lipper et al., 2014). Critics point out that many agricultural practices are being "rebranded" as climate-smart, although they may not actually be addressing climate change issues (GRAIN, 2015; Ewbank, 2015).

Overall few studies have measured success of CSA practices and studies have begun an examination of synergistic practices in an attempt to increase the cost-effectiveness and efficiency of CSA (e.g., Chandra et al., 2016). Other studies conclude that CSA should involve an assessment of the optimal mixes of synergistic adaptation-mitigation practices (Lipper et al., 2014) that minimize negative trade-offs. Yet synergies and trade-offs between agrifood systems, climate change responses, and development are interconnected and complex. Detailed knowledge of local social-ecological contexts and transformational processes (Thornton and Comberti, 2013) affecting the implementation of adaptation and mitigation practices within agrarian societies is limited (Chandra et al., 2017).

Scientific understanding of the characteristics of CSA, evaluating practices and processes contributing to agriculture resiliency remains underdeveloped. Gender, social equity, and economic costs—benefits of CSA interventions have been poorly researched in literature (Beuchelt and Badstue, 2013). Similarly, Bryan et al. (2013) conclude from their research in Kenya that promoting and adopting CSA and triple-win strategies is a major challenge given access, costs, lack of incentives and investments for market-oriented smallholder systems. A vast majority of recent studies identify the need to research policy and institutional dimensions of CSA. Theoretical studies provide general principles of integration (Tompkins et al., 2013) while empirical cases can go a long way in validating synergies, triple-win, or cobenefits of CSA across scales (Steenwerth et al., 2014).

institutions, universities, national research institutions in agriculture, and international organizations. Placing CSA on climate change political agendas in ASEAN and the recent increase of research activities though the CGIAR raises two fundamental questions: First, what constitutes CSA in the smallholder farming landscapes that bear limited historical responsibility for the increasing GHG emissions? Second, if agriculture becomes climate-smart, who benefits? Framed this way, CSA as a solution to challenges confronting smallholder agriculture has been met with criticism from farmer groups and nongovernmental organizations (NGOs) because of the lack of clarity on what constitutes CSA for smallholder farmers and their lack of engagement within its governance at the global level. For example, some NGOs have questioned whether the global partnerships or the concept is an equitable solution to smallholder farmer vulnerability, adaptation/mitigation needs, and farming practices (Neufeldt et al., 2013; Sugden, 2015).

A new research and policy agenda that emphasizes the strengthening of smallholder farmers, civil society and local governments is therefore needed. We argue this point by assessing comparatively whether the approaches and practices of community-based programs in Southeast Asia are likely to prompt adaptation, mitigation, and food security at subnational levels. As examples, we investigate two community-based adaptation (CBA) programs in Philippines and Timor-Leste that are implemented by farmer organizations and civil society organizations. Drawing primarily on descriptive analysis of semistructured interviews, focus group discussions (FGDs), and field observations, this chapter presents the similarities and differences between key characteristics of CSA in the two case study countries. We begin by asking what CSA means to smallholder farmers and identify five key characteristics of such practices (i.e., institutional, finance and market). We then follow with a comparative analysis of three of the five characteristics of CSA that were found in the two case studies. Finally we discuss the implications of our research findings and explore prospects for the integration of adaptation and mitigation strategies.

13.2 METHODS

13.2.1 Sampling and Data Collection

Our study specifically compares and explains the success, failure, and challenges of implementing CSA approaches across CBA programs in the Philippines and Timor-Leste. The following information outlines the criteria for selection of the research sites and recruitment of research informants (Chandra et al., 2016, 2017), mutually developed with NGO and community-based research partners:

1. *Stakeholders*: case study areas were supported by the CBA program partners. All farming communities had ongoing collaboration with community-based organizations (CBOs) on climate change and agricultural livelihood initiatives;
2. *Enabling conditions*: multiple farming practices (e.g., land management, water resources management, sustainable agriculture, conservation of watersheds and traditional knowledge) were in use. They were also focus areas of national/regional government support in terms of resource management and disaster risk reduction (DRR);
3. *Researcher access*: sites were outside of restricted, social conflict or "rido" in the Philippines (war) zones and were accessible to the researcher;
4. *Climate vulnerability*: sites had a history of and exposure to climate related events that caused loss/damage to livelihoods as these factors related to the research questions—environmental settings where the challenges associated with adaptation and mitigation practices in agricultural settings and research could therefore be investigated;
5. *Assets and entitlement*: farms were located across a range of temporal ecosystem and land tenure types—upland (mountain), mid, and coastal lowland (watershed areas);
6. *Unique or critical characteristics*: communities had unique environmental or cultural resources such as indigenous knowledge, practices, and endemic biotic resources; and
7. *Common governance*: The sites were contained within a common regional governance framework, which greatly reduced the variability among the cases in terms of sociocultural and institutional development paths.

In January 2013 and September 2015 fieldwork was conducted and data was collected by the lead author assisted by the CBOs and NGOs who implement these programs across the seven research sites in two countries. We employed a mixed-method approach (Creswell, 2014) to allow for triangulation in gathering qualitative data and verifying results on smart agricultural practices. Two CBA programs implemented in the Philippines and Timor-Leste constitute the research focus for this study. A purposeful sampling technique was used to select research sites: In the Philippines, five municipalities across three provinces: Sultan Kudarat (Esperanza, Bagumbayan), North Cotabato (Alamada,

TABLE 13.1 Methodological Strategies and Data Instruments

Method	How It Was Used in the Study	No. of Informants and Units of Analysis	
		Philippines	Timor-Leste
Key informant interviews	Different stakeholder groups comprised of farmers, CBOs, local NGOs, international NGOs, government and municipal councils, private sector, and donors. These stakeholders have a stake and interest in the climate change or agriculture issues in the Philippines and Timor-Leste	77	17
Field observations and transect walks	Field observations were undertaken across the farms in the case study sites, and photographs/notes were taken to capture climate/disaster impacts, climate-smart practices, and off-farm markets and trading areas	7	4
Participant observations	CSA practices used by farmers were observed, including participation in community gardening, climate-resiliency field schools, NGO workshops, international workshops/seminars, and collection of climate data at national level	5	4
Focus group discussions	Community, NGO and local government focus groups were conducted to understand impacts of climate change and disasters, characteristics of CSA and practices implemented by stakeholders	14	4
Secondary data analysis	To triangulate and verify CSA practices and institutional/policy measures identified in FGDs and interviews, content analysis of organizational documents/reports. Organizational program plans, program and policy reports, national level policies, workshop proceedings, and negotiation strategies (limited to climate change, agriculture) were analyzed. Coding categories were derived directly from the text analysis, interpreting similarities and differences between CSA practices and impacts	~75	~55

Pigcawayan), and Agusan del Norte (Jabonga); in Timor-Leste two districts (Viqueque and Oecusse). Data collection included key informant interviews, document analysis, field visits and observations, participant observation, and a literature review, as summarized in Table 13.1.

This research drew heavily upon semistructured interviews and FGDs with farmers, NGOs, the private sector, local and national government agencies, donors and international research institutes involved in agriculture, climate change, and international development. During the interviews and FGDs, participants discussed priorities for climate change in agriculture in the respective case studies, using a set of semistructured questions—they focused on key characteristics of CSA, specific examples from farms, policy cases, the support required and practical steps needed to effectively upscale CSA practices. With only two cases, it is not expected that major generalizations of CSA strategies will emerge from this research. However, the contrast between the Philippines and Timor-Leste cases provide detailed features specific to Southeast Asian farming communities. Given the research focus on tropical agriculture, it is likely that CSA strategies emerging from this research could differ for other agriecological zones and ecosystems.

13.2.2 Data Analysis

A diverse set of indicators exist on effectiveness of adaptation, mitigation, and food security practices. However, no one framework addresses all the three pillars of CSA. Globally there is a lack of a common approach to measure and evaluate success of climate resiliency, in particular that of smallholder agriculture systems (UNFCCC, 2014a). Several monitoring and evaluation frameworks exist on effectiveness of adaptation and mitigation measures and are being implemented at various levels by different organizations. To understand the effectiveness of CSA farming practices, nine qualitative indicators relating to adaptation, food production, and mitigation were identified from literature review, trialled in Timor-Leste, and workshopped and validated with research informants in the Philippines (see Chandra et al., 2016 for results on the effectiveness of CSA practices).

The interviews and FGDs were transcribed to understand the characteristics of CSA. This qualitative data was then analyzed using a grounded theory analysis approach. Themes emerging during the coding process (Robson, 2002) were analyzed using NVivo 10 (version 10; QSR, 2012). Our coding system was designed to examine farmer and NGO experiences on the characteristics identified in the interviews and FGDs that are presented in the results section. The community-level risk and hazard specific data were triangulated with other expert interviews and secondary data gathered from different stakeholder organizations in the two countries. This allowed us to compare the findings and identify similarities and differences in CSA practices implemented by smallholder farmers.

13.3   RESULTS: CHARACTERIZES, SIMILARITIES, AND DIFFERENCES OF CSA PRACTICES

13.3.1   Characteristics of Climate-Smart Agriculture

While diverse aspects of sustainable agriculture have been described in the literature as climate-smart (Neufeldt et al., 2013; Campbell et al., 2014; Lipper et al., 2014), there is a general acceptance that the characteristics of CSA are not universally standardized. Agricultural communities in Southeast Asia face a number of drivers of change, including market access, population growth, access to land, and extreme natural disaster events. In this section, we present data that explore the characteristics of and motivations for adopting CSA that are important to female and male smallholder farmers. Interviews and FGDs with farmers, civil society and local government officials identified five important characteristics and 20 microlevel elements, as illustrated in Fig. 13.1.

Although Fig. 13.1 illustrates characteristics and elements spanning from socioeconomic and political factors at the national level, the global discourse about CSA was rather perceived by community and NGO informants as a set of technical actions. For the approach to be useful to smallholder farmers, case study respondents emphasized the importance of integrating social empowerment through multipronged participatory approaches. Participants across all stakeholder groups in both countries said that efforts to integrate adaptation and mitigation in the agriculture sector are still in the formative stage. The key reason for adopting CSA was the need to address climate vulnerability and disaster risks, not to pursue mitigation (later described as a cobenefit of adaptation actions). An important outcome of CSA identified by informants was social empowerment of women and indigenous farmers. Smallholder farmers described two factors as being important in empowerment: (1) Inclusion: At-risk female and male farmers should be provided with and have access to the necessary incentives, political space, knowledge, resources, authority, responsibilities, and accountabilities for resilient livelihoods; and (2) Accountability: State and nonstate actors should be accountable to farmers regarding adaptation, mitigation, and food security decisions.

The climate change literature lists a variety of characteristics as determinants of farm-level climate adaptation measures that may shape the adaptive capacity and resilience of smallholder farmers in developing countries (Adger, 2003; Pelling and High, 2005; Bryan et al., 2009; Deressa et al., 2009; Nabikolo et al., 2012; Nhemachena et al., 2014). Among the determinants, institutional, finance, and market mechanisms are particularly critical for influencing small-scale subsistence farmers with limited resources, to respond to climate change impacts. One limitation has been studying the determinants individually rather than sufficiently exploring their linkages with national and local stakeholder organizations. While knowledge and technology to support social learning and adaptation has received attention in literature (IAASTD, 2009; Scherr et al., 2012; UNFCCC, 2014a,b) fewer studies have unpacked institutional, finance, and market characteristics within smallholder farming landscapes (Davidson, 2016). Although factors associated with the three characteristics, such as participation, representation, decision-making, empowerment, resource availability, and agriculture trade policies, interact at various levels, few empirical researches have examined their interactions (Rosenzweig and Tubiello, 2007; Engle and Lemos, 2010). Even fewer examples of cases exist on how these interactions build on the practices implemented under CBA programs and community-based stakeholders. Therefore, based on the general feedback on CSA elements and characteristics, we followed with a closer look at empirical examples and approaches to CSA implemented under the CBA programs in the Philippines and Timor-Leste focusing on institutional, finance, and market characteristics. The next sections illustrate three of the five characteristics of CSA and analyze the similarities and differences between the two case studies within the framework outlined in Fig. 13.1.

13.3.2   Institutional: Climate-Resiliency Field Schools Versus Consortium Approach

Two distinct forms of multiscale institutional arrangements for farmer movement building emerged in our investigation: Philippine's climate-resiliency field schools (CrFS); and Timor-Leste's NGO consortium approach. In Timor-Leste, the consortium approach to CBA incorporated a range of NGOs with diverse skill sets and experience in agriculture and

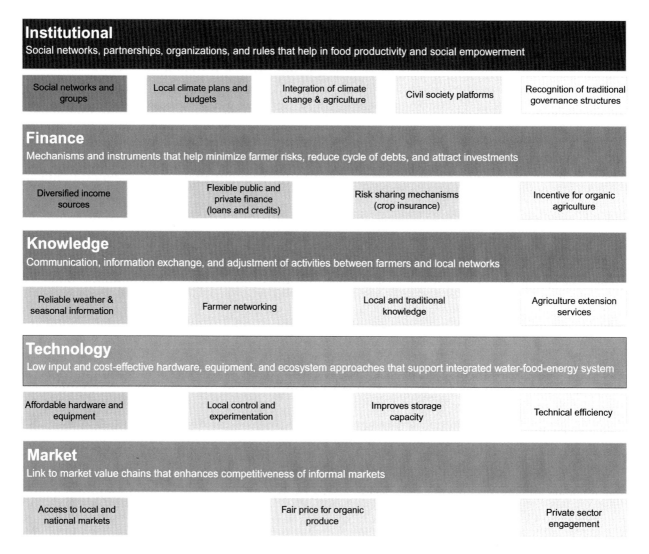

Institutional
Social networks, partnerships, organizations, and rules that help in food productivity and social empowerment

| Social networks and groups | Local climate plans and budgets | Integration of climate change & agriculture | Civil society platforms | Recognition of traditional governance structures |

Finance
Mechanisms and instruments that help minimize farmer risks, reduce cycle of debts, and attract investments

| Diversified income sources | Flexible public and private finance (loans and credits) | Risk sharing mechanisms (crop insurance) | Incentive for organic agriculture |

Knowledge
Communication, information exchange, and adjustment of activities between farmers and local networks

| Reliable weather & seasonal information | Farmer networking | Local and traditional knowledge | Agriculture extension services |

Technology
Low input and cost-effective hardware, equipment, and ecosystem approaches that support integrated water-food-energy system

| Affordable hardware and equipment | Local control and experimentation | Improves storage capacity | Technical efficiency |

Market
Link to market value chains that enhances competitiveness of informal markets

| Access to local and national markets | Fair price for organic produce | Private sector engagement |

FIGURE 13.1   This figure drawn from informant interviews and FGDs, shows the characteristics of CSA in smallholder farming systems, as assessed in the Philippines and Timor-Leste. The characteristics are defined at the broad level followed by elements unpacked under each of the characteristic identified as important to foster synergistic mitigation and adaptation actions in the agriculture sector.

natural resources management. The key challenges in implementing climate-resilient agricultural practices in Timor-Leste are the lack of capacity for programming climate change (Chandra et al., 2016), and farmer access to agricultural extension services. However, with a consortium of three international (Oxfam, Caritas, and Catholic Relief Services) and 11 local NGOs, stakeholders were able to implement the CBA program in 23 villages (*sucos*) in four districts. Interviews with government and consortium staff indicated that the consortium approach allowed different agencies to share knowledge, skills, and resources, avoiding duplication of program activities. The consortium approach addressed capacity gaps by building an alliance of NGOs to provide technical support to rural farming communities over a large geographic area, targeting remote communities to sustainably apply farming practices and using climate risk assessments and existing program relationships to improve livelihood strategies. At the subnational level, the interviews indicated that building upon existing community relationships at subdistrict and village levels enabled NGOs to directly engage with leaders of traditional villages and subvillages (*aldeia*) and agencies of governance. In many cases, village leaders were involved in the improved livelihood activities, or had community action plans presented to them for approval.

The consortium model also opened up new spaces for political engagement and joint national action. Local government and NGO staff described aid and development community in Timor-Leste as a "crowded landscape." Generally, a project-based environment is favored by government and donors, and current efforts on agriculture and climate change lack long-term programmatic and community-based approaches. A handful of donors dominate the climate change landscape, and formal platforms (or policy spaces) for NGO engagement are lacking. There is also a continued dependence on external donor finance for environmental and climate change programmes, despite the government receiving considerable revenue from the resources sector (e.g., petroleum and other extractive industries such as mining).

In the absence of dedicated formal platforms for NGOs in Timor-Leste, the consortium and Australian Aid partnerships provided multiple benefits of directly engaging with communities and local NGOs. Joint agriculture actions with the Ministry of Agriculture, Forestry, and Fisheries (MAFF) extension staff, the *Programa Nasional Dezenvolvimentu Suku* or National Program for Village Development (PNDS) staff, and NGOs were developed, defining what successful and smart agriculture adaptation practices look like. Practical and analytical climate skills of local NGO partners, community members, and (to a lesser degree) local government stakeholders were developed by international NGOs. The consortium approach also led to the scaling out of agriculture livelihoods activities from Oecusse to other districts under the CBA program. Joint actions also helped to promote adaptation approaches at the national level. For example, the consortium members linked village leaders and MAFF extension staff, and mainstreamed successful practices into the PNDS program. However, the long-term effects of mainstreaming are difficult to measure.

In contrast, the Philippines CrFS was institutionally embedded within local communities that brought together NGO and government organizations to work directly with farmers (Fig. 13.2). Unlike the Timor-Leste case, CrFS is inscribed in the Philippine Strategy on Climate Change Adaptation (2010−22) as an institutional means for information, education, and communication on climate change and DRR in the agriculture sector (DENR, 2010). CrFS in the national strategy enabled a unique polycentric multilevel, multiactor governance model to emerge at the subnational level. CrFS was implemented in partnership with local farmers, NGOs, Local Government Units (LGU), the Philippine Atmospheric, Geophysical, and Astronomical Services Administration (PAGASA), and national government agencies. As a consequence of formal prescribed climate policies, CrFS was characterized by strong collaboration with local government. Dedicated platforms that ran parallel to CrFS helped with the deployment of regular technical assistance and extension services from the local and national governments. For example, NGO informants said that in Pigcawayan, strong advocacy for local CrFS governance through the local committee on agriculture favored budget allocation and support for organic agriculture from the LGU. Farmer organizations and LGU staff (agriculture extension workers and

FIGURE 13.2 Mixed-crop farming in Mindanao, The Philippines.

technicians) said that CrFS enabled them to obtain regular training and early warning assistance from the NGOs and PAGASA (i.e., use of climate information, data capture, interpretation, and reporting). NGO experience in climate risk assessment, technical support related to agricultural livelihoods and women's leadership complemented the planning support from LGU.

Training in the CrFS was found to be effective, but farmer-to-farmer learning sessions and workshops were described by all informants as the most effective way to raise awareness about improved practices. Key informant responses indicated that the CrFS workshops directly responded to the needs of local farmers and mobilized immediate assistance from government to respond to reoccurring droughts and floods. Workshops in the CrFS enabled farmers, facilitators and NGO and LGU partners to reflect on seasonal calendars, consolidate experiences and learn from each other. Similarly, workshops in the villages (barangays) helped refine community action plans and traditional governance structures, roles, and tasks under the Barangay/Municipal Disaster Risk Reduction Management Council.

13.3.3 Finance: Bottom-Up Budgeting Versus Microfinance

Climate finance in both cases relies heavily on international aid, and much of this aid comes from bilateral and multilateral donor agencies (Barnard et al., 2015). Much of the debate in climate finance literature for the agriculture and forestry sectors has centered on country readiness, sustaining finance from donors, and institutional access to funds from multilateral and bilateral donors (Streck, 2012; Dugumaa et al., 2014; Ye et al., 2015). However, mechanisms that channel climate finance to subnational and vulnerable communities have been poorly demonstrated (Siedenburg et al., 2012). As such, informants from both case studies said that reaching the most vulnerable farmers was a challenge that can best be overcome by continuous risk assessments, local government involvement, and mainstreaming within local development plans. Two distinct ways to finance low-cost community-level adaptation and mitigation practices in the agriculture sector were identified: Bottom-up Budgeting (BuB) and microfinance.

In the Philippines, funding for climate-resilient agriculture was secured by mainstreaming risk reduction and adaptation activities into local development plans and budgets. Informants stated that BuB introduced by the national government in 2012 has helped local communities and organizations to directly participate in the budgeting process. They described the BuB process as inclusive, empowering, participatory, and said that it directly engaged grassroots communities. Inclusive and transparent consultation platforms and CSO assemblies in the municipalities helped local farmers and organizations articulate and set budget priorities. To ensure that important issues were represented, a variety of stakeholders participated in the BuB process. These included farmers, women's groups, civil society and local governments. The community priorities identified through the BuB process included education, poverty reduction, health, and support for organic agriculture, which were then integrated in the government's local development and poverty reduction plans. The identified priorities were then prescribed budgets that were combined with community-based projects to establish the collective operating budget for a municipality.

To ensure that climate change, DRR, and agriculture issues formed part of the BuB priorities, consultations in some municipalities occurred back-to-back with participatory climate change and disaster risk assessments. The risk assessment validation sessions coincided with the municipal BuB process, helping secure LGU budgets for priority adaptation and DRR activities. In some instances, NGOs lobbied and farmers exchanged experiences on increasing climate risks to escalate livelihood priorities in the BuB processes. Through the BuB process, LGUs in the Philippines case study have started budgeting for agriculture adaptation actions identified in the risk assessments. For example, LGU officials from Alamada said that in the BuB process, about 200,000 pesos (USD 4248) was allocated to purchase an automatic weather station and related instruments, 35,000 pesos (USD 753) was provided to fund season-long CrFS training, and 100,000 pesos (USD 2153) was set aside to promote organic agriculture. LGUs of Pigcawayan and Bagumbayan similarly allocated funds for automatic weather stations, and support for organic agriculture and DRR activities. These examples indicate the pivotal role played by stakeholder advocacy and devolved budgeting processes in stimulating local government planning and investment for CSA.

In contrast, the Timor-Leste case study used microfinance as a catalyst to facilitate climate-smart actions. Microfinance savings and credit schemes are a traditional financing instrument used by government and development agencies to reduce loss and damage resulting from disasters and climate change (Linnerooth-Bayer and Hochrainer-Stigler, 2015). Despite the potential for fostering successful adaptation and mitigation, little is actually known about how loans and credits function at the local level. In Timor-Leste, the NGOs developed a saving and credit system (UBSP) to diversify farmer livelihood initiatives, help spread risks associated with seasonal and extreme drought events and encourage local ownership of project investments. Interviews with NGOs suggest that in total, program partners facilitated the establishment of about 30 UBSP groups across 12 subvillages (aldeias). The UBSP schemes formed the basis of social and economic safety nets at the household level, ensuring that financial resources were available to households during extreme drought events. The UBSP schemes were developed within farming communities and additional support was provided by NGOs to strengthen off-farm activities, help households recover from a range of climate shocks, and diversify agricultural income. Male and female farmers in the subvillages had regular access to cash for household needs, business opportunities, and training. UBSP in Timor-Leste targeted the poor rural farmers—both males and females.

Informant interviews elicited three interesting aspects of loan and savings programs that spread risk by diversifying agricultural livelihoods. First, strengthening women's groups increased female participation and raised the community profile of the UBSP mechanism. Interviews with farmers and NGOs indicated that in the subvillages, female farmers were organized into groups who shared responsibility for administration, payments, and support to other groups who needed help. The savings groups were formed to encourage farmers to save money and provide funds to buy agricultural supplies and food during extreme seasonal shocks. Female farmers were effective UBSP managers who contributed to community cohesion and led a mutual support system for their households during difficult times. Second, NGO stakeholders said that a combination of direct and indirect capacity-building support and life skills was useful in delivering the loans and savings microfinance support scheme. NGOs noted that community-led UBSP schemes within farming communities were successful because they were accompanied by crop diversification, financial/literacy training, and food processing activities. Complementary training of women in business management, literacy, finance, and profit-sharing mechanisms imparted basic administration skills. This training helped develop various income-generating activities such as small community-led business enterprises involving the production of beverages, cashews, animal husbandry, vegetable gardens, and traditional weaving.

Finally, integration of DRR and agricultural adaptation was made possible through the microfinance scheme by diversifying a range of different livelihood assets and activities. For example, UBSP groups in Oecusse were trained to grow and process local crops into marketable products. Loans from UBSP made investment in farming assets possible, including equipment, livestock, houses, and start-up business capital. By expanding their assets and activities, farmers received routine income and replenished the loan and saving transactions in the UBSP. Overall, the UBSP helped to create community independence from farming by diversifying the income base.

13.3.4 Market: Climate Risks Versus Market Shocks

Some issues with access to market may be unique to the two cases but others are similar to those observed by CBA programs focusing on rural farms. The two cases, consistent with the literature (Biagini et al., 2014; Ensor et al., 2015), found that adaptation projects that target highly vulnerable communities tend to be geographically remote, are likely to lack access to central markets, and face high transaction costs when transporting produce to market. In the case of the Philippines, the project was located in remote and conflict-prone areas of Mindanao Island, where the main markets are in the capital city of Davao. Conversely, in the Timor-Leste case, project sites in the highland areas, particularly in Oecusse District, had limited access to the Dili markets (although they were closer to Indonesia, transportation and economic activities are restricted by political boundaries). The importance of supporting smallholder farmer access to markets within climate change projects is clear through the absence of such support in the Philippines and Timor-Leste cases.

Making smallholder farmers more resilient to climate change by using adaptation and mitigation practices was described by some government and NGO stakeholders as rather inward-looking, where the broader influence of market policies and practices were absent from the program's scope. Both case studies used participatory risk assessments and climate scenarios as tools to guide adaptation pathways and selection of CSA and livelihood activities. Interviews with donors, NGOs, and government stakeholders suggested that the risk assessments overly focused on disaster and climate risks, and underweighted market risk. The market risks identified in our case studies include access to formal markets, market suitability of diversified and organic produce, and demand and supply of mixed farm produce. The majority of farmers indicated that a range of market practices and policies are needed when promoting livelihood diversification, farming intensification, or the introduction of new crop varieties (see Table 13.2). These market factors are well-known from other agriculture production chains, and as such this study reinforces the role of these influential elements on market access and impact on smallholder farmers.

Farmer responses indicated that organic crops brought to the market had to compete with nonorganic produce or were less attractive to consumers, and therefore had to be sold at lower prices. Where the farmers traveled to larger markets outside of their villages, factors such as transportation cost, the higher cost of produce and food spoilage resulted in lower profit margins. Female farmers in the Philippines reported difficulties finding space in local markets and negotiating fair deals with local traders. Similarly, male and female interviewees in both cases reported that

TABLE 13.2 Factors Influencing Access to Markets and Impacts on Smallholder Farmers

Market Factors	Impact on Smallholder Farmers
Institutional coordination and power relations	• Participation in a number of groups like NGOs and local government CrFS, school, senior citizen groups, church, women's organizations, competes with time available for family and farming—multiple function of farmers and reduces farm productivity • Training helps enhance knowledge and awareness of market demand and supply • NGO and local government seminars, trainings, meetings
Private lending and borrowing	• High interest rates create a cycle of debt and dependency on loans • Private loans are used to finance education (and other household needs) and high interest rates are charged by private lenders • Frequent harvest loss due to droughts, disasters and infestation (e.g., rat infestation) • Forced to sell produce at a lower price to traders compared to input costs and interest rates
Marketing and trading skills	• Lack of education gives women fewer opportunities and options for off-farm livelihoods • Women in particular have low negotiation skills and training
Government bureaucracy and red-tape	• Requirements like registration of women's organization (e.g., fiduciary standards, accreditation), for registration limit's women's access to government finance and projects • No budget from government for farmer associations • Government needs to empower the association by training and seminars
Low profit margins of organic agriculture	• Costly farm inputs results in high prices for produce and low profit margins • Crop price controlled by traders; price manipulation results in lower value of organic products, i.e., economic loss • Lack of profits results in loans and cycle of debts
Market incentives	• Conflicting government policies on subsidies to support organic agriculture • Farmers lack adequate legal protection and capital assets to borrow from banks and as a consequence they take loans from traders at high interest rates • Farmers lack understanding of rules and requirements for crop insurance
Product standards and certification	• Costly product quality requirements, e.g., packaging, standards, and labeling • Costly government requirements for product accreditation • Product value-adding reduces market competitiveness of smallholder farmers
Transportation	• High cost of transporting produce to markets increases product price • Poor road conditions, e.g., flooding, delays the transport of fresh quality farm produce to markets

wealthier farmers and larger agribusinesses had more leverage in the main market due to their access to credit and loans, government support, and market stall space. Smallholder farmers described that it was difficult for them to compete individually with larger farmers who had more resources and produced crops using inorganic fertilizers. NGO interviews described how local government agricultural policies were commodity focused, and subsidies were not incentivizing sustainable organic agriculture practices.

Both case studies also found a close relationship between knowledge dissemination, technology diffusion, and market access. Farm technology use is closely linked to crop, type of agriculture, and the commercial availability of hardware and equipment (Tambo and Abdoulaye, 2012). Although the Philippines and Timor-Leste case studies introduced food processing technologies, there was little evidence of its broad adoption and visible signs of access to larger markets in the capital cities. In both instances, the link between technology selection and market factors (Table 13.2) was weak, and there was little evidence of private sector engagement with smallholder farmers. As one NGO expert explained, the selection of technology for building climate resilience was heavily focused on addressing environmental risks and only weakly incorporated local economic shocks. For example, in the Philippines, crop diversification and organic agriculture technologies focused on increasing rice yield, vegetables and other crops. However, it was not clear if the local markets could support increased yields, new crop varieties, and/or higher prices of organic produce. Project staff explained that the short time frame of the programs, budget restrictions, limited NGO skills relating to market analysis, and donor focus on addressing climate risk were key factors limiting opportunities for market integration in the case study sites. While increasing yield and building smallholder resilience requires application of new practices, both cases illustrate that climate-smart technology choices should be closely linked to market value chains. A clear strategy that includes market risk and opportunities would help accelerate adoption decisions for climate-smart strategies and technologies in rural smallholder farms.

13.4 DISCUSSION: IMPLICATIONS FOR THEORY AND PRACTICE

This comparative analysis between the Philippines and Timor-Leste illustrates three characteristics that we believe are necessary to integrate adaptation and mitigation measures to drive climate-smart strategies in smallholder agriculture settings. These are: institutional strategies, devolved financing mechanisms, and market conditions. These characteristics are consistent with previous research on adaptive capacity (Gupta et al., 2010; Lemos et al., 2013), implementing GHG mitigation options (Smith et al., 2007; Jakob et al., 2014), and enhancing agricultural productivity (Lipper et al., 2014). However, our analysis of context-specific processes and examples from the perspective of smallholder farmers and civil society organizations is also useful in informing three arguments on CSA theory, science, and practice.

First, CSA in smallholder systems is not a set of separate practices or individual activities. Rather, it is a combination of "no-regrets" adaptation, resilient livelihoods, and food security practices. Our case studies illustrated that CSA has multiple definitions that are informed in vulnerable developing countries by adaptation actions implemented at the farm level (a key development priority). At the community level in the agriculture sector, different adaptation and mitigation options are less arbitrary and favor practices that offer more opportunities for livelihoods, food security, nutrition, and education. CSA practices should ideally aim to achieve acceptable profit and sustained production during periods of climate variability, while minimizing the negative effects of intensive farming by preserving ecosystem services. The potential for mutual benefits makes a strong and logical case for the adoption of CSA practices. To this end, we assert that farming practices that integrate adaptation and DRR approaches and technologies can and will have significant mitigation cobenefits.

When combined with NGO participation, varied and targeted agroecological practices and technologies (e.g., physical structures, equipment, and hardware) help to embed CSA within communities and ensure local ownership and an optimal mix of approaches and technologies; both critical for sustainability. Other participatory considerations for embedding CSA practices and technologies include capturing local relevance, organizational capacity, and experimentation or testing across different farm sites. Our analysis found that combining adaptation approaches and technologies provides multiple cobenefits and positive outcomes in food security, nutrition, livelihood diversification, education, income, and local development (Chandra et al., 2016). For instance, adaptation options such as reforestation, rice intensification, crop management, and soil and water conservation complemented mitigation priorities (sequestration of GHG emissions) and local development (improved food production and livelihoods diversification). To gain these multiple cobenefits, programs in Timor-Leste and the Philippines integrated socio-economic aspirations of farming communities with technical interventions. Local-level participatory risk assessments in the farming communities ensured that the underlying factors that drive climate vulnerability were addressed in the two projects.

Second, while there was evidence for the dissemination of climate-smart practices in some farms and districts, our two case studies suggest that adoption behavior and cobenefits varied amongst farmers. The adoption behavior of farmers in our cases was strongly influenced by the availability of institutional support from stakeholders, financial opportunity, knowledge, technology, and access to local markets. Structural factors (power relationships, inequality, and justice) sustain and perpetuate vulnerability within smallholder communities (Yates, 2014) and influence the choice of climate-smart practices. An approach that is "climate-smart" for one farmer or community is not necessarily the appropriate solution for another; it is different based on local vulnerability and mitigation/adaptation options available to farmers. In our view, the adoption behaviors and climate-smart choices of smallholder farmers can be better informed by offering a diversity of flexible CSA practices that maximize synergies and cobenefits and minimize trade-offs between adaptation, mitigation, food security, and local development. For instance, smallholder farmers were more likely to integrate climate-smart options in farming when they were offered multiple on-farm and off-farm livelihood opportunities. Off-farm measures such as insurance, social protection, and market access complement risk reduction, adaptation, and mitigation (Harvey et al., 2014) and the necessary components for community-based CSA interventions. This is not a new finding, as recent studies have argued that a livelihood resilience approach to climate change projects helps tackle a range of shocks and stresses, including food security, conflict, and disasters (Tanner et al., 2014). Agriculture and climate change investments are more likely to be successful if they address systemic off-farm policy challenges such as market barriers, product standards, climate risks to value chains, and access to loans and credits. We recommend that projects targeting vulnerable smallholder farmers in developing countries tailor the choice of CSA practices to local risks, livelihood priorities, and development objectives.

Third, we posit that the integration of adaptation and mitigation strategies in the agriculture sector is more likely to be achieved if policy at the national level is improved through interlinked and coordinated policies. As Candel and Biesbroek (2016) suggest, policy integration is a process. In our case studies, five process-based characteristics (see Fig. 13.2) have influenced the integration of adaptation and mitigation and CSA more generally. Further, we found at the community-level that there were less disparities between adaptation and mitigation measures while at the national level they were treated differently within policies, practices, and funding mechanisms. Scholarly research suggests the effectiveness of integrated adaptation and mitigation strategies in addressing policy and institutional barriers to agriculture resilience, including transforming the forestry, natural resources, and infrastructure sectors (see Suckall et al., 2014; Inderberg et al., 2015). In contrast to theory, integration of adaptation and mitigation in practice is polarized as a consequence of perceptive and political factors. Lessons from the Philippines and Timor-Leste case studies suggest that CSA in developing countries depends upon adaptation actions, with mitigation described as a cobenefit. Addressing vulnerability to climate and natural disasters is the key priority that is reinforced in policies and by institutions. The integration of adaptation and mitigation is impeded by poor coordination between government agencies (Departments of environment and agriculture), siloed discussions within policy groups (e.g., National Adaptation Plans, Nationally Appropriate Mitigation Actions, Nationally Determined Contributions, and efforts to Reduce Emissions from Deforestation and forest Degradation or REDD +), inconsistent agriculture sector policies, and the weak participation of agriculture departments in climate change policy negotiations.

The aforementioned factors give rise to the existing dichotomy within national policies that impedes the adoption of CSA. Policy dichotomy shaped by the historically siloed negotiation processes of adaptation and mitigation in the international climate policy (Harvey et al., 2014) have led to separate policy agendas and funding mechanisms. While efforts are underway in the two case study countries to mainstream adaptation and mitigation in the agriculture sector by government and donors agencies, this is in the early stages. The focus instead is on institutionalizing risk assessment, hazard identification, and building capacity at the national level. In the Philippines, the divide between mitigation and adaption in the agriculture sector is addressed by the Department of Agriculture through system-wide adaptation and mitigation initiatives in its agriculture program. Its strategic objective is to increase agricultural productivity potential by accelerating CSA development and building adaptive capacity. The program is an example of how agriculture policies can be reformed by removing barriers to climate-smart solutions across political boundaries. Further coordinated effort is needed to address the specific issues that confront smallholder farmers and to develop a diverse portfolio of funding opportunities that encourage integrated approaches in agriculture.

13.5 CONCLUSION AND RECOMMENDATIONS

The evolution of the global agriculture system will require smallholder food systems to transition to CSA. This need raises two important questions: "Which climate-smart processes can optimize smallholder farming?" and "How can these processes maximize engagement with organizations they represent including people, civil society, and local government?" Answers to these questions require CSA strategies that support broad transformational and livelihood resilience agendas that acknowledge inclusion and collaboration. In practice, the comparative examples from the Philippines and Timor-Leste have showed that civil society organizations and local government have raised awareness and implemented a mix of science, local knowledge, technology, and flexible financing opportunities to support successful climate-smart interventions.

Our comparative analysis yields three key conclusions. First, most studies of CSA have viewed the need to adopt specific technical and scientific practices under the aegis of adaptation, mitigation, and food production. Attempts are underway to identify global criteria and metrics for monitoring integrated strategies and CSA, and as a consequence there is an urgent need for studies on successful empirical cases. In our view, smallholder farmer interventions are climate-smart when they directly address local climate risks, support a combination of adaptation, food security, and livelihood strategies, and empower at-risk/marginalized populations. It is important to recognize that *global* CSA strategies emphasize integrated technical adaptation and mitigation measures, while in developing countries *local* CSA strategies should be characterized by a combination of "no-regrets" adaptation measures that support technology, knowledge, finance, policy, and institutional and market measures. Reinforcing policies that directly address community and country needs is crucial for policy relevance and CSA success.

Second, empirical research from the two case studies confirms the effectiveness of CSA interventions through multiactor partnerships with grassroots organizations in local knowledge production, planning, and implementation.

These partnerships ideally include farmer networks, women's groups, civil society and local government organizations. Cross-scale collaborative partnerships maximize synergy between adaptation, mitigation, and food security, balance trade-offs, reduce potential conflicts between practices, strengthen local institutions, pool organizational expertise, mobilize extension services, generate local knowledge, and test technology options. We recommend that CSA initiatives focus on increasing collaboration at the subnational level, strengthening local production and integrating knowledge.

Third, we conclude that CBOs are uniquely positioned at the subnational level to coordinate services with farmers, generate new knowledge, and implement climate-smart farming solutions. Civil society, community-based groups, and organizations can cooperate, collaborate, and work together spanning different administrative levels and expert boundaries to strengthen resilience and reduce climate risks.

ACKNOWLEDGEMENTS

The authors thank the Oxfam Philippines and Timor-Leste Community-based Climate Change Program teams, especially Julia Kalmirah, Ana Maria Caspe, Dante Dalabajan, Anefel Granada, Christopher Dable, Maricar Jaro and Linda Tan Joan Lopez, who helped organize community consultations. We acknowledge Dr. Paul Dargusch for his advice and Daniel Benns for graphic design. This research was supported through assistance from Oxfam Philippines and the University of Queensland Graduate School International Travel Awards.

REFERENCES

Adger, W.N., 2003. Social capital, collective action, and adaptation to climate change. Econ. Geogr. 79 (4), 387–404. Available from: https://doi.org/10.1111/j.1944-8287.2003.tb00220.x.

Barnard, S., Nakhooda, S., Caravani, A., Schalatek, L., 2015. Climate Finance Regional Briefing: Asia. Overseas Development Institute, London, https://www.odi.org/sites/odi.org.uk/files/odi-assets/publications-opinion-files/10061.pdf (accessed 18.08.16).

Beuchelt, T.D., Badstue, L., 2013. Gender, nutrition- and climate-smart food production: opportunities and trade-offs. Food Sec. 5 (5), 709–721. Available from: https://doi.org/10.1007/s12571-013-0290-8.

Bhatta, G.D., Aggarwal, P.K., 2016. Coping with weather adversity and adaptation to climatic variability: a cross-country study of smallholder farmers in SouthAsia. Clim. Develop. 8 (2), 145–157. Available from: https://doi.org/10.1080/17565529.2015.1016883.

Biagini, B., Bierbaum, R., Stults, M., Dobardzic, S., McNeeley, S.M., 2014. A typology of adaptation actions: a global look at climate adaptation actions financed through the Global Environment Facility. Global Environ. Change 25, 97–108. Available from: https://doi.org/10.1016/j.gloenvcha.2014.01.003.

Bryan, E., Deressa, T.T., Gbetibouo, G.A., Ringler, C., 2009. Adaptation to climate change in Ethiopia and South Africa: options and constraints. Environ. Sci. Policy 12 (4), 413–426. Available from: https://doi.org/10.1016/j.envsci.2008.11.002.

Bryan, E., Ringler, C., Okoba, B., Koo, J., Herrero, M., Silvestri, S., 2013. Can agriculture support climate change adaptation, greenhouse gas mitigation and rural livelihoods? Insights from Kenya. Clim. Chang. 118 (2), 151–165. Available from: https://doi.org/10.1007/s10584-012-0640-0.

Campbell, B.M., Thornton, P., Zougmoré, R., van Asten, P., Lipper, L., 2014. Sustainable intensification: what is its role in climate smart agriculture? Curr. Opin. Environ. Sustain. 8, 39–43. Available from: https://doi.org/10.1016/j.cosust.2014.07.002.

Candel, J.J.L., Biesbroek, R., 2016. Toward a processual understanding of policy integration. Policy Sci. 49 (3), 211–231. Available from: https://doi.org/10.1007/s11077-016-9248-y.

Chandra, A., Dargusch, P., McNamara, K.E., 2016. How might adaptation to climate change by smallholder farming communities contribute to climate change mitigation outcomes? A case study from Timor-Leste, Southeast Asia. Sustain. Sci. 11 (3), 477–492. Available from: https://doi.org/10.1007/s11625-016-0361-9.

Chandra, A., McNamara, K.E., Dargusch, P., Caspe, A.M., Dalabajan, D., 2017. Gendered vulnerabilities of smallholder farmers to climate change in conflict-prone areas: a case study from Mindanao, Philippines. J. Rural Stud. 50, 45–59. Available from: https://doi.org/10.1016/j.jrurstud.2016.12.011.

Creswell, J.W., 2014. Research Design: Qualitative, Quantitative, and Mixed Methods Approaches. Sage Publications, Thousand Oaks, CA.

Davidson, D., 2016. Gaps in agricultural climate adaptation research. Nat. Clim. Chang. 6, 433–435. Available from: https://doi.org/10.1038/nclimate3007.

DENR, 2010. Philippine Strategy for Climate Change Adaptation 2010–2022. Department of Environment and Natural Resources, Quezon City, The Philippines.

Deressa, T.T., Hassan, R.M., Ringler, C., Alemu, T., Yesuf, M., 2009. Determinants of farmers' choice of adaptation methods to climate change in the Nile Basin of Ethiopia. Glob. Environ. Chang. 19 (2), 248–255. Available from: https://doi.org/10.1016/j.gloenvcha.2009.01.002.

Dugumaa, L.A., Wambugua, S.W., Minanga, P.A., van Noordwijkb, M., 2014. A systematic analysis of enabling conditions for synergy between climate change mitigation and adaptation measures in developing countries. Environ. Sci. Policy 42, 138–148. Available from: https://doi.org/10.1016/j.envsci.2014.06.003.

Engle, N.L., Lemos, M.C., 2010. Unpacking governance: building adaptive capacity to climate change of river basins in Brazil. Glob. Environ. Chang. 20 (1), 4–13. Available from: https://doi.org/10.1016/j.gloenvcha.2009.07.001.

Ensor, J., Boyd, E., Juhola, S., Castán Broto, V., 2015. Building adaptive capacity in the informal settlements of Maputo: lessons for development from a resilience perspective. In: Inderberg, T.H., Eriksen, S., O'Brien, K., Sygna, L. (Eds.), Climate Change Adaptation and Development: Transforming Paradigms and Practices. Routledge, London.

Ewbank, R., 2015. Climate-Resilient Agriculture: What Small-Scale Producers Need to Adapt to Climate Change. Christian Aid, London, UK.

FAO, 2010. Climate-Smart Agriculture: Policies, Practices and Financing for Food Security, Adaptation and Mitigation. Food and Agriculture Organization of the United Nations, Rome.

GRAIN, 2015. The Exxons of Agriculture. GRAIN, Barcelona, Spain.

Gupta, J., Termeer, C., Klostermann, J., Meijerink, S., Van den Brink, M., Jong, P., et al., 2010. The adaptive capacity wheel: a method to assess the inherent characteristics of institutions to enable the adaptive capacity of society. Environ. Sci. Policy 13 (6), 459–471. Available from: https://doi.org/10.1016/j.envsci.2010.05.006.

Harvey, C.A., Chacón, M., Donatti, C.I., Garen, E., Hannah, L., Andrade, A., et al., 2014. Climate-smart landscapes: opportunities and challenges for integrating adaptation and mitigation in tropical agriculture. Conserv. Lett. 7, 77–90. Available from: https://doi.org/10.1111/conl.12066.

IAASTD, 2009. Agriculture at a Crossroads. The Global Report of the International Assessment of Agricultural Knowledge, Science, and Technology. Island Press, Washington, DC.

Inderberg, T.H., Eriksen, S., O'Brien, K., Sygna, L. (Eds.), 2015. Climate Change Adaptation and Development: Transforming Paradigms and Practices. Routledge, London, UK.

IPCC, 2007. Climate Change 2007: Mitigation. Working Group III Contribution to the Fourth Assessment Report of the IPCC. Cambridge University Press, Cambridge.

IPCC, 2014. Chapter 7: food security and food production systems. Climate Change 2014: Impacts, Adaptation, and Vulnerability. IPCC Working Group II Contribution to AR5. IPCC Secretariat, World Meteorological Organization, Geneva.

Jakob, M., Steckel, J.C., Klasen, S., Lay, J., Grunewald, N., Martínez-Zarzoso, I., et al., 2014. Feasible mitigation actions in developing countries. Nat. Clim. Chang. 4, 961–968. Available from: https://doi.org/10.1038/nclimate2370.

Lemos, M.C., Agrawal, A., Eakin, H., Nelson, D.R., Engle, N., Johns, O., 2013. Building adaptive capacity in less developed countries. In: Asrar, G. R., Hurrell, J.W. (Eds.), Climate Science for Serving Society: Research, Modelling and Prediction Priorities. Springer Netherlands, Dordrecht, pp. 437–457.

Linnerooth-Bayer, J., Hochrainer-Stigler, S., 2015. Financial instruments for disaster risk management and climate change adaptation. Clim. Chang. 133, 85. Available from: https://doi.org/10.1007/s10584-013-1035-6.

Lipper, L., Thornton, P., Campbell, B.M., Baedeker, T., Braimoh, A., Bwalya, M., 2014. Climate-smart agriculture for food security. Nat. Clim. Chang. 4, 1068–1072. Available from: https://doi.org/10.1038/nclimate2437.

Nabikolo, D., Bashaasha, B., Mangheni, M.N., Majaliwa, G.M., 2012. Determinants of climate change adaptation among male and female headed farm households in eastern Uganda. Afr. Crop Sci. J. 20, 203–212.

Neufeldt, H., Jahn, M., Campbell, B.M., Beddington, J.R., DeClerck, F., De Pinto, A., et al., 2013. Beyond climate-smart agriculture: toward safe operating spaces for global food systems. Agric. Food Sec. 2 (12), 1–6.

Nhemachena, C., Hassan, R., Chakwizira, J., 2014. Analysis of determinants of farm-level adaptation measures to climate change in Southern Africa. J. Develop. Agric. Econ. 6 (5), 232–241. Available from: https://doi.org/10.5897/JDAE12.0441.

Pelling, M., High, C., 2005. Understanding adaptation: what can social capital offer assessments of adaptive capacity? Glob. Environ. Chang. 15, 308–319. Available from: https://doi.org/10.1016/j.gloenvcha.2005.02.001.

QSR, 2012. NVivo Qualitative Data Analysis Software (Version 11). QSR International Pty Ltd, London, UK.

Robson, C., 2002. Real World Research: A Resource for Social Scientists and Practitioner-Researchers. Blackwell, Oxford, UK.

Rosenzweig, C., Tubiello, F.N., 2007. Adaptation and mitigation strategies in agriculture: an analysis of potential synergies. Mitigat. Adapt. Strat. Glob. Chang. 12, 855–873.

Scherr, S.J., Shames, S., Friedman, R., 2012. From climate-smart agriculture to climate-smart landscapes. Agric. Food Sec. 1, 12.

Siedenburg, J., Martin, A., McGuire, S., 2012. The power of "farmer friendly" financial incentives to deliver climate smart agriculture: a critical data gap. J. Integrat. Environ. Sci. 9 (4), 201–217. Available from: https://doi.org/10.1080/1943815X.2012.748304.

Smith, P., Martino, D., Cai, Z., Gwary, D., Janzen, H., Kumar, P., et al., 2007. Policy and technological constraints to implementation of green house gas mitigation options in agriculture. Agric. Ecosyst. Environ. 118 (1–4), 6–28. Available from: https://doi.org/10.1016/j.agee.2006.06.006.

Steenwerth, K.L., Hodson, A.K., Bloom, A.J., Carter, M.R., Cattaneo, A., Chartres, C.J., et al., 2014. Climate-smart agriculture global research agenda: scientific basis for action. Agric. Food Sec. 3 (11), 1–39. Available from: https://doi.org/10.1186/2048-7010-3-11. 2014.

Streck, C., 2012. Financing REDD + : matching needs and ends. Curr. Opin. Environ. Sustain. 4 (6), 628–637. Available from: https://doi.org/10.1016/j.cosust.2012.10.001.

Suckall, N., Tompkins, E., Stringer, L., 2014. Identifying trade-offs between adaptation, mitigation and development in community responses to climate and socio-economic stresses: evidence from Zanzibar, Tanzania. Appl. Geogr. 46, 111–121.

Sugden, J., 2015. Climate-Smart Agriculture and Smallholder Agriculture. The Critical role of Technology Justice in Effective Adaptation. Practical Action Publishing, Rugby, UK.

Tambo, J.A., Abdoulaye, T., 2012. Climate change and agricultural technology adoption: the case of drought tolerant maize in rural Nigeria. Mitigat. Adapt. Strat.r Glob. Chang. 17 (3), 277–292. Available from: https://doi.org/10.1007/s11027-011-9325-7.

Tanner, T., Lewis, D., Wrathall, D., Bronen, R., Cradock-Henry, N., Huq, S., et al., 2014. Nature Climate Change 23−26. Available from: https://doi.org/10.1038/nclimate2431.

Thorn, J.P.R., Friedman, R., Benz, D., Willis, K.J., Petrokofsky, G., 2016. What evidence exists for the effectiveness of on-farm conservation land management strategies for preserving ecosystem services in developing countries? A systematic map. Environ. Evid. 5 (13), 1−29. Available from: https://doi.org/10.1186/s13750-016-0064-9.

Thornton, T.F., Comberti, C., 2013. Synergies and trade-offs between adaptation, mitigation and development. Clim. Change 1−14. Available from: https://doi.org/10.1007/s10584-013-0884-3.

Tompkins, E.L., Mensah, A., King, L., Long, T.K., Lawson, E., Hutton, C., et al., 2013. An Investigation of the Evidence of Benefits From Climate Compatible Development. Centre for Climate Change Economics and Policy, Leeds, UK.

UNFCCC, 2014a. Report on the meeting on available tools for the use of indigenous and traditional knowledge and practices for adaptation, needs of local and indigenous communities and the application of gender-sensitive approaches and tools for adaptation. Bonn, Subsidiary Body for Scientific and Technological Advice, UNFCCC Secretariat.

UNFCCC, 2014b. Technologies for Adaptation in the Agriculture Sector. The Technology Executive Committee, United Nations Climate Change Secretariat, Bonn.

WB, 2009. World Development Report 2010: Development and Climate Change. The World Bank, Washington, DC.

Yates, J.S., 2014. Power and politics in the governance of community-based adaptation. In: Ensor, H., Berger, R., Huq (Eds.), Community-based Adaptation to Climate Change. Practical Action Publishing, Rugby, UK, pp. 15−34.

Ye, W., Vasileiou, I., Förch, W., 2015. Climate Finance for Agricultural Adaptation. CGIAR Research Program on Climate Change, Agriculture and Food Security (CCAFS), Copenhagen, Denmark, Working Paper no. 155.

Chapter 14

Challenges in Building Climate-Resilient Quality Energy Infrastructure in Africa

Ashbindu Singh, H. Gyde Lund and Jane Barr
Environmental Pulse Institute, Trent, SD, United States

Chapter Outline

14.1 INTRODUCTION

Climate trends are expected to have long-term gradual impacts on the energy system through increasing air and water temperatures, sea level rise, and other changes in weather patterns as well as impacts caused by extreme events, such as heat waves, droughts, heavy precipitation, and storms (IEA, 2016).

Changes in weather and climate extremes will occur globally and in all regions, albeit unevenly, according to projections in the Intergovernmental Panel on Climate Change (IPCC) Fifth Assessment Report (IPCC, 2014c). These impacts, as well as combinations of these impacts that can intensify their effect, will have significant implications for business as usual in the energy sector (IEA, 2016).

Energy demand is expected to change, potentially dramatically in some areas, as a result of increasing temperatures and changing weather patterns, affecting heating and cooling demands. While demand for heating may decrease, demand for space cooling will increase in all parts of the world (IEA, 2016). *Energy supply* will face changing conditions, including reduced efficiency of thermal plants, cooling constraints on thermal and nuclear plants, and pressure on transmission systems; electricity generation from hydro, wind, and other renewable and biofuel production will also be affected (IEA, 2016). *Energy infrastructure* could be exposed to sea level rise, glacial melt, as well as more frequent and intense extreme weather events (EWEs) including increased wind speeds and ocean storminess. These may threaten coastal power generation infrastructure, onshore transmission and distribution infrastructure, as well as offshore installations and pipelines and could ultimately lead to various interruptions of energy delivery systems (IEA, 2016).

Changing demographics may also influence energy demand. Africa's population is increasing and the African's are also increasingly affluent. Economic growth in Sub-Saharan Africa is projected to reach 2.6% in 2017 and is expected to continue in 2018−19 (World Bank Group, 2017).

14.2 POTENTIAL IMPACT OF CLIMATE CHANGE IN AFRICA

Climate change will increasingly impact Africa due to many factors, including the proximity of many African countries to the equator. These impacts are already being felt and will increase in magnitude if action is not taken to reduce global carbon emissions. The impacts include higher temperatures, drought, rising sea levels, changing rainfall patterns, and increased climate variability. These conditions have a bearing on energy production and consumption. Droughts in 2015−16 (Anyadike, 2017) in many African countries, which has been linked to climate change, adversely affected both energy security and economic growth across the continent. With increasing population and corresponding energy demand, energy security must be addressed because energy is crucial for sustainable development (Besada and Sewankambo, 2009).

As the most vulnerable continent to climate variability and change, access to energy services in African countries also ranks among the lowest in the world. Currently, over 645 million Africans do not have access to electricity, 700 million have no access to clean cooking energy, and 600,000 die each year from indoor pollution due to their reliance on biomass for cooking. Installed grid-based power generation capacity in Africa reached 170 GW in 2014. Per capita electricity use in Africa averages 181 kWh, compared to about 13,000 kWh in the United States and over 6500 kWh in Europe. Africa's poorest pay some of the highest energy costs in the world, about 60−80 times per unit more in Northern Nigeria than residents in cities like New York and London (AfDB, 2016). Energy sector bottlenecks and power shortages are estimated to cost Africa about 2%−4% of GDP annually (AfDB, 2016).

It is anticipated that climate change in Africa will result in aridification, decreased runoff, increased air temperatures, and increased extreme weather conditions such as floods, droughts, and high winds. Water resources, agriculture, human health, forestry, rangelands, biodiversity, fishing, forestry, and tourism are all sectors that will be affected. The most vulnerable energy subsector is biomass fuels, which are used by the largest and most vulnerable category of consumers—poor women (Annecke, 2002).

Climate-related factors will make it harder for African countries to tackle extreme poverty in the future for three reasons:

- *Warming is unavoidable* as a result of past emissions of greenhouse gases, which will cause the loss of cropland, a decline in crop production, worsening undernourishment, higher drought risks, and a decline in fish catches
- *Further warming may materialize*, which will have disastrous consequences for the region in the form of heat extremes, increased risk of severe drought, crop failures every 2 years, a 20% reduction in major food crop yields, and, by the end of the 21st century, up to 18 million people affected by floods every year
- *Considerable uncertainty on what the warming impact will be on local weather patterns and hydrological cycles*, which poses formidable challenges for development planning, and for the design of projects related to water management such as irrigation and hydropower, and more generally climate-sensitive infrastructure such as roads or bridges (World Bank, 2016).

14.3 ADAPTATION OPTIONS IN ENERGY SECTOR

Changes in climate attributes (temperature, precipitation, windiness, cloudiness, etc.) will affect energy sources and technologies in Africa differently. Gradual climate change will progressively affect the operation of energy installations and infrastructure over time. Possible changes in the frequency and intensity of EWEs as a result of climate change represent a different kind of hazard for them. Table 14.1 provides an overview of the most important impacts and adaptation options in both categories.

The African continent is home to many of the world's fastest growing economies and populations. It holds substantial untapped energy resources. Maintaining economic growth for these countries requires access to reliable, affordable, and sustainable energy. Yet power generation capacity is often inadequate (MacWilliams, 2014).

Climate change will affect the major sources of Africa's energy, although the literature is still sparse on the subject. It will affect the various energy sources and technologies differently, depending on the resource (water, wind, solar, nuclear, fossil, biomass, and/or waste), the technological processes involved, and the location. Changes in temperature, precipitation, windiness, cloudiness, etc., and the frequency and intensity of EWEs will increasingly affect energy operations. The main climate-related problems for thermal and nuclear power plants are changes in the availability and temperature of water for cooling and the potential decline in thermal conversion efficiency. There is also concern for climate change's potential to influence the integrity and reliability of pipelines and electricity grids, which may require changes in construction and operation design standards (IPCC, 2014a). Due to Africa's high dependence on biomass, climate change will no doubt have adverse effects on the sector, since potential declines in rainfall and increasing drought can lead to vegetation loss and ultimately to desertification. Other human impacts on the environment, such as land use change and deforestation, will exacerbate climate change effects (Habtezion, 2012). In a recent study on the effects of climate change on the potential for expanding wind and solar energy resources in southern Africa for future planning purposes, the authors looked at the long-term mean of wind speed and Global Horizontal Irradiance, which are reliable indicators of change in the potential for electricity production. They found that changes in wind and solar potential by 2050 are expected to be small (Fant et al., 2016; AfDB, 2017).

TABLE 14.1 The Main Projected Impacts of Climate Change and Extreme Weather Events on Energy Supply and the Related Adaptation Options

Technology	Changes in Climatic or Related Attributes	Possible Impacts	Adaptation Options
Thermal and nuclear power plants	Increasing air temperature	Reduces efficiency of thermal conversion.	Siting at locations with cooler local climates where possible
	Changing (lower) precipitation and increasing air temperature increases temperature and reduces the availability of water for cooling	Less power generation; annual average load reduction by 0.1%–5.6%, depending on scenario	Use of nontraditional water sources (e.g., water from oil and gas fields, coal mines and treatment, treated sewage); Reuse of process water from flue gases (can cover 25%–37% of the power plants cooling needs), coal drying, condensers (dryer coal has higher heating value, cooler water enters cooling tower), flue-gas desulphurization; Using ice to cool air before entering the gas turbine increases efficiency and output, melted ice used in cooling tower; Condenser mounted at the outlet of cooling tower to reduce evaporation losses (by up to 20%). Alternative cooling technologies: dry cooling towers, regenerative cooling, heat pipe exchangers; Costs of retrofitting cooling options depends on features of existing systems, distance to water, required additional equipment, estimated at US$250,000–500,000 MW^{-1}
	Increasing frequency of extreme hot temperatures	Exacerbates impacts of warmer conditions: reduces thermal and cooling efficiency; limits cooling water discharge; overheats buildings; causes self-ignition of coal stockpiles	Cooling of buildings (air-conditioning) and of coal stockpiles (water spraying)
	Drought: reduced water availability	Exacerbates impacts of warmer conditions, reduces operation and output, causes shutdown	Same as reduced water availability under gradual climate change
Hydropower	Increase/decrease in average water availability	Increased/reduced power output	Schedule release to optimize income
	Changes in seasonal and interannual variation in inflows (water availability)	Shifts in seasonal and annual power output; floods and lost output in the case of higher peak flows	Soft: adjust water management Hard: build additional storage capacity, improve turbine runner capacity
	Extreme precipitation causing floods	Direct and indirect damage to dams and turbines (by siltation, debris carried from flooded areas), lost output due to releasing water through bypass channels	Soft: adjust water management Protect vegetation in watershed Remove debris Hard: increase storage capacity
Solar energy	Increasing mean temperature	Improve performance of thermal heating (TH) (especially in colder regions), reducing efficiency of photovoltaic (PV) and concentrating solar power (CSP) with water	

(Continued)

TABLE 14.1 (Continued)

Technology	Changes in Climatic or Related Attributes	Possible Impacts	Adaptation Options
		cooling; PV efficiency drops by ~0.5%/1°C temperature increase for crystalline Si and thin-film modules as well, but performance varies across types of modules, with thin-film modules performing better; Long-term exposure to heat causes faster aging	
	Changing cloudiness	Increased unfavorable (reduced output), decreased beneficial (increased output) for all types, but evacuated tube collectors for TH can use diffuse insolation. CSP more vulnerable (cannot use diffuse light)	Apply rougher surface for PV panels that use diffuse light better; optimize fixed mounting angle for using diffuse light, apply tracking system to adjust angle for diffuse light conditions; Install/increase storage capacity
	Hot spells	Material damage for PV, reduced output for PV and CSP; CSP efficiency decreases by 3%–9% as ambient temperature increases from 30 to 50°C and drops by 6% (tower) to 18% (trough) during the hottest 1% of time	Cooling PV panels passively by natural air flows or actively by forced air or liquid coolants
	Hail	Material damage to TH: evacuated tube collectors are more vulnerable than flat plate collectors; Fracturing as glass plate cover, damage to photoactive material	Flat plate collectors: using reinforced glass to withstand hailstones of 35 mm (all of 15 tested) or even 45 mm (10 of 15 tested); only 1 in 26 evacuated tube collectors withstood 45 mm hailstones; Increase protection to current standards or beyond them
Wind power	Windiness: total wind resource (multiyear annual mean wind power densities).	Change in wind power potential	Site selection
	Wind speed extremes: gust, direction change, shear	Structural integrity from high structural loads; Fatigue, damage to turbine components; reduced output	Turbine design, LIDAR-based protection

Source: IPCC, 2014a. Climate Change 2014: Impacts, Adaptation, and Vulnerability. Part A: Global and Sectoral Aspects. Contribution of Working Group II to the Fifth Assessment Report of the Intergovernmental Panel on Climate Change. Cambridge, United Kingdom and New York, NY, USA: Cambridge University Press. Retrieved July 14, 2016, from http://hcl.harvard.edu/collections/ipcc/docs/4__WGIIAR5-PartA_FINAL.pdf (IPCC, 2014a).

Climate change, however, can have mixed consequences. The warming of the Mediterranean region appears to be behind an increase in rainfall in the Sahel region (Park et al., 2016). The increase in rainfall leads to more vegetation. However, Africa's growing population may offset an advantage gained by the greening of the Sahel (Brandt et al., 2017).

14.3.1   Climate Adaptation in Hydropower

The "Enhancing the Climate Resilience of Africa's Infrastructure" (ECRAI) study (Cervigni et al., 2015), conducted by the World Bank and the United Nations Economic Commission for Africa (UNECA), found that failure to integrate climate change in the planning and design of power and water infrastructure could entail, in the driest climate scenarios, losses of hydropower revenues of between 5% and 60% (depending on the basin); and increases in consumer expenditure for energy up to three times the corresponding baseline values. In the wettest climate scenarios, business-as-usual infrastructure development could lead to foregone revenues in the range of 15%−130% of the baseline, if the larger volume of precipitation is not used to expand the production of hydropower. In irrigation, the largest loss in revenue is in the 10%−20% range for most basins. In wet scenarios, the largest foregone gains are in the range of 1%−4%, with the exception of the Volta basin, where they are estimated to be one order of magnitude higher (World Bank, 2015a,b).

The IPCC reports that the impact of climate change in Africa on future water scarcity will be modest relative to other drivers, such as population growth, urbanization, agricultural development, and land use change. There will be regional and subregional variations, however, with already water-stressed regions such as northern Africa and parts of southern Africa projected to become drier (IPCC, 2014b). In East Africa, climate change has already contributed to varying rainfall patterns with impacts on hydropower. Severe droughts reduce water supplies for hydroelectric power generation, while excessive flooding contributes to silt deposits in hydropower reservoirs, affecting the amount of water available for electricity generation (Habtezion, 2012). By way of example, Ghana's electricity demand is increasing by 10% a year while declining precipitation and higher temperatures due to climate change are reducing hydropower output leading to a national power crisis (REN21, 2015; AfDB, 2017) (Box 14.1).

BOX 14.1 The Impacts of Climate Change on the Energy Economy

Some research suggests that repeated droughts are creating a power crisis in East Africa, which derives almost 80% of its electric supply from hydropower (Habtezion, 2012). For example, hydroelectricity accounts for about half of all electricity generated in Uganda. Lake Victoria is very sensitive to rainfall variability and the impacts of climate change. The IPCC reports that the rise in global temperatures is already affecting the lake's water levels (IPCC, 2014b). Rainfall variability led to nearly a 2 m drop in Lake Victoria's water level over 3 years in the mid-2000s, reducing the amount of power generated at the Kira and Nalubale hydropower stations. The shortage of power and need for alternative sources increased operation costs and affected industry in particular, which in turn led to a drop in GDP growth and increased power tariffs (AfDB, 2017).

The Kariba Dam, the world's largest man-made reservoir, is located in the Kariba Gorge of the Zambezi River Basin between Zambia and Zimbabwe. Constructed between 1956 and 1959, this double curvature concrete arch dam supplies water to two underground hydropower plants, one on the north (left) bank in Zambia and the other on the south (right) bank in Zimbabwe, with a total capacity of 1830 MW and generating more than 10,035 GWh of electricity The impacts of climate change will affect the reservoir's reliability. World Bank projections indicate that under the driest scenarios, hydropower generation could fall by more than 60%, and in the Zambezi basin, unmet irrigation demand could decrease by more than 25%. Under wetter scenarios, hydropower production could increase by up to 25% and irrigation water provision potential would rise by a few percent, illustrating how climate change is an important factor in water and power infrastructure performance in the Zambezi (AfDB, 2014). Power from the Kariba dam helped to make Zambia's economy one of the fastest growing in Africa. The country relies on hydropower for 95% of its electricity and the dam usually generates more than 40% of its power. A severe drought in 2015, exacerbated by climate change, led to near-record lows in water levels and blackouts that hurt the economy. Impacts were especially significant in the copper and steel industries, which suffered from increased production costs, declines in production capacity and prices, and lost jobs. This power crisis focused attention on the contribution climate change makes to natural temperature and rainfall variations that affect lake levels (Fig. 14.1) over time (Onishi, 2016). More recently, water levels appear to be increasing (AfDB, 2017).

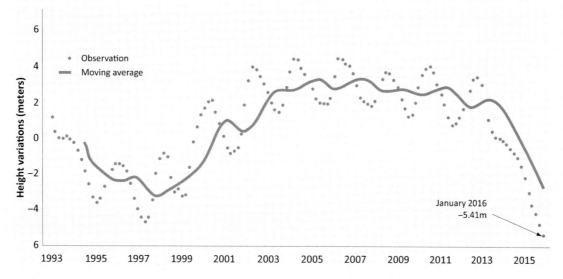

FIGURE 14.1 Variations in Lake Kariba's water levels, 1993−January 2016. *NASA Earth Observatory, 2016. The Decline of Lake Kariba. Retrieved September 23, 2016, from NASA Earth Observatory. http://www.earthobservatory.nasa.gov/NaturalHazards/view.php?id = 87485 (NASA Earth Observatory, 2016).*

14.4 RESPONSE

To tackle the climate challenge in collaboration with African governments and a variety of regional and international partners, the Africa Climate Business Plan focuses on increasing adaptation through a dozen priority areas grouped into three clusters;

- *Strengthening resilience*, which includes initiatives aimed at boosting the continent's natural capital (landscapes, forests, and oceans), physical capital (cities and transport infrastructure), and human and social capital, including improving social protection for the more vulnerable against climate shocks and addressing the climate-related drivers of migration;
- *Powering resilience*, which includes opportunities to increase low-carbon energy sources as societies with inadequate energy sources are more vulnerable to climate shocks; and
- *Enabling resilience* by providing essential data, information, and decision-making tools for promoting climate-resilient development across sectors through strengthening hydromet systems at the regional and country level, and through building the capacity to plan and design climate-resilient investments (World Bank, 2016).

Adapting to today's climate is not same as adapting to future climate change. Climate variability and climate change are dynamic. This makes adaptation targets shifting goals to be achieved—which necessitate making continual efforts and struggles. The transition from vulnerability to resilience is to be determined by conscious and concerted actions through designing and implementing adaptation measures. Some of the crucial conditions include: improving climate change governance; mainstreaming climate change in development; enhancing knowledge systems; and forging strategic partnerships and cooperation to support such a transition (Gabriel, n.d.).[1]

Most national governments are initiating governance systems for adaptation (Table 14.2). Disaster risk management, adjustments in technologies and infrastructure, ecosystem-based approaches, basic public health measures, and livelihood diversification are reducing vulnerability, although efforts to date tend to be isolated (IPCC, 2014c).

TABLE 14.2 Examples of African Intended Nationally Determined Contribution (INDC) Energy Sector Adaptation Actions (summarized from www.unfcc.org)

INDC Energy Sector Adaptation Actions

Energy Sources-Related Actions

- Promote renewable energy and other energy sources.
- Promote use of biofuels for lighting and cooking replacing fossil-based fuel.
- Sustainably manage traditional energy sources (forests, firewood, and charcoal).
- Increase the efficiency in the use of biomass in the traditional energy sector such as using briquettes as substitute for firewood and charcoal.
- Promote an energy mix that moves people away from use of biomass to avoid deforestation.
- Expand electric power generation from geothermal, wind, and solar sources to minimize the adverse effects of droughts on predominantly hydroelectric energy sector.

Infrastructure-Related Actions

- Promote the use of energy-efficient technologies and behavior. Implement energy-efficient and electricity-saving strategies.
- Climate proofing investments in electricity power sector. Strengthen existing energy infrastructure, in part through early efforts to identify and implement all possible "no regrets" actions. Move critical infrastructure out of flooding/storm surge risk areas. Include increased protective margins in construction and placement of energy infrastructure (i.e., higher standards and specifications).
- Increase the resilience of current and future energy systems. Increase the efficiency in the modern energy sector, mainly of electricity. Strengthen transmission and distribution infrastructures for public utilities to ensure climate resilience (i.e., flooding).
- Ensure the best use of hydropower by careful management of the water resources. Construct storage dams for hydropower generation. Enhance integrated basin catchment and upstream land management for hydro sources. Protect water catchments around hydropower sources.
- Build institutional and technical capacities of different units in the energy sector in climate change issues. Keep the options for diversifying electricity sources under continual review. Develop mini hybrid networks for rural electrification.
- Develop and diversify secure energy backup systems to ensure both civil society and security forces have access to emergency energy supply.

(Continued)

---

1. n.d. = not dated.

TABLE 14.2 (Continued)

INDC Energy Sector Adaptation Actions

Research-Related Actions

- Support research and technological development to enable the electricity sector to deal properly with climate change.
- Conduct comprehensive studies to assess the impact of climate change on hydroelectricity, mining, and the energy sector.
- Propose appropriate adaptation measures and estimate the economic cost of the proposed adaptation measures.
- Determine the safe locations for the construction of power generation projects.
- Explore and invest in energy diversification systems including the application of renewable energy technologies.
- Undertake risk assessment and risk reduction measures to increase resilience of the energy sector.

Technological-Related Actions

- Acquire low-carbon transport modes including hybrid and electric means of mass transportation.
- Promote new, clean and energy-saving technologies in industry and construction.
- Promote solar PV and use of energy efficient bulbs and the thermal insulation of buildings.
- Promote the production and distribution of improved cook stoves.
- Recover methane from the solid wastes and used water.

Source: C2ES, 2016. Submitted Intended Nationally Determined Contributions (INDCs). Arlington, VA: Center for Climate and Energy Solutions. Retrieved March 22, 2017, from https://www.c2es.org/international/2015-agreement/indcs (C2ES, 2016).

Africa has experienced economic growth of more than 5% per annum during the past decade, but to sustain this growth, investment in infrastructure is fundamental. In recognition of this fact, the Program for Infrastructure Development in Africa (PIDA), endorsed in 2012 by the continent's heads of state and government, has laid out an ambitious, long-term plan for closing Africa's infrastructure gap. In the water and power sector, PIDA calls for an expansion of hydroelectric power generation capacity by more than 54,000 MW and of water storage capacity by 20,000 km^3 (Cervigni et al., 2015).

14.5 CONCLUSIONS

Policy-makers in Africa are recommended to recognize the potential role of renewable energy in meeting the energy challenges being faced by the region, and assume a proactive role in implementation. There is an urgent need to assume an integrated and coordinated approach at a regional level to scale up the deployment of renewable energy technologies so as to increase access to modern energy services and increase energy security to support economic and social development (UNIDO, 2009).

Policy directions for change as articulated in Uganda's Renewable Energy Policy (March 2007) could serve as a model for wider applicability (Besada and Sewankambo, 2009):

- Ensure the legal and regulatory framework is responsive to the development of renewable energy sources and facilitate their promotion.
- Establish an appropriate financing and fiscal policy framework for renewable energy technology investments, including targeted subsidies, tax rebates/exemptions, favorable feed-in tariffs, risk mitigation, and credit enhancement mechanisms.
- Mainstream poverty eradication, equitable distribution, social services, and gender issues in renewable energy strategies.
- Promote research and development, international cooperation, technology transfer and adoption, and standards in renewable energy technologies.
- Promote and enhance the sustainable production and utilization of biofuels, including developing the necessary legislation.
- Promote and encourage the conversion of municipal waste biomass to energy, in particular, municipal and industrial waste.

Priorities for governmental intervention to ensure improved energy supply—to improve the situation, African governments could:

- Introduce reforms to liberalize energy supply and attract private sector investment. Such reforms could include breaking the monopoly of vertically integrated utilities to allow more players in the sector, and introducing an independent regulation.
- Encourage small-scale renewable energy generation (up to 20 MW) by creating incentives for private sector investment. An example might be a published feed-in tariff: a minimum amount that anyone exporting electricity to the grid will be paid for each unit exported, a scheme specifically designed to make these investments financially viable.
- Set efficiency improvement targets for utilities, such as loss reduction in the system, to be achieved in each business center, i.e., in generation, transmission, and distribution, respectively.
- Establish standard legal documents for entities entering into contractual arrangements in bulk energy in order to reduce transaction time.
- Establish a regulator with powers to prevent market abuse and ensure good service and transparency.

Building climate-resilient infrastructure will bring an additional layer of complexity in terms of cost and engineering specifications to Africa's development and growth challenges, requiring long-term sustainability to be included as a core consideration from the earliest stages of infrastructure project planning. But developing quality infrastructure that is resilient to climate change also offers the continent considerable opportunities, including ensuring that growth and development is both inclusive and sustainable (ICA, n.d.). By transitioning to a lower carbon economy, jobs can be created and skills improved, benefiting Africa and its people.

REFERENCES

AfDB, 2014. Project: Kariba Dam Rehabilitation Project: Environmental and Social Impact Assessment Summary. African Develompment Bank Group.

AfDB, 2016. The Impact of Energy and Climate Change on Growth in Africa. African Development Bank (AfDB) Group, Retrieved March 22, 2017, from: https://www.afdb.org/fileadmin/uploads/afdb/Documents/Generic-Documents/Annual_Meetings/AfDB_Annual_Report_2015_-_Extract_2_of_3_-_Energy_and_climate_change.pdf.

AfDB, 2017. Atlas of Africa Energy Resources. African Development Bank (AfDB), Abidgan, Côte d'Ivoire, Retrieved March 25, 2017, from: http://hdl.handle.net/20.500.11822/20476.

Annecke, W., 2002. Climate change, energy-related activities and the likely social impacts on women in Africa. J. Glob. Environ. Issues 2 (3/4), 207−222. Retrieved March 22, 2017, from: http://www.inderscience.com/info/inarticle.php?artid=2400.

Anyadike, O., 2017, March 15. Farmers, traders and consumers across East and Southern Africa are feeling the impact of consecutive seasons of drought that have scorched harvests and ruined livelihoods. IRIN. Retrieved July 21, 2017, from: https://www.irinnews.org/feature/2017/03/17/drought-africa-2017.

Besada, H., Sewankambo, N.K., 2009. Climate Change in Africa: Adaptation, Mittigation and Governance Challenges. Centre for International Governance Innovation (CIGI), Waterloo, ON, Retrieved March 22, 2017, from: https://www.unicef.org/esaro/Climate_Change_in_Africa.pdf.

Brandt, M., Rasmussen, K., Penuelas, J., Tian, F., Shurgers, G., Verger, A.V., et al., 2017. Human population growth offsets climate-driven increase in woody vegetation in sub-SaharanAfrica. Nat. Ecol. Evol. 1 (0081), Retrieved July 21, 2017, from: https://www.nature.com/articles/s41559-017-0081.

C2ES, 2016. Submitted Intended Nationally Determined Contributions (INDCs). Center for Climate and Energy Solutions, Arlington, VA, Retrieved March 22, 2017, from: https://www.c2es.org/international/2015-agreement/indcs.

Cervigni, R., Liden, R., Neumann, J., Strzepek, K., 2015. Enhancing the Climate Resilience of Africa's Infrastructure - The Power and Water Sector. Agence Française de Développement and the World Bank, Washington, DC. Available from: https://doi.org/10.1596/978-1-4648-0466-3.

Fant, C., Schlosser, C.A., Strzepek, K., 2016. The impact of climate change on wind and solar resources in southern Africa. Appl. Energy 161, 556−564. Available from: https://doi.org/10.1016/j.apenergy.2015.03.042.

Gabriel, A.H., n.d. Climate Resilience, Development and Adaptation in Africa. African Union Commission. Retrieved March 25, 2017, from: http://www.uneca.org/sites/default/files/uploaded-documents/Climate/CCDA1/ccda1_2-climate_resilience_development_and_adaptation_en.pdf.

Habtezion, S., 2012. Gender and Energy. Gender and Climate Change: Capacity development series Africa Training Module 3, United Nations Development Programme, New York. Retrieved January 15, 2016, from: http://www.undp.org/content/dam/undp/library/gender/Gender%20and%20Environment/TM3-Africa_Gender-and-Energy.pdf.

ICA, n.d. Climate-Resilient Infrastructure. The Infrastructure Consortium for Africa (ICA). Retrieved March 25, 2017, from: https://www.icafrica.org/en/topics-programmes/focal-points/climate-resilient-infrastructure/.

IEA, 2016. Resilience of the Energy Sector to Climate Change. International Energy Agency, Paris, Retrieved March 26, 2017, from: http://www.iea.org/topics/climatechange/subtopics/resilience/.

IPCC, 2014a. Climate Change 2014: Impacts, Adaptation, and Vulnerability. Part A: Global and Sectoral Aspects. Contribution of Working Group II to the Fifth Assessment Report of the Intergovernmental Panel on Climate Change. Cambridge University Press, Cambridge, United Kingdom and New York, NY, USA, Retrieved July 14, 2016, from: http://hcl.harvard.edu/collections/ipcc/docs/4__WGIIAR5-PartA_FINAL.pdf.

IPCC, 2014b. Climate Change 2014: Impacts, Adaptation, and Vulnerability. Part B: Regional Aspects. Contribution of Working Group II to the Fifth Assessment Report of the Intergovernmental Panel on Climate Change. Cambridge University Press, Cambridge, United Kingdom and New York, NY, USA.

IPCC, 2014c. Climate Change 2014: Synthesis Report. Intergovernmental Panel on Climate Change (IPCC), Geneva, Retrieved March 27, 2017, from: http://ar5-syr.ipcc.ch/ipcc/ipcc/resources/pdf/IPCC_SynthesisReport.pdf.

MacWilliams, J.J., 2014, September 24. Financing Clean Energy Infrastructure in Africa. *Energy.Gov*. Retrieved from https://energy.gov/articles/financing-clean-energy-infrastructure-africa.

NASA Earth Observatory, 2016. The Decline of Lake Kariba. Retrieved September 23, 2016, from NASA Earth Observatory. http://www.earthobservatory.nasa.gov/NaturalHazards/view.php?id = 87485.

Onishi, N., 2016, April 16. Climate Change Hits Hard in Zambia, an African Success Story. The New York Times.

Park, J.-y, Bader, J., Matei, D., 2016. Anthropogenic Mediterranean warming essential driver for present and future Sahel rainfall. Nat. Clim. Chang. 6, 941−945. Available from: https://doi.org/10.1038/nclimate3065.

REN21, 2015. Renewables 2015 Global Status Report. Renewable Energy Policy Network for the 21st Century. Paris: REN21 Secretariat. Retrieved January 16, 2016, from: http://www.ren21.net/wp-content/uploads/2015/07/REN12-GSR2015_Onlinebook_low1.pdf.

UNIDO, 2009. Scaling Up Renewable Energy in Africa. United Nations Industiral Development Organization (UNIDO), Vienna, Retrieved March 22, 2017, from: http://www.unido.org/fileadmin/user_media/Services/Energy_and_Climate_Change/Renewable_Energy/Publications/Scaling%20Up%20web.pdf.

World Bank, 2015a. A Facility for Climate Resilient Investment Planning in Africa. The World Bank, Washington, DC, Retrieved March 25, 2017, from: https://www.afdb.org/fileadmin/uploads/afdb/Documents/Events/COP21/Abstracts/08_Dec_2015_-_Facility_for_Climate_Resilient_Investment_Planning_in_Africa.pdf.

World Bank, 2015b. The Kariba Dam Rehabilitation Project: Fact Sheet. Retrieved September 23, 2016, from The World Bank Group. http://www.worldbank.org/en/region/afr/brief/the-kariba-dam-rehabilitation-project-fact-sheet.

World Bank, 2016. Africa Climate Business Plan. The World Bank Group, Washington, DC, Retrieved March 25, 2017, from: http://www.worldbank.org/en/programs/africa-climate-business-plan.

World Bank Group, 2017, April. Africa's Pulse No. 15. Africa's Pulse (15). Retrieved July 21, 2017, from: http://hdl.handle.net/10986/26485.

Tools and Approaches

Chapter 15

Ethics, Communities, and Climate Resilience: An Examination by Case Studies

Kerry W. Bowman, Alan Warner and Yousef M. Manialawy

University of Toronto, Toronto, ON, Canada

Chapter Outline

15.1 BACKGROUND AND SIGNIFICANCE: THE NEED FOR CLIMATE RESILIENCE

Reducing anthropogenic greenhouse gas emissions to stable and sustainable levels ought to be a key political, economic, and social priority for nations around the world. Failure to adequately address the imperatives of climate change through a coherent and effective policy framework represents a harm of extreme magnitude that will have potentially devastating effects for all of humanity, with those already impoverished standing to lose the most. Inaction in this domain is an abrogation of social responsibility. Although the ethical implications of failing to address climate change are monumental, governments around the world have indeed failed to implement a policy framework that adequately reflects the urgent need for a comprehensive climate change strategy. Efforts thus far to mitigate the impact of anthropogenic activities have been spotty, inconsistent, minimal, or absent entirely. In turn, much of the center stage discourse is diluted with endless debates on the ins and outs of carbon trading and on arguments about which forms of lifestyle changes are most warranted. Many people put great stock in the scientific development of future technological discoveries that may help to solve some of the problems that emerge as a result of climate change, but the effectiveness of scientific intervention will likely be limited and is by no means a certainty. In the absence of discernible, visible risks, governments are hesitant to devote resources to addressing a threat that may not fully present itself during their tenure. Virtually all contemporary political structures are focused on clear and deliverable results during their political term. Mitigation and resilience strategies for climate change fall outside of this mandate.

Yet we are living in the real world and change is upon us. We hold an obligation to protect human populations and the environment from the climatic threats that are here now, and ethical arguments for reducing emissions as well as building climate resilience can be enormously beneficial. For example, consider the analysis of resilience for climate change through the lens of "justice." Extreme weather and above average temperatures are "already undermining the realization of a broad range of internationally protected human rights" (International Council on Human Rights Policy,

Resilience. DOI: https://doi.org/10.1016/B978-0-12-811891-7.00015-3

2008, p. 1). In general, developing countries, minorities, the poor, and indigenous populations are more susceptible to environmental changes and climate-sensitive health outcomes induced by climate change (Jäger and Kok, 2007).

Addressing the current state of affairs is by no means a simple task. Overshadowing the massive scientific challenge of building climate change resilience is finding within ourselves the ability to cultivate adaptive, resilient behavior; adaptation is as much a social as a scientific process. A problem of this magnitude, however, requires both short-term and long-term initiatives. It is beyond the scope and mandate of this chapter to fully articulate all resilience-building initiatives and moral positions related to the need for climate change strategies; rather, our task is to review the ethical arguments, challenges, and at times contradictions inherent in the building of global capacity for climate resilience. Through case studies this chapter examines what factors are necessary to effectively build this capacity, and how those factors can be harnessed and nurtured. We seek to explore whether there is an alchemy of ethical, cultural, and sociopolitical factors that ultimately promotes adaptation towards climate resilience.

One factor that has proven to be common of successful examples of climate resilience adaptation is the notion of social capital. Social capital is often defined as the connections between people and the reciprocity and motivation to help in the face of crisis or change. The concept can be categorized into distinct strains of *bonding*, *bridging*, and *linking* social capital. Bonding social capital entails relationships amongst those who are similar under classifications such as age, ethnicity, socioeconomic status, and education, whereas bridging social capital is characterized by relationships between people who demonstrably differ in these areas. Linking social capital defines and captures the degree to which individuals establish effective relationships with those people and institutions that have relative power over them (Hawkins and Maurer, 2010). Although social capital has been discussed broadly in much environmental literature, it is rarely looked at through the lens of adaptive climate resilience or moral motivation. As such we know little about the relationship between social capital and adaptation to accommodate changing environmental conditions. However, there are many indications that the development of social capital can significantly contribute to productivity and resilience, as well as social justice.

It is important to note that many community-based initiatives do not define social capital as an overt strategy yet clearly the elements of social capital are present. One study that specifically focused on well-defined social capital in Burundi found that promoting social capital adaptations had the effect of positively changing livelihood policies, institutions, and processes, particularly in places with poor governance (Vervisch et al., 2013). Furthermore, nurturing the development and expansion of social capital frequently transcends gender differences, lining up with a large body of research suggesting that targeting women for community-based education and initiatives can be effective. In the Limpopo Province of South Africa, a team of researchers conducted a study to monitor intervention groups that offered women enhanced social capital through microfinance training and health education for more than two years (Pronyk et al., 2008). Beyond educational and economic gains, women in the program reported higher levels of solidarity in times of crisis, exhibited higher levels of collective action, were more likely to be members of social groups, and believed they had the support of the community to back them.

In many cases, such community progress is not brought about through "top-down" initiatives introduced by state systems. In wealthy and developed nations, governments are rarely able to see beyond short-term economic priorities. On the flipside, in many struggling tropical nations the absence of effective state institutions and programs leaves climate change resilience strategies undeveloped and unexplored. Consequently, this chapter frequently focuses on "bottom-up" initiatives emerging from communities and small groups as these often serve as the primary conduit for local adaptation. Such initiatives come from two sources: they may be a natural adaptive response of creative and observant communities to changing environmental factors such as crop failures, different weather patterns, and increased levels of fire, drought, or flooding; they may also originate from community leaders or small nongovernmental organizations (NGOs). On both fronts a patchwork of community-based initiatives emerges.

Before proceeding to "bottom-up" examples, however, we review one "top-down" approach that has demonstrated success in promoting climate resilience as a state initiative—the Bangladesh Climate Change Resilience Fund (BCCRF). Hatched in the wake of the United Kingdom–Bangladesh Climate Change Conference in 2008, the BCCRF was established as a multidonor trust fund under the World Bank to address the increasingly frequent and severe cyclones and floods that plagued the Bay of Bengal (Nasiruddin, 2013). The fund symbolizes a natural progression in Bangladesh's efforts to invest in adaptation measures such as livelihood diversification, disease risk surveillance, and new coastal infrastructure ever since a devastating cyclone claimed the lives of over 140,000 people in 1991. The projects chartered by the Fund so far have yielded remarkable outcomes, the most notable of which is the development of a series of cost-effective shelters that housed over 40,000 people and 4300 livestock when Cyclone Mahasen landed in May 2013; in stark contrast to the events of 1991, a mere 17 casualties were reported. When not being used for shelter, the structures double as schools, tree plantations, and food storage facilities. In addition to the shelter construction project, the initiative includes significant investments in agriculture adaptation, afforestation, rural electrification, and renewable energy sources.

The success of the BCCRF can likely be attributed to its creative and multifaceted approach in addressing such a logistically complex state of affairs. It is primarily worth noting that unlike previous development projects, the allocation of funds came with no special conditions, thereby sparing impoverished Bangladesh the hardship of repaying any debt. Furthermore, the flexibility of funding has enabled NGOs already working in target areas to receive financial assistance under the so-called "Community Climate Change Project" organized by the Palli Karma-Sahayak Foundation. Incorporating NGOs effectively takes advantage of existing resources and allows an almost simultaneous bottom-up/top-down intervention to develop. The recruited organizations have already contributed to a wide variety of activities such as road repair, construction of grain banks, livestock housing, promotion of flood-resilient agriculture, and exchange visits between members of different villages to build resilience (and, in turn, social capital). The Fund is also noteworthy in that it incorporates analytical specialists to research and monitor critical aspects of the projects such as eco-engineering and climate-sensitive disease impacts, thereby enabling continuous adaptation of project strategies. This approach allows for a much broader view of climate resilience to include and protect human health. What truly distinguishes the BCCRF, however, is that it consistently aims to educate and empower local communities that are most affected by the consequences of climate change. In contrast to simply punishing those defying climate regulations, bestowing responsibility upon the inhabitants allows simple projects to take on the character of a movement, ultimately cultivating social capital through community purpose. As explained by a pair of geographers regarding coastal communities: "Only when the coastal residents would participate directly in such initiatives they would develop a sense of ownership of the future of coastal Bangladesh" (Rasid and Paul, 2014).

Despite the success that this initiative has demonstrated in meeting its targets (World Bank, 2016), a great deal of work remains to be done if Bangladesh is to truly become climate-resilient. Annual flooding brought about by seasonal monsoon rains continues to ravage many vulnerable populations in low-lying regions; the practically unprecedented floods in the summer of 2017 reportedly submerged almost a third of the entire country, killing 142 people and affecting almost 9 million (George, 2017). Additionally, over 700,000 homes were partially or completely destroyed (Al Hasnat, 2017). Thus, there remains an urgent need to develop infrastructure that would serve to curtail the damage inflicted on villages and crops against an inevitable barrage of future floods. While it is worth commending the provision of shelters that housed thousands of displaced villagers during the events, there lies a danger in the complacency of pushing climate resilience strategies when prevention of further climate change as a whole remains crucial to long-term survival and stability; the Sundarban mangrove forests that straddle the coastal border of Bangladesh and India are a case in point.

The Sundarbans comprise the largest continuous mangrove ecosystem, harboring an impressive level of biodiversity in addition to supporting the livelihoods of over 4 million people (Mahadevia and Vikas, 2012). The forests additionally serve as a natural barrier against the cyclones and storm surges that plague the region (Uy and Shaw, 2012). The sheer importance of the Sundarbans therefore makes their decline especially alarming. Evaluation under a multitude of established climate models have indicated a steady temperature increase in the mangrove forests (Agrawala et al., 2003). Warmer air over the area reduces air pressure over the land, enhancing the levels of rainfall during the monsoon season that contribute to more severe flooding and land erosion in a region that is already quite vulnerable to such events. This is compounded by a net rise in sea levels that diminishes the forests' capacity to withstand regional cyclones, in addition to increasing salinization of Sundarban freshwater systems via coastal flooding. The threat to the forests is further exacerbated by the burden of a rising population; the ecosystem is plagued by harmful practices such as illegal poaching and logging (as well as legal practices such as damming and overcultivation of shrimp) (Mahadevia and Vikas, 2012). Thus, while it is beneficial to invest in adaptive measures for local populations, it must be done so in a manner that does not forfeit the continued existence of the Sundarbans and other systems like it.

It is worth noting that the Bangladeshi government invested in a "Gorai River Restoration Project" (GRRP) that ran from 1998 to 2014, with the aim of restoring freshwater flow to the Sundarbans via dredging of the Gorai (a key distributary of the Ganges river). While the project brought about positive impacts on food production, employment, and living standards of surrounding communities, the project failed to reduce water salinity to desired levels (Talukder, 2014). Despite failing to reach desalinization targets, at the very least the GRRP—like the BCCRF—demonstrates the potential in top-down approaches when properly executed. However, effectively addressing wide-ranging human threats such as illegal logging and overfishing likely remains outside the reach of impersonal, large-scale initiatives, mandating more personalized approaches like those devised by Health in Harmony and Blue Ventures, both of which are outlined later in the chapter.

While the BCCRF and the GRRP were heavily reliant on the support of a handful of Western nations, it is unfair to assume that effective climate resilience strategies are exclusively dependent on the assistance of wealthy states. As an example, several environmental scientists and researchers in North Korea (DPRK) have launched a range of environmental initiatives including mass tree planting and agroforestry (Bowman, 2014). In Ethiopia, a desire to build climate resilience has assumed the form of multiple projects in and around the rift valley, adapting by substituting commonly used crop seeds with alternatives better aligned to a faster yield cycle; farmers matched this with small but constant experimentation with afforestation and agroforestry (Bowman, 2010). However, many such "entrepreneurs" lack the social capital to extend their adaptive practices to the greater populace. Despite knowledge of growing climate challenges, those with the influence to enact necessary changes (and many of those who elect them in democratic institutions) lack the motivation or perspectives needed to promote empathetic and forward-thinking behavior. It is to these ethical challenges that we now turn.

15.2   ETHICS, SOCIAL CAPITAL, AND CLIMATE RESILIENCE

A useful definition of resilience is "the ability of a system to cope with a disturbance in ways that keep up its functions and identity and maintains its capacity for adaptation and transformation" (IPCC, 2014). The great challenge is to both *learn* from and *adapt* to changes as they emerge, in turn building resilience. Promoting resilient practices rests on a thorough understanding of the ethics that drive such behavior. Although there is significant variation on theory, application, and the ordering of priorities, ethics essentially explores questions of justice, equity, and respect for autonomy (which can be viewed individually, collectively, or even culturally). Environmental ethics deviates only slightly by focusing on the preservation, protection, and promotion of the integrity of interdependent life.

The nurturing and expansion of social capital is of great benefit to building climate resilience strategies, but remains associated with significant challenges. False information may well flow through the very structures of social capital, potentially creating more problems than it solves. Furthermore, while not absolute and somewhat tractable, "us and them" thinking is an unfortunate reality of human perception and priorities. This creates a considerable challenge for nurturing social capital (especially bridging social capital) as it depends on building and maintaining relationships between people facing similar difficulties but who occupy distinct social groups. There are many examples of the expansion of social capital in conditions of disaster, war, and emergency. However, the challenge resides in conveying the magnitude of slow and often obtuse threats of climate change so as to trigger such a response. As such, there are times when clear ethical arguments that resonate with people can be powerful in eliciting change; the nature of these arguments and perspectives will vary greatly depending on culture and circumstances. Unfortunately, for many countries trust at the municipal, regional, or national level may not exist, and in some cases for good reason. Ethical perspectives will have far greater power if they emerge from local communities and are tied to cultural and practical perspectives. Even so, trying to motivate people toward an expanding moral responsibility may be truly difficult (Fig. 15.1).

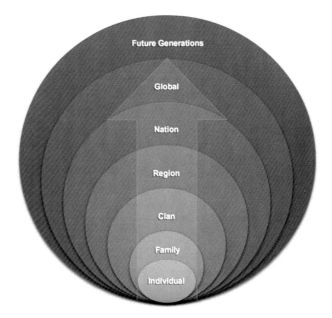

FIGURE 15.1   The expanding levels of moral responsibility. In general, it is difficult to motivate moral thinking beyond the more immediate levels.

Certain arguments from an evolutionary point of view claim that our moral perspectives and psychology have evolved to deal only with immediate community and relationships (Persson and Savulescu, 2016), the theory being that we are poorly equipped to consider moral obligations the further we extend from such parties. Considering the needs of people living in the future is considered a remote concept. Although respect for ancestors and even ancestor worship transcends virtually all human cultures, beyond having children we so far don't seem to be able to project the same sentiments forward. Yet research into motivating and understanding human morality in such realms remains in its infancy. Some of the subsequent case studies suggest social capital beyond immediate relationships can be nurtured and expanded.

The ethical principle of justice comes into sharp focus as we look to future generations.

15.3 ETHICS, THE PRECAUTIONARY PRINCIPLE, AND THE NEED FOR CLIMATE RESILIENCE

In an environmental context, the precautionary principle can be defined as such: when an identified threat of serious or irreversible damage to the environment or human health exists, a lack of full scientific knowledge about the situation should not be used to delay remedial steps if the balance of potential costs and benefits justifies action (Resnik, 2003). It goes beyond the notion of prevention, which seeks solely to minimize the impact of identified risks. The adoption of the principle in the medical field (Goldstein, 2001) emerged later. Although it now has far greater resonance in health care, the principle often remains ignored by decision-makers as reason for climate change mitigation or adaptation. This is ironic considering that the precautionary principle was originally created in the mid-1970s as part of Germany's environmental protection policy (Gignon et al., 2013).

Can the precautionary principle be commandeered to build support for the adoption of climate resilience? Whether stated or not, the very acceptance, commitment to, and development of climate resilience is grounded in the principle of justice. Its main element revolves around a general rule of public policy action to be applied in situations of potentially serious or irreversible threats to health and wellbeing, where there is a need to act to reduce potential hazards before there is strong, definitive *proof* of harm, taking into account the likely costs and benefits of both action and inaction. It is important to clarify that building resilience is not an abandonment of the precautionary principle; it is an extension of it. The principle lies at the heart of adaptation and resilience initiatives, and encompasses several sub-elements such as the proportionality principle, in which the costs of actions to prevent or minimize hazards should not be disproportionate to the likely benefits.

Prevention, awareness, and mitigation has often been used to mobilize government interventions in the face of public health threats. Epidemiology (the study of the incidence, distribution, cause, control, and prevention of disease) is not just an instrument for definitive *proof* of the distribution and determinants of disease, but also serves as a tool for early detection and warning of preventable risk. Generally, there is considerable societal acceptance within the world of promoting health without enormous political and ideological pushback. Why then is it so tough to gain the same acceptance in the domain of the environment? At the level of the state, the very core of democracy is the responsibility of governments and in turn their public health authorities to seek and protect the common good. The answers to this are not fully understood, but appear to include an array of factors including political ideology, social attitudes, costs, and public perception of science. A greater movement of climate resilience towards evidence-based practice in combination with a strong emphasis of the health effects of climate change may well go a long way towards public acceptance of the need for resilience. This in combination with the language of ethics may indeed be one of the best ways to move forward and solidify the need for action.

These health-centered moral quandaries run so deep that they may well determine the fate of scientific solutions to climate instability. After two decades of paralysis and false starts by the international community to respond adequately to the threat of global climate change, discussions of the possibility of geoengineering a cooler climate have recently proliferated. Alongside the considerable optimism that these technologies have generated, there has also been wide acknowledgement of significant ethical concerns. Risks are potentially substantial as well as poorly understood. Another set of complex questions emerges when planning for the cessation of climate engineering. The wide range of geoengineering technologies currently being discussed makes it prudent that each technique should be evaluated individually. However, in light of the uncertainty that surrounds the impacts of geoengineering, its ultimate acceptance or rejection may well be more influenced by ethics than science.

15.4 CAN ETHICS STRENGTHEN CLIMATE RESILIENCE?

Perhaps the greatest and most glaring practical and ethical criticism of building climate resilience is that it is only responding to the problem but not *dealing* with its cause. In essence this criticism holds true, yet one does not necessarily negate the other. In fact, there is a moral relationship between the two and climate resilience must be undertaken along with, not instead of, climate change prevention, meaning addressing the root causes—to the best of our knowledge and ability—of climate change. Although few community-based programs set out to bring global change, an expanding patchwork of projects that begin to form a web of resilience will.

Building climate resilience sits at the cusp of both immediate need and future planning. Knowing, as we do, that the present feeds and penetrates the future, our actions now, in this domain, have moral significance that can resonate for years to come. Much of the future is beyond our grasp and many decisions made by public officials now related to building climate resilience can either expand or limit our options in the future. We have an obligation to avoid closing such options that could affect the survival and well-being of current and future generations in the face of potential calamities.

For a good part of the 19th and 20th century, social scientists believed that man's greatest asset—his cognitive capacity—was the perfect conduit to communicating hard data related to risk. The thinking was that "the public will act in their own best interests if given discernible statistics and logic" (Ropeik, 2010). The unfortunate truth is that we tend to assess risks through a series of subjective, emotionally formulated calculations often occurring well below the conscious mind. This is why developing practical, resilient strategies with an eye to long-term sustainability but based on the immediate needs of people may well be the best way to begin the process of change and resilience building.

15.5 INDIGENOUS PEOPLE: VULNERABILITY AND KNOWLEDGE

The term indigenous refers to individuals in sovereign countries whose social, political, cultural, and economic conditions are distinct from mainstream society on account of their ancestry and genealogical linkages to the populations which inhabited the country (Indigenous and Tribal Peoples Convention, 1989). Indigenous populations are presently some of the most environmentally vulnerable groups of people in the world. Their cultures are often broad and complex, covering a range of perspectives and interactions with the environment. The knowledge and perspectives of these groups may therefore turn out to be great sources of building climate resilience.

Critics of climate change validity build an argument that *all* naturally occurring fluctuations in weather are now viewed through a lens of media-driven perspectives, in turn leading to a general overreaction to naturally occurring climate-related events. However, indigenous groups in remote regions of the Amazon provide evidence against this claim. Many of the inhabitants lack access to any form of global media, and would therefore have no knowledge of Western news cycles; in some cases, the communities are so remote that they remain uncontacted by outsiders. A constant concern expressed in these indigenous communities is that weather patterns are changing, or that trees are not flowering or fruiting when they should be (Bowman, 2014, 2015). The annual flood waters (a massive, naturally occurring cycle in the western Amazon) have been disrupted, making the timing and severity of shifting water levels unpredictable to those who have lived alongside them for countless generations. There are remarkably similar sentiments in the Eastern Democratic Republic of Congo (DRC) (Bowman, 2016). People living far from towns and villages (including the indigenous Batwa and Mobuti pygmies) speak greatly of the challenge and consequences of changing weather patterns. An additional challenge of climate change is that it exacerbates other environmental and geopolitical problems. In eastern Congo, these changes interplay with forest loss, radical biodiversity decline, and military conflict that all magnify the threat to survival. Yet within both regions, nascent initiatives have emerged to adapt to these changes; much can be learned from them.

Indigenous peoples' lives are often directly woven into the environment. The interactions of these groups with the environment can take many forms with millennia of collective knowledge about the ecology of their surroundings. Indigenous land titles can therefore have a massively beneficial effect on the environment. For example, research indicates that the effect of titling land to indigenous groups in the Peruvian Amazon reduces clearing by more than three-quarters and forest disturbance by roughly two-thirds (Blackman et al., 2017). With such depth of knowledge and experience, even minute changes in water cycles, wildlife, soil, and weather are readily apparent.

Unfortunately, the same closeness to the land that has given indigenous peoples early warning about climate change also means that they suffer the consequences of it to a far greater degree than others. Historical or contemporary subjugation and oppression have left many indigenous people living on land that is already marginal, so even relatively small changes in temperature or rainfall have large implications. Although the term indigenous is challenging and sometimes contentious in an African context, the case of the Maasai is representative of the plight of indigenous groups worldwide. The Maasai—who originally grazed their cattle on verdant savannah—were pushed by colonization and dominant societies onto semiarid scrubland where their herds can barely find enough food under only the best of climatic conditions. Drought can be, and is, catastrophic for them. Yet many prove adaptive and resilient; some Maasai in southern Kenya are beginning to herd camels rather than cattle, this despite a cultural taboo grounded in a belief that camels bring drought.

Indigenous peoples are not only suffering the consequences of climate change more acutely; they are facing them sooner. The problems faced by these populations are harbingers of what all peoples will face eventually. Indigenous peoples offer an important alternative viewpoint of the world, one that sees natural processes as cyclical, one that reflects respect for the earth and, most of all, one that considers all our actions in the context of future generations. Resilience strategies are now emerging from indigenous communities that we need to learn from, respect, model after, and find inspiration from.

15.5.1　Detailed Case Studies

15.5.1.1　Health in Harmony—Indonesia

Health in Harmony is a nonprofit organization committed to promoting human well-being and protecting biodiversity by primarily gleaning insight from the words of local people and facilitating community-based solutions. It was founded in 2005 by Dr. Kinari Webb in West Kalimantan, Indonesia based on community awareness of economic need to access health services. Dr. Webb had come to appreciate the rich landscape of the local rainforests in previous trips, but she also observed how inadequate access to health care in West Kalamantan's rural communities promoted an unsustainable relationship with the environment. At the time, illegal logging was one of the few ways to efficiently acquire income to pay for necessary health services. As an example, the felling of 60 trees would provide enough money necessary for one Cesarean section (Horton, 2016). Such circumstances forced locals to decide between the health of their family and the health of the rainforest in Indonesia.

Health in Harmony, by its very practice, reinforces important ethical messages about building environmental resilience and an obligation to the provision of environmental justice. It specifically analyzes the complex relationship between poverty, logging, and health care affordability (Bussard, 2016). The main approach behind programs run by Health in Harmony is the concept of "radical listening." Under this idea, Dr. Webb held more than 400 hours of talks with local communities on how to adequately address their needs (Horton, 2016). With an already strongly established community network of bonding social capital, the root causes of poverty, poor health, and destruction of the natural environment were explored before implementing specific solutions. Illegal logging was identified as a consequence of a lack of economic alternatives which would have enabled purchase of necessary health care services. It was primarily through logging that communities were able to procure enough income to access health facilities for their families. However, continuous illegal logging has only served to reinforce the cycle of poverty, habitat and biodiversity loss, and subsequently increased flooding and disease via vector-borne routes such as mosquitoes.

15.5.1.2　Alam Sehat Lestari (ASRI)—"Healthy Nature Everlasting"

One of the central programs of Health in Harmony is ASRI, started in 2007. It provides clinical practice service, medical and public health training, and rainforest conservation. The clinic is meant to be a permanent feature for positive long-term change and capacity building of the community (Bessinger, 2012). Health challenges in the community include tuberculosis, malaria, diabetes, heart disease, and diarrheal illnesses. Patients may receive health care irrespective of the ability to pay with cash; alternative options include payment in seedlings, manure for organic farming, traditional hand crafts, and assistance with labor to reduce clinic costs. Since the opening of the clinic more than 14,500 patients have been cared for and have received regular education about disease prevention (Shetty, 2009). Impact surveys of 1300 households between 2007 and 2012 showed that infant death and incidences of fever, diarrhea, or cough for more than 3 days (which is especially pertinent in this region due to tuberculosis) all declined significantly and continue to do so (Saving lives by saving trees, Kinari Webb, TEDxRainier, 2016). In the same time period, there was a decline from 1350 logging households to 450 and to date that number has further declined to 180 (Rulistia, 2016). In West Kilimantan alone where Health in Harmony operates, it is estimated that over 1500 orangutans have been saved due to drastic reduction in illegal logging.

Local communities have demonstrated a keen interest in the protection of the forest, but the lack of alternative economic opportunity often meant that illegal logging of the forest was a quick way to increase income. This important community insight played a key role in the creation of the Sustainable Farming program to provide training in sustainable organic farming—a much more effective and environmentally responsible form of agriculture. Before this initiative, the prevailing method in the region was "slash-and-burn" (Salisbury, 2017). However, this technique is only effective when population density is low and the forest cover is sufficient enough to allow forests to regenerate between cycles. Over time, the combination of this practice with illegal logging came to drastically upset the delicate nutrient balance of the soil. Since implementing the new farming program, farmers have learned how to improve soil quality with inexpensive and local ingredients. Many of these conservation methods are being shared and implemented in primary schools as well. The success of the farming program additionally helped spur the founding of the "Goats for Widows" program to support the most disadvantaged in the community. Through "radical listening" the widows in the community were identified as having the least economic opportunity but the most need. Today, the program has distributed several hundred goats which provide a necessary source of income, including much-needed manure to sell to farmer groups (Bussard, 2016).

The success of Health in Harmony is largely owed to harnessing the strong bonding social capital of the community members, where growing levels of trust have rapidly brought about positive outcomes, civic engagement, and efficacy as described in the examples above. The organization is also characterized by a theme of bridging social capital across external networks; the groups benefit from such connections when engaging in collective action across multiple regions, giving access to untapped resources in other spheres. For example, many physicians from excellent medical institutions in the U.S. provide free training for Indonesian doctors who go on to further their knowledge in institutions across Indonesia. Looking to the future, Health in Harmony's long-term goals include: expansion across Indonesia with a strong focus on community-led initiatives, establishing a global Health Training Center in Borneo, and creating mobile applications for disseminating health-conservation education that includes the radical listening methodology (Health in Harmony, 2017).

15.5.1.3 Blue Ventures—Madagascar

Blue Ventures is a science-based, social nonprofit founded in 2003 in Madagascar. The organization's philosophy focuses on placing responsibility for fisheries management in the hands of local communities. This is particularly necessary in Madagascar's coastal communities due to limited central capacity and infrastructure for marine conservation. In 2013, fish provided 3.1 billion people with almost 20% of their intake of animal protein worldwide. Additionally, dependence on fish is usually higher in coastal than inland areas as well as small islands (FAO, 2016). Marine ecosystems and traditional coastal livelihoods are facing unprecedented pressures from overfishing and climate change, with many fish stocks being rapidly depleted. Blue Ventures has several interrelated conservation initiatives, two of which will be explored here. The first revolves around rebuilding the fisheries, while the second focuses on community health.

15.5.1.4 Rebuilding Fisheries

Blue Ventures began with surveying coral reefs in the Mozambique Channel. Here, they encountered the seminomadic Vezo community on the South West coast of the neighboring island of Madagascar, who expressed concerns over the noticeable effect of overfishing on octopus stock—one of their primary food sources as well as one of the most significant sources of small-scale ocean revenue. For this community, the decline in stocks was particularly alarming as they depend entirely on the sea for food and sources of income. In collaboration with Vezo communities, Blue Ventures implemented a temporary closure of an octopus fishing area as an experiment for a few months. When the closure reopened, local communities experienced a huge increase in octopus and subsequent fisheries income. Some areas have experienced an increase in octopus size by as much as 41% when compared to controls (Benbow et al., 2014). Positive results like these have inspired what is now a community-led change in fisheries management in Madagascar. Further growth and economic opportunity has contributed to the creation of the country's first Locally Managed Marine Area (LMMA) governed by a small network of fishing villages. The method of temporary closure has since spread to the southern, western, and northern coastlines of Madagascar—greatly helping to improve food security and income.

The model of temporary closure harnesses and nurtures bonding social capital in rebuilding coastal fisheries. Social capital was best illustrated by community involvement and cooperation in determining which fishing grounds were most appropriate to implement closure, the placement of the closures to prevent fishing, and finally the monitoring of these areas during closure. Its expansion is particularly successful as it provides economic incentive to protect marine biodiversity loss, food security, cultural identity, income sources, and future generations while building socioecological resilience to climate change.

15.5.1.5 Community Health

At the request of Vezo community members in 2007, Blue Ventures incorporated family planning into its program of rebuilding fisheries. Due to their isolation, many people lack access to even the most basic health services. Interviews with

Vezo community members indicated many couples wanted to use contraceptives but would have to go several kilometers to the nearest clinic by foot in order to access them (Blue Ventures, 2014) As a result, more children were born than they would choose or could adequately support. Since the introduction of Safidy (which roughly translates to "the freedom to choose"), reproductive health services have expanded to serve a total of 20,000 people (Robson et al., 2017). It has greatly helped to establish a balance between family size, reproductive health, and food security. Local women are supported and offered community-based family planning and other health services, integrated closely with Madagascar's public health system. These community-driven health programs uphold the reproductive rights of women to choose freely the number and spacing of their births (Robson et al., 2017). None of the successes seen here in building integrated resilience for both the environment and public health would have been possible without the strong contributions of bonding, bridging, and linking social capital networks available to these communities. The Safidy program helped to greatly address the unmet family planning and health needs that many members of the Vezo community faced and has—among many positive outcomes—allowed women to secure jobs, enabling them to afford to send their children to school.

15.5.1.6 Reduced Emissions from Deforestation and Degradation (REDD+) Kasigau Corridor— Kenya

The UN-backed "Reduced Emissions from Deforestation and Degradation" (REDD+) initiative in the Kasigau corridor is part of the Wildlife Works Organization. It encompasses approximately 500,000 acres of dryland forest in southeastern Kenya that forms a corridor between Tsavo East and Tsavo West national parks. The corridor is home to over 50 large mammal species, over 300 species of birds, and important populations of endangered species (CodeREDD, 2017). The mission is to protect endangered wildlife while balancing the need for resources and work for rural Kenyan communities of approximately 150,000 people who share the same space. A strong reason for this initiative stems from the impacts that illegal logging, mining, and poaching have had on communities and the environment. Slash and burn agriculture in the dryland forest area has proven to be an additional threat with the gradual expansion of local communities; coupled with climate change effects, the result is an overall negative impact on agricultural productivity in an already semiarid region. Community members were often forced into these destructive environmental practices to make sufficient income to feed their families, purchase health care services, and educate their children, as the harsh conditions of the area adversely affect crop yield and by extension their earning potential.

The REDD+ project brings value to the natural capital of forest through the implementation of sustainable economic solutions that provide an income for rural communities, allowing them to increase food security and educate their families without damaging the environment (Kizaka, 2013). Based in southeastern Kenya, the project entails a Voluntary Emission Reductions (VER) initiative that has been planned to run for 5 years. It is expected to avoid the emission of over 30 million tonnes of CO_2 during its course (CodeREDD, 2017). Corporations seeking to offset their carbon footprint can purchase VERs equivalent to 1 tonne of CO_2 emissions from REDD+ projects. These transactions pay local communities living in project regions to conserve biodiversity, while also generating income for sustainable development in areas such as education and health care. The six key elements of REDD+ that make this initiative such a strong example of resilience include job creation, agricultural intensification, physical protection and monitoring of forests, fuel wood substitution, agroforestry, and social benefits (CodeREDD, 2017).

In the Kasigau corridor, land is primarily respected as ancestral soil. The REDD+ project initiated in this area provides an opportunity to protect communal land, generate income, improve agricultural productivity, and stimulate investment to positively influence sustainable development in the region. Money generated from the project can be given back to the community for education initiatives such as bursaries for high school. A substantial percentage of earnings has been geared towards education, with the remainder funneled into community projects. New classrooms have been constructed for school children, which include rainwater harvesting systems to provide a supply of clean water. Other notable benefits include training for wildlife rangers to protect local wildlife and biodiversity.

This UN-backed carbon offset project with economic opportunities has dovetailed with the strong bonding capital of the rural Kenyan community and has succeeded to become acceptable both in the marketplace and rural communities. This level of trust via linking social capital through REDD+ facilitates an improvement in trust in regional, national, and international decision-makers. The outcomes of the Kasigau corridor project represent a strong example of building environmental resilience and have started to demonstrate substantial results in supporting communities, forest, and wildlife.

15.5.1.7 South Kivu Corridor—Democratic Republic of Congo

Climate resilience strategies may very well develop as part of deep-rooted spiritual beliefs in tight-knit communities. One such initiative has emerged from the desire to protect the now critically endangered eastern lowland gorilla.

Deforestation, the bushmeat trade, weak governance, and ongoing conflict continue to put many of the people and the environment of eastern DRC at growing risk. The Burhinyi Community Forest Project in the South Kivu province of eastern DRC seeks to address these problems by highlighting the importance of building environmental resilience through blending of indigenous and modern scientific knowledge. Burhinyi is a tropical forest that lies between Kahuzi-Biega National Park and the Itombwe Natural Reserve, home of the Eastern lowland gorilla (Ortolani, 2017). Socioeconomic surveyors, primatologists, and botanists studied the great apes and examined community norms, interviewing over 800 community members to learn how locals sustain the forest and its wildlife. Important insights showed that community chiefs were instrumental in the following: designating sacred areas in the forest to prevent poaching of great apes, reinforcing a precolonial ban on hunting pregnant animals, and prohibiting clearing of forests by burning (Ortolani, 2017). The Batwa (pygmy) communities hold a wealth of forest, resilience, and food security knowledge and were central to this initiative. The blended knowledge approach to conservation harnesses and enhances the strong bonding social capital evident in these communities and creates opportunities to protect the environment and build resilience. Unfortunately, waiting for a top-down federal management strategy to build environmental resilience in DRC is unrealistic; although the present government has supported many environmental initiatives, the state lacks the resources to initiate, monitor, and maintain such programs.

This initiative nurtures the concept of solidarity, which in turn nurtures the development of social capital. More specifically, it has not only promoted bonding social capital within communities but has also enabled bridging social capital by categorizing us as part of our biological community (human and nonhuman), our local community, and our national community in achieving our goals. This is especially highlighted by the direct involvement of the Batwa (pygmy) communities that are so often excluded from wide-scale environmental initiatives.

The Burhinyi Community Forest Project does not push formal conservation laws, but rules from community chiefs and the indigenous Batwa act as guidelines. This forest model ensures that "communities have the rights of management and governance of their traditional lands, in alignment with conservation goals" (Ortolani, 2017). A deeper understanding of local conservation views in the Burhinyi community forest began in 2012 under the guidance of Dominique Bikaba, who grew up in close contact with the local wildlife and primate species in particular. Follow-up research in 2014 by Yale forest and environmental conservationist Sarah Tolbert showed that in terms of the use of natural resources by local communities in Burhinyi, primate bushmeat practically vanished from the food market as a result of a community-led effort to protect primate species (Ortolani, 2017). These findings help to demonstrate how influential traditional chiefs can be over their community, and how instrumental they are in increasing the potential to protect great apes and, by extension, biodiversity.

The Burhinyi community forest is a clear example of how environment and habitat cannot be thoroughly protected with a top-down strategy, but must be inclusive of local people. This is an increasingly important lesson echoed by conservation NGOs in recent years. Governments ought to consider, draw from, and respect the authority and belief systems of local communities when passing laws, creating policies, and establishing protected lands. DRC's government has received an overall positive response to the passing of a 2014 law that allows local communities to apply for land tenure. As part of the new initiative, the Burhinyi Community Forest will become the first nonprotected area in South Kivu Province where local communities will be allowed to apply scientifically grounded conservation policies in concert with traditional knowledge and practices in order to protect great apes. The connection between the government and the South Kivu population demonstrates the positive impact of linking social capital, and the potential for unity in linking scientific knowledge with local beliefs. Community-based conservation strategies focused on the long-term preservation of habitats and endangered species in Burhinyi community forest have thus proven pivotal to the harmonious coexistence between gorillas and the indigenous local population.

These case studies reinforce the importance of the expanding levels of moral responsibility to help build resilience for climate change. While it may seem initially difficult to tangibly motivate moral thinking beyond the more immediate levels of personal relation and identity, Health in Harmony, Blue Ventures, REDD + , and the South Kivu Corridor all present real success stories that should act as a nidus for resilience to climate change. Across each case study, the harnessing of varying degrees of bonding, bridging, and linking social capital is evident. Each example demonstrates how resilience to climate change expands where growing levels of trust have rapidly brought about positive outcomes, civic engagement, efficacy, and increased sustainable income streams while restoring the environment. Health in Harmony initiatives have decreased infant morbidity and mortality, while at the same time reducing the number of logging households, protecting wildlife habitats, and providing sustainable alternative revenue streams. The Vezo community in Madagascar benefited from economic incentives to protect marine biodiversity and bolster climate resilience that in turn improved food security, reproductive health, cultural identity, income sources, and the well-being of future generations. The REDD + project has brought value to the natural capital of forest through the implementation of sustainable income for rural communities, allowing them to increase food security and educate their families without damaging the environment. The South Kivu Corridor aided local communities in applying scientifically grounded conservation policies in concert with traditional knowledge and practices in order to protect great apes. In summary, all of these cases demonstrate that we cannot expand moral circles without interconnected communities with strong social cohesion—to proceed otherwise is incongruent with the lessons learned in pursuing climate resilience.

15.6 CONCLUSION

Many would say that focusing on climate resilience strategies rather than mitigation is fundamentally wrong, as it is an acknowledgement of defeat and sends the wrong message to an already apathetic public. Yet a problem of this magnitude and complexity requires both short- and long-term strategies. To choose not to develop climate resilience strategies when they hold the potential to protect life and well-being is far more ethically problematic. Developing climate resilience does not negate the need for mitigation. To repeat an often made but salient ethical point, Western nations have contributed far more to climate change and in turn hold an obligation to the most vulnerable in developing climate resilience infrastructure. Although autonomy alone as an ethical concept holds cultural bias, the concept of self-determination, if pursued in a respectful manner and often on a community level, holds tremendous promise in developing climate resilience strategies. As idealistic as it may sound, cultivating a sense of solidarity that transcends the many levels of group inclusion across the expanding levels of the moral circle is essential to building effective climate resilience strategies.

What is the common ground among the case studies reviewed? Allowing solutions and initiatives to emerge grounded in the *perspectives and needs of the people* is critically important. The obvious and great challenge is to see whether these initiatives can broadly expand, merge, or complement each other and prove effective on a larger scale. For example, in the case of Health in Harmony, providing health care in exchange for rainforest protection and alternative forms of payment could be limited by location and culture, yet the concept of "radical listening" holds the potential to promote transferability of the initiative considering that it is always intended to address the needs of the local community. From this standpoint, incorporating indigenous perspectives as well as protecting indigenous territories and lands may well be very helpful in building climate resilience. Far more attention needs to be paid to how we can perceive and plan for risk, as well as to how we might bridge divisions at the individual, community, national, and even global levels. Perspectives from political science, psychology, and ethics are all needed. Furthermore, a spirit of democracy is called for on a level never before attempted.

REFERENCES

Agrawala, S., Ota, T., Ahmed, A., Smith, J., van Aalst, M., 2003. Development and climate change in Bangladesh: focus on coastal flooding and the sundarbans. Org. Econ. Co-operat. Develop. 41–47.
Al Hasnat, M., 2017. No respite for flood victimsDhaka Tribune Retrieved from . Available from: http://www.dhakatribune.com/bangladesh/2017/10/.
Benbow, S., F Humber, F., TA Oliver, T.A., Oleson, K.L.L., Raberinary, D., Nadon, M., et al., 2014. Lessons learnt from experimental temporary octopus fishing closures in south-west Madagascar: benefits of concurrent closures. Afr. J. Marine Sci. 36 (1), 31–37.
Bessinger, C.M., 2012. Yale Doctors Around the World. Yale School of Medicine Institutional Planning and Communication 1–3.
Blackman, A., et al., 2017. Titling indigenous communities protects forests in the Peruvian Amazon. Proc. Natl. Acad. Sci. USA 114 (16), 4123–4128.
Blue Ventures, 2014. Empowering Communities to Live with the Sea,. Retrieved from: https://blueventures.org/conservation/community-health/.
Bowman, K.W., November 2010. Observation Made in Research Log in Bale Mountains, Ethiopia.
Bowman, K.W., August 14, 2014. Field Log Recording, DPRK Site Visit.
Bowman, K.W., August 25, 2014 & September 4, 2015. Field Log Recording, Javari Valley, Brazil.
Bowman, K.W., February 15, 2016. Dominique Bakaba and Batwa Interview Records, Democratic Republic of Congo.
Bussard, M., 2016. From the Forest to the Health Clinic. Carnegie Council for Ethics in International Affairs. New York, NY.
CodeREDD, 2017. Wildlife Works Carbon/Kasigau Corridor, Kenya.
FAO, 2016. The State of World Fisheries and Aquaculture 2016. Contributing to food security and nutrition for all. Rome, 200, pp. 4.
George, S., 2017. A third of Bangladesh under water as flood devastation widens CNN. Retrieved from https://www.cnn.com/2017/09/01/asia/bangladesh-south-asia-floods/index.html.
Gignon, M., et al., 2013. The precautionary principle: is it safe. Eur. J. Health Law 20 (3), 261–270.
Goldstein, B.D., 2001. The precautionary principle also applies to public health actions. Am. J. Public Health 91 (9), 1358–1361.
Hawkins, R., Maurer, K., 2010. Bonding, bridging and linking: how social capital operated in new orleans following Hurricane Katrina. Bri. J. Social Work 40, 1777–1793.
Health in Harmony, 2017. Scaling. Retrieved from: https://healthinharmony.org/programs/scaling-up/.

Horton, R., 2016. Offline 2016: some lessons to consider. Lancet 388 (10063), 2970.

Indigenous and Tribal Peoples Convention, (1989). International Labour Organization C-169 Part 1. General Policy Article 1 section (b).

International Council on Human Rights Policy, 2008. Climate Change and Human Rights: A Rough Guide.

Jäger, J., Kok, M.T.J., 2007. Vulnerability of people and the environment: challenges and opportunities. In: UNEP (Ed.), Global Environmental Outlook 4: Environment for Development. Earthscan, Nairobi.

Kizaka, C.P., 2013. Kasigau, Corridor District. REDD Talks.

Mahadevia, K., Vikas, M., 2012. Climate change-impact on the sundarbans: a case study. Int. Sci. J. 2, 7—15.

Nasiruddin, M., 2013. Bangladesh Climate Change Resilience Fund (BCCRF) Annual Report 2013 (No. 93959) (pp. 1—84). The World Bank.

Ortolani, G., 2017. Indigenous traditional knowledge survival helps conserve great apes. Mongabay January 20, 2017. Glenn Scherer.

Persson, I., Savulescu, J., 2016. Unfit for the future: the need for moral enhancement. J. Moral Philos. 13 (6), 751—754.

Pronyk, P.M., et al., 2008. A combined microfinance and training intervention can reduce HIV risk behaviour in young female participants. AIDS 22 (13), 1659—1665.

Rasid, H., Paul, B., 2014. Climate Change in Bangladesh: Confronting Impending Disasters. Lexington Books, Plymouth, pp. 199—200.

Resnik, D.B., 2003. Is the precautionary principle unscientific? Stud. History Philos. Sci. Part C 34 (2), 329—344.

Robson, L., Holston, M., Savitzky, C., Mohan, V., 2017. Integrating community-based family planning services with local marine conservation initiatives in southwest madagascar: changes in contraceptive use and fertility. Stud. Family Plan. 48 (1), 73—78.

Ropeik D., (2010). How Risky Is It Really? Why Our Fears Don't Always Match Facts, McGrawHill Companies. Rulistia, N., 2016. Healthy Forests for Healthy Lives. Jakarta Post (Tue Feb 23rd, 2016).

Rulistia, N., 2016. Healthy Forests for Healthy Lives. Jakarta Post Tuesday, February 23, 2016.

Salisbury, C., 2017. Paying for healthcare with trees: win-win for orangutans and communities. Mongabay Series. Great Apes. Glenn Scherer .

Saving lives by saving trees, Kinari Webb, TEDxRainier, 2016.

Shetty, P., 2009. Kinari Webb: saving lives and saving rainforests. Lancet 374 (9705), 1882.

Talukder, S., 2014. Mid-term Evaluation of Gorai River Restoration Project, Phase-II (Final Report). Bangladesh Ministry of Planning, pp. 1—46.

Uy, N., Shaw, R., 2012. Ecosystem-based Adaptation. Emerald Group Publishing Limited, Bingley, pp. 75—76.

Vervisch, T., Titeca, K., Vlassenroot, K., Braeckman, J., 2013. Social capital and post-conflict reconstruction in Burundi: the limits of community-based reconstruction. Dev. Change 44, 147—174. Available from: https://doi.org/10.1111/dech.12008.

World Bank, 2016. Bangladesh Climate Change Resilience Fund (BCCRF) Annual Report 2016 (English). World Bank Group, Washington, D.C. Available from: http://documents.worldbank.org/curated/en/194721498048042073/Bangladesh-Climate-Change-Resilience-Fund-BCCRF-annual-report-2016.

PHOTOGRAPHIC SOURCES

Annual damages caused by large-scale natural disasters. Source: United Nations International Strategy for Disaster Reduction. http://www.unisdr.org/files/25831_20120318disaster20002011v3.pdf.

    List of Small Island Developing States: UN Members and Non-UN Members https://sustainabledevelopment.un.org/topics/sids/list.

    Relative Contribution of Tourism to Selected Caribbean Countries.

Worrell, D., Belgrave, A., Grosvenor, T., Lescott, A., 2011. An analysis of the tourism sector in Barbados. Econ. Rev. Volume 37 (1), 64. March.

FURTHER READING

Aaheim, A., Amundsen, H., Dokken, T., Wei, T., 2012. Impacts and adaptation to climate change in European economies. Glob. Environ. Chang. 22 (4), 959—968. Available from: https://doi.org/10.1016/j.gloenvcha.2012.06.005.

Andreou, C., 2006. Environmental damage and the puzzle of the self-torturer. Philos. Public Affairs 34 (1), 95—108. Available from: https://doi.org/10.1111/j.1088-4963.2006.00054.x/full. Retrieved from: http://onlinelibrary.wiley.com.

Balbus, J.M., Malina, C., 2009. Identifying vulnerable subpopulations for climate change health effects in the United States. J. Occupat. Environ. Med. 51 (1), 33—37.

Barrett, S., 2011. Rethinking climate change governance and its relationship to the world trading system. World Econ. 34 (11), 1863—1882. Available from: https://doi.org/10.1111/j.1467-9701.2011.01420.x.

Basher, R., 2006. Global early warning systems for natural hazards: systematic and people-centred. Philos. Trans. Royal Soc. 364 (1845), 2167—2182. Available from: https://doi.org/10.1098/rsta.2006.1819. August 15th.

Burke L., Reytar K., Spalding M., Perry A., 2011. Reefs at Risk Revisted. Word Resources Institute, pp. 6, 7.

Cutter, S., Boruff, B., Shirley, W., 2003. Social vulnerability to environmental hazards. Soc. Sci. Quart. 84 (2), 242—261. Retrieved from: http://doi.wiley.com/10.1111/1540-6237.8402002.

DARA, 2010. Climate Vulnerability Monitor 2010: The State of the Climate Crisis.

Epstein, J., 2006. Friendship: An Exposé. Houghton Mifflin, Boston.

Gebresenbet, F., Kefale, A., 2012. Traditional coping mechanism for climate change of pastoralists in South Omo, Ethiopia. Indian J. Tradit. Knowledge 11 (4), 573—579. Oct.

Harris Paul, G., 2009. World Ethics and Climate Change: From International to Global Justice., 22. Edinburgh University Press, pp. 112—114.

Hobson M., Campbell L., 2012. How Ethiopia's Productive Safety Net Programme (PSNP) is responding to the current humanitarian crisis in the Horn. Humanitarian Practice Network, Issue 53. Retrieved from: http://www.odihpn.org/humanitarian-exchange-magazine/issue-53/how-ethiopias-productive-safety-net-programme-psnp-is-responding-to-the-current-humanitarian-crisis-in-the-horn.

Kasperson, R.E., et al., 2005. In: Hassan, R., Scholes, R., Ash, N. (Eds.), United Nations Environmental Program: Vulnerable Peoples in Ecosystems and Human Well-Being: Current State and Trends, 1. Island Press, Washington, DC, pp. 143–164.

Korchinsky, M., 2010. Wild Life Works Carbon Kasigau Corridor, Kenya an African REDD Project, pp. 1–23.

Macchi, M., Oviedo, G., Gotheil, S., Cross, K., Boedhihartono, A., Wolfangel, C, Howell, M., 2008. International Union for Conservation of Nature (IUCN). Indigenous and Traditional Peoples and Climate Change, pp. 15–22.

Mercer, J., Kelman, I., Taranis, L., Suchet-Pearson, S., 2010. Framework for integrating indigenous and scientific knowledge for disaster risk reduction. Disasters 34 (1), 214.

Moncel, R., Joffe, P., McCall, K., Levin, K., 2011. Building the Climate Change Regime: Survey and Analysis of Approaches.

National Geographic New Watch: Tribes in Historic "Cultural Crossroads" of Ethiopia's OmoVallet Endangered by Man and Land Grabs, April 23, 2013. Joanna Eede, Survival International. Retrieved from: https://blog.nationalgeographic.org/2013/04/23/tribes-living-in-historic-cultural-cross-roads-of-ethiopias-omo-valley-endangered-by-dam-and-land-grabs/.

Nossiter, A., 2012. Cholera Epidemic Envelops Coastal Slums in West Africa. New York Times August 22, 2012.

Oxfam America, 2011. An ounce of prevention: preparing for the impact of a changing climate on us humanitarian and disaster response. Bri. J. Sports Med. Retrieved from: http://bjsm.bmj.com/content/39/6/312.short.

Pham, P., Vinck, P., 2012. Technology, conflict early warning systems, public health, and human rights. Health Human Rights Int. J. 14, E106–E117.

Report of The United Nations Conference on Environment and Development, Rio de Janeiro, 3–14 June 1992. Principle 15. http://www.un.org/docu-ments/ga/conf151/aconf15126-1annex1.htm.

Salick Jan and Anja Byg Indigenous Peoples and Climate Change. Tyndall Centre for Climate Change Research, Oxford, 2007 pp 4, 5,7, 13,14.

Singer, P., 1972. Famine, affluence, and morality. Philos. Public Affairs 1 (1), 229–243.

Survival International, 2013. The Omo Valley Tribes. Cited on a website for Survival International at http://www.survivalinternational.org/tribes/omovalley

The Organization for Economic Co-operation and Development (OECD), 2009. Poverty and Climate Change: Reducing the Vulnerability of the Poor through Adaptation Executive Summary, pp 9.

The Barbados Advocate Focus on Early Warning Systems, January 12th, 2013.

The World Bank, 2010. Climate-Smart Agriculture: A Call to Action, pp. 2–24.

United Nations Development Program (UNDP), 2012. International Human Development Indicators from the Human Development Report Office as of 15 October 2012. http://hdrstats.undp.org/en/indicators/103106.html.

United Nations Environment Programme, 2007. Vulnerability of People and the Environment: Challenges and Opportunities published in Global Environmental Outlook 4: Human Dimensions of Environmental Change (Chapter 7), pp. 304–355.

United Nations Envrionment Programme, 2007. Environmental Emergencies News. Issue 6. April 2007, pp. 2–3. Retrieved from: http://www.unep.org/DEPI/PDF/EEsnewsletterissue6.pdf

United Nations News & Media, October 16th, 2012. Barbados warns of impact of climate change and natural disasters, Speaker Maxine McClean. http://www.unmultimedia.org/radio/english/2012/10/184152/.

United Nations University, 2013. Traditional Knowledge and Climate Science: Traditional Knowledge and Climate Science Toolkit, pp. 9.

United Nations Office for Disaster Risk Reduction, 2006. Global Survey of Early Warning Systems.

Webster, M., Ginnetti, J., Walker, P., Coppard, D., Kent, R., 2008. The Humanitarian Costs of Climate Change (December). Retrieved from: http://www.indiaenvironmentportal.org.in/files/FeinsteinTuftsclimatechange.pdf.

Werz, M., Conley, L., 2012. Climate Change, Migration, and Conflict: Addressing Complex Crisis Scenarios in the 21st Century.

World Food Program, 2013. Livelihoods, Early Assessment and Protection. Full citation found at: http://www.wfp.org/disaster-risk-reduction/leap.

World Health Organization, 2013. Health Topics: Epidemiology. Full citation found at: www.who.int/topics/epidemiology/en/

World Health Organization, 2013. Climate Change and Human Health: Climate Change Adaptation to Protect Human Health. http://www.who.int/glo-balchange/projects/adaptation/en/index1.html.

Worrell, D., Belgrave, A., Grosvenor, T., Lescott, A., 2011. An analysis of the tourism sector in Barbados. Econ. Rev. 37(1), pp 49, 64.

Chapter 16

A Framework for Assessing the Effectiveness of Ecosystem-Based Approaches to Adaptation

Hannah Reid[1], Amanda Bourne[2], Halcyone Muller[2], Karen Podvin[3], Sarshen Scorgie[2] and Victor Orindi[4]

[1]*International Institute for Environment and Development, London, United Kingdom,* [2]*Conservation South Africa, Cape Town, South Africa,* [3]*Regional Office for South America, Quito, Ecuador,* [4]*National Drought Management Authority, Nairobi, Kenya*

Chapter Outline

16.1 INTRODUCTION

The global climate is changing rapidly, and nations must plan accordingly. International, bilateral, and national sources of finance are increasingly available to meet global adaptation needs. But how should international climate finance be allocated? And how should countries developing and funding plans to address climate change, independent of international or bilateral finance, prioritize amongst the various different adaptation options available to them?

The main approach to climate change adaptation has tended to involve investment in engineered interventions, such as sea walls or irrigation infrastructure (Jones et al., 2012). There is growing realization, however, that ecosystem-based adaptation (EbA) may sometimes provide the optimal adaptation solution, particularly for poorer countries where people are more dependent on natural resources for their lives and livelihoods. A growing number of organizations and countries are adopting EbA as a means for climate adaptation, especially at the community level and in the context of disaster risk reduction. For example, 101 of the 166 Nationally Determined Contributions (covering 195 countries) submitted to the United Nations indicate ecosystem-orientated actions for adaptation, of which 68 include EbA (Seddon et al., 2017).

EbA involves the restoration and protection of biodiversity and ecosystem services to help people adapt to the adverse effects of climate change (CBD, 2009, 2010). Examples include: restoring coastal ecosystems to lower the energy of tropical storms and protect local communities against erosion and wave damage; wetland and floodplain management to prevent floods and maintain water flow and water quality in the face of changing rainfall patterns; conservation and restoration of forests and natural vegetation to stabilize slopes and prevent landslides, and regulate water flows preventing flash flooding; and establishment of diverse agroforestry systems to help maintain crop yields under changing climates. Box 16.1 describes some of the defining characteristics of effective EbA, derived from a review of relevant literature (taken from Seddon et al., 2016).

If properly implemented, EbA should bring many cobenefits and can meet objectives under all three Rio Conventions (Seddon et al., 2016). For example, EbA often involves maintaining the ability of natural ecosystems to control water cycles, or supports effective management regimes for dry areas, and thus aligns with the goals of the United Nations Convention to Combat Desertification (UNCCD). EbA also involves maintaining or restoring natural ecosystems and increasing habitat connectivity which improves conservation prospects for biodiversity and thereby helps countries meet their commitments under the Convention on Biological Diversity (CBD). Many EbA activities

Resilience. DOI: https://doi.org/10.1016/B978-0-12-811891-7.00016-5

BOX 16.1 Defining Characteristics of Effective Ecosystem-Based Approaches to Adaptation (EbA)

1. *Is human-centric.* EbA emphasizes human adaptive capacity or resilience in the face of climate change.
2. *Harnesses the capacity of nature to support long-term human adaptation.* It involves maintaining ecosystem services by conserving, restoring, or managing ecosystem structure and function, and reducing nonclimate stressors. This requires an understanding of ecological complexity and how climate change will impact ecosystems and key ecosystem services.
3. *Draws on and validates traditional and local knowledge.* Humans have been using nature to buffer the effects of adverse climatic conditions for millennia. Traditional knowledge about how best to do this should thus be drawn upon when implementing EbA.
4. *Is based on best available science.* An EbA project must explicitly address an observed or projected change in climate parameters, and as such should be based on climatic projections and relevant ecological data at suitable spatial and temporal scales.
5. *Can benefit the world's poorest,* many of whom rely heavily on local natural resources for their livelihoods.
6. *Is community-based and incorporates human rights-based principles.* Like community-based adaptation, EbA should use participatory processes for project design and implementation. People should have the right to influence adaptation plans, policies, and practices at all levels, and should be involved with framing both the problem and identifying solutions. EbA initiatives should be accountable to those they are meant to assist and not simply those providing support (i.e., donors or governments). EbA should consistently incorporate nondiscrimination, equity, the special needs of the poor, vulnerable and marginalized groups, diversity, empowerment, accountability, transparency, and active, free, and meaningful participation.
7. *Involves cross-sectoral and intergovernmental collaboration.* Ecosystem boundaries rarely coincide with those of local or national governance. Moreover, ecosystems deliver services to diverse sectors. As such, EbA requires collaboration and coordination between multiple sectors (e.g., agriculture, water, energy, transport) and stakeholders. EbA can complement engineered approaches, e.g., combining dam construction with floodplain restoration to lessen floods.
8. *Operates at multiple geographical, social, planning and ecological scales.* EbA can be mainstreamed into government processes (e.g., national adaptation planning) or management (e.g., at the watershed level), provided that communities remain central to planning and action.
9. *Integrates decentralized flexible management structures* that enable adaptive management.
10. *Minimizes trade-offs and maximizes benefits with development and conservation goals* to avoid unintended negative social and environmental impacts. This includes avoiding maladaptation, whereby adaptation "solutions" unintentionally reduce adaptive capacity.
11. *Provides opportunities for scaling up and mainstreaming* to ensure the benefits of adaptation actions are felt more widely and for the longer-term.
12. *Involves longer-term "transformational" change* to address new and unfamiliar climate change related and the root causes of vulnerability, rather than simply coping with existing climate variability and "climate-proofing" business-as-usual development.

Sources: Travers et al. (2012), Jeans et al. (2014), Faulkner et al. (2015), Reid (2014a,b), Girot et al. (2012), Ayers et al. (2012), Anderson (2014), Andrade et al. (2011), GEF (2012), ARCAB (2012), Bertram et al. (2017), Reid et al. (2009).

---

also sequester carbon and some prevent the greenhouse gas emissions that would be emitted from hard infrastructure-based approaches to adaptation. In this way EbA helps countries meet the mitigation targets of the United Nations Framework Convention on Climate Change (UNFCCC). Also, by increasing the resilience of vulnerable communities to extreme events such as flooding and landslides, EbA helps countries to meet the goals of the Sendai Framework for Disaster Risk Reduction (Renaud et al., 2013).

Despite its strong theoretical appeal, many positive anecdotes from around the world, and acknowledged multiplicity of cobenefits, EbA is not being widely or consistently implemented, or sufficiently mainstreamed into national and international policy processes (e.g., in South Africa see Pasquini et al., 2015). Relative to hard infrastructural options, EbA currently receives a small proportion of adaptation finance (Chong, 2014) and it is often poorly represented in climate policy (but see Seddon et al., 2017, and note that some countries like South Africa are exceptions to this). There are four major explanations for this (Biesbroek et al., 2013; Ojea, 2015; Vignola et al., 2009, 2013; Seddon et al., 2016).

1. First, there is uncertainty around how best to finance EbA. International climate finance, through mechanisms such as the Green Climate Fund or the Adaptation Fund, is one possibility. This will not provide enough to address adaptation challenges at the scale required to meet the needs of the world's poorest. Payments for Ecosystem Services (PES) may provide an alternative source of funding, or large-scale government social protection, employment generation, or environmental management programs. In the context of providing finance for adaptation, however, both are in their infancy.

2. Second, many climate change impacts will be long term, but this does not sit well with what are usually short-term political decision-making processes often based on standard electoral cycles. Photogenic engineered adaptation solutions with immediate but inflexible benefits are thus often favored over the long-term flexible solutions offered by EbA under which benefits may only be apparent in the future.
3. Third, the evidence base for the effectiveness of EbA (especially its economic viability) is currently weak. Evidence on the ecological, social, and economic effectiveness of EbA projects relative to alternative approaches is needed (Doswald et al., 2014; Travers et al., 2012; Reid, 2011, 2014a; UNEP, 2012).
4. The final major challenge to EbA relates to issues around governance. EbA necessitates cooperation and communication across multiple sectors and varying administrative or geographical scales. Most governance models struggle to support this.

This chapter describes work aimed at addressing the last two challenges in the above list. It describes a framework (Table 16.1) for assessing EbA effectiveness that was recently developed as part of the "Ecosystem-based approaches to adaptation: strengthening the evidence and informing policy" project (Reid et al., 2017; Seddon et al., 2016), and emerging observations from framework application in selected countries.

16.2 HOW DO WE KNOW IF EBA IS EFFECTIVE?

Effective EbA is defined as "an intervention that has restored, maintained, or enhanced the capacity of ecosystems to produce services. These services in turn enhance the well-being, adaptive capacity or resilience of humans, and reduce their vulnerability. The intervention also helps the ecosystem to withstand climate change impacts and other pressures" (Reid et al., 2017 based on Seddon et al., 2016). Seddon et al. (2016) explain that this definition generates two overarching questions that need to be addressed in order to determine whether a particular EbA initiative is effective:

1. *Did the initiative allow human communities to maintain or improve their adaptive capacity or resilience, and reduce their vulnerability, in the face of climate change, while enhancing cobenefits that promote well-being?*
2. *Did the initiative restore, maintain or enhance the capacity of ecosystems to continue to produce services for local communities, and allow ecosystems to withstand climate change impacts and other stressors?*

By definition, EbA should also be financially and/or economically viable, and for benefits to materialize it needs support from local, regional, and national governments, and to be embedded in an enabling policy, institutional, and legislative environment (Seddon et al., 2016; Reid et al., 2017). This leads to two further overarching questions:

3. *Is EbA cost-effective and economically viable?*
4. *What social, institutional, and political issues influence the implementation of effective EbA initiatives and how might challenges best be overcome?*

These questions encompass much important detail regarding how to assess and compare effectiveness in ecological, social, and economic terms. They lead to a further set of nine more specific questions (Table 16.1)[1] that reflect the growing consensus around the key characteristics of effective EbA (Box 16.1).

16.3 APPLYING THE FRAMEWORK—IS EBA EFFECTIVE?

The framework can be used to help climate change policy makers, amongst others, recognize when EbA is effective and then, where appropriate, integrate EbA principles into national and international climate adaptation policy and planning processes. The framework can also be used to explore the policy and institutional obstacles to, and opportunities for, improving EbA implementation on the ground and its uptake in national adaptation planning and policy making.

Practitioners and researchers who are planning and implementing new EbA projects can use the framework to develop indicators, which can help measure initiative effectiveness, and support adaptive project management over time. These indicators will differ by location. The framework can also be used to assess whether a project is indeed

---

1. These questions derive from: a distillation of literature on evidence and evidence gaps relating to EbA; criteria for identifying best practices and lessons learned on adaptation effectiveness collated under the Nairobi Work Programme of the UNFCCC and the Convention on Biodiversity; and, the IUCN Learning Framework, developed to capture some common lessons from the portfolio of IUCN EbA projects. Seddon et al. (2016) provides more detail on the background behind these questions and Reid et al. (2017) details subquestions and options for answers in a more detailed framework for assessing EbA effectiveness.

TABLE 16.1 Framework for Assessing EbA Effectiveness

1. **Effectiveness for Human Societies**

Did the initiative allow human communities to maintain or improve their adaptive capacity or resilience, and reduce their vulnerability, in the face of climate change, while enhancing cobenefits that promote long-term well-being?

1. Did the EbA initiative improve the resilience and adaptive capacity of local communities, and help the most vulnerable (e.g., women, children, and indigenous groups)? If so, over what timeframes were these benefits felt, and were there trade-offs (or synergies) between different social groups?
2. Did any social cobenefits arise from the EbA initiative, and if so, how are they distributed and what are the trade-offs between different sectors of society?
3. What role in the EbA initiative did stakeholder engagement through participatory processes and indigenous knowledge play? Did/does the use of participatory processes support the implementation of EbA and build adaptive capacity?

2. **Effectiveness for the Ecosystem**

Did the initiative restore, maintain, or enhance the capacity of ecosystems to continue to produce adaptation services for local communities, and allow ecosystems to withstand climate change impacts and other stressors?

4. What were/are the factors threatening local ecosystem(s)? How did/do these pressures affect the resilience of the ecosystem(s) to climate change and other stressors and their capacity to deliver ecosystem services over the long term?
5. After the EbA initiative, which ecosystem services were restored, maintained, or enhanced, and did the resilience of the ecosystem change? Over what geographic scale(s) and time frame(s) were these effects felt, and were there trade-offs (or synergies) between the delivery of different ecosystem services at these different scales?

3. **Financial and Economic Effectiveness**

Is EbA cost-effective and economically viable over the long term?

6. What are the general economic costs and benefits of the EbA initiative? How cost-effective is it, ideally in comparison to other types of interventions, and are any financial or economic benefits sustainable over the long term?

4. **Policy and Institutional Issues**

What social, institutional, and political issues influence the implementation of effective EbA initiatives and how might challenges best be overcome?

7. What are the key policy, institutional, and capacity barriers to, or opportunities for, implementing EbA at the local, regional, and national levels over the long term?
8. What, if any, opportunities emerged for replication, scaling up, or mainstreaming the EbA initiative or for influence over policy, and how?
9. What changes in local, regional, and/or national government or in donor policies are required to implement more effective EbA initiatives?

genuine EbA or whether it is a "rebranded" conservation or natural resource management initiative (Reid, 2015; Martin, 2016; Bertram et al., 2017).

The framework is being applied under the "Ecosystem-based approaches to adaptation: strengthening the evidence and informing policy" project, which aims to consolidate and compare evidence from 13 existing EbA project sites across Asia, Africa, and Central and South America in order to better assess EbA effectiveness. In each country, relevant project literature is being reviewed, and interviews with a range of project stakeholders held (key national level policy and decision-makers, local authority level officials, project implementing partners, and local communities). At the time of writing, data from all project sites is still being collected, so this chapter can only describe selected and qualitative results from early framework application in four countries (see Table 16.2).

16.3.1 Effectiveness for Human Societies

EbA can help people maintain or improve their adaptive capacity or resilience, and reduce their vulnerability in the face of climate change (Doswald et al., 2014; Jeans et al., 2014; UNFCCC, 2017). For example, agroforestry (a key EbA activity) has been shown to increase crop yields in times of drought and improve livelihood resilience by providing alternative sources of both income and fruit, fodder, or fuel (Charles et al., 2013; Garrity et al., 2010). Early project results reinforce this view. For example, Tari et al. (2015) explain how in 2014 Kenya experienced a long dry season.

TABLE 16.2 Selected Project Countries and Partners Under the "Ecosystem-based Approaches to Adaptation: Strengthening the Evidence and Informing Policy" Project

Country	Project Partner Organization	Project Name or Activities
Kenya	Adaptation Consortium/Kenya Drought Management Authority	Supporting Counties in Kenya to Mainstream Climate Change in Development and Access Climate Finance
South Africa	Conservation South Africa	Climate Resilient Livestock Production on Communal Lands: Rehabilitation and Improved Management of Dryland Rangelands in the Succulent Karoo
Uganda	IUCN	Ecosystem-Based Adaptation in Mountain Ecosystems Programme (Uganda)
Chile	IUCN	Ecosystems Protecting Infrastructure and Communities (EPIC), South America geographical component

Despite similarly low rainfall amounts recorded across the region, Isiolo County did not reach drought "alarm" level warnings, whereas neighboring counties did. This uncharacteristic decoupling of drought and its impacts on humans has been attributed to good practices in local natural resource management, including EbA activities, in Isiolo (Tari et al., 2015).

Beyond helping people adapt to the adverse effects of climate change, many EbA projects report a wide range of social cobenefits (UNEP, 2015; CBD, 2009, 2016; Doswald et al., 2014; Rizvi et al., 2014). These include food and water security, disaster risk reduction, livelihood diversification, and reduction in conflict over scarce resources (GIZ, 2013; Rao et al., 2013; Reid and Alam, 2017). For example, the EPIC project in Chile showed that investing in ecological infrastructure—namely trees through improved forest management—could reduce disaster risk (from avalanches and falling rocks/debris) (IUCN, 2017; Rizvi et al., 2014). In Uganda, there is more money for sending local children to school than before the Mountain EbA Project, and the stable supply of water and food, better nutrition, and reductions in indoor air pollution have significant health benefits (Rossing and Nyman, 2015). Interviewees reported improvements in knowledge and capacity levels, and social cohesiveness at several project sites. Capturing the full value of these cobenefits, however, remains challenging. Many are hard to measure and quantify and are thus overlooked.

It is the world's poorest who are often most vulnerable to climate change impacts and most reliant on natural resources (Reid, 2014a). However, few studies closely examine exactly who benefits from EbA initiatives (Doswald et al., 2014; Reid, 2011). EbA was able to benefit the poorest and most vulnerable people in South Africa. Here, rehabilitation of critical rangeland and wetland ecosystems benefitted pastoralists in the Northern Cape—one of South Africa's poorest provinces—by improving access to water and reducing the risks of stock loss during droughts (Black et al., 2016). The poorest and most vulnerable communities rely on pastoralism so particularly benefitted. Meanwhile, in Kenya most EbA-related investments under the Isiolo County Adaptation Fund (CAF) up until 2014 focused on clean water (e.g., from the rehabilitation and fencing of water pans and accompanying water governance activities)—an issue of particular importance to poor rural women whose role it is to fetch water (Kenya National Drought Management Authority, 2014).

Much of the published literature on EbA emphasizes its benefits, and EbA is frequently characterized as "win-win" for society and the environment (Ash et al., 2009) or "no regrets" (Rizvi et al., 2014). Less is known about costs and trade-offs in terms of who benefits (Doswald et al., 2014; Reid, 2015; Munroe et al., 2012). In South Africa, e.g., rangeland rehabilitation may require the exclusion of livestock or grazing restrictions for roughly 5 years, to the detriment of those with stock (De Villiers, 2013; Bourne et al., 2017). People growing vegetables on the banks of the Sipi River in Uganda also lost benefits from crops due to tree-planting and riverbank rehabilitation activities. In Chile, interviews revealed that some tourism sector stakeholders felt that the study's emerging recommendations could disrupt elements of the local economy that depended on tourism.

Benefits can also take time to accrue. For example, sand dams in Kenya take a few years to accumulate sufficient water. At some project sites, incentives helped ameliorate the short-term losses associated with trade-offs or built support for EbA until benefits accrued. Conclusions from the Mountain EbA project sites in Peru, Uganda, and Nepal reiterate the importance of such incentives (UNEP, 2015). In Uganda these included a specific community environment conservation fund, local service procurement, market outlets for produce, a gravity flow water scheme, and provision of water stands (UNEP, 2015).

A key attribute of effective EbA is the use of participatory processes, especially where adaptation benefits target poor communities (Bertram et al., 2017; Reid, 2014b). Box 16.1 explains how participatory processes should be applied so that people have the right to influence adaptation planning, implementation and policies at all levels (Reid et al., 2009). Many project interviewees felt that the more participatory engagement in EbA project planning and implementation was, the better project outcomes were. Indeed, EbA investments in Isiolo, Kenya, included the provision of support for planning meetings and the operational costs of four Borana community customary range management institutions (dedhas). Dedhas are the repositories for much traditional knowledge of climate, water management and rangeland management (Victor Orindi pers. comm., 2016). Likewise in South Africa, support for climate resilient planning was provided to the local agricultural cooperative in the Kamiesberg Local Municipality (Amanda Bourne pers. comm., 2017).

16.3.2   Effectiveness for the Ecosystem

If ecosystems are unable to maintain their structure and function in the face of climate change and other stresses—such as overgrazing and invasive species—they will not be able to provide adaptation benefits to people (UNEP, 2015; Jeans et al., 2014). Little is known about the effect of EbA on ecosystem resilience, but there are indications that if properly implemented, EbA can improve ecosystem resilience and restore, maintain, or enhance the capacity of ecosystems to continue to produce ecosystem services for local communities (EC, 2013; Epple and Dunning, 2013; CBD, 2016; Reid and Alam, 2017). EbA in some ecosystems (such as grasslands/savanna, mountains, and marine biomes) or relating to some components of biodiversity (such as genetic diversity) is particularly poorly understood (Doswald et al., 2014).

There may, in some instances, be a minimum ecosystem size needed to secure ecosystem resilience (Ekins et al., 2003). The boundaries that influence ecosystem resilience in the context of EbA are poorly understood (Doswald et al., 2014), although EbA project stakeholders in Chile and Uganda felt that the watershed was an appropriate level to focus on in order to ensure ecosystem/landscape level resilience in addition to comprehensive and sustainable EbA benefits and local level political support from district level governments and protected area managers (UNEP, 2015). Operating according to natural boundaries, however, can be challenging as they often do not match administrative boundaries.

There may also be thresholds or tipping points (induced by climate change or other stresses) beyond which ecosystems cannot function effectively or recover from once crossed (TEEB, 2010; CBD, 2009; MEA, 2005). Little is known about these thresholds in the context of EbA (Munroe et al., 2012). For example, appropriate stocking rates and the best way to restore degraded rangeland (or whether this is indeed possible) in some dryland areas in Africa are poorly understood (Bourne et al., 2017; Van der Merwe et al., 2011; James et al., 2013). It is also unclear at what stage shifts in avalanche patterns will be seen at the EPIC project site in Chile as a result of snowmelt.

As with social benefits, cobenefits relating to ecosystem integrity and service provision are known to accrue from EbA, particularly downstream from EbA project areas in the same water catchment (Rossing and Nyman, 2015). Beekeeping activities in Sironko, Mount Elgon, Uganda, also helped pollination and thus may have increased crop productivity more broadly (Rossing and Nyman, 2015).

There can be trade-offs too, however, e.g., where one ecosystem service is more strongly emphasized than another or where ecosystem services in one area are restored at the expense of another (Jiménez Hernández, 2016). For example, sand dams in Kenya could reduce downstream water availability (Victor Orindi pers. com. 2016). In addition, some ecosystem services take longer to materialize than others, e.g., tree-planting to protect vulnerable riverbanks in Uganda (UNEP, 2015) or dryland ecosystem restoration in South Africa (Bourne et al., 2017).

16.3.3   Financial and Economic Effectiveness

Economic justifications for action invariably have most traction with decision-makers, and while there is some evidence to suggest EbA can be cost-effective, robust financial and economic arguments for and against adopting EbA approaches are in short supply (Reid, 2011, 2014a; Rao et al., 2013). Financial viability refers to the project operational costs and benefits, or the business case for EbA. Economic evaluations consider broader additional costs and benefits such as avoided (or increased) losses due to disaster risk reduction, opportunity costs, land value increases, and local income enhancement.

Project evidence varies from EbA site to site and is rarely comparable, but in all four sites EbA is, or is perceived by interviewees as, economically beneficial. In Uganda, a project cost−benefit analysis showed that EbA practices were viable over a 15-year period (MWE, 2015). Investing in EbA was financially worthwhile for local farmers in that the benefits from this exceeded costs. And in Chile, IUCN (2017) argue that the protective role played by healthy forests against avalanches and falling debris in mountain ecosystems could lead to significant economic savings. But not all EbA projects were financially viable. Cost—benefit and cost-effectiveness analyses in the Namakwa District Municipality, South Africa (Bourne et al., 2017; Black et al., 2016), showed rangeland and wetland rehabilitation to be risky and expensive, with benefits taking many years to materialize. This study did not, however, include wider economic benefits, which may have altered results. Likewise, the historical systematic undervaluation of pastoral production systems, also identified in Kenya, has evolved in part because it is not easy to measure and quantify the full economic benefits that pastoralism provides.

EbA projects studied were generally characterized by a multitude of economic cobenefits, most of which were hard to quantify. For example, interviewees reported land value increases in Uganda, and integrating EbA into publicly funded Expanded Public Works Programmes in South Africa (such as Working for Water) showed huge potential for employment creation (Bourne et al., 2017).

As with social and ecosystem-related benefits, sites exhibited trade-offs in terms of where and when financial/economic costs and benefits accrued and who accrues them. In contexts such as rangeland rehabilitation in the drylands of South Africa, where the benefits of EbA interventions will take years to materialize, immediate incentives sometimes helped to offset short-term opportunity costs.

16.4 POLICY AND INSTITUTIONAL ISSUES

The ultimate success of EbA initiatives hinges on the institutional, governance and policy context in which they operate. This is true both at the local level—where capable local institutions are needed to make decisions and ensure active community participation—and also for the high-level institutional and policy arena in which initiatives sit, on the understanding that it is not actors at the local level alone that define the success of a local adaptation initiative (Faulkner et al., 2015; Reid, 2015; Pasquini et al., 2015).

Project sites revealed a wide range of policy-related and institutional factors that contributed to EbA project success. These included strong institutions such as dedha and Ward Climate Change Planning Committees in Kenya, and the presence of "champions" and government prioritization of EbA at various levels. Appropriate incentives given to people to compensate for short-term losses were also important at some sites. These included incentives for sustainable land management from the Community Environment and Conservation Fund in Uganda (UNEP, 2015), and direct employment for monetary gain in the rangeland and wetland rehabilitation and management work in South Africa, along with skills development and livestock improvement schemes (Amanda Bourne pers. comm., 2017). Many interviewees felt that a supportive policy and legislative context was important (indeed, most of the countries looked at had a strong national policy framework that recognized the role of ecosystems in addressing climate change).

Operationalizing these policies and translating them into action, however, proved more challenging. EbA is cross-sectoral in nature (Baudoin and Ziervogel, 2016), which requires collaboration in planning and management at local, provincial, and national levels. Most government planning, however, is highly sectoral. For example, housing climate change within the Department of Environmental Affairs in South Africa inhibited EbA implementation at times because the issue was seen as an additional responsibility that stakeholders from other departments had to undertake without extra funding or support. This gave it a negative reputation. Similarly, the national office for managing emergencies and disasters in Chile currently focuses mostly on disaster response rather than prevention, although it envisages incorporating various types of territorial planning units relating to ecosystems or water basins into its work in the future (ONEMI, 2014).

Other challenges listed by interviewees included a lack of capacity, technical or knowledge constraints, unclear institutional mandates, and limited financial resources. Most local and provincial level authorities in the countries studied faced obstacles regarding budgeting and funding, and lacked the necessary capacity to implement EbA.

The case studies revealed a number of opportunities for mainstreaming EbA practices into wider planning and management, scaling them up in other ways for greater impact, and financing EbA at scale. These included embedding EbA into the Expanded Public Works Programmes in South Africa and funding EbA at scale through the Isiolo County CAF, which supports bottom-up climate change responses at scale in Kenya's drylands. In Uganda, work to bundle watershed services into credits to sell to international or national buyers (such as the government-owned National Water and Sewerage Corporation of Uganda) showed that PES models could finance EbA. Over 200 EbA-related tools also exist to support EbA replication and use elsewhere.[2]

---

2. See https://www.iied.org/call-for-feedback-inventory-tools-support-ecosystem-based-adaptation

16.5 CONCLUSION

This chapter has outlined a recently developed framework for looking at EbA and answering the question "is it working?" Applying this framework involves assessing whether EbA can strengthen human adaptive capacity or resilience, whether it can support ecosystem resilience and service provision, and whether it makes financial sense to invest.

The chapter has also described some early emerging qualitative results from applying this framework in selected countries under the "Ecosystem-based approaches to adaptation: strengthening the evidence and informing policy" project to date. EbA has been able to help the poorest and most vulnerable people at the four project sites assessed, and strong participation has often played a key role in this context. Evidence from the sites assessed so far supports the commonly held view (e.g., Rossing and Nyman, 2015; Doswald et al., 2014) that EbA can provide a multitude of social and ecosystem-related cobenefits. Indications are that it made financial sense to invest in EbA at the four sites, but that the broader economic benefits of interventions needed to be factored into planning at some sites (e.g., South Africa) for this to hold true. These broader economic benefits are not easy to quantify. Project sites illustrated a multitude of social, environmental, and economic trade-offs that needed to be acknowledged and addressed during implementation. Providing incentives sometimes helped ameliorate the short-term losses associated with these trade-offs or built support for project activities until benefits accrued.

This project will now collate and consolidate the results of framework application in all 13 project sites to provide a stronger evidence base for answering the question of whether EbA works or not. Framework application by researchers outside the project would also serve to strengthen the evidence base for EbA effectiveness, which could then help inform wider planning.

A number of political, policy-related, institutional, and capacity challenges influenced EbA effectiveness at the project sites studied. The cross-sectoral collaboration needed at local, provincial, and national levels to translate policy and plans into genuine benefits proved challenging at most sites. This research has begun identifying some of these challenges at specific sites, but greater understanding of how to identify and capitalize on opportunities to make EbA effective, and overcome related policy and institutional challenges is needed. Improving knowledge in this area is important if EbA is to go beyond the project level and reach the millions of poor people vulnerable to climate change who need help.

ACKNOWLEDGMENT

This chapter presents work conducted under an International Climate Initiative (IKI) project "Ecosystem-based approaches to adaptation: strengthening the evidence and informing policy" implemented by IIED, IUCN, and UNEP-WCMC in collaboration with 13 in-country partner organizations in Bangladesh, China, Nepal, Kenya, South Africa, Uganda, Burkina Faso, Peru, Chile, Costa Rica, El Salvador, and Senegal. The German Federal Ministry for the Environment, Nature Conservation, Building and Nuclear Safety (BMUB) supports this initiative on the basis of a decision adopted by the German Bundestag. The authors received help with writing and collecting results described in this chapter from Liaquat Ali, Anu Adhikari, Edmund Barrow, Evelyne Busingye, Ced Hesse, Charlotte Hicks, Xiaoting Hou-Jones, Val Kapos, Michelle Kimeu, Sophie Kutegeka, Rob Munroe, Marta Perez de Madrid, Ali Raza, Dilys Roe, Fiona Roberts, Natalie Seddon, and Sylvia Wicander.

REFERENCES

Anderson, S., 2014. Getting Ahead of the Curve: When Climate Adaptation Has to Get Radical. IIED briefing paper. November 2014.

Andrade, A., Córdoba, R., Dave, R., Girot, P., Herrera-F, B., Munroe, R., et al., 2011. Draft Principles and Guidelines for Integrating Ecosystem-based Approaches to Adaptation in Project and Policy Design: A Discussion Document. IUCN-CEM, CATIE, Turrialba, Costa Rica.

ARCAB, 2012. Action Research for Community Adaptation in Bangladesh: Monitoring and Evaluation Framework Paper. Final report.

Ash, N., Ikkala, N., Parker, C., 2009. Ecosystem-based Adaptation (EbA). Contribution to the Fifth session of the UNFCCC Ad Hoc Working Group on Long-Term Cooperative Action under the Convention (AWG-LCA), Bonn, 29 March to 8 April 2009. IUCN, Gland.

Ayers, J., Anderson, S., Pradha, S., Rossing, T., 2012. Participatory Monitoring, Evaluation, Reflection and Learning for Community-based Adaptation: A Manual for Local Practitioners. CARE International.

Baudoin, M., Ziervogel, G., 2016. What role for local organisations in climate change adaptation? Insights from South Africa. Reg. Environ. Change 16 (7). Available from: https://doi.org/10.1007/s10113-016-1061-9.

Bertram, M., Barrow, E., Blackwood, K., Rizvi, A.R., Reid, H., von Scheliha-Dawid, S., 2017. Making Ecosystem-based Adaptation Effective: A Framework for Defining Qualification Criteria and Quality Standard. FEBA (Friends of Ecosystem-based Adaptation) technical paper developed for UNFCCC-SBSTA 46. GIZ, Bonn, Germany, IIED, London, UK, and IUCN, Gland, Switzerland.

Biesbroek, G.R., Termeer, C.J.A.M., Klostermann, J.E.M., Kabat, P., 2013. On the nature of barriers to climate change adaptation. Reg. Environ. Change 13, 1119−1129.

Black, D., Turpie, J.K., Rao, N., 2016. Evaluating the cost-effectiveness of ecosystem-based adaptation: Kamiesberg Wetlands case study. South Afr. J. Environ. Manage. Sci. 19 (5), 702–713.

Bourne, A., Muller, H., de Villiers, A., Alam, M., Hole, D., 2017. Assessing the efficiency and effectiveness of rangeland restoration in Namaqualand, South Africa. Plant Ecol. 218 (0), 7–22.

CBD, 2009. Connecting Biodiversity and Climate Change Mitigation and Adaptation. Report of the Second Ad Hoc Technical Expert Group on Biodiversity and Climate Change. CBD Technical Series No. 41. Secretariat of the Convention on Biological Diversity, Montreal, Canada.

CBD, 2010. Convention on Biological Diversity 2010: Decision adopted by the Conference of the Parties to the Convention on Biological Diversity at its 10th Meeting. X/33. Biodiversity and climate change. UNEP/CBD/COP/DEC/X/33.

CBD, 2016. Synthesis Report on Experiences with Ecosystem-Based Approaches to Climate Change Adaptation and Disaster Risk Reduction. CBD Technical Series no. 85. CBD, Montreal.

Charles, R.L., Munishi, P.K.T., Nzunda, E.F., 2013. Agroforestry as adaptation strategy under climate change in Mwanga District, Kilimanjaro, Tanzania. Int. J. Environ. Protect. 3 (11), 29–38.

Chong, J., 2014. Ecosystem-based approaches to climate change adaptation: progress and challenges. Int. Environ. Agreem. Politics Law Econ. 14 (4), 391–405.

De Villiers, A., 2013. Ecosystem-based Adaptation to climate change in Namaqualand, South Africa: cost-effectiveness of rangeland rehabilitation for erosion control. Conservation South Africa, Cape Town.

Doswald, N., Munroe, R., Roe, D., Giuliani, A., Castelli, I., Stephens, J., et al., 2014. Effectiveness of ecosystem-based approaches for adaptation: review of the evidence-base. Clim. Dev. 6 (2), 185–201.

EC, 2013. Thematic Issue: Ecosystem-Based Adaptation. Science for Environment Policy. Issue 37. Science Communication Unit, Bristol.

Ekins, P., Simon, S., Deutsch, L., Folke, C., De Groot, R., 2003. A framework for the practical application of the concepts of critical natural capital and strong sustainability. Ecol. Econ. 44 (2–3), 165–185.

Epple, C., Dunning, E., 2013. Ecosystem resilience to climate change: what is it and how can it be addressed in the context of climate change adaptation? Technical report for the Mountain EbA Project. UNEP-WCMC, Cambridge.

Faulkner, L., Ayers, J., Huq, S., 2015. Meaningful measurement for community-based adaptation. In: Monitoring and evaluation of climate change adaptation: a review of the landscape New Directions for Evaluation (eds. D. Bours, C. McGinn and P. Pringle), pp. 89–104.

Garrity, D.P., Akinnifesi, F.K., Ajayi, O.C., Weldesemayat, S.G., Mowo, J.G., Kalinganire, A., et al., 2010. Evergreen Agriculture: a robust approach to sustainable food security in Africa. Food Sec. 2, 197–214.

GEF, 2012. Operational Guidelines on Ecosystem-Based Approaches to Adaptation. LDCF/SCCF Council meeting, 15 Nov 2012, Washington DC.

Girot, P., Ehrhart, C., Oglethorpe, J., Reid, H., Rossing, T., Gambarelli, G., et al., 2012. Integrating community and ecosystem-based approaches in climate change adaptation responses. ELAN, unpublished.

GIZ, 2013. Saved health, saved wealth: an approach to quantifying the benefits of climate change adaptation. Practical application in coastal protection projects in Viet Nam. GIZ, Bonn.

IUCN, 2017. Chile and Nepal - Forest Protecting against Snow Avalanches. EPIC project blog. See http://www.epicproject.net/?page_id = 8.

James, J.J., Sheley, R.L., Erickson, T., Rollins, K.S., Taylor, M.H., Dixon, K.W., 2013. A systems approach to restoring degraded drylands. J. Appl. Ecol. 50, 730–739.

Jeans, H., Oglethorpe, J., Phillips, J., Reid, H., 2014. The role of ecosystems in climate change adaptation: lessons for scaling up. In: Schipper, E.L.F., Ayers, J., Reid, H., Huq, S., Rahman, A. (Eds.), Community Based Adaptation to Climate Change: Scaling it up. Routledge, London, pp. 253–265.

Jiménez Hernández, A., 2016. Ecosystem-based Adaptation Handbook. IUCN NL, Amsterdam.

Jones, H.P., Hole, D.G., Zavaleta, E.S., 2012. Harnessing nature to help people adapt to climate change. Nat. Clim. Change 2 (7), 504–509.

Kenya National Drought Management Authority, 2014. Isiolo County Adaptation Fund: Activities, Costs and Impacts after the 1st Investment Round. Project Report: June 2014.

Martin, S., 2016. EbA Revisited, Part 1: disentangling misconceptions about nature and adaptation. Blog 14 June 2016. http://www.climateprep.org/stories/2016/6/14/eba-revisited-part-1-disentangling-misconceptions-about-nature-and-adaptation.

MEA, 2005. Millennium Ecosystem Assessment. Ecosystems and Human Well-being: Synthesis. Island Press, Washington, DC.

Munroe, R., Roe, D., Doswald, N., Spencer, T., Möller, I., Vira, B., et al., 2012. Review of the evidence base for ecosystem-based approaches for adaptation to climate change. Environ. Evid. 1, 13.

MWE, 2015. Natural Resource Economic Assessment for Ecosystem Based Adaptation in the Mt. Elgon Ecosystem. Ministry of Water and Environment, Republic of Uganda.

Ojea, E., 2015. Challenges for mainstreaming Ecosystem-based Adaptation into the international climate agenda. Curr. Opin. Environ. Sustain. 14, 41–48.

ONEMI, 2014. Política Nacional para la Gestión de Riesgo de Desastres. Ministerio del Interior y Seguridad Pública, Santiago de Chile.

Pasquini, L., Ziervogel, G., Cowling, R.M., Shearing, C., 2015. What enables local governments to mainstream climate change adaptation? Lessons learned from two municipal case studies in the Western Cape, South Africa. Clim. Dev. 7 (1), 60–70.

Rao, N.S., Carruthers, T.J.B., Anderson, P., Sivo, L., Saxby, T., Durbin, T., et al., 2013. An Economic Analysis of Ecosystem-Based Adaptation and Engineering Options for Climate Change Adaptation in Lami Town, Republic of the Fiji Islands. Secretariat of the Pacific Regional Environment Programme (SPREP), Apia, Samoa.

Reid, H., 2011. Improving the evidence for ecosystem-based adaptation. IIED Opinion: lessons from adaptation in practice series. IIED, London, November 2011.

Reid, H., 2014a. A natural focus for community-based adaptation. In: Ensor, J., Berger, R., Huq, S. (Eds.), Community-based Adaptation to Climate Change: Emerging Lessons. Practical Action Publishing, Rugby, pp. 35−54. , 2014.

Reid, H., 2014b. Ecosystem- and Community-based Adaptation: Learning From Natural Resource Management. IIED Briefing. IIED, London.

Reid, H., 2015. Ecosystem- and community-based adaptation: learning from community-based natural resource management. Clim. Dev. . Available from: https://doi.org/10.1080/1756552920151034233.

Reid, H., Alam, S.S., 2017. Ecosystem-based approaches to adaptation: evidence from two sites in Bangladesh. Clim. Dev. 9 (6), 518−536.

Reid, H., Cannon, T., Berger, R., Alam, M., Milligan, A., 2009. Community-based adaptation to climate change. Participatory Learning and Action 60. IIED, London.

Reid, H., Seddon, N., Barrow, E., Hicks, C., Hou-Jones, X., Kapos, V., et al., 2017. Ecosystem-based Adaptation: Question-based Guidance for Assessing Effectiveness. IIED, London.

Renaud, F.G., Sudmeier-Rieux, K., Estrella, M., 2013. The Role of Ecosystems in Disaster Risk Reduction. UN University Press, Tokyo/New York/ Paris.

Rizvi, A.R., Barrow, E., Zapata, F., Cordero, D., Podvin, K., Kutegeka, S., et al., 2014. Ecosystem based Adaptation: Building on No Regret Adaptation Measures. Technical Paper prepared for the 20th session of the Conference of the Parties to the UNFCCC and the 10th session of the Conference of the Parties to the Kyoto Protocol, Lima, Peru, 1-12 December 2014. IUCN, Gland.

Rossing, T., Nyman, N.I., 2015. Generating multiple benefits from Ecosystem-based Adaptation in Mountain Ecosystems. Global ecosystem-based adaptation in mountains programme learning brief. UNDP, New York.

Seddon, N., Reid, H., Barrow, E., Hicks, C., Hou-Jones, X., Kapos, V., et al., 2016. Ecosystem-based Approaches to Adaptation: Strengthening the Evidence and Informing Policy: Research Overview and Overarching Questions. IIED, London.

Seddon, N., Daniels, E., Davis, R., Harris, R., Hou-Jones, X., Huq, S., et al., 2017. Ecosystem-based adaptation and the Paris Agreement: global recognition that ecosystems are key to human resilience in a warming world (manuscript in preparation).

Tari, D., King-Okumu, C., Jarso, I., 2015. Strengthening Local Customary Institutions: A Case Study in Isiolo County, Northern Kenya. IIED, London.

TEEB -The Economics of Ecosystems and Biodiversity (TEEB), 2010. The Economics of Ecosystems and Biodiversity: Mainstreaming the Economics of Nature: A Synthesis of the Approach, Conclusions and Recommendations of TEEB. Progress Press, Malta.

Travers, A., Elrick, C., Kay, R., Vestergaard, O., 2012. Ecosystem-based Adaptation Guidance: Moving From Principles to Practice. UNEP, New York.

UNEP, 2012. Making the Case for Ecosystem-based Adaptation: Building Resilience to Climate Change. UNEP, Nairobi.

UNEP, 2015. Making the Case for Ecosystem-Based Adaptation: The Global Mountain EbA Programme in Nepal, Peru and Uganda. UNEP, New York.

UNFCCC, 2017. Adaptation planning, implementation and evaluation addressing ecosystems and areas such as water resources. Synthesis report by the secretariat for the Subsidiary Body for Scientific and Technological Advice, 46th session, Bonn, 8−18 May 2017.

Van der Merwe, H., van Rooyen, M.W., 2011. Life form and species diversity on abandoned croplands, Roggeveld, South Africa. Afr. J.Range Forage Sci. 28 (2), 99−110.

Vignola, R., Locatelli, B., Martinez, C., Imbach, P., 2009. Ecosystem-based adaptation to climate change: what role for policy-makers, society and scientists? Mitig. Adapt. Strateg. Global Change 14 (8), 691−696.

Vignola, R., McDaniels, T.L., Scholz, R.W., 2013. Governance structures for ecosystem-based adaptation: using policy-network analysis to identify key organizations for bridging information across scales and policy areas. Environ. Sci. Policy 3, 71−84.

Chapter 17

The Global Framework for Climate Services Adaptation Programme in Africa

Janak Pathak and Filipe D.F. Lúcio

World Meteorological Organization, Geneva, Switzerland

Chapter Outline

17.1 INTRODUCTION

While countries attempt to adapt to climate variability and change, including extreme weather and climate events, climate services are becoming an important part of public and private decision-making process. There is a growing need to improve our understanding of climate variability, predictions, and use of climate information for a better decision making. In response to the growing need for better coordination and science-informed climate services, the Global Framework for Climate Services (GFCS)[1] was established at the World Climate Conference-3 (WCC-3) in 2009 with the aim to guide the development and use of science-based climate services for decision-making in climate-sensitive sectors. Climate Services are defined by the GFCS of World Meteorological Organization (WMO) as, "providing climate information in a way that assists decision-making by individual and organizations. A service requires appropriate engagement along with an effective access mechanism and must respond to user needs."

The GFCS aims to strengthen and coordinate existing initiatives and develop new infrastructure where needed in order to: (1) establish and/or improve access to climate services needs in all countries; (2) enhance capacity to deal with climate-related risks; (3) improve availability and quality of climate data; (4) support better interaction between users and providers; and (5) improve climate service needs to better meet user requirements. These objectives are based on the five challenges identified through a widespread consultation both at and subsequent to the World Climate Conference-3 in 2009 (WMO, 2014). The GFCS provides a worldwide mechanism for coordinated actions to codesign, coproduce, disseminate, and use quality and quantity climate services as a potential for climate-resilience development, both for adaptation and mitigation.

For successful implementation of climate services, the GFCS implementation plan has outlined five pillars (further details in Box 17.1 and Fig. 17.1) (WMO, 2014). These represent the different stages of the production, delivery, and application of climate information and services.

---

1. http://www.wmo.int/gfcs/.

Resilience. DOI: https://doi.org/10.1016/B978-0-12-811891-7.00017-7

BOX 17.1 Pillars of the Global Framework for Climate Services

1. *UIP (User Interface Platform)*: a structured means for users, climate researchers, and climate information providers to interact at all levels; providing the functions of feedback, dialogue, monitoring and evaluation, and outreach;
2. *CSIS (Climate Services Information System)*: the mechanism through which information about climate (past, present, and future) will be routinely collected, stored, and processed to generate products and services that inform often complex decision-making across a wide range of climate-sensitive activities and enterprises;
3. *Observations and Monitoring*: to ensure that climate observations and other data necessary to meet the needs of end-users are collected, managed, and disseminated and are supported by relevant metadata;
4. *Research, Modeling, and Prediction*: to foster research towards continually improving the scientific quality of climate information, providing an evidence base for the impacts of climate change and variability, and for the cost-effectiveness of using climate information;
5. *Capacity Development*: to address the particular capacity development requirements identified in the other pillars and, more broadly, the basic requirements for enabling any Framework-related activities to occur.

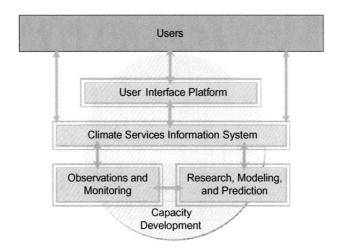

FIGURE 17.1   A schematic illustration of the five pillars of Global Framework for Climate Services and their linkages.

17.2   THE GFCS ADAPTATION PROGRAMME IN AFRICA: PHASE I

The "Global Framework for Climate Services Adaptation Programme in Africa — Building Resilience in Disaster Risk Reduction, Food Security, and Health through Climate Services" is a pilot project of the GFCS, implemented in Tanzania and Malawi from January 2014 to May 2017, with financial support from the Norwegian Government. The project aimed to increase the resilience of people most vulnerable to the impact of weather and climate-related hazards in the climate-sensitive sectors of Disaster Risk Reduction (DRR), food security, and health.

   This pilot project is a model of how agencies can work together under the GFCS umbrella, and the programme has a strong communications component to enable partners to reach different target audiences for the uptake of weather and climate information for planning and decision making both at national and local levels. A key characteristic of the Programme is the partnership approach, involving seven different international agencies and research institutes focusing on food security and agriculture, health, and DRR. They are World Food Programme (WFP), World Health Organization (WHO), and International Federation of Red Cross and Red Crescent Societies (IFRC) including Norwegian Red Cross and Red Cross/Red Crescent Climate Centre. The Tanzanian Meteorological Agency (TMA) and the Malawian Department of Climate Change and Meteorological Services (DCCMS) led the design, production, and dissemination of weather and climate information services in their respective countries.

   The research partners CCAFS (CGIAR Research Program on Climate Change, Agriculture and Food Security), CICERO (Center for International Climate and Environmental Research) in Oslo, and CMI (the Chr. Michelsen

Institute) with the local research partners University of Dar es Salaam and Lilongwe University of Agriculture & Natural Resources conducted baseline study at the start of the project in 2014 and at the end of the project in 2016. They assessed the effectiveness of the project interventions, such as WFP interventions for food security. Other research partners assessed users' awareness and satisfaction of climate services in the project districts Longido, Kondoa, and Kiteto in Tanzania, and Chikwawa, Lilongwe, Balaka, Nsanje, and Zomba in Malawi. Furthermore, the research partners with their local research partners assessed the use of Indigenous Knowledge (IK) by the local communities in weather prediction and explored the role of IK in weather and climate prediction. WMO and the GFCS Office led the overall coordination and implementation of Phase I in close consultations with the partner organizations.

To ensure the sustainable and effective provision of climate information services through TMA and DCCMS in the project districts, the scoping study for project design and planning, and the baseline assessment, methodologies, and approaches were developed in consultation with the project partners. Intermediaries, such as Red Cross volunteers, agriculture extension workers, and health professional played active roles in the uptake of weather and climate information needed by the users and this has helped to narrow the gaps between providers and users of weather and climate information. For the sustainable provision of high-quality climate services and continuation of key activities needed by the users in the project districts, the development and operationalization of National Framework for Climate Services (NFCS) in Tanzania and Malawi is underway in close consultation with high-level government bodies. Overall, Phase I tried to address how climate and weather information services best meet the decision-making needs of farmers and pastoralists and tried to overcome challenges that influence sustainable production, dissemination, uptake, and use of weather and climate information.

17.3   ENGAGEMENT FOR THE IMPLEMENTATION OF CLIMATE SERVICES

The participation of National Meteorological and Hydrological Services (NMHS), relevant government ministries/departments, local governments/communities, research partners (both international and national), implementing partners, volunteers, and extension workers (intermediaries) in the research, coproduction, dissemination, and uptake of science-informed weather and climate information in health, agriculture, and DRR sectors were prominent factors enabling adoption of climate services into these sectoral planning and decision-making both at the national and district levels.

Two mechanisms governed the design and implementation of the GFCS Adaptation Programme in Africa.

1. A PSC (Programme Steering Committee) consisted of representatives from each partner organization and the donor. PSC is chaired by WMO and was responsible for the overall management oversight of the programme and strategic decision-making.
2. PDTs (Project Delivery Teams) established in each country consisted of local representatives from each partner organization supporting the implementation of the project on the ground. The membership of the PDTs is expanded to include government officials from the Ministries of Agriculture, Health, Disaster Management Departments (DMDs), and other departments. PDTs were responsible for the planning and implementation of joint activities on the ground.

WFP led activities on agriculture and food security, WHO on health, and IFRC on DRR as an interagency collaboration. The GFCS Office led the overall coordination for the implementation of activities in close consultations with all the partners including TMA in Tanzania and DCCMS in Malawi.

WMO helped to build the capacity of TMA and DCCMS staffs for the production of science-based climate information services based on users' needs and supported the development of "Climate Services Information System" of TMA and DCCMS. The technical support provided by WMO helped TMA and DCCMS to produce weather and climate information services needed by the users in the project districts. More importantly, WMO in consultation with the project partners took a lead for the establishment of "User Interface Platform (UIP)" and operationalization of the climate services in both countries. TMA and DCCMS chaired the PDTs in their respective countries and facilitated the participation of the national and local institutions in a coordinated manner. Thus, this has helped to address local issues and challenges during the implementation of the activities on the ground in the project districts.

Operationalization of climate services and its UIP require a proper participation and high-level engagements of providers and users of climate services in all aspects of climate service production, delivery, and uptake. For the establishment of climate services in both countries, two major activities were in the plan: (1) Consultation for the establishment of UIP; and (2) Development of NFCS. UIP aims to guide and provide strategic direction for the operationalization of climate services in both countries. This involved ad-hoc and planned meetings of UIPs both at the higher level and on the ground. For example, an annual PSC meeting aimed to provide strategic direction for the implementation of the project activities. Similarly, every three months PDTs exchanged experiences and discussed challenges in the coproduction, dissemination, and use of weather and climate information services.

Implementation of Phase I activities in close consultation with PDT members helped to mainstream climate services into sectoral planning and development. PSC members from each sector consulted with their PDT members and relevant ministries to support the implementation of sector-specific activities on the ground. More importantly, engagement of government representatives from the Vice-President's and Prime Minister's Offices and relevant ministries showed the importance and priorities and supported the establishment and operationalization of climate services in both countries.

The research institutes (CICERO and CMI) consulted with the partners (both local and international partners) and the communities to establish a baseline for monitoring and evaluation of user satisfaction of climate services with the focus on the programme-targeted districts in Tanzania and Malawi. At the end of Phase I, the research institution conducted second round data collection to compare with the baseline data and to assess the interventions in Longido, Kiteto, and Kondoa in Tanzania and Nsanje, Lilongwe, Chikwawa, Zomba, and Balaka in Malawi. Qualitative focused group discussions, in-depth interviews, key informant interviews, and quantitative farmer-to-farmer communication methods were used. In addition, the research partners with their local research partners (University of Dar es Salaam and Lilongwe University of Agriculture and Natural Resources in Tanzania and Malawi, respectively) assessed the role of IK in weather and climate prediction and its uses in day-to-day lives, especially for weather prediction.

The Programme appropriately designed the communication of climate information involving extension workers, Red Cross volunteers, and health professionals who were trained to ensure the uptake and benefit of using climate information for the local communities in project districts.

17.4 OPERATIONALIZATION OF CLIMATE SERVICES

Several consultation sessions outlining the modality to operationalize climate services in both countries resulted in setting up institutions to engage researchers, users, and producers of climate services and to facilitate effective coordination, prioritization, and planning for the coproduction, dissemination, and use of climate services into sectoral planning, and contribute to the National Adaptation Plan (NAP) process. For the operationalization of climate services, the project has started consultation with the governments. Part of the operationalization process involved building national ownership of the project and continuation of the activities needed by the communities through regular financial and institutional support from the national governments.

17.4.1 Tanzania

For the operationalization of Climate Services, Tanzania has drafted its NFCS and shared with Tanzanian Disaster Relief Committee (TANDREC) for endorsement. TANDREC is anchored at the Prime Minister's Office (PMO) DMD and therefore DMD provides the Secretariat role to TANDREC. The DMD is also responsible for coordinating disaster issues in the country and as such plays a crucial role in GFCS DRR implementation. The Vice-President's Office (VPO) is a coordinating entity of climate issues including the NAP process, both TANDREC and PDT have VPO representatives involved in the implementation of Phase I.

To facilitate effective implementation of Phase I activities in DRR, and to establish a constant exchange between TENDREC and PDT, a representative from the PMO took part in the discussion and helped to coordinate DRR activities in health and food security.

17.4.2 Malawi

In Malawi, PDT drafted Terms of Reference (TOR) of the Technical Working Group for Climate Services (TWGCS) for DCCMS to review and finalize in consultation relevant government's ministries. The aim is to establish a Technical Working Group for Climate Services as a permanent structure under the National Technical Committee on Climate Change (NTCCC) responsible for the operationalization of climate services in Malawi. DCCMS shared the draft TOR with NTCCC for NTCCC to consult with the National Steering Committee for Climate Change (NSCCC) for endorsement. Malawi is considering drafting a Malawian Framework for Climate Services under the TWGCS.

17.5 THE INCEPTION OF PHASE I

During the inception phase at the national level, greater participation between all the stakeholders helped effective cross-sectoral planning for the coproduction and application of climate services including mapping crosscutting activities, dialogue, and creation of agreed "Roadmaps" to develop frameworks for climate services at the national level.

The baseline studies (Coulibaly et al., 2015b) identified the user needs of weather and climate information for agriculture and food security, health, and DRR in the project districts of Tanzania and Malawi. Understanding users' needs showed that weather and climate information should be accessible and timely, understandable, and useful for users' specific needs. Household-level baseline surveys on the use of climate services in the project districts in Malawi and Tanzania led by CCAFS (Coulibaly et al., 2015a,b) showed their reliance more on IK, personal experiences, and the traditional cropping calendar than on weather and climate information for their farm decision-making. In addition, these studies informed TMA and DCCMS on farmers' needs and access to climate information services.

Phase I made an effort and tried to take into account the different needs of men and women in the delivery of climate services. It was noticed that access to means (radio/radio programme, SMS) for receiving weather and climate information by men and women varied. With this in consideration, Phase I focused on the delivery of weather and climate information based on the means available to all stakeholders to access weather and climate information in the project districts.

The findings of these studies were very useful in the design and delivery of weather and climate information services. For example, the seasonal forecasts for the project districts were downscaled for 2015/16 season in Malawi, and for the October–December season in Tanzania as one of the requested information by the communities.

17.6 PROGRESS BASED ON GFCS FIVE PILLARS

Phase I has made progress in the coproduction, dissemination, and uptake of weather and climate information. There is a need for continued production of weather and climate information as needed by the users to reduce weather- and climate-related risks and increase the resilience of vulnerable communities in the project districts. In addition, the project also helped to build awareness and strengthened all the pillars in the countries.

Although many of the foundational capabilities and infrastructure of the GFCS pillars for climate services already exist and are being used in these countries, proper coordination of programs and institutions that have addressed individual aspects of climate services is generally weak. Therefore, the operationalization of climate services in both countries aims to bring all the players to share and discuss their interventions in climate services to overcome this challenge. All these come under an umbrella called operationalization of climate services through user–provider partnerships in countries. The GFCS Framework helps to facilitate and strengthen already existing pillars in countries to coordinate and operate climate services nationally.

17.6.1 User Interface Platform

It is a platform where users, climate researchers, and climate information providers interact in person or meet remotely. UIPs were formed on an ad hoc basis as the need arose during the implementation of the project. Functioning UIPs at the national level existed before the project in both countries. For example, Climate Service Platform in Tanzania and Climate Service Forum in Malawi are the platforms where the Met Services are providing advisories to ministries. However, these UIP were not planned and coordinated appropriately and were mainly for receiving advisories from the Met Services.

The level of interaction and participation of stakeholders depend on the scope and the purpose of activity. These UIPs are formed ad hoc or well-planned, temporarily or long-term (project period and/or beyond project lifetime). UIP is an important pillar of the Framework that can help to coordinate, establish, and guide other pillars and can lead the operationalization of climate services in countries by providing overall strategic direction for research, production, dissemination, and uptake of climate services into national planning and development. UIP helps to develop appropriate products based on user needs, identify capacity development needs, and influence the direction of observational investments and research for climate services (WMO, 2014). Appropriate functioning of UIP at different levels can help in the design, coproduction, dissemination, and uptake of weather and climate information into national adaptation and mitigation strategies and climate-related policy decisions based on high-quality climate information. It can also play a vital role in the development and contribute to the implementation of the NAPs. For example, the use of weather and climate information to address adaptation needs is identified in the NAPs.

During the implementation of Phase I, UIPs at different levels were operational for health, food security, and DRR. An interactive UIP for a health activity needing weather and climate information generally has representatives from WHO national and international offices, a representative from a ministry of health, and a representative from Met Service. The participation of other stakeholders is based on the nature of the activity. If an activity has crosscutting issues then other ministries, PDTs, and other project partners, including intermediaries and extension workers, will be involved to interact with Met Services. This means, during the implementation of the project, numerous ad hoc and pre-planned UIPs were envisioned for the consultations and implementation of project activities. Sometimes full participation of all stakeholders was limited in this project, especially the Met Services, and relevant ministries, due to their mandates and commitments to their national priorities, and partly due to limited human and financial resources.

UIPs at project level were set up during the inception of the project. These are PSC and PDTs (as mentioned earlier). PSC members consult with their PDTs to understand the weather and climate information needed in the project districts to implement health, food security, and DRR interventions. Then they shared the information received from their PDTs at the PSC meetings. PDTs formed in both countries were led by TMA and DCCMS in Tanzania and Malawi, respectively. The meetings of PDTs every three months, chaired by the directors of TMA and DCCMS in their respective countries, helped to share information and challenges among local researchers, users, and climate information providers acting on the ground. In terms of the functioning of the PDTs, the focus was more on updating the implementation of project activities, and not much on the needs and improvement of sector-specific tailored weather and climate services. Sometimes it was challenging for TMA and DCCMS to fulfill the requests for weather and climate information by the users (both the partners implementing sector-specific activities and the ministries representing in the PDTs). Moreover, fulfilling national mandates given to TMA and DCCMS by their governments and at the same time products and services required by other projects with their limited technical capacity, human and financial resources put them in difficult situations to fulfill all the requests.

UIPs at the community level involved interactions between local researchers, climate information providers, intermediaries/health professionals, agriculture extension workers, and Red Cross volunteers with the districts' communities including farmers and pastoralists. At the community level, the interactions were focused on disseminating and supporting the intake of weather and climate information, and at the same time receiving feedback from the farmers and pastoralists. In addition, the focus was also to understand users' needs and satisfaction of using weather and climate information (Daly et al., 2016).

Climate services also require the well-defined participation of civil society and private sector for the codesign, coproduction, sharing, and uptake of climate services in adaptation to climate change interventions. A proper and well-defined collaboration among public and private sectors and their engagement should come into place in the implementation of Phase II.

At the national level, a multistakeholder consultative process to develop a framework for climate services and action plans in Tanzania and Malawi is an ongoing process. The drafted Tanzanian Framework for Climate Services is to be endorsed by the TANDREC. Following the endorsement, TMA is expected to lead the process for establishing UIP in close consultation with TANDREC and the development of action plans for the implementation of climate services into sectoral planning. In Malawi, the aim is to establish a TWGCS as a high-level institution to integrate weather and climate information into sectoral planning and national policy. DCCMS has prepared a draft TOR for a TWGCS to be established as a permanent structure under the NTCCC. After the endorsement of the TOR by the NSCCC, the TOR acts as a basis for the establishment of UIP at the national level to guide in the development of Malawian framework for climate services and action plans.

Other UIPs functioning to support dialogue between users, climate researchers, and climate information providers were WFP-led Planning and Review Days, Radio Listening Hubs, TMA- and DCCMS-led coproduction of advisories/forecasts for radio programmes and feedback from users, and RC volunteer community exchanges/feedback loops.

WHO led a consultative process involving relevant members from the Ministry of Health, WHO representatives, health professionals, and climate service providers to assess climate-related health vulnerability. The Health Vulnerability and Adaptation Assessments prepared by WHO identified health risks and vulnerabilities due to extreme weather and climate variability including information gaps and the required actions to reduce risk of cholera and malaria by understanding and incorporating needed weather and climate information on the El Nino Southern Oscillation, seasonality/seasonal forecast, and long-term climate change and projections.

Phase I involved trained intermediaries, such as Red Cross volunteers, extension workers, and health professionals, interacting with climate information providers and users on the ground. They helped to disseminate weather and climate information and to bridge the gap between service providers and the practical application of weather and climate information by the local communities in the project districts.

17.6.2 Climate Services Information System

Climate Service Information System (CSIS) is "the principal mechanism through which information about climate (past, present, and future) will be routinely collected, stored, and processed to generate products and services that inform often complex decision-making across a wide range of climate-sensitive activities and enterprises" (WMO, 2014). It resembles a proper functioning of Met Services with needed infrastructure and experts to develop, generate, and distribute a wide range of climate information and services for the use of products and services at different levels. TMA and DCCMS are leading CSIS in Tanzania and Malawi.

TMA and DCCMS support the provision of weather information and climate services for their countries by using data and information from the regional and global climate centers and closely interact with national, local, and international partners. TMA and DCCMS have produced downscale seasonal forecasts, weather information for Farm SMS and radio programs, assisted in developing a malaria maproom, provided agrometeorological bulletin in local language (Swahili), supported the production of health and DRR bulletins, produced a crop weather calendar, and issued monthly weather bulletins to the district level with advisories for agriculture, health, and DRR.

Both Met Services have qualified and dedicated staff. However, the number of staff was limited in comparison to the growing need for weather and climate information. In addition, both human and financial resources are not sufficient to meet the needs of the users and to coordinate and standardize operations and products. The seasonal forecast and its uses are well understood by TMA and DCCMS but there is a need for further training, including validation on the use of the seasonal forecast.

17.6.3 Observation and Monitoring

Observation and monitoring are main foundations for the production of science-informed end-user weather and climate information services. High-quality weather and climate data from past and real-time observations are required to assess climate variability and change. TMA and DCCMS are leading weather and climate observation and data exchange as a basis for the coproduction of weather and climate information services. The collection, processing, and use of in situ (surface-based) and ex situ (satellite-based) observations have helped in the coproduction of end-user weather and climate information. The quality of climate data was enhanced through the establishment of Enhancing National Climate Services (ENACTS). ENACTS aims to improve availability, access, use of climate data, and helps to translate climate data into information and products that are tailored to the needs of the users for decision-making[2]. In Malawi, ENACTS helped to improve climate data and integrate malaria and climate data by creating a malaria maproom to produce bulletins showing the status of malaria risk. In Tanzania, the integration of weather, climate, and malaria data helped to alert the communities. TMA is exploring technical and financial supports to expand the ENACTS maproom to serve in the agriculture sector.

Both countries lack a proper monitoring coverage of weather and climate data and a proper operational database management system. The "Observation and Monitoring" pillar needs a baseline assessment of existing hydrometeorological monitoring stations, both temporal and spatial coverage, including the parameters monitored and other data needs. This will help to identify gaps in monitoring and observation and ultimately help to meet the end-users' needs. Infrastructure to manage and operate weather and climate data of TMA and DCCMS is underdeveloped due to not enough budget, challenges keeping the capable technical experts, and past data records are still available only on paper. There is an urgent need to upgrade and deploy a proper weather and climate data management system, and initiate data rescue to digitize all the past paper data in both countries.

The engagement of TMA and DCCMS is critical for the observation and monitoring of weather and climate parameters as they are the nationally mandated organizations leading the observation and monitoring of weather and climate data, and the compilation, management, and assimilation of meteorological and climate information. Therefore, there is a need for a reliable data storage system in an appropriate database and building capacities of TMA and DCCMS staffs.

2. http://iri.columbia.edu/resources/enacts/.

17.6.4 Research, Modeling, and Prediction

CICERO and CMI with the local partners University of Dar se Salaam in Tanzania and Lilongwe University of Agriculture & Natural Resources in Malawi assessed users' satisfaction of climate services at national, district, and local levels.

In addition, CICERO and CMI with their local research partners assessed the role of IK in weather and climate prediction and how the local communities have been using IK in their day-to-day lives and methodologies for the integration of IK into weather and climate prediction in Tanzania and Malawi.

According to their findings,[3] awareness of and access to climate information and services have improved among the communities in the project districts and the uptake of weather and climate information has increased. Those who received the seasonal forecast generally expressed satisfaction with and utility of the weather forecast and advisories for decision-making. A community in Longido where people mainly rely on IK where IK was more preferred and expressed their desire to receive weather information based on science. Some communities relied on both IK and scientific knowledge. The study also found that not all the IK used by the communities is useful due to the impact of climate change. Some IK may not be available and other may not be useful in a rapidly changing climate. Therefore, there is a need to assess the reliability and validity of the IK used by the communities for weather prediction before integrating IK into weather and climate prediction.

17.6.5 Capacity Development

For a proper use of climate services, strengthening capacity in all areas of the Framework is essential, educating users on the benefit of climate services, and enhancing the capacity of climate information producers and their understanding of the user needs including observations, data management systems, and research capabilities (WMO, 2014). The program has enhanced the capacity of the TMA and DCCMS with respect to the CSIS pillar. The project has enhanced the capacities of TMA and DCCMS to respond to user needs with high quality tailored weather and climate information. Trained the staffs in climate modeling, downscaling seasonal forecasts; training of TMA, DCCMS and partner-institutions' staffs in packing and disseminating tailored climate information and products, collecting feedback from users and exposure of Met Service staffs to regional centers.

The project has trained intermediaries including agricultural extension workers and Red Cross volunteers to communicate climate information, risk attenuation options to vulnerable communities, and establish direct delivery of weather and climate information to communities through SMS and radio. The project has also trained vulnerable communities to access and use weather and climate information to maximize yields, to minimize loss of crop and loss of life, to increase food security, and to improve health in regions heavily impacted by seasonal climate variations and extremes.

17.7 PROJECT OUTCOME/BENEFITS

The 3-year intervention of the Programme has raised awareness of and access to climate information and services among the users. It has strengthened the capacity of the vulnerable communities in the project district to uptake and use weather and climate information, and helped to build resilience to the impact of weather and climate-related hazards. Phase I has done the groundwork and prepared mechanisms for the establishment and operationalization of climate services in Tanzania and Malawi. For example, TANDREC and TWGCS in Malawi are envisioned to act as high-level institutions to lead the overall development of climate services. Users' satisfaction of climate services at national, district, and local levels showed that the awareness of climate services and its use have increased. Downscaled seasonal forecasts were highly appreciated by the users, who requested their continued production and dissemination even after the project. Furthermore, the methodologies for the integration of IK into weather and climate prediction were developed.

In Tanzania, a total of *4190* farmers and livestock keepers (37% women, 63% male) from Kiteto, Longido, and Kondoa districts, and 88 farmer groups from Balaka and Nsanje districts in Malawi with a total of *1992* farmers (66% women, 34% men) received weather and climate information. In total the project has reached *over 25,000 people* directly and indirectly.

The process is already underway to institutionalize the activities that are needed by the users beyond the project life span. Several UIPs were established for each of the three climate-sensitive priority sectors (health, food security, and

---

3. The report to be published: Assessment of the project interventions based on the baseline study and the second round data collection.

DRR). These included the WFP Planning and Review Days, the Radio Listening Hubs, the Red Cross volunteer community, and the climate and health core groups at the Ministries of Health. TMA and DCCMS interactive sessions to disseminate weather and climate information and receiving feedback from the communities in the project districts has helped to understand the essential role of users in developing climate services.

The implementation of the GFCS in the two countries has made progress across the five GFCS pillars. Phase I as "a proof of concept," as tested by WMO and the Met Services (TMA and DCCMS) for the sustainable provision of high-quality climate services, has led to the development and operationalization of a NFCS. In Tanzania and Malawi, a process for developing a NFCS has been initiated and expected to adopt by the national government during the implementation of Phase II.[4]

TANDREC and the TWGCS in Malawi as high-level institutions are envisioned to coordinate climate services development among the climate-sensitive line ministries and national stakeholders, with the aim of providing coordination in identifying and meeting needs for weather and climate information in sectoral planning and national policy processes.

The project has enhanced the capacity of the TMA and DCCMS. In Malawi, DCCMS staffs were trained to downscale seasonal forests at a district level. The ENACTS database and maproom installed at the DCCMS office has facilitated the use of the regional and global datasets and helped to develop a 10-year climate scenario. Other capacity building activities helped DCCMS to issue the agro-meteorological bulletins and crop weather calendar. DCCMS were able to use historical data in modeling and projections, and developed and disseminated weather information via SMS. The establishment of the Malaria Maproom enabled training of health professionals to use climate data to assess malaria risk. DCCMS forecasters were able to use area-specific short- and medium-range weather forecasts.

Similarly, in Tanzania, TMA staff trained to downscale seasonal forests were able to issue seasonal forecasts in the project districts. The ENACTS database and maproom installed have helped to use the regional and global datasets for weather prediction and the production of weather and climate information. TMA has developed a climatological zone using the datasets to help in sectoral planning. Other capacity building activities helped TMA to use historical climate data and issue weather information for radio and FarmSMS. TMA are now able to use historical data in modeling and projections, develop and issue weather information via SMS, and issue the agro-meteorological bulletin in Swahili. Similarly, the establishment of the Malaria Maproom enabled the training of health professionals to use climate data to assess malaria risk.

Strengthening of the Observation and Monitoring pillar has been through other programme initiatives in both countries. For example, the DFID WISER[5] project has helped to improve the meteorological infrastructure in Tanzania by investing in weather stations to enhance the monitoring and observation network. Furthermore, the CIRDA[6] project in Malawi is helping to install monitoring equipment to support the development of an early warning system in Malawi. The Green Climate Fund project[7] in Malawi further aims to invest heavily in the country's observation network and modernization of the national hydrometeorological facilities. These agencies have started to work with partners to improve meteorological data collection and management, downscaled climate products, including the seasonal forecast, and issued monthly weather bulletins at the district level with advisories for agriculture, health, and DRR.

The project supported the training of Met staff in downscaling seasonal forecasts in Malawi and Tanzania. Seasonal forecasts in both countries at district level were very useful, especially for early planning in agriculture. Training of intermediaries, including agricultural extension workers and Red Cross volunteers, enabled the communication of weather and climate information as well as risk attenuation options to vulnerable communities, and established direct delivery of weather and climate information to communities through SMS and FarmRadio. Tailored technical training for academics, health professionals, and analytical approaches using climate and weather information helped the dissemination and uptake of weather and climate information by the users.

---

4. Five steps to establishing a National Framework for Climate Services: (1) Baseline Assessment; (2) Initial National Consultation Workshop; (3) National Action Plan on Climate Services joint development with sectoral users; (4) High-level endorsement of the National Action Plan on Climate Services by all entities; (5) Launch of the National Climate Service Framework, and implementation of National Climate Service Action Plan priorities, including rigorous monitoring and evaluation.

5. http://www.metoffice.gov.uk/about-us/what/international/projects/wiser.

6. http://adaptation-undp.org/projects/programme-climate-information-resilient-development-africa-cirda.

7. http://www.greenclimate.fund/-/scaling-up-of-modernized-climate-information-and-early-warning-systems-in-malawi.

17.8   CONCLUSION

Providing timely and accurate climate and weather information has helped to take early measures to reduce the impact of climate-related hazards and has supported the livelihoods of farmers and pastoralists in the project districts. However, more and continued investment in climate services in developing regions is needed to help sustain the generation of new information, and to compile lessons on the effective production, delivery, and uptake of climate information to improve sectoral decision-making.

Although there is a growing realization among development practitioners, governments, private sectors, and providers and users of climate services of the benefit of weather and climate information, there is still a need to reach out to the wider community, such as civil society and private sectors.

Climate services could potentially contribute to the adaptation needs identified in the NAP process and the emission reduction through the Nationally Determined Contribution, e.g., climate information needed by renewable energy sectors, such as hydropower, wind, and tidal energy. A properly functioning national UIP can potentially contribute to NAPs and other national policies in countries. UIP at a regional and local level can help tailored weather and climate services to reach local communities and local businesses. The scope of climate services and the role of Met Services need to be recognized by locals, private sectors, governments, and donors to encourage them to invest more in climate services.

NMHS should be given a clear mandate for their leading role in "Climate Service Information System," and "Observation and Monitoring." NMHS need to understand the existing potential market and demand for climate services, and develop a proper business model. This market potential of weather and climate services is largely untapped by the NMHSs.

There is an urgent need for data rescue, monitoring, and observation of weather and climate parameters. Based on the users' needs, a reliable online data storage system in an appropriate database, as well as, operational implementation of quality control and homogenization of monitoring and observation of weather and climate would fulfill the users' needs for weather information and climate services.

Additionally, a prominent question is how IK is affected by climate change, and to what extent IK is reliable including its temporal range and legitimacy. Therefore, there is a need to assess IK in weather forecasting and climate services, especially a comprehensive documentation of IK indicators that are still valid, and how locally-specific IK can help to improve scientific forecasting in particular localities.

For a proper operationalization of climate services in countries and its sustainability, a regular budget from national government is a must, as is consistency in donor funding. There is an opportunity to enhance new and existing investments in climate services. Operationalization of NFCS is not possible without a proper resources commitment from governments and donors, and active and well-defined participation of relevant government departments, research institutes, private sectors, and Met Services.

More investment in knowledge management, sharing, and learning events on weather and climate information services are need to ensure benefits reach vulnerable regions like sub-Saharan Africa and South Asia.

REFERENCES

Coulibaly, Y.J., Kundhlande, G, Amosi, N, Tall, A., Kaur, H., Hansen, J., 2015a. What Climate Services Do Farmers and Pastoralists Need in Tanzania? Baseline study for the GFCS Adaptation Program in Africa. CCAFS Working Paper no. 110. CGIAR Research Program on Climate Change, Agriculture and Food Security (CCAFS), Copenhagen.

Coulibaly, Y.J., Kundhlande, G., Tall, A., Kaur, H., Hansen, J., 2015b. What Climate Services Do Farmers and Pastoralists Need in Malawi? Baseline Study for the GFCS Adaptation Program in Africa. CCAFS Working Paper no. 112. CGIAR Research Program on Climate Change, Agriculture and Food Security (CCAFS), Copenhagen.

Daly, M.E., West, J.J., Yanda, P.Z., 2016. Establishing a Baseline for Monitoring and Evaluating User Satisfaction with Climate Services in Tanzania. <https://brage.bibsys.no/xmlui/handle/11250/2382516>.

WMO (World Meteorological Organization), 2014. Implementation Plan of the Global Framework for Climate Services. GFCS, Geneva.

Chapter 18

Supporting Farmers Facing Drought: Lessons from a Climate Service in Jamaica

John Furlow[1], James Buizer[2], Simon J. Mason[1] and Glenroy Brown[3]

[1]*International Research Institute for Climate and Society, Columbia University, New York, NY, United States,* [2]*Institute of the Environment, University of Arizona, Tucson, AZ, United States,* [3]*Meteorological Service of Jamaica, Kingston, Jamaica*

Chapter Outline

18.1 INTRODUCTION

In 2014, Jamaica entered one of the worst droughts in past 40 years. The drought was associated with the El Niño of 2014−15. Losses to agriculture and to farmers' livelihoods were substantial, but not all farmers suffered equally. This chapter describes a seasonal drought forecast service that was developed by the Jamaican Meteorological Service (JMS) and the Rural and Agricultural Development Agency (RADA) to help farmers anticipate and prepare for drought. The service, which integrated new technical scientific information, interactive farmers' forums, and various ways of communicating the information, grew directly out of Jamaica's stakeholder-driven climate policy process. Jamaica's success offers a useful example for how high-level planning, such as National Adaptation Plan processes and Nationally Determined Contributions can trigger actions that offer tangible benefits to vulnerable actors critical to sustaining key components of a country's economy.

18.1.1 Policy Development Process

Jamaica's new drought service was not a one-off project designed to solve the problem of drought; it grew out of a broader Jamaican effort to systematically address climate risks across its economy. The drought service went online in 2014, but its roots go back to 2012, when Jamaica's then-Prime Minister Portia Simpson-Miller, asked the international development community to help her country consolidate several unfinished climate change policies into one that would cover the entire economy, support the national development strategy, and address greenhouse gas mitigation opportunities as well as needs for adapting to the impacts of a changing climate. She asked the United States Agency for International Development (USAID) to lead the effort, supporting the newly established Climate Change Division (CCD) and the Planning Institute of Jamaica (PIOJ).

To launch the effort to develop a climate policy framework, USAID and the CCD convened a workshop that would place climate risks in the context of Jamaica's social and economic aspirations and build a sense of ownership of the

Resilience. DOI: https://doi.org/10.1016/B978-0-12-811891-7.00018-9

climate challenge well beyond the CCD. The CCD and the Ministry of Finance and the Public Service jointly invited representatives from across the government, the private sector, and civil society to participate. Most international development agencies active in Jamaica were invited and many participated. More information about that workshop is available in the report, "Climate Change: Towards the Development of a Policy Framework For Jamaica" (Climate Change, 2012).

At the workshop, participants identified climate stresses that undermine priority economic sectors identified in Jamaica's national development strategy, Vision 2030. USAID and the CCD designed the workshop following USAID's Climate Resilient Development Framework, a "development-first" approach that emphasizes sought-after development outcomes rather than starting with potential changes in climate. At the workshop, working groups addressed economic sectors identified in Vision 2030. Each working group identified development outcomes, necessary inputs and conditions for achieving the outcomes, climate and nonclimate stresses and constraints that undermine the use of the inputs, possible solutions to alleviate the stresses and constraints, and actors necessary to plan and implement the solutions. The information was recorded on flip-charts that look something like a results framework.

The agriculture working group identified a lack of information about the weather and climate as a significant constraint to production. In fact, the lack of weather and climate information was commonly identified as a constraint across many of the sector working groups (Fig. 18.1).

Climate Services involve the production, translation, transfer, and use of climate knowledge and information in climate-informed decision making and climate-smart policy and planning. Climate services ensure that the best available climate science is effectively communicated with agriculture, water, health, and other sectors, to develop and evaluate adaptation strategies.

USAID (and others) saw the need for information as an opportunity to provide support, drawing on other US government and academic expertise in the area of climate services. A team from USAID and the International Research Institute for Climate and Society (IRI) at Columbia University conducted a capacity assessment of the JMS and

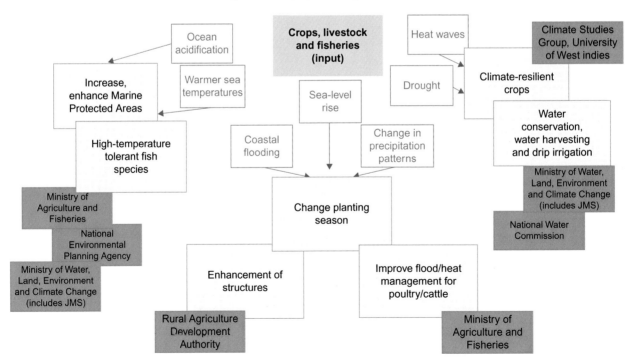

Agriculture input, stresses, solutions

FIGURE 18.1 Participants in the agriculture working group identified crops, livestock, and fisheries as key inputs. Climatic stresses are shown in red. Potential solutions are shown in pink boxes, and key actors are shown in peach. The Jamaica Met Service was part of the Ministry of Water, Land, Environment, and Climate Change.

facilitated discussions between the JMS and several of its "clients" in other government agencies, such as RADA, agricultural research, ports authority, and the water authority.

18.1.2 Developing a Climate Service

The team found that the JMS is comprised of talented, dedicated staff who are overwhelmed by demands for their skills. The director of JMS at the time said that so many requests had come in over the years that they had simply reverted to producing a monthly bulletin and trying to keep up with that and demands such as supporting the UNFCCC process, IPCC, and other international requirements. By 2013, when USAID and IRI got involved, JMS seemed reluctant to produce tailored information for any particular user group for fear that others who had also requested support would complain that they were being left out.

The discussions with "clients," users from other agencies and from outside the government, revealed a similar level of frustration. A number of clients said they no longer looked to the JMS for support because past requests had gone unanswered. In some cases, better-funded agencies were developing their own monitoring networks to do work that would be more appropriately conducted by the meteorological service.

The JMS described being stuck in a vicious cycle of constrained resources and limited service delivery. The limited resources prevented JMS from meeting all the demands for services, but failure to meet that demand prevented the rest of the government from wanting to increase JMS's budgets. JMS needed help breaking that cycle: USAID offered to provide funding to help JMS deliver one tool for one powerful stakeholder with the intention that if it did so, that powerful stakeholder would help JMS secure better budgets in the future (Fig. 18.2).

USAID proposed convening a workshop at which the JMS and its clients would discuss the potential that climate services offer to support decision-making. The workshop, co-convened by the JMS, USAID, and IRI, brought together leaders from the national water authority, the agriculture ministry, the fisheries, environment, and forestry departments, and the disaster management office, as well as the University of Technology of Jamaica. Participants discussed the decisions they make, or that they support, and the role that weather and climate play in their decision-making. They discussed times when better information would have enhanced their decision-making. The JMS explained its capacity and its limitations. Toward the end of the workshop, the organizers challenged the participants to work with the JMS to (1) identify an important sector or group of actors that are at risk from climate impacts; (2) commit to working with JMS to codevelop an information product that would support decision making in that sector. The tool had to be finished by December 2013 so that it could be demonstrated at the Third International Conference on Climate Services (2013) in Jamaica. While everyone in the room volunteered, all agreed that RADA's proposal to help farmers anticipate drought would help an important economic sector and an at-risk group of decision makers.

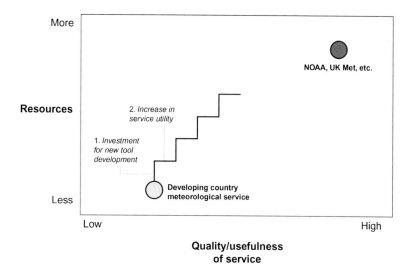

FIGURE 18.2   Underresourced weather agencies struggle to provide useful services to the range of potential users. Without more resources, they cannot improve delivery of services; but, without better service delivery, there is reluctance to provide more resources. The Jamaican government and USAID are experimenting with improving resources and services, one service at a time.

JMS and RADA committed to work together. USAID agreed to support the development of the tool; if it proved useful, the Jamaican government would fund its maintenance and USAID would try to support additional information products. The intention was that if the JMS could demonstrate its value to better-funded, more powerful ministries and agencies, those ministries would come to see JMS as a valued partner in achieving Jamaica's development objectives, particularly in the face of weather and climate variability. The challenge will be that the benefits and cost savings of better preparedness will accrue to actors other than those providing the climate service. As recognition of the value of climate services grows, it is important that serious thought is given to funding models that are both fair and sustainable.

With USAID funding, IRI hosted representatives from JMS and RADA, as well as the Caribbean Institute for Meteorology and Hydrology (CIMH) and ACDI/VOCA (an NGO implementing the Jamaica Rural Economy and Ecosystems Adapting to Climate Change project (JaREEACH) for USAID) in New York, where they jointly identified the challenges facing farmers and designed the information products to address those challenges. Each group brought its own expertise to the effort: JMS developed the data products, while RADA and JaREEACH brought their understanding of how farmers receive and make use of information. The resulting product is what came to be known as the Jamaican Drought Service. The information component of the drought service was shared in December 2013 and first went online in January 2014. The Jamaica information, and information developed by the CIMH were used at the 2014 Caribbean Regional Climate Outlook Forum, held in Kingston in May, 2014 (see workshop report at: http://www.cimh.edu.bb/pdf/CariCOF_Meeting_Report2014.pdf).

Throughout 2014, the JMS, RADA, and ACDI/VOCA's JaREEACH project (https://www.usaid.gov/documents/1862/jamaica-reech) disseminated the information from the drought forecast to farmers via farmer field schools, text messages, online bulletins, and other media. The goal was not just to make information available, but to help farmers understand that information and how to respond to it.

The information we received from the Met office gave us drought forecasts in terms of probabilities. We still decided to plant because we were fortunate to have access to the river and could fill up water drums ahead of time in anticipation of the drought.

Melonie Risden, farmer in Crooked River, Clarenon Parish.

By the end of 2014, the Jamaican government estimated economic losses of 30% in the agriculture sector, compared to 2013. In August of 2014, Minister Robert Pickersgill reported that "to date, the estimated loss to the [agriculture] sector as a result of the drought is 2,190 hectares of crops valued at over J$953.5 million. In this sector, some 18,309 farmers have been affected" (Statement of Robert Pickersgill). The drought continued in 2015. By early 2015, the JMS began hearing stories from farmers who had used the drought service and greatly reduced their losses. Farmers were storing water, switching crops, and taking other steps to prepare for and manage for the dry conditions.

The question the JMS, RADA, and USAID asked was, given these dramatic losses and the stories of farmers taking steps to manage their losses, how much credit could be given to the drought service, and how many of the actions would have occurred anyway? If losses were 30% with the service in place, might they have been greater, lesser, or the same without the service?

There are three components to the drought service: the drought forecast itself; farmer forums to educate farmers about climate and the forecast; and, communication of the information to farmers, primarily by text message. The rest of this chapter discusses the technical aspects of the drought forecast and an evaluation produced by the University of Arizona under contract by USAID (Rahman et al., 2016).

18.2 TECHNICAL DETAILS OF THE DROUGHT INFORMATION

18.2.1 The Standardized Precipitation Index

The Standardized Precipitation Index (SPI; McKee et al., 1993) is a widely-used measure of meteorological drought (Mishra and Singh, 2010) that is used in both monitoring (Hayes et al., 1999) and forecasting (Mishra and Desai, 2005; Cancellicre et al., 2007; Moreira et al., 2008). The popularity of the SPI as a measure of drought is partly a reflection of its relative mathematical simplicity compared to some alternatives, which facilitates interpretation (Guttman, 1998). Its simplicity also contributes to the SPI being a good indicator of drought onset and of intensity (Keyantash and Dracup, 2002). It can be used at a wide range of timescales (data availability permitting (Wu et al., 2005), and so is highly flexible. It is also closely related to soil moisture deficits (Sims and Raman, 2002; Hirschi et al., 2011), and therefore can be a good indicator of possible crop impacts (Quiring and Papakryiakou, 2003). In the context of

Jamaica, the SPI is therefore an appropriate measure of meteorological drought for indicating possible impacts on agriculture (Richards et al., 2013).

The Index measures deficits (or surpluses) in accumulated precipitation (hereafter, referred to as rainfall; other forms of precipitation are negligible) over a specified period (typically a few months). These deficits are corrected for effects of seasonality and location so that the index can be compared easily from place to place. The number of months over which the rainfall accumulation is measured is typically indicated; e.g., "SPI6" represents a 6-month index. The index expresses the rainfall accumulation not in millimeters nor as percentages of average, but in the form of a unit-less number that measures the severity of drought in terms of how unusual the accumulation is (Mishra and Desai, 2005).

18.2.2 Drought Forecasting Procedure

The JMS has been producing drought monitoring information using the SPI since the late 2000s (Sims and Raman, 2002). Similarly, the JMS has been issuing seasonal rainfall forecasts operationally for approximately 10 years. These rainfall forecasts are submitted monthly to the CIMH as input to a region-wide seasonal rainfall forecasting initiative that began in 2010 (Gerlak et al., 2017). In designing a drought forecasting system for the country, technical specifications were, of course, requisite, but to build upon these existing operational procedures and systems also was an important criterion. Implementing a drought forecasting system that is entirely consistent with existing rainfall forecasts would ensure that there is no contradiction between the forecasts, and might also facilitate regional upscaling. Apart from avoiding inconsistencies, working with an already-accepted drought measure would help avoid unnecessary confusion.

Consistency amongst these various products and procedures can be retained by calculating a common rainfall forecast, which can then be expressed in SPI units as well as in the standard tercile-based probabilistic format (Mason et al., 1999) that is used by CIMH and JMS for the regional and national seasonal forecasts, respectively. The JMS derive their probabilistic forecast from a deterministic forecast of accumulated seasonal rainfall in millimeters using the Climate Predictability Tool (CPT) (Mason and Tippett, 2017), while the probabilities can be calculated from the prediction error variance (Mason and Mimmack, 2002).

An option was implemented into CPT to calculate an SPI forecast that was perfectly compatible with the standard rainfall forecast. The JMS could therefore continue to monitor the 3-month SPI, e.g., and then predict the coming 3-month SPI. However, especially in situations of prolonged and severe drought an additional capability is desirable to address questions such as what will the 3-month SPI be in 1 month's time, or what will the 6-month SPI be by the end of the dry season given that we are already part way into the dry season? Answering such questions requires the ability to combine recent observations with forecasts of rainfall for the immediate next month or more. Therefore, an additional option was implemented in CPT to combine the forecast with recent observations to answer such questions.

18.3 FROM TOOL TO SERVICE: COMMUNICATING DROUGHT INFORMATION TO FARMERS

In addition to developing the drought information, that information had to be delivered to farmers in ways they could and would use it. This provision function was led by the JMS, with support from RADA and the JaREEACH project. JMS, RADA, and JaREEACH took three approaches to provision. Beginning in early 2014, the information was provided to more than 300 farmers through the mechanisms of farmers' forums—daylong training events organized by RADA and the JaREEACH project—and cell phone text messages; both the farmers' forums and text message communications continued over the course of June 2014—June 2015 (Fig. 18.3).

The first is farmer forums. RADA organized 12 farmer forums in the different parishes, 7 held in 2014 and 5 in 2015. The target audience was 50 farmers, but participation varied from parish to parish. The farmer forums educated farmers on the weather and climate terminology used by the JMS in its various information products. The forums also enabled farmers to better understand the information being produced by JMS so they could make sense of it and of subsequent updates to be delivered via text message. JMS explained the maps it used as one means of communicating the forecast. JMS also previewed the text messages and signed up farmers to receive them.

The second component of the provision was text messages delivering updates to the forecast. Seasonal texts went out either monthly or every 3 months; updates could be requested by farmers or extension agents. The information was also posted on a JMS-managed website (http://www.jamaicaclimate.net/farmers-bulletin/).

The third component was RADA's extension agents, who already had relationships with many farmers and who took part in the farmer forums. Their knowledge and relationships reinforced the other two components of provision.

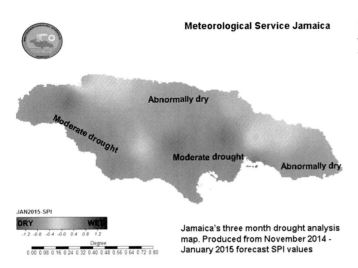

Meteorological Service Jamaica

FIGURE 18.3 Map from the Farmer Bulletin produced by the JMS showing conditions in November 2014—January 2015. Available at http://www.jamaicaclimate.net/farmers-bulletin/.

Jamaica's three month drought analysis map. Produced from November 2014 - January 2015 forecast SPI values

18.4 IMPACT EVALUATION

18.4.1 Evaluating the Impact of the Drought Information Service on Agricultural Production

The timing of the release of the drought service was fortuitous: drought began in Jamaica soon after the drought services became available. As with many tools, the ability of the drought service to reduce losses and suffering would depend on the quality of the information and the extent to which people used that information to make better decisions that yielded better outcomes. Personal stories from farming communities suggested that the losses in agricultural production might have been much greater if not for the provision of the new seasonal drought forecast information produced by the JMS. The economic value of the drought information service would derive from the notion that the recipient farmers were able to make better farming decisions compared to nonrecipient farmers, hence minimizing the adverse productivity impacts of the drought (Hirshleifer and Riley, 2002).

While the positive personal anecdotes were encouraging, USAID and the JMS wanted more solid evidence of the value of the drought service. In order to support the anecdotal evidence, in 2015, USAID commissioned the University of Arizona to conduct an evaluation of the impact of the drought information service on agricultural output by Jamaican farmers. While the primary objective of the study was to estimate the economic impact of the seasonal drought information service (information service, henceforth) received by more than 300 farmers during June 2014—June 2015, the overarching goal was to provide a comprehensive evaluation of the drought information service. First, the relative impacts of the various components of the information service were identified and estimated; second, actions taken by farmers armed with the information were identified; and third, some insights into the demand for climate forecast information services in Jamaica were offered.

18.4.2 Approach

The primary question asked in this study: was the average loss in agricultural production by the farmers exposed to the drought information service statistically smaller than the average loss in agricultural production of the farmers who were not recipients of the information?

To arrive at an answer to this question, in August 2015, the University of Arizona team surveyed 453 farmers from 10 parishes in Jamaica. Of these, 204 had participated in farmer forums and/or received drought information via text messages (the "drought information service"). An additional 249 farmers who had not participated in the drought information service (but may have received information through other mediums, e.g., radio and TV) were also surveyed. Data was collected through phone interviews, primarily because detailed information about farmers' location, addresses, and availability was not readily available.

Through the survey, information was collected about households' demographic and socioeconomic characteristics, sources of income and livelihood, challenges and constraints to decision-making and management, sources and utilization of climate information, participation in the information service, farming decisions and management practices before and after the information was received, reported agricultural output in the pre- and postdrought periods, attitude and perceptions about the information, obstacles to utilization of the information, among other data.

18.4.3   Findings

The drought had substantial impact on the agricultural production and income status of Jamaican farmers. Among surveyed farmers, lack of water, finances, and the uncertainty of "water," "rainfall," and/or "drought" (uncertainty of WRD, henceforth) are the most frequently reported challenges and constraints. Lack of access to climate information was reported as a constraint by a small number of farmers. The average reported percent loss in agricultural production (by volume) from 2014 to 2015 is 57%. (Note that the drought began in 2014, and production and incomes in 2014 were lower than in 2013. The survey asked for comparisons between 2014 and 2015.) Farmers reported that the value of agricultural production declined 31% in the same period. Among farmers who had identified uncertainty of WRD as a constraint, losses were 25% higher than the average (71% vs 57%).

Provision of climate information services, especially to farmers, is based on the assumption that it can lead to better farming decisions, by reducing climate uncertainty, raising awareness, and improving adaptation capacity. However, this will work only when there is demand for such information. Demand for climate information will exist if lack thereof is perceived by the farmer as an obstacle to better farming decisions and management, or if it can reduce climate uncertainty. Is the lack of climate information an obstacle to farmers in Jamaica? Is the uncertainty about "water," "rain," and "drought" a challenge or constraint to agricultural decisions and management in Jamaica? To answer these questions, the farmers were asked through open-ended questions to list the challenges and constraints to their agricultural decisions and management. They were allowed enough time to think about the question before listing their challenges.

The following specific findings stand out from the analysis of the data:

1. Farmers in Jamaica have attained a relatively high educational level compared to their counterparts in Africa, Asia, and Latin America, making them better equipped to receive and apply probabilistic climate information.
2. Jamaican farmers' income and livelihood sources are not very diversified, beyond agricultural-based activities. On average, agriculture accounts for over 60% of household incomes. Further, within agriculture, the on-farm activities are not very diversified. These results suggest high economic vulnerability to climate variability and change.
3. The impact of drought on agricultural production during June 2014−June 2015 is substantial. The average reported percent loss in the volume of agricultural production relative to production in the preceding year is 57%. The self-reported overall income status of Jamaican farmers was much worse in June 2015 relative to the income status in June 2014.
4. Lack of water, lack of finances, and the uncertainty of "water," "rain," and/or "drought" are amongst the most frequently reported challenges and constraints faced by the farmers. Limited access to finances is a concern for another reason; it may inhibit a farmer's ability to act upon new climate information.
5. The uncertainty of WRD has a substantial adverse effect on agricultural production. For the group of farmers faced with the uncertainty of WRD, the loss in the agricultural production was on average 25% larger relative to the mean loss of 57%.
6. However, in the WRD group, the reported loss in agricultural production declines with the increasing degree of exposure to the information service. In other words, the losses in agricultural production for the farmers faced with the constraint of "water," "rain," and "drought" would have been much greater if not for the provision of the information service. Losses among farmers identifying WRD as a concern who only attended a farmer forum were 46%; losses among farmers attending a farmer forum and receiving text messages were 39% (Rahman et al., 2016).
7. TV, radio, and the agricultural extension services provided by RADA are the most common sources of climate information for the farmers. They are also the most reliable and trustworthy sources of climate information as identified by the farmers. This suggests a relatively low level of awareness about the activities of the JMS, implying a potential obstacle to the utilization of climate information disseminated by the JMS.
8. Not all of the components of the information service were effective. The information service provided through farmer forums and phone text messages were the most effective mechanisms of information dissemination.

9. The information service contributed to the agricultural production by influencing the agricultural decisions and management of the farmers. It influenced the planting and sowing time, choice of crops, harvesting time, amount of land cultivated, mulching practices, chemical and fertilizer use, and irrigation. 79% of the farmers who participated in the climate service indicated that the information impacted their decisions. Of these farmers, the following were the actions they indicated they took: (a) planting and sowing time (71%); (b) choice of crops (61%); (c) amount of land cultivated (32%); (d) harvesting time (27%); (e) mulching practices (27%); (f) chemical and fertilizer use (25%); (g) irrigation (19%).

10. There is a strong demand for the continued provision of the climate information service. More importantly, the farmers would be willing to pay for timely, relevant, and accurate seasonal climate forecast information. 97% of respondents indicated that they would like to receive timely and useful climate forecast information in the future, and 84% of respondents indicated that they would sign up for climate information service even if it required payment for the service.

18.4.4 Limitations of the Study

In this study, a preliminary analysis of the economic impact of drought on Jamaican farmers, as well as an estimation of the impact of the drought information service was conducted.

A randomized controlled experiment is a more desirable sample design to study the impacts of an intervention (or program, or treatment), particularly in contexts where (1) comparison-treatment and control-groups can be well defined and (2) the treatment is randomly assigned over the comparison groups. In such cases, it is relatively straightforward to identify and estimate the treatment effects.

However, the drought information service considered here was not conceptualized as a randomized experiment and was not intended for rigorous impact evaluation. Therefore, the evaluators lacked baseline data and "well-defined" comparison groups. Although only approximately 300 farmers were part of the treatment group, the drought information was available on the JMS website and hundreds of bulletins were circulated to farmers across Jamaica. Therefore, it was difficult to identify a comparable group of farmers who had no access to the seasonal forecast information, whose agricultural outcomes can be compared with the outcomes of 300 treated farmers, for estimating the causal impact of the information service.

For logistical reasons, data was collected via phone interviews in August 2015. It should be noted that phone interview as a method for collecting data suffers from many limitations including low participation rate, lack of clear communication, and difficulty of earning and keeping the trust of respondents, which are desirable for obtaining accurate information, among others. Also, it limits the ability of researchers to ask as many questions as they would like to obtain detailed information.

In light of the above limitations, the impact of the information service was estimated by comparing the distribution of agricultural outcomes of subgroups of the farmers. Here subgroups of farmers, e.g., refer to farmers who identified or did not identify uncertainty of "water," "rainfall," or "drought" as challenges or constraints to their agricultural decisions and management, among others.

18.5 RECOMMENDATIONS REGARDING CLIMATE SERVICES

1. Climate information services must be demand-driven, requiring *baseline assessments* for the identification of needs, constraints, and effective mediums of information dissemination.

2. Climate information services usually do multiple things (e.g., farmer forums, information dissemination), and use more than one means to communicate information (e.g., JMS, RADA).

3. Identification and estimation of contributions of the respective components in such cases are impossible, unless impact evaluation is embedded in the program, and the program has been designed accordingly.

4. An information service aimed at providing information that reduces management production constraints (in the case of this experiment, the uncertainty of "water," "rain," and "drought" by farmers) is unlikely to be utilized, if the uncertainty is not perceived as a constraint to his/her agricultural decisions. In other words, providing information to a farmer can potentially improve his/her agricultural decisions; thereby minimize loss in agricultural production, only if he/she needs the information.

5. New funding models need to be found to support the development and maintenance of climate services. The Jamaica experience demonstrates that well designed services can reduce losses, which can result in savings for farmers and agencies that would otherwise be expected to provide relief in bad years. The cost of developing and maintaining the service falls on the met service, while the savings accrue to the ministry of agriculture, the disaster management agency, and individual farmers. Work is needed to develop effective funding models.

18.6 CONCLUSIONS

Jamaica's effort to implement a drought forecast service tailored to the needs of farmers shows that through collaboration the strengths of multiple actors can be brought together in a single service. The drought service links the needs of famers directly to the national climate policy, and demonstrates a clear role for the JMS in both helping economic actors and supporting the implementation of the national policy. The service also shows that the provision of information via trusted intermediaries is as important as the information itself. Efforts should be taken to design such services so that rigorous evaluations can be conducted. And efforts should be taken to find sustainable funding models to support useful climate services.

ACKNOWLEDGMENT

This research was supported by the United States Agency for International Development (USAID) under the Climate Change Resilient Development Task Order No. AID-OAA-TO-11-00040, under the Integrated Water and Coastal Resources Management Indefinite Quantity Contract (WATER IQC II Contract No, AID-EPP-I-00-04-0024), and related funding from the National Oceanic and Atmospheric Administration (grant NA13OAR4310184) under the International Research and Applications Project, a joint effort by the University of Arizona and Columbia University's International Research Institute for Climate and Society. One of the authors worked for USAID at the time of the Jamaica project. The views expressed herein are those of the authors and do not necessarily reflect the views of the United States

REFERENCES

Cancelliere, A., Mauro, G.D., Bonaccorso, B., Rossi, G., 2007. Drought forecasting using the standardized precipitation index. Water Resour. Manage. 21 (5), 801−819.

Climate Change: Towards the Development of a Policy Framework for Jamaica, Kingston, Jamaica, July 26−27, 2012. Available at: < http://pdf. usaid.gov/pdf_docs/PA00JM6B.pdf > .

Gerlak, A.K., Guido, Z., Vaughan, C., Rountree, V., Greene, C., Liverman, D., et al., 2017. Progress and challenges of the Caribbean Climate Outlook Forum: building a framework for process-oriented evaluation of Regional Climate Outlook Forums. Weather Climate and Society. Available at: < https://journals.ametsoc.org/doi/abs/10.1175/WCAS-D-17-0029.1 > .

Guttman, N.B., 1998. Comparing the palmer drought index and the standardized precipitation index. J. Am. Water Resour. Assoc. 34, 113−121.

Hayes, M.J., Svoboda, M.D., Wilhite, D.A., Vanyarkho, O.V., 1999. Monitoring the 1996 drought using the standardized precipitation index. Bull. Am. Meteorol. Soc. 80 (3), 429−438.

Hirschi, M., Seneviratne, S.I., Alexandrov, V., Boberg, F., Boroneant, C., Christensen, O.B., et al., 2011. Observational evidence for soil-moisture impact on hot extremes in southeastern Europe. Nat. Geosci. 4 (1), 17−21.

Hirshleifer, J., Riley, J.G., 2002. The Analytics of Uncertainty and Information. Cambridge University Press, Cambridge.

https://www.usaid.gov/documents/1862/jamaica-reeach.

Keyantash, J., Dracup, J.A., 2002. The quantification of drought: an evaluation of drought indices. Bull. Am. Meteorol. Soc. 83 (8), 1167−1180.

Mason, S.J., Mimmack, G.M., 2002. Comparison of some statistical methods of probabilistic forecasting of ENSO. J. Clim. 15, 8−29.

Mason, S.J., Tippett, M.K., 2017. Climate Predictability Tool version 15.5.10. Columbia University Academic Commons. < https://doi.org/10.7916/D8G44WJ6 > .

Mason, S.J., Goddard, L., Graham, N.E., Yulaeva, E., Sun, L., Arkin, P.A., 1999. The IRI seasonal climate prediction system and the 1997/98 El Niño event. Bull. Am. Meteorol. Soc. 80, 1853−1873.

McKee, T.B., Doesken, N.J., Kleist, J., 1993. The relationship of drought frequency and duration to time scales. Preprints, Eighth Conference on Applied Climatology. American Meteorological Society, Anaheim, CA, pp. 179−184.

Mishra, A.K., Desai, V.R., 2005. Drought forecasting using stochastic models. Stoch. Environ. Res. Risk Assess. 19 (5), 326−339.

Mishra, A.K., Singh, V.P., 2010. A review of drought concepts. J. Hydrol. 391 (1), 202−216.

Moreira, E.E., Coelho, C.A., Paulo, A.A., Pereira, L.S., Mexia, J.T., 2008. SPI-based drought category prediction using loglinear models. J. Hydrol. 354 (1), 116−130.

Quiring, S.M., Papakryiakou, T.N., 2003. An evaluation of agricultural drought indices for the Canadian prairies. Agric. Forest Meteorol. 118 (1), 49−62.

Rahman, T., Buizer, J.B., Guido, Z., 2016. The Economic Impact of Seasonal Drought Forecast Information Service in Jamaica, 2014−15. A report to USAID by the University of Arizona.

Richards, J., Madramootoo, C.A., Goyal, M.K., Trotman, A., 2013. Application of the standardized precipitation index and normalized difference vegetation index for evaluation of irrigation demands at three sites in Jamaica. J. Irrig. Drain. Eng. 139 (11), 922−932.

Sims, A.P., Raman, S., 2002. Adopting drought indices for estimating soil moisture: a North Carolina case study. Geophys. Res. Lett. 29 (8), 1183.

Statement of Robert Pickersgill, Robert D. Pickersgill M.P., Minister of Water, Land, Environment and Climate Change. Available at: < http://jis. gov.jm/joint-ministerial-statement-effects-drought-schools-agriculture/ > .

The Third International Conference on Climate Services was Held in Montego Bay in December, 2013. More information is available at: < http://www.climate-services.org/iccs/iccs-3/ > .

USAID Climate-Resilient Development: A Framework for Understanding and Addressing Climate Change. Available at: < http://pdf.usaid.gov/ pdf_docs/pbaaa245.pdf > .

Vision 2030 Jamaica: National Development Plan. Available at: < http://www.vision2030.gov.jm/National-Development-Plan > .

Wu, H., Hayes, M.J., Wilhite, D.A., Svoboda, M.D., 2005. The effect of the length of record on the standardized precipitation index calculation. Int. J. Climatol. 25 (4), 505−520.

Chapter 19

Forecast-Based Financing and Climate Change Adaptation: Uganda Makes History Using Science to Prepare for Floods

Eddie Wasswa Jjemba[1], Brian Kanaahe Mwebaze[2], Julie Arrighi[1], Erin Coughlan de Perez[1,3,4] and Meghan Bailey[1,5]

[1]Red Cross/Red Crescent Climate Centre, The Hague, The Netherlands, [2]Uganda Red Cross Society , Kampala Uganda, [3]VU University Amsterdam, Amsterdam, The Netherlands, [4]Columbia University, Palisades, NY, United States, [5]School of Geography and the Environment, University of Oxford, Oxford, United Kingdom

Chapter Outline

19.1 INTRODUCTION

Uganda is a low-income country with a GDP per capita of USD714.6 (2011−2015)[1] and high population growth rate of 3.3%. The country's economy is still tied to agricultural production employing more than 70% of the population. Since 1897 Uganda has suffered losses from several disasters some human-induced while others are related to natural hazards. Nature-related disasters include droughts, floods, landslides, earthquakes, storms, and epidemics.

Climate change projections for the region have a lot of uncertainty but the last IPCC report indicated that East Africa is likely to see an increase in heavy precipitation. Similarly, some studies suggest an increase in maximum floods in rivers in the area (Nohara et al., 2006).

In the last 30 years, Uganda has lost USD4,171,000, and USD1,600,000 to riverine floods and drought respectively.[2] Meanwhile, a news feed from UNISDR shows that more than 70 percent of natural hazards in Uganda are associated with extreme hydro meteorological events.[3] In 2007, the Teso region located in the eastern part of Uganda experienced heavy rains, leading to devastating floods affecting more than 100,000 people who were also recovering from internal conflicts. To reduce suffering, Uganda Red Cross Society applied for more than USD9,000,000 from IFRC's Disaster Relief and Emergency Fund to provide temporal shelter, basic water, and sanitation among others.[4]

Humanitarian organizations and development partners are increasingly searching for ways of reducing human suffering at affordable and sustainable costs. This in part explains the growing investment and interest in early action based on sound science (IFRC, 2009). The Forecast-based Financing (FbF) approach piloted in northeastern Uganda seeks to

---

1. http://data.worldbank.org/indicator/NY.GDP.PCAP.CD
2. http://emdat.be/country_profile/index.html
3. https://www.unisdr.org/archive/47105
4. http://www.ifrc.org/docs/appeals/07/MDRUG006rev.pdf

Resilience. DOI: https://doi.org/10.1016/B978-0-12-811891-7.00019-0

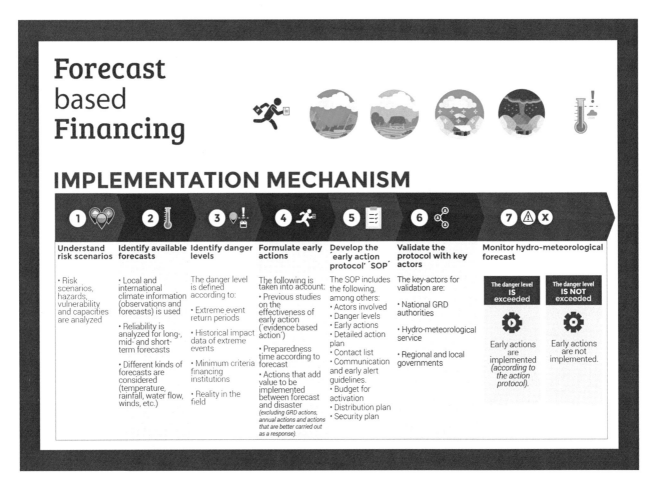

FIGURE 19.1   FbF has six major steps that can vary depending on country context. Source: *Peru Red Cross Brochure.*

contribute to this effort. The approach involves disbursement of funds for preparedness actions after release of the forecast but before a disaster occurs.

In Uganda, FbF is implemented as part of Integrated Climate Change Adaptation project, supported by the German Red Cross. It is envisaged that FbF will strengthen local early warning systems and subsequently reduce the impacts floods that often ravage the area. As the name suggests, the success of a Forecast-based Finance system largely depends on the quality of the forecasts. However, since our target area does not have a locally-calibrated hydrological model, this pilot uses the Global Flood Awareness System (GloFAS) a product of the European Commission and the European Centre for Medium-Range Weather Forecasts (ECMWF) to forecast floods (Alfieri et al., 2013). Accessing forecasts however does not immediately translate into saving lives, but it rather calls for a well thought through process to turn this scientific information into a locally useful resource that is actionable (Haile, 2005). To this end, the project team facilitated an iterative process of matching forecasts with appropriate actions. The selected actions were then compiled into Standard Operating Procedures (SOPs) detailing responsibilities, where to act, when to act, along with a specified budget allocated for each action. However, developing SOPs is just one of the six steps of the FbF process. See Fig. 19.1.

19.2   DEVELOPING THE SOPS

The FbF approach has two main components: (1) developing Standard Operating Procedures (SOPs) that link specific forecast triggers to specific early actions; and (2) committing resources necessary to implement those actions when a triggering forecast is issued. Guided by selected questions (see Box 19.1), the process to develop SOPs involved a series of consultative meetings facilitated by the Red Cross Red Crescent Climate Centre (RCCC). These consultative

BOX 19.1 Key Guiding Questions During the Consultative Process

- How can we use the available scientific information (hydro met forecasts) to reduce impacts of disasters in communities?
- When should we take action?
- Who should be involved?
- What happens if we take action and the floods do not occur?
- How do we make sure that we target the most vulnerable?
- What role can other partners play in this process?

At the end of the consultations, a workshop was held with Uganda Red Cross (URCS) staff, NGO representatives, and technical experts from Uganda Meteorology Authority to agree on the danger levels, set the triggers (the point at which action is taken based on a forecast), and deliberate on early actions for the different lead times (Coughlan de Perez et al., 2016). The result of the workshop was a draft SOP that was tested and refined through a simulation conducted in target communities. Local partners were invited to be part of the simulation as another way to enhance participation and ownership. Once finalized, the SOPs were approved by the Secretary General of Uganda Red Cross just in time for the October to December short rains of 2015.

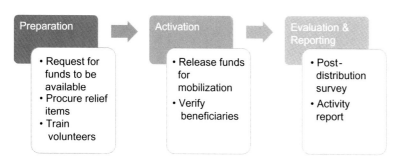

FIGURE 19.2   SOPs are organized into three phases: Preparation, Activation, and Reporting.

meetings brought together climate scientists, practitioners and opinion leaders from the local councils, communities, districts, and national level.

19.3 OVERVIEW OF SOPS

Standard Operating Procedures for FbF explain the steps that should be taken to reduce the impact of an expected hazard. In the Uganda pilot, the SOPs focused on three areas that reduce the impact of floods: (1) enhancing water purification and storage to minimize diarrheal infections; (2) encouraging food storage to minimize post-harvest losses; and (3) digging drainage channels (trenches) to reduce damage to houses and gardens. Each of the activities was completed in three steps starting with the preparation phase, followed by activation and finally reporting (see Fig. 19.2 above). The preparation phase takes place at the beginning of the year and involves procurement of relief items. Whereas the activation phase is triggered within 15 days of the expected occurrence of the hazard. However, the evaluation and reporting varies between 30−60 days after the activation depending on the activity implemented. To ensure timely implementation, the German Red Cross provided a dedicated fund for the actions identified.

19.4 THE TRIGGERS

Earlier on, community representatives and other actors agreed to take action wherever there is a 50% probability of flooding in target areas. As shown in Fig 19.3, on November 9, 2015, the GloFAS model indicated that the threshold set for digging trenches and distributing food storage bags was exceeded. For the very first time in history, a humanitarian organization took preventative action in a forecast-based financing system. See Box 19.2 for details on how the model works.

19.5 DISTRIBUTING RELIEF ITEMS

Following a quick beneficiary verification assessment, the Red Cross team distributed nonfood items to 370 households in the villages of Okoboi, Omatai, Apedu, and Akulonyo in anticipation of flooding. Each household in the target

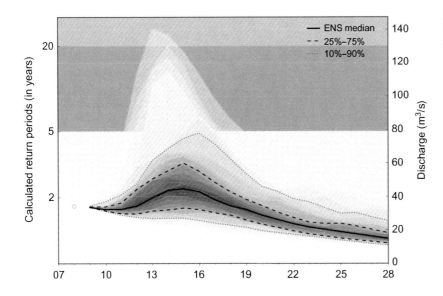

FIGURE 19.3 ERA-1 Forecast hydrograph showing heightened probability of flooding in Kapelebyong (33.9E;2.2N) from November 12–28.

BOX 19.2 How Does the GloFAS Model Work

The Global Floods Awareness System (GloFAS) is a model that simulates how water will flow once rain falls and hits the ground. This hydrological model maps the entire world. First, it divides the world into pieces of about 10 km^2, and then uses a rainfall forecast to see how much rain will fall in each piece. Based on the hills and valleys of the region, the model then shows how the water will probably flow into rivers, and how much water will end up in the big rivers. It then tells us whether this amount of water is unusually high for the rivers. When we see a signal that the water in our project areas is likely to be unusually high, a level that only happens every 2–3 years, then we trigger action. However, because the model does not have local observations included, it cannot tell for sure whether there will be a flood. When we trigger based on this forecast, there is about a 50% chance of real impacts on the ground. See Alfieri et al. (2013) for more information on the model.

villages received a 30–day supply of water treatment tablets, two bars of soap, two jerry cans for storing clean water, shovels for digging drainage trenches, and food bags for storing their harvest. In the week that followed, more people within the region were adversely affected by floods to the extent that Uganda Red Cross applied for emergency funds to support 50,000 affected households in the Teso region.

19.6 PREPARING TO ASSESS IMPACTS

To assess the impacts of the forecast-based actions, comparison communities were identified in advance. These communities were in close proximity to the targeted beneficiary communities so that they would be likely to experience similar environmental conditions. They were also similar in other characteristics, including household size and livelihoods. Full lists of households in both the targeted beneficiary communities and nearby comparison communities were compiled, giving each household a unique identification number. From these lists, a randomized representative sample of households was selected to be interviewed following a flood.

19.7 IMPACT RESULTS

During early stages of the FbF pilot, stakeholders agreed to distribute water purifying tablets to vulnerable households as a way of reducing diarrhea resulting from drinking contaminated water. The impact evaluation comparing the beneficiaries communities to comparison communities was conducted four months after triggering the forecast-based actions. The results indicated that 83% of respondents from communities benefiting from FbF used water purifying tablets compared to 6% from comparison villages. Similarly only 5% of FbF beneficiary communities drank unpurified water compared to 52% in the neighboring communities. However, the difference in diarrhea rates between communities that received water purifying tablets and those that didn't was minimal, with both communities experiencing very high rates. Seventy-six percent of households who received water purification tablets reported a household member with diarrhea, compared to 84% of households affected who received no tablets (statistically significant at 90% confidence level). This highlights the need to consider other causes of diarrhea.

On April 30, 2016, a second trigger was released indicating more than 50% chance of experiencing floods in the target areas. As in the previous trigger, the information was verified against in-country official forecasts by Uganda National Meteorology Authority officials. Consequently, URCS distributed purification tablets, two jerry cans, five storage-sacks, and two bars of soap to over 2000 people within the target area. In addition, URCS partnered with other stakeholders including the local government and WATESO—a local NGO working on health sector. Local government officials mobilized communities while WATESO conducted malaria tests, and sensitized communities about managing water borne diseases including diarrhea. However, this time the flood did not materialize, and as such the actions were considered to have been done *in vain*. The concept of *acting in vain* forms part of the FbF methodology—it is understood that forecasts are never entirely certain and so occasionally acting in anticipation of a disaster event that does not materialize is an accepted risk. Interestingly, community consultation conducted to obtain people's perceptions about *acting in vain* revealed that both beneficiaries and Red Cross staff found it worth taking action. Communities alluded to the fact that the actions taken and items distributed were still useful even when the forecast is not realized. On the other hand, the field-based team had a new-found appreciation of forecasts as probabilities that have possibilities of not being realized. The team however, emphasized the advantages of implementing "no regret actions" to keep the Red Cross' reputation in the face of a false alarm.

19.8 KEY LESSONS

Both the first and second triggers provided URCS and the wider humanitarian community an opportunity to learn from a realized forecast and a false alarm.

- It is noteworthy that in the first trigger GloFAS was able to pick up the floods in the target areas. Although the trigger came when waters had already risen quite close to the danger level, the experience indicated that global models can often be an alternative in areas with limited local climate data to facilitate early response. However, the thresholds need to be tested and revised. For example, Ngariam (located in Katakwi District) one of the other target areas experienced floods, but was not triggered by GloFAS indicating a challenge in the threshold set.
- Even with imperfect forecasts, having prior deliberation of actions and a well-defined funding mechanism improved the response by saving time and money. Just under UGX90 million (USD25,000) was used to support close to 400 households.
- FbF should be complemented with other ongoing initiatives from government and nongovernment actors. For example, in Teso the Integrated Climate Change Adaptation project (ICCA) invested in other adaptation options such as providing improved seeds to beneficiary communities. Similarly, during the 2015 short rains the government of Uganda conducted a countrywide awareness campaign on El Nino and its impacts. These two initiatives most likely enhanced flood preparedness in both target and comparison communities and hence it was hard to directly attribute the results to FbF.
- Acting in vain: Because forecasts are never perfect, there is always a possibility of taking action without realizing a flood. This creates a false alarm and consequently people can lose trust in the entity/organization coordinating FbF. To manage potential negative impacts of *acting in vain*, entities wishing to invest in FbF should communicate the inherent uncertainties of forecasts to the target communities. In addition, partnerships and networks should be explored to select low regret actions that offer benefits at an affordable cost even in the case of failed forecasts (Coughlan de Perez et al., 2014).
- Assessing impacts of forecast-based actions using comparison communities and representative samples of the population is important for institutional learning—in this case, learning which actions produce the expected impacts, which in turn informs future SOPs in the program area. Although, we did not conduct a cost−benefit analysis for the actions undertaken, we highly recommend one for other pilots.
- One of the key outstanding lesson from this pilot was on selection of beneficiary communities. This first pilot preselected villages, but a new FbF approach is being suggested that will take a larger-scale approach and intervene in the villages that are forecasted to have the greatest impact (not preselected).

19.9 CONCLUSION

Climate change is changing the risk of extreme weather events that threaten the survival of millions of vulnerable people. Uganda Red Cross' experience in piloting the Forecast-based Finance approach shows that the approach has great potential to enhance climate change adaptation by strengthening early warning systems and disaster financing in developing countries. Having an operational protocol will allow timely response to these extreme events as they come. As the experimentation with scientific models continues to improve flood forecasting, it is imperative that FbF focuses on a wider geographical scope, including at national and regional levels. Selecting appropriate and effective early actions was quite challenging in this pilot project and the final decision resonated so much with the usual Red Cross Red Crescent activities of distributing nonfood items but this could be resolved by introducing cash transfer instead. The potential of cash transfer is huge given the increased penetration of mobile money services within East Africa. Uganda's experience shows that the approach should not be constrained by local administrative boundaries but rather should focus on a larger area such as a river basin. For continuity, future FbF initiatives should be integrated into disaster risk management strategies of hosting organizations or even better, the government institutions responsible for disaster management. As next steps in this direction, Uganda Red Cross Society will continue to promote FbF understanding at subnational level while garnering national-level support. At the local level, URCS will revise the Standard Operating Procedures from this pilot in light of the lessons learnt so far. Considerable effort will also focus on engaging external partners to appreciate and collaborate on the Forecast-based Financing concept as a key approach in disaster risk management.

REFERENCES

Alfieri, L., Burek, P., Dutra, E., Krzeminski, B., Muraro, D., Thielen, J., et al., 2013. GloFAS – global ensemble streamflow forecasting and flood early warning. Hydrol. Earth Syst. Sci. 17 (3), 1161–1175. Available from: https://doi.org/10.5194/hess-17-1161-2013.

Coughlan de Perez, E., van den Hurk, B., van Aalst, M., Jongman, B., Klose, T., Suarez, P., 2014. Forecast-based financing: an approach for catalyzing humanitarian action based on extreme weather and climate forecasts. Nat. Hazards Earth Syst. Sci. Discuss. 2 (5), 3193–3218. Available from: https://doi.org/10.5194/nhessd-2-3193-2014.

Coughlan de Perez, E., van den Hurk, B., van Aalst, M.K., Amuron, I., Bamanya, D., Hauser, T., et al., 2016. Action-based flood forecasting for triggering humanitarian action. Hydrol. Earth Syst. Sci. Discuss. . Available from: https://doi.org/10.5194/hess-2016-163.

Haile, M., 2005. Weather patterns, food security and humanitarian response in sub-Saharan Africa. Philosophical Transactions of the Royal Society B: Biological Sciences 360 (1463), 2169–2182.

IFRC: World Disasters Report 2009: Focus on Early Warning, Early Action, 2009.

Nohara, Daisuke, Kitoh, Akio, Hosaka, Masahiro, Oki, Taikan, 2006. Impact of climate change on river discharge projected by multimodel ensemble. J. H 7 (5), 1076–1089.

Chapter 20

Managing Risks from Climate Change on the African Continent: The African Risk Capacity (ARC) as an Innovative Risk Financing Mechanism

Ekhosuehi Iyahen[1] and Joanna Syroka[2]

[1]Policy & Advisory Services, African Risk Capacity, Johannesburg, South Africa, [2]Research & Development, African Risk Capacity, Johannesburg, South Africa

Chapter Outline

20.1 INTRODUCTION

Although their exact timing and magnitude are uncertain, most weather events are predictable. Agricultural production in many parts of Africa is affected by natural climate variability and is likely to be significantly compromised by climate change through the higher incidence of drought, erratic rainfall, and damaging high temperatures.

As currently structured, the system for responding to natural disasters is not as timely or equitable as it should, or could be, with much of the cost borne by farmers and vulnerable households. International assistance through the traditional appeals system is secured on a largely ad hoc basis after disaster strikes, and forcing governments to reallocate funds in national budgets from essential development activities to crisis response. Only then can relief be mobilized toward the people who need it most—and it is often too late. Lives and livelihoods are lost, assets are depleted, and development gains reversed—forcing more people into chronic hunger, malnutrition, and destitution across the continent.

Given the present disaster risk financing framework, in which African governments and donors are already struggling to meet year on year humanitarian response needs let alone preparing for and working to mitigate risks in the future, African leaders have been exploring innovative and diverse ways to address the challenge of providing funding for disaster response and climate adaptation across the continent.

In 2012 the African Union Commission took a bold step to establish the African Risk Capacity (ARC) Specialized Agency of the African Union with a mandate to help Member States improve their capacities to better plan, prepare, and respond to extreme weather events and natural disasters, therefore to protect the food security of their vulnerable populations. ARC's mission is to combine modern finance mechanisms such as risk pooling and risk transfer with early warning and contingency planning to create pan-African response systems that enable African countries to meet the needs of people impacted by natural disasters in a more timely, cost-effective, objective, and transparent manner.

One of the value drivers of ARC for its Member States is the concept of risk pooling which underpins its sovereign insurance pool. Through risk pooling the risk of, say, a drought occurring across several countries can be combined into

Resilience. DOI: https://doi.org/10.1016/B978-0-12-811891-7.00020-7

a single pool to take advantage of the natural diversity of weather systems across Africa. That pool then takes on the risk profile of the group rather than the risk profile of each individual country. Since it is unlikely that droughts will occur in the same year in all parts of the continent, not every country participating in the pool will receive a payout in a given year. Further, because a continental risk pool's exposure to covariant drought risk is, in a relative sense, significantly smaller than a given country's or region's exposure to the same risk, an ARC pool can manage drought risk with fewer funds than if each country financially prepared for its own worst-case drought scenario individually. This makes risk pooling not only a powerful political statement of solidarity, but also a financially efficient contingency funding mechanism. As more risks, such as floods or cyclones, and countries across Africa are added to the pool, the efficiency benefits of risk pooling increase further.

As a development finance institution, ARC's role is to provide both the financial tools and the necessary infrastructure and capacity building services to help African countries manage their natural disaster risk and adapt to climate change through leveraging the products and services offered by ARC. Currently consisting of 33 Member States, ARC has in place an established capacity building programme around participation in its risk pool where countries (i) define their risk profile through *Africa RiskView*, a satellite weather surveillance software platform developed by ARC, to estimate the impact of drought (and soon other risks) on vulnerable populations—and the response costs required to assist them (ii) prepare a contingency plan for implementing ARC insurance payouts in times of disaster, (iii) determine risk transfer parameters to trigger insurance payouts from ARC's insurance pool when events occur. When countries have satisfactorily completed this process they are awarded a certificate of Good Standing (CGS) from the ARC Agency Board which serves as the basis upon which a government can secure insurance coverage from the ARC risk pool. This Pre-planning and capacity building work is aimed at ensuring that potential payouts are used quickly and effectively and that ARC funds reach the most vulnerable population in a timely manner. The evidence suggests that one dollar spent on early intervention through a mechanism like ARC saves nearly four and a half dollars spent after a crisis is allowed to evolve.[1]

ARC Agency launched its initial drought risk insurance product for member states through its financial affiliate, ARC Insurance Company Limited (ARC Ltd.) in 2014.

ARC Ltd. is a specialist hybrid mutual insurance company and Africa's first ever disaster insurance pool, aggregating risk by issuing insurance policies to participating governments and transferring it to the international market. ARC Ltd. uses *Africa RiskView*, as the basis of the placement of the index-based insurance contracts and associated triggering of payouts at or before harvest time if the rains are poor.

With a USD 200 million initial capital commitment provided by the governments of Germany (KfW) and the United Kingdom (Department for International Development, DFID), ARC Ltd. has issued drought insurance policies totalling over USD 400 million for a total premium cost of USD 53 million in its first four years of operation.

Eight African governments have participated in these drought risk pools—Kenya, Mauritania, Niger, Senegal, The Gambia, Burkina Faso, Mali, and Malawi—with premium funding coming directly out of national budgets. In just three years of operation, ARC Ltd. has made payouts of over USD 36 million to four countries as a result of poor rainfall seasons: the governments of Senegal, Niger, and Mauritania as a result of a poor 2014 agricultural season and the government of Malawi as a result of the poor 2015/2016 agricultural season respectively. More recently, in February 2018 the Government of Mauritania received a payment of USD 2.4 million as a result of a poor 2017 season. These payouts have been used to assist over 2.1 million food insecure people and provide over 900,000 cattle with subsidized feed in the affected countries. Work is ongoing to add flood insurance and cyclone insurance for ARC Member States.

Table 20.1 below presents more detail of ARC's implemented activities in the four countries in which payouts have been made to date. It and demonstrates that, the payouts were used to scale up existing social safety net systems with the aim of quickly respond to those affected.

The innovation in ARC is its linking of early warning, contingency planning, and modern financial mechanisms (specifically insurance), to one offering to enable governments to provide targeted responses to disasters in a more timely, cost-effective, objective, and transparent manner, thereby reducing response costs but also protecting the lives and livelihoods of those affected in order to avoid further costs down the line. ARC is based on the fundamental principle that responding earlier to a disaster, before it develops into a crisis, is financially efficient, more economical, and significantly more effective in protecting development and resilience gains.

ARC is changing the way African governments and their partners are managing weather risks across the continent. The value proposition ARC offers its Member States, however, could be threatened should drought and other insured

---

1. Boston Consulting Group (BCG), 2012, African Risk Capacity Cost Benefit Analysis.

TABLE 20.1 ARC Payouts and Implementation I

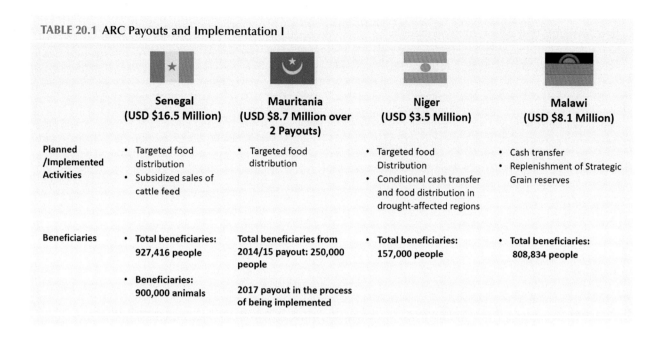

	Senegal (USD $16.5 Million)	Mauritania (USD $8.7 Million over 2 Payouts)	Niger (USD $3.5 Million)	Malawi (USD $8.1 Million)
Planned /Implemented Activities	• Targeted food distribution • Subsidized sales of cattle feed	• Targeted food distribution	• Targeted food Distribution • Conditional cash transfer and food distribution in drought-affected regions	• Cash transfer • Replenishment of Strategic Grain reserves
Beneficiaries	• Total beneficiaries: 927,416 people • Beneficiaries: 900,000 animals	Total beneficiaries from 2014/15 payout: 250,000 people 2017 payout in the process of being implemented	• Total beneficiaries: 157,000 people	• Total beneficiaries: 808,834 people

risks significantly increase from their current baselines. In order to help ensure the sustainability of the risk management systems that they are building by investing in ARC as described above, ARC must also help its countries secure the funds required to adapt to a changing climate. The Extreme Climate Facility (XCF) is ARC's response to this challenge.

20.2 INNOVATIVE FINANCING FOR CLIMATE CHANGE ADAPTATION: THE EXTREME CLIMATE FACILITY (XCF)

While African countries, through participating in ARC, are taking action to better manage today's weather, climate change, particularly an increase in extreme events and their intensity, threatens the scalability and sustainability of ARC's growth targets and the risk management infrastructure African countries are building, by potentially increasing the premiums countries will need to pay or making the payouts they receive insignificant (see Box 20.1).

Significant additional investments will be required to offset the predicted negative impacts of Africa's future climate and, at the very least, maintain the current status quo where insurance, together with other risk management and mitigation measures, can be a financially effective tool for managing weather risks (Clarke and Hill, 2013). The capital required for such climate change adaptation investment in Africa is substantial however, funds have not been forthcoming to the scale required.

Experts estimate an adaptation investment cost need of USD 14−17 billion per year over the period 2010−2050 for sub-Saharan African countries to adapt to an approximately 2°C warmer climate forecast for 2050. This is particularly threatening to the future of African agriculture, which impacts global food security and the economic livelihoods of hundreds of millions of Africans.

To support the level of international funding available, that will be made available as well as countries' own investment in resilience and adaptation, the ARC is now developing a new financial mechanism that will track extreme climate shocks and will pay out to countries, already managing their weather risk through ARC Ltd., in the case that extreme event frequency and/or intensity increases. This facility will utilize both public and private sector funds and will facilitate direct access to climate adaptation finance for African governments.

While already within its mandate, the ARC Agency was specifically requested by the African Union Conference of Ministers of Finance in March 2014 to develop a proposal for a mechanism by which African states can gain access to financing to respond to the impacts of increased climate volatility.

BOX 20.1 Analyzing Past Climate Shifts in the Sahel

In the late 1960s the Sahel experienced a pronounced drop in annual precipitation, reaching its nadir in the early 1980s following the devastating droughts in 1973, 1974, 1983, and 1984. Only now, some 40 years later, have annual precipitation levels recovered to the levels of the decades prior to the 1970s. While subject to much analysis in the climate science community, the leading cause of this drying is believed to have been aerosol emissions, otherwise known as pollution, from the northern hemisphere which lead to changes in sea surface temperatures. Notably however some scientists believe increasing greenhouse gases have contributed to the amelioration of the rainfall in recent years. As an example of a dramatic and rather rapid climate shift on the continent, analysing what would have happened had ARC Ltd. been operational before, during, and after the drying is an important exercise in understanding the sustainability of ARC's insurance value proposition in the light of potential future climate shifts.[2]

Fig. 20.1 shows the modeled drought response cost estimates for each year in Niger—a current ARC Ltd. member—from 1931, in terms of today's population and costs. ARC Ltd. insurance contracts are indexed to these modeled responses costs. The figure shows that what was once a 1-in-10-year drought event at the start of the period, becomes a 1-in-5-year drought event, or even more frequent, by the mid-1980s. A drought insurance policy that would have cost 10% in 1931 ends up costing 25% for the same policy by the late 1980s. The increase in the frequency and magnitude of extreme drought events would have impacted Niger in one of two ways had they been members of ARC Ltd. during this period: the country would have either had to pay the increased premium, which at some point would no longer be affordable on an ongoing basis, or it would have had to select less coverage, or protection that triggered less frequently, meaning less funds available for an effective and meaningful response for future disasters. The figure clearly illustrates that to make ARC sustainable—with insurance products that are effective and affordable for member states in the long-run—and to protect the growth and resilience goals ARC has for the future, ARC wants and needs to help its countries secure the funds required to adapt to a changing climate and to help insure the sustainability of current risk management systems.

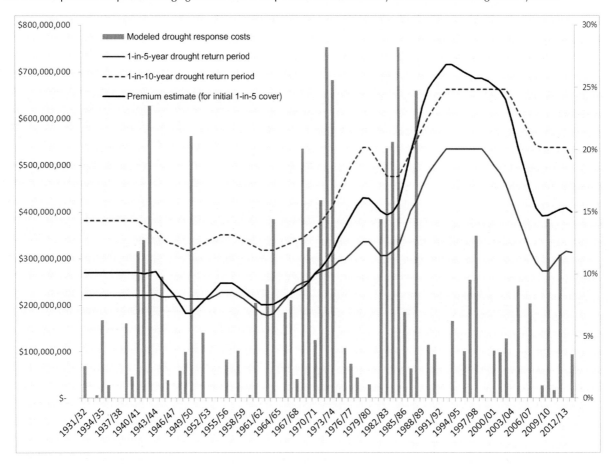

FIGURE 20.1   Insurance Estimates for Niger, 1931–2013.

2. E.g. Booth, B. B. et al., Aerosols implicated as a prime driver of twentieth-century North Atlantic climate variability, Nature, 2012, doi:10.1038/nature10946 found at http://www.nature.com/nature/journal/vaop/ncurrent/abs/nature10946.html.

In 2015 ARC formally launched the Research & Development phase for its Extreme Climate Facility (XCF). The XCF is envisioned as a data-driven, multiyear financial vehicle that will track the frequency and magnitude of extreme weather events in Africa and will provide additional financing should these statistics of extreme weather events increase.

Payments to countries would be entirely data-driven over a predetermined monitoring period such as five years. If there are no significant increases in the frequency or magnitude of extreme weather events over climatology during that period, then no payments would be made. Where payments are made, countries would use those funds to invest in climate change adaption measures specified in predefined country-level adaptation plans. Payment size would increase with extreme event number and magnitude over and above a specified threshold, corresponding to the degree of confidence that extreme events are increasing, the climate over the monitoring period compared to its baseline climatology has changed, and that intensified adaptation is needed.

In order to attract private capital in particular, XCF payments would be based on an objective, multihazard Extreme Climate Index (ECI). This index would be based on meteorological data, specified by climatic region and designed to capture the severity and frequency of heat, drought, flood, and other extreme weather events important to particular regions, such as cyclones. Similar to the insurance contracts issued by ARC Ltd. that are based on Africa RiskView, the XCF payment triggering mechanism would be index-based. In contrast to ARC Ltd. insurance products, which pay one country based on a specified weather event trigger, XCF payments would be given to all eligible countries in a region, irrespective of whether the triggering event(s) happened in a country's territory should the ECI in a given year exceed a predefined threshold, indicating an increase of severe weather across that region. The threshold would be set to identify extremes in the ECI time series, subsequent breaches of which during the monitoring period could indicate a potential shift to a new climate regime with a heightened risk of intense weather events occurring.

Payments from the XCF would start small, and increase in size with subsequent breaches of the threshold growing alongside increasing evidence, of observed deviations in the ECI from the current baseline climatology period. Payments would not be linked to the underlying losses of those events—those would be covered by ARC Ltd. insurance contracts with each participating country—but rather would be set at a meaningful level to directly support a country's climate adaptation plan pipeline.

Specifically designed to access private capital, the XCF could utilize a number of risk transfer strategies to secure risk capital from the markets. For example, XCF could be structured along the lines of a catastrophe bond programme, where the XCF's financial obligations over a series of 5-year financing windows would be securitized, issued as a bond and financed by capital provided through private bond investors. Donors would support the annual bond coupon payments to investors, thereby leveraging public capital to access larger private funds.

Should no ECI-based payment events occur during the bond's tenor, the capital provided would be returned to investors at the bond's maturity, in addition to the yield collected through the annual coupon payments; should an ECI-based payment event occur, some or all of the bond's capital—depending on the frequency and severity of the triggering event or events—would be triggered and channelled to participating countries to implement adaptation plans (see Fig. 20.2).

Should extreme events continue to occur and payments from the XCF become very likely, reflected by a decrease in premium affordability or interest from the private sector for the risk transfer program, donor capital may be required to directly support XCF payments. Therefore, in this model, the XCF would leverage private capital to fund the uncertain risk, particularly in the early years of the program, with public money reserved to fund the more certain and frequent risk should it be required in later years.

20.3   XCF RESEARCH & DEVELOPMENT

To be effective, the following three elements are central to XCF's design: (1) implementation-ready country-level adaptation plans conforming to a set standards for climate-resilient investments, or, at the very least, a credible channel very least, a credible channel through which these additional climate adaptation funds could be directed to participating countries; (2) a data-driven mechanism to track extreme weather events across the continent in an objective manner over time, with established indices, thresholds and criteria for triggering XCF payments to regions; (3) an efficient financial vehicle that could finance XCF's obligations to African governments over time.

20.4   CLIMATE ADAPTATION PLANS

For XCF to achieve its objective of enabling countries to improve their adaptation and risk management systems in light of a changing climate, and protect the effectiveness of ARC's value proposition to its Member States, there must be some mechanism in place to support countries to put the XCF funds, if triggered, to good use.

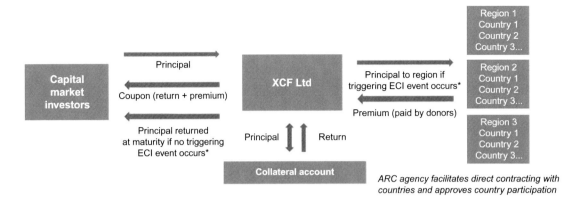

A similar principle underlies the ARC's insurance products, which require that countries submit detailed contingency plans which clearly explain how a potential payout from the facility would be managed and spent. Currently, in the event that the country is to receive an ARC insurance payout, they must submit to the ARC's Governing Board a Final Investment Plan, which, if and when approved, allows the payout to be released for implementation.

Despite their similarities, there are fundamental differences between ARC's insurance products and the XCF, which require that the standards and guidelines against which countries' XCF plans are assessed be adjusted to reflect the nature of the new mechanism. In order to start developing such standards and guidelines, in 2015 ARC partnered with Climate Knowledge Development Network (CDKN) to commission a study in order to:

- Explore whether and to what extent investment-ready adaptation and DRR plans exist in a group of ARC Member States, and enumerate adaptation and DRR "good practices" which can form the basis for the standards and guidelines;
- Determine how countries absorb climate finance through "direct access" modalities, and consider which fiduciary standards should be fulfilled to receive funds;
- Offer recommendations regarding how countries can best and most easily access XCF funds to maximize their potential for resilience and adaptation building.

Findings from this analysis were used to inform preliminary standards and guidelines for XCF to use in assessing investment plans received from prospective members.

Moreover the team designed a three-step process to adaptation planning for countries to utilize, ensuring they can successfully meet the set standards and guidelines and access funding, should it be triggered, at the end of the 5-year risk financing window, while making them eligible to participate, and without overly burdening countries with in-country adaptation activities, at the start of the risk financing window (see Fig. 20.3).

These proposed standards and three-step approach were developed around the principles of simplicity, flexibility, and complementarity with other international climate finance mechanisms.

20.5 EXTREME CLIMATE INDEX

A key building block of XCF is the Extreme Climate Index (ECI). To meet XCF's requirements for detecting changes in extreme weather event activity in Africa over time, and to create an index that could be confidently used to underpin potentially large transactions with the international risk and capital markets, the ECI needs to satisfy the following criteria.

- Be multihazard, targeting the extreme events that are likely to have the gravest impact on Africa's vulnerable populations and economic growth potential;
- Be able to capture individual extreme events and be suitable for monitoring changes in extreme climate event frequency and intensity over a 30-year or so time frame;
- Be standardized, so that it could be aggregated and compared across larger geographical regions;
- Have a well-defined parametric probability distribution function, to allow for straightforward tests of hypotheses, and other statistical diagnostics, related to changes, such as the direct computation of return times;

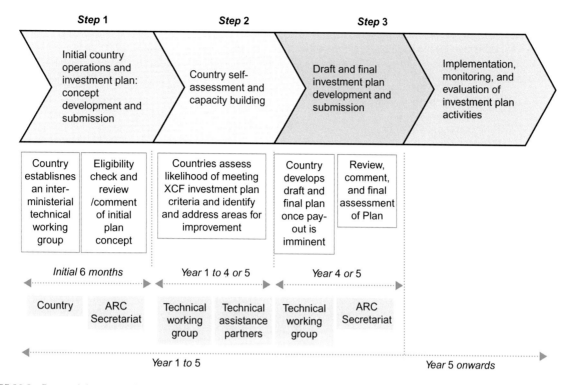

FIGURE 20.3   Proposed three-step adaptation investment plan development process.

- Be able to reflect the large-scale climate picture of a region;
- Be constructed from data satisfying risk transfer criteria, i.e., data with a consistent, sufficiently long, high-quality historical record that will also be produced objectively, reliably, and consistently in near real time going forward.

A prototype ECI was fully developed in 2016 and has been designed to satisfy these requirements (See Developing the Index Below). The index can be used to track extreme weather events, their frequency and changes thereof, and the definition of the ECI builds on a review of existing indicators already adopted for climate monitoring in Africa. The prototype ECI is defined by two components:

1. Its intensity, which indicates whether an event is extreme or not;
2. One or more angular components, which represent the contribution of each single hazard to the overall intensity of the index.

The main hazards covered by ECI are extreme dry, wet, and heat events, with the possibility of adding other region-specific risk events. The ECI is standardized across broad geographical regions, so that extreme events occurring under different climatic regimes in Africa can be compared (see Figs. 20.3 and 20.4). In order to support the triggering of payouts, a rigorous methodology has also been developed to identify statistically significant changes in the ECI index in a region over a five-year monitoring period when compared to a previous 30 year climatology baseline (see Fig. 20.5 and discussion in Developing the Index).

Research is currently underway at the Council for Scientific and Industrial Research (CSIR) and Southern African Science Service Center for Climate Change and Adaptive Land Management (SASSCAL) in southern Africa, to verify and further develop the ECI and the proposed triggering mechanism for application in African countries.

The focus will be on finding the most appropriate definitions of extremely wet and dry conditions in Africa, in terms of their impact across a multitude of subregional climates of the African continent. ECI calculations will be based on remotely sensed precipitation, with verifications performed using the SASSCAL weather station network. Changes in the ECI under climate change will subsequently be projected, using detailed regional projections generated by the CSIR and through the Coordinated Regional Downscaling Experiment (CORDEX). This work will be concluded by the development of a web-based climate service informing African Stakeholders on climate extremes and the real-time status of the ECI.

Developing the Index

While the work to define the ECI methodology and thresholds is still ongoing, Fig. 20.4 shows an example of how Africa could be broken down into climate regions for which a regional ECI would be calculated historically[3] and in real time at monthly intervals during the financing window.

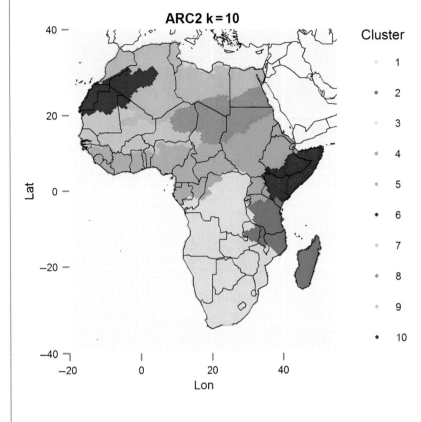

ARC2 k = 10

Cluster

1
2
3
4
5
6
7
8
9
10

FIGURE 20.4 The example above show the technical results of a clustering analysis performed on the Standardized Precipitation Index (SPI) — one of the key inputs into the ECI — across Africa, constraining the analysis to define only ten aggregation regions, k. This cluster analysis identifies areas that exhibit similar SPI characteristics and move together in terms of SPI variations over time. The SPI in this example is constructed using the ARC2 satellite-based rainfall estimate dataset[4] which covers Africa daily from 1983 with a resolution of 0.1 degrees latitude by 0.1 degrees longitude. This roughly translates to an area of 10 x 10 kilometres per pixel. The XCF R&D team is currently working to understand the impact of modifying such clusters for practical considerations (e.g. covering wholes countries in each climate region) and how sensitive clusters are to underlying datasets and which parameters (e.g. rainfall, SPI, ECI, temperature) are used to perform the clustering.

(*Continued*)

20.6 FINANCIAL & LEGAL STRUCTURE

To be viable the XCF will need an efficient financial structure that can use both public and private funds to finance its obligations to African governments. The ARC Group offers a unique opportunity to quickly set up a new facility, leveraging the experience and significant work that went into establishing the existing ARC institutional structure. This approach will allow XCF to leverage ARC's existing expertise and infrastructure and to enable it quickly take advantage of market conditions in a cost-effective and flexible manner.

Mirroring the ARC Agency and ARC Ltd. arrangement whereby ARC Agency acts as the gateway for countries to participate in the ARC Ltd. insurance pool by enforcing certain standards and best practices, it is envisaged that ARC Agency would play the same role for XCF with respect to approval of climate adaption plans, determining a country's eligibility to participate in the new facility and facilitating country participation. Minimizing XCF establishment costs will allow ARC to focus interactions with countries on the more substantive elements of their XCF engagement, namely the development of robust and meaningful adaptation plans.

---

3. Going back to 1983 given the input data available.
4. Climate Prediction Center (CPC) Africa Rainfall Climatology Version 2.0 (ARC2). Climate Prediction Center, College Park, Maryland, USA. Available Online: http://www.cpc.ncep.noaa.gov/products/fews/AFR_CLIM/arc2_201303_final.pdf (technical description).

(Continued)

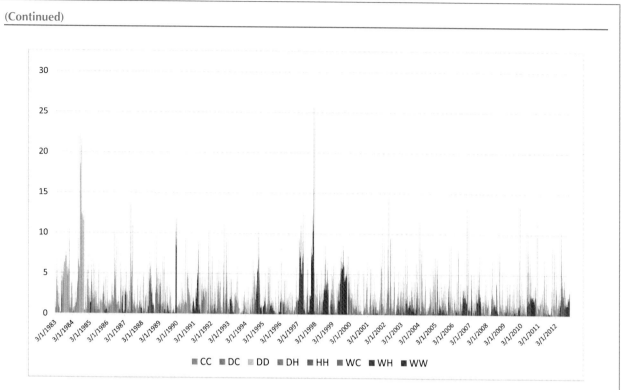

FIGURE 20.5   An example of the ECI aggregated for pixels in a Sahel cluster (this example uses a different input rainfall dataset and therefore represents a different geographic area than the light orange areas show in Figure 3 from 1983–2012. This example ECI only has two components: a three month SPI to represent short-term/seasonal dryness (D) or wetness (W) and a daily Standardised Heat Index (SHI) that has been developed by the XCF R&D team to track the number of events where daily maximum temperatures exceed 35°C for three or more consecutive days. The SHI can be used to track the presence (H, i.e. hot) and absence (C, i.e. cold) of such events. All deviations in the ECI index from any of these components are represented as positive deviations on the y-axis which represents standardised ECI units. The greater the deviation on the y-axis, the stronger and more anomalous the underlying weather event. The nature of these deviations are represented by colours as shown by the legend and well-known past events are visible in the chart such as: the severe drought of the 1983/1984 (denoted as dry (DD) and dry and hot (DH)) and the extreme wet events (WW) that results in flooding in the late 1990s.

The concept is that climate regions will be defined by contiguous areas that exhibit correlated variations in weather variables and therefore would be expected to behave similarly under large scale climate boundary forcing, such as the El Nino Southern Oscillation peak events or increased global temperatures. Work is still on-going to investigate the extent to which these climate regions should/could be modified to also take other considerations, e.g. country boundaries, into account.

Fig. 20.5 shows an initial prototype ECI index for one of these regions. In this example the index considers four different weather hazards: short-term/seasonal dryness (as measured by a three-month SPI); short-term/seasonal wetness (as measured by a three month SPI); extreme heat (as measured by the SHI); and anomalously cold temperatures (as measured by the SHI).[5] It is clear that there is significant variability in the regional ECI from month to month, with peaks in the time series denoting particularly strong underlying weather events from any hazard within the composite index. Thresholds will be set for each region and the XCF will track events that exceed the ECI threshold in any month. Multiple and severe exceedance events over the five year financing window, denoting extreme event activity that is significantly different to the previous in history (where history is defined by the previous 30 year climatology), would be aggregated and if a pre-specified payout trigger on this aggregate is breached, XCF would trigger a payout to countries in the region at the end of the five year period. Therefore at the end of the five year financing window XCF may have to pay out to countries in just one, several or—in a worst case scenario—all climate regions depending on deviations in the regional ECIs with respect to their thresholds. Should the trigger be reached before the end of the five-year period, it would be possible to make earlier payouts to countries.

(Continued)

5. The XCF R&D team is working to add longer-term (i.e. multi-year) drying and wetting components, as measured by a 12 or 24 month SPI, into the ECI.

(Continued)

As the R&D work is still on-going it is too early to estimate the potential expected loss, i.e. the expected payout size at the end of the five-year period, for XCF. However, given that numerous regional ECIs would be used to trigger potential payouts—leading to potentially higher expected losses than if a continental approach was taken—it is clear that from a financing perspective the XCF will need to have the flexibility to enter into various risk financing instruments with the market to fully finance its obligations to XCF-participating countries as cost-effectively as possible.[6]

20.7 CONCLUSION

As ARC's insurance program continues to grow, the next phase in the XCF research and development program is to complete the necessary technical work to move it into the operational phase so that it can start offering protection against future shifts in climate to countries insuring their current weather risks with ARC.

Leveraging ARC's existing infrastructure, the XCF is an innovative opportunity through which funding could be effectively channeled to support adaptation efforts on the continent in an objective manner, ensuring that African countries—and the international community—properly monitor climate shocks and are financially prepared not only to responds to weather events but to undertake greater adaptation measures should their frequency and intensity increase.

An instructive example for the global climate finance community on how to tackle risk management across multiple time scales, with ARC and XCF Africa is leading the way in innovative climate finance, diversifying the sources and increasing the amount of international funding available for both disaster response today and climate adaptation for the continent, as and when required, for tomorrow.

REFERENCES

Besson et al., 2016: Risk Financing for Climate Adaptation and Resilience: Extreme Climate Facility (XCF) (accessible online at http://cdkn.org/wp-content/uploads/2015/05/Briefing-paper-Risk-Financing-for-Climate-Adaptation-and-Resilience_07092016_final.pdf).

Clarke, D. and R. Hill, 2013: Cost-Benefit Analysis of the African Risk Capacity Facility, IFPRI Discussion Paper 01292, Markets, Trade and Institutions Division, International Food Policy Research Institute, Washington, DC, USA (accessible online at http://www.ifpri.org/sites/default/files/publications/ifpridp01292.pdf).

World Bank, 2010. Economic of Adaptation to Climate Change: Synthesis Report. World Bank, Washington, DC, USA (accessible online at http://documents.worldbank.org/curated/en/2010/01/16436675/economics-adaptation-climate-change-synthesis-report).

FURTHER READING

Bordi, I., Sutera, A., 2012. Drought Assessment in a Changing Climate, Department of Physics. Sapienza University, Rome, Italy.

---

6. In risk financing terms, such coverage is known as term aggregate cover, where losses (i.e. ECI threshold exceedance events) throughout the period ('term') aggregate to lead to a final loss estimate at the end of the term.

Chapter 21

Climate Change Adaptation in Ethiopia: Developing a Method to Assess Program Options

Karyn M. Fox[1], Suzanne Nelson[2], Timothy R. Frankenberger[2] and Mark Langworthy[2]

[1]TANGO International, Washington, DC, United States, [2]TANGO International, Tucson, AZ, United States

Chapter Outline

21.1 INTRODUCTION

Increasingly, communities are coping with unprecedented impacts of a changing climate. Weather patterns have become more erratic and climate-related events more extreme. As a result, there is a growing impetus among development organizations to enhance resilience to climate change and integrate adaptation measures in their food and livelihood security programs. This requires understanding the climate-related risks that affect the communities where they work and identifying effective, context-appropriate interventions. While there are numerous climate change adaptation (CCA) strategies, how do donors and implementing agencies determine how to invest limited resources to achieve the greatest impact (Somda et al., 2011; Bours et al., 2013)?

The United States Agency for International Development (USAID) Adaptation, Thought Leadership, and Assessments (ATLAS) project is supporting efforts to answer this question and develop more robust approaches and tools to inform decisions about how to help people manage climate-related shocks and meet desired development outcomes. The ATLAS study presented in this chapter centers on a set of analytical tools pilot tested in rural Ethiopia to assess various adaptation strategies.[1] A climate change "hotspot," Ethiopia is expected to experience extreme climate variability, with negative effects on agricultural and pastoral livelihoods (Deressa et al., 2009; Conway and Schipper, 2010; Golrokhian et al., 2016).

An initial review of decision-making tools in climate science identified a set of four complementary qualitative and quantitative assessment tools (TANGO International, 2015b). The team adapted and implemented the tools using a multilevel approach intended to engage stakeholders at the community and regional levels, and facilitate a process of

---

1. This research was made possible through the support of USAID and the ATLAS project, but the contents of this chapter are the sole responsibility of the authors and do not necessarily reflect the views of USAID or the United States Government.

Resilience. DOI: https://doi.org/10.1016/B978-0-12-811891-7.00021-9

validation across these levels of inquiry. Researchers piloted the approach in late 2015 in chronically poor and vulnerable communities in Ethiopia where two USAID Feed the Future projects, GRAD and PRIME, are working to improve food security and livelihoods, and strengthen resilience to climate change (Box 21.1). For each project, the team conducted research in two different regions (four regions total) to capture a range of agroecological and livelihood diversity. These regions were SNNPR (Sidama Zone) and Tigray (Debubawi Zone) for GRAD and Oromia (Borena Zone) and Somali (Jijiga Zone) for PRIME.

BOX 21.1 Assessing Climate Change Adaptation Options in Ethiopia: The GRAD and PRIME Projects

The Graduation with Resilience to Achieve Sustainable Development (GRAD) project seeks to enhance livelihood options of chronically food insecure households in Ethiopia's rural highlands, improve community and household resilience to climate change, and strengthen the enabling environment to promote scale up and sustainability.

Pastoralist Areas Resilience Improvement Through Market Expansion (PRIME) focuses on strengthening resilience to climate change of pastoralist and agropastoralist households in Ethiopia's arid and semiarid lands and to increase household incomes through market linkages, particularly livestock value chains.

The analysis generated two sets of distinct but interrelated findings, presented in this chapter. The first focused on the effectiveness of adaptation options implemented in the GRAD and PRIME projects. While the findings may be of interest primarily for implementers and stakeholders in the Ethiopian context, they highlight characteristics of effective CCA strategies that may be generalized to other food and livelihood security programs. Second, the study reports on the appropriateness and effectiveness of the analytical tools themselves. Here, we offer recommendations to maximize the tools' usefulness for determining which adaptation options are most effective in helping households and communities manage climate variability. The field testing of the tools and subsequent analysis informed practical guidance for applying and testing the analytical tools in other contexts.

This chapter begins with a brief discussion of key concepts related to resilience and CCA. We then describe the study methodology, followed by a discussion of the findings at two levels, both for the CCA strategies implemented in the Ethiopia projects and the effectiveness of the assessment tools. The final section offers concluding remarks and recommendations for implementing and assessing CCA strategies.

21.2 CONCEPTS FOR CCA OPTIONS ANALYSIS

21.2.1 Climate-Related Shocks, Climate Change Adaptation, and Resilience

Designing programs to build resilience in the face of climate change requires context-specific knowledge of the types of climate-related shocks people face and how they experience these shocks. In Ethiopia, as elsewhere, climate-related hazard projections include higher temperatures, increased incidence of drought and other extreme events, and more erratic and unpredictable rainfall (ATLAS, 2016). For resource-poor farmers who lack access to robust weather information, e.g., the effects of these hazards can have devastating impacts on their production and the food and livelihood security of their families, their aspirations and willingness to make livelihood investments, and, in the worst situations, result in relocation or loss of life, both human and valued livestock (Webb et al., 1992; Deressa et al., 2009; Kosec and Mo, 2017).

In identifying climate-related shocks and the impact of these events, it is important to consider the characteristics of the shock, in terms of frequency, duration, and severity, as well as periodicity and seasonal aspects. Droughts, e.g., may be an annual occurrence. Severe droughts may occur two to three times in a 10-year period, while severe and persistent drought (e.g., multiple seasons or years) may be less common, but have far more extreme and compounded impacts on households and communities.

With increasing attention to climate change and its potential impacts, particularly on agriculture, CCA has been widely promoted as a key programming strategy by donors, development actors, and governments. The growing interest in resilience, particularly in relation to climate-related shocks and stresses, has blurred the lines about where CCA ends and resilience building begins. This is perhaps not surprising, as a key component of CCA involves building resilience, particularly to climate-related shocks and stresses, and building resilience involves consideration of climate change and climate variability.

In a development context, resilience is the capacity that ensures adverse stressors and shocks do not have long-lasting adverse development consequences (Constas et al., 2014). Building resilience involves strengthening three capacities—an *absorptive* capacity that involves using risk management strategies that moderate, or help people cope with, the impacts of shocks and stresses; an *adaptive* capacity reflected in the ability to make forward-looking decisions and changes in behavior based on past experience and knowledge of future conditions; and a *transformative* capacity that promotes enabling environments, which in turn support absorptive and adaptive capacity (e.g., good governance, infrastructure, formal and informal social protection mechanisms, basic service delivery, policies/regulations) (Béné et al., 2015). These three capacities are not mutually exclusive and exist at multiple levels (e.g., individual, household, community, national).

In broad terms, CCA is the capacity to make decisions and take actions to minimize the negative effects of climate change and to take advantage of potential positive opportunities it may present. The International Panel on Climate Change (IPCC) defines adaptive capacity in relation to climate change impacts as the "ability of a system to adjust to climate change (including climate variability and extremes), to moderate potential damages, to take advantage of opportunities, or to cope with the consequences" (IPCC, 2007). Likewise, the European Community defines CCA as "anticipating the adverse effects of climate change and taking appropriate action to prevent or minimize the damage they can cause, or taking advantage of opportunities that may arise" (EC, 2017).

Although there is diversity in how the concepts are articulated (e.g., "adaptive capacity in relation to climate change" vs "climate change adaptation"), there is tremendous overlap between the ability to deal with climate change (including climate variability) and resilience capacity. Both concepts involve the ability to cope with climate change and its impacts. Thus, CCA, resilience capacity, and strategies for coping with climate change are mutually reinforcing concepts and do not necessarily represent distinct CCA, resilience, or development interventions per se (Table 21.1). Rather, it has been argued that effective adaptation options derive from existing good development practice (Vermeulen et al., 2013). CARE's Climate Vulnerability and Capacity Analysis study in Ethiopia suggests that climate change "cannot be tackled in isolation from livelihoods and food security" and that CCA is tightly linked to sustainable management of natural resources (Dazé, 2014:19), an important element of building resilience.

Where differences may be more apparent is in how CCA and resilience are measured. CCA is primarily measured as uptake of improved technologies and practices, and access and use of information in decision-making related to household production or income. Resilience is measured in terms of how people respond to a specific shock or stress and how that response did or did not translate into improved well-being (e.g., food security, poverty, health and nutrition status) (Béné et al., 2015).

TABLE 21.1 Illustrative Program Interventions to Promote CCA and Resilience Capacities

CCA	Resilience Capacities		
	Absorptive	**Adaptive**	**Transformative**
• Improve supplementary/small-scale irrigation • Support adoption of early-maturing/drought-tolerant crops • Enhance livelihoods diversification • Strengthen natural resource management • Promote water conservation practices (rain-water harvesting, ponds, dams) • Promote soil conservation practices (terracing, bunds) • Promote participatory scenario planning	• Promote DRR/DRM • Increase cash savings • Strengthen social capital (bonding) • Support access to informal safety nets • Increase asset ownership • Promote hazard insurance • Develop early warning systems • Promote water conservation practices • Promote soil conservation practices	• Improve access to information • Strengthen human capital • Strengthen social capital (bridging & linking) • Enhance livelihoods diversification • Increase asset ownership • Improve access to financial services • Support use of drought-tolerant crops/livestock breeds • Improve access to irrigation	• Support formal safety nets • Increase access to: Infrastructure Basic services Ag extension services Natural resources Markets • Strengthen empowerment of women, children, elderly • Promote transparent/equitable governance • Strengthen social capital (bridging & linking)

21.2.2 What Are CCA Options?

The range of CCA options can vary widely across contexts. Examples include use of scarce water resources more efficiently, building flood defenses, using drought-tolerant crops, choosing tree species and forestry practices that reduce vulnerability to storms and fires, improving income-generating ability through value-chain engagement, and setting aside land corridors to help species migrate (EC, 2017). These strategies can contribute to strengthened resilience capacities (absorptive, adaptive, transformative) and in so doing improve food and livelihood security outcomes, illustrated in Table 21.1.

Most CCA interventions address seasonal or annual variations around a long-term mean (e.g., variability in temperature and/or rainfall, including extreme events). Using drought-tolerant crops or livestock breeds, water conservation strategies, and even flood mitigation structures all address climate variability more than they address the potential impact of a change in the long-term mean value of temperature or rainfall, given the uncertainty in predicting just what that change will be. Thus, they reflect the possibility that there might be more flooding than drought or dry spells, or vice versa, in the current year or next, but not 30 years in the future.

Access to climate information provides another example; smallholder farmers need access to weather predictions (seasonal or annual predictions of temperature and/or rainfall), not necessarily future climate predictions. The predicted average temperature in the year 2050 as determined through one or various global models is of limited use to the annual decision-making processes of smallholder farmers or pastoralists who are more concerned with making informed decisions about what or when to plant—or where to graze livestock—this year or next.

21.3 METHODOLOGY

21.3.1 Selection of Adaptation Options for Analysis

To assess the effectiveness of household and community-level strategies to adapt to climate variability, the study team first identified adaptation options promoted by the GRAD and PRIME projects. Options shown to have some traction in the project communities based on rates of adoption reported in project documents were selected for further analysis (see TANGO International, 2015a). In selecting adaption interventions, the team also prioritized those that were central to the project's causal pathway or approach (e.g., alternative livelihoods, savings, Participatory Scenario Planning processes), represented a mix of household and community-level activities, or had a gender focus (e.g., fuel-efficient stoves) (Table 21.2).

21.3.2 Selection and Application of the Research Tools

In recent years there has been a proliferation of approaches and methodologies in CCA assessment, monitoring and evaluation, to promote learning and facilitate informed decision-making for CCA investments at project, subnational,

TABLE 21.2 GRAD and PRIME Climate Change Adaptation Options in Ethiopia

GRAD CCA Options	PRIME CCA Options
1. Planting early-maturing, drought-tolerant, or short season crops	1. Gully treatment
2. Moisture-conserving practices	2. Water point rehabilitation/upgrade
3. Water harvesting	3. Fodder production-hay making
4. Fuel-efficient stoves	4. Herd diversification
5. Rope and washer pumps	5. Postharvest storage bags
6. Information for decision-making	6. Information for decision-making
7. Alternative and diversified livelihoods	7. Alternative and diversified livelihoods
8. Reforestation	8. Management of dry and wet season grazing
9. Community-based upland management	9. Participatory Scenario Planning
10. Savings	10. Savings
11. *Livestock diversification*	
12. *Weather-indexed crop insurance*	

Note: Italicized options added during fieldwork.

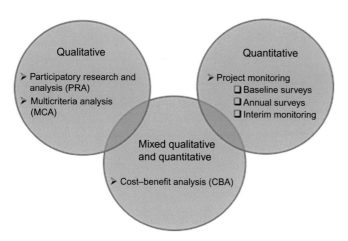

FIGURE 21.1   CCA options assessment tools.

and national levels (Deressa et al., 2009; Beaulieu, 2010; World Bank, 2010; Somda et al., 2011; UNFCC, 2011; Ayers et al., 2012; Grafakos and Olivotto, 2012; IISD, 2012; Bours et al., 2013; GIZ, 2013; Dinshaw et al., 2014). In a synthesis and review of several frameworks and approaches to monitoring and evaluation for CCA, Bours et al. (2013:59) found that the evidence base informing CCA is "fragmentary and nascent," lacks practical applicability to reduce vulnerability to climate risks, and in some cases is overly simplistic. Following on from Brooks et al. (2011), the authors call for more nuanced approaches in CCA assessment, monitoring and evaluation that encompass an array of measures and capture local specificities across different temporal and spatial scales.

In the next phase of the ATLAS study, the research team reviewed and synthesized this literature to identify methodologies that could address this gap in CCA measurement approaches. The team selected four complementary tools: (1) participatory research and analysis (PRA); (2) multicriteria analysis (MCA); (3) cost−benefit analysis (CBA); and (4) quantitative analysis of survey data (Fig. 21.1; TANGO International, 2015b). The tools use both qualitative and quantitative methods and, using a modified-Delphi technique (described below). They could be combined to collect information at multiple levels of inquiry, with a focus on community and regional levels (Table 21.3).

The qualitative tools of PRA and MCA (OECD, 2015; Watkiss et al., 2014) were piloted through fieldwork conducted at GRAD and PRIME project sites using focus group discussion (FGD) formats and key informant interviews (KIIs). In the PRA activities, participants discussed the primary climate shocks and stresses they experience, identified adaptation strategies in rank order, and explored the conditions, benefits, and challenges associated with these strategies. For the MCA component, participants agreed on a set of decision-making criteria they use to determine the effectiveness and/or appropriateness of a CCA intervention. FGD participants then weighted each criterion (e.g., "lifesaving" tends to carry more weight than "low risk") and scored it against each option to establish a list of priority interventions. The quantitative project monitoring tool focused on analyzing data from project monitoring and evaluation activities, including baseline and annual surveys and interim project monitoring activities. Using mixed qualitative and quantitative methods, the cost−benefit tool was used to collect and analyze data on the project and household costs associated with various options in relation to estimated economic benefits.

21.3.3   Research Process: Validation and Consensus Building

The field research design incorporated a modified-Delphi technique. The Delphi technique is a methodological approach designed to build consensus across a range of stakeholders (Helmer, 1967; Hsu and Sanford, 2007). The process allows participants to reassess their opinions and consider feedback from other participants. In the Ethiopia pilot study, the technique was adapted and implemented at three levels of knowledge or "expertise" in each of the four study regions including: (Level 1) female and male FGDs and KIIs at the kebele (village) level; (Level 2) a mixed-gender community validation team (CVT) acting as "community representatives" or "experts" at the kebele level to refine and validate Level 1 findings; and (Level 3) regional (or zonal) validation team (RVT) of technical or programming representatives from NGOs, government, technical experts, and other relevant stakeholders who offered new insights from "higher" geographic and institutional levels. The latter findings were compared with kebele level findings during the analysis phase. Overall, the process served to condense and verify findings and proved to be an effective method for reaching agreement with focus group participants around key CCA options.

TABLE 21.3 Strengths and Limitations of CCA Options Assessment Tools

Tool	Strengths	Limitations
Participatory research and analysis	• Guided exploration provides key contextual insights and change over time (e.g., previous 10 years) • Participants define CCA options and respective costs/benefits in their own terms • Enables study of differential impacts of CC and uptake of CCA options among social groups (e.g., men, women, elderly, children) • Use of modified-Delphi technique facilitates consensus-building and/or efficient data collection from multiple stakeholder groups across multiple levels	• Time constraints, FGDs and KIIs: Implementing multiple qualitative tools (FGD, CBA, MCA) into participatory FGD format requires trade-offs in coverage • Researcher choices about what information to emphasize during data collection may introduce bias into the study (e.g., some CCA options or criteria may be "dropped" from the ranking and scoring exercises) • At the RVT level, potential for sectoral bias based on participants' expertise
Multicriteria analysis	• Facilitates focused debate and negotiation around range of options • Process of ranking and scoring promotes prioritization and complements more open-ended discussions • Discussions help to identify key decision-making criteria	• Time constraints limit the range of options that can be ranked and scored in the FG setting; risks becoming a "mechanical" scoring activity • Criteria can be difficult to elicit and define; linking to CBA can help determine relevant criteria • Ranking and scoring (i.e., making "hard" choices) CCA options proved difficult for severe or persistent climate shocks • Ranking of individual options does not adequately capture the perceived benefits of clustering and timing/sequencing activities
Cost–benefit analysis	• Provides logically rigorous framework for assessing effectiveness and impacts of CCA options and their externalities • Provides insight into locally-perceived trade-offs associated with each option, and constraints or opportunities for uptake • Qualitative discussions highlight sequencing and clustering interventions to optimize uptake and effectiveness	• Limited data on the monetary costs and benefits of the CCA options, at the project and local levels as well as for externalities • Time constraints limit exploration of qualitative costs and benefits in FGD/KI formats
Quantitative analysis	• Quantitative data from appropriately designed surveys provide statistically representative and generalizable results that can be useful both for regular project management and decision-making, as well as for assessments of the effectiveness of alternative CCA options and strategies	• Project annual monitoring surveys do not collect important descriptive information about household characteristics and outcomes, thereby preventing analysis of associations between uptake and effectiveness of CCA options and household characteristics

21.3.3.1   Key Study Questions

Using the GRAD and PRIME projects, the study explored two sets of key questions. The first set of questions focused on the *appropriateness and effectiveness of the CCA interventions*:

• Has the chosen intervention delivered the intended change for the investment?
• Do certain interventions appear effective across the range of assessment tools?
• Would it be possible to generate more benefits using alternative interventions?
• In the future, should we improve on existing interventions or apply new ones to maximize benefit?

The second set of key questions centered on the *effectiveness of the tools* themselves:

• Do the tools effectively and efficiently generate meaningful findings to guide decision-making for investment in CCA?
• Are the chosen tools complementary? Collectively, do they enhance breadth and depth of understanding around CCA?

- How are the tools best adapted to a particular research context?
- What skill sets and information/inputs are required to implement the tools?

21.3.3.2 Analysis

Following from the study questions, the analysis was carried out at two levels. With regard to the effectiveness of the CCA options, the analysis used both qualitative and quantitative data to, first, compare and contrast the prioritization of CCA options by FGDs, KIIs, the CVT, and RVT *within each agroecological zone* (two per program for a total of four), how they differ (or not), and why. Second, the analysis compared the rankings of the two RVTs in different agroecological zones *within the same program area*. Finally, the rankings of the RVTs *across the programs* were compared (two RVTs in GRAD and two RVTs in PRIME).

Review of preliminary findings generated during fieldwork was ongoing and iterative, to facilitate the process of refining and validating the data. Full analysis of qualitative and quantitative data was conducted post fieldwork. The analysis of the tools themselves considered both the implementation process and the quality of the data generated by the tools. For the qualitative component, each of the tools was piloted in different contexts and adapted in real time based on feedback from study participants and the research team.

21.4 FINDINGS

21.4.1 Findings on GRAD and PRIME Adaptation Options

Here, we present a summary of the findings on priority CCA options generated by piloting each of the four tools, in relation to drought (Box 21.2). Results should be considered in light of limitations to the analysis, particularly regarding gaps in available data for the CBA and quantitative tools.

BOX 21.2 Climate Shocks and Stresses in the Study Area: A Focus on Drought

The findings focus on drought and the effectiveness of the various options for dealing with shorter-term climate shocks and stresses resulting from climate variability, rather than those resulting from longer-term climate change. Across the research sites, drought was considered the most common and most severe climate shock, occurring two to three times over the previous decade, with severe droughts occurring once or twice—including the year of the study (2015). Successive or progressively severe droughts are particularly difficult for communities to deal with; their ability to cope is often much lower the second year of a drought than the first year. Participants also highlighted the challenges of erratic or untimely rainfall. Across the study sites, insufficient rainfall and high temperatures trigger pest outbreaks, plant and animal disease, fires and, in severe years, crop failure and livestock death. In multiple FGDs, participants commented that weather has become increasingly unpredictable and severe in the last 3−5 years.

21.4.1.1 Participatory Research and Analysis

Participatory research conducted with focus groups and key informants showed differences between programs and agroecological regions. While this finding is not surprising, it highlights the importance of tailoring interventions to specific contexts. With that caveat, the findings do indicate some clear priorities within projects and/or regions.

GRAD: Findings from GRAD study sites show some common priorities as well as differences between the regional and local levels, particularly with regard to the scale of interventions. At the regional level, participants gave high priority to clustering interventions implemented at the community or watershed scale. In Tigray, participants agreed that these options should be combined as a single intervention of "integrated watershed management." In SNNPR, participants categorized interventions at community and household levels and ranked these two "parallel lines," emphasizing the importance of addressing CCA at both levels simultaneously. Across both regions, the priority interventions were water harvesting, soil and moisture conservation, and community-based and integrated watershed management.

At the local level, there was a great deal of variation in ranks across groups, particularly in SNNPR. Overall, however, a set of priorities across both regions emerged, namely savings, water harvesting/soil and moisture conservation, livelihood diversification, and improved seeds. In both GRAD regions, at the local level in particular, participants highlighted the value of collective formation through Village Economic and Social Associations (VESAs) for gaining knowledge and skills related to CCA activities, supporting uptake of options, exchanging information related to climate change, and expanding beneficial social networks—all of which enhance household and community resilience in the face of climatic variability.

PRIME: In PRIME study sites, there was fairly strong agreement among all groups (and across regions) regarding the top ranked CCA options for dealing with drought. In these arid and semiarid pastoralist regions of Ethiopia, water resources are critical for dealing with drought. Access to information, particularly predictions of upcoming weather and livestock market prices, is also considered very important for helping households and communities deal with drought. Households reported that information allows them to plan ahead, to shift what and when to cultivate, sell and/or purchase livestock, migrate or evacuate in response to weather predictions.

For PRIME, the top three ranked CCA options were water point rehabilitation/upgrade, information for decision-making (including, as perceived by study participants, Participatory Scenario Planning), and fodder-hay production. When comparing Borena and Jijiga, the most notable difference is the ranking of dry/wet season grazing. Privatization of land in the kebeles studied in Jijiga limits—if not completely eliminates—dry/wet season grazing as a viable option for dealing with drought in that area, thus its low rank in this study site. In contrast, dry/wet season grazing is the primary livelihood strategy (i.e., mobile pastoralism) used by pastoralists in the Borena region and is highly ranked. In both study zones, participants described their social networks as more important than savings for helping them deal with drought. They suggested that cultural barriers have so far limited uptake of savings as a key CCA strategy. As stated in one FGD and affirmed across the PRIME study area, "Somali's do their saving through the pockets of their relatives."

21.4.1.2 Multicriteria Analysis

In the MCA exercise, participants identified specific selection criteria they use to decide what CCA option to apply. CVT and/or RVT participants then scored program-specific CCA options against each of the selected criteria (which, as noted above, were assigned weights). Thus, the MCA exercise generated a rank-order list of CCA options, prioritized according to these program-specific criteria. The selection criteria and priority options are presented here, by program.

GRAD: The results on decision-making criteria for GRAD show that there is greatest agreement across groups with respect to options that are considered: low risk, provide short-term or timely benefits, or are multipurpose, meaning that a single intervention is well-integrated with other complementary activities or that it addresses multiple constraints simultaneously. At the regional level, the most important criteria for CCA option selection reflected a regional and operational perspective and included availability of local knowledge and materials, technical feasibility, potential for scale up or "generalizability," and low initial investment cost. At the community-level, participants shared these regional-level criteria, along with those that reflect household and community-level perspectives: tested and proven (i.e., "I've seen it work on my neighbor's farm"); aspirations (i.e., to achieve what a neighbor achieves); and of benefit to the community.

For the GRAD project overall, the MCA scoring and ranking exercises showed similar results to those generated by PRA ranking activities. The top CCA options that emerged from the MCA results were water harvesting, livelihood diversification, improved seeds, pumps and small-scale irrigation, savings, and soil and moisture conservation. Reforestation was also highly ranked in Tigray using the MCA tool. In the semiarid Tigray study sites, participants reported multiple benefits of revegetation and closures in their communities, including increased availability of forest products and forage, improved soil and moisture conservation, and increased local rainfall attributed to tree cover.

PRIME: For PRIME research sites, participants across sites identified "need" as a primary criterion, i.e., whether or not a proposed CCA option was perceived as needed by the community. This is important for both donors and project developers, as buy-in from potential beneficiaries is critical to long-term success of development programs and projects. In comparisons between community and regional levels, some differences did emerge. Kebele-level groups tended to emphasize criteria that result in immediate effects, such as whether or not a CCA option saves lives (human or animal), what costs are incurred and to whom, and how much time or labor is required from the community or household. RVTs, or technical experts, included such criteria but also emphasized longer-term impacts, such as whether a CCA option promoted resilience, sustainability, and food security. Across RVTs in Borena and Jijiga, the top four ranked CCA options identified in the MCA analysis included soil conservation/gully treatment, dry/wet season grazing, water point rehabilitation, and herd diversification.

21.4.1.3 Cost–Benefit Analysis

The CBA methodology assigns monetary values to both direct and indirect costs and benefits associated with any CCA investment or activity. Quantitative analysis was limited by a lack of monetary values for costs and benefits—particularly for indirect benefits and externalities—associated with many of the options. For this reason, three CCA options were selected for CBA: improved seeds, moisture-conserving practices, and fuel-efficient stoves. Findings indicated that, in terms of generating net benefits, investment in all three strategies provides similar returns. Available data was used to derive a Benefit–Cost Ratio, or the amount of net benefits provided to adopters of each of the three practices per dollar spent by the project. It provides some insight into the relative effectiveness of these alternative intervention options.

Overall the Benefit–Cost values were very high and similar in magnitude across the three practices—approximately US$12 of net revenue generated per each dollar spent on promoting the activities by the projects. The values are high for two important reasons. First, the estimates of the net benefits to adopters are probably quite optimistic, and second, the costs are only the direct costs of promoting these specific interventions, and project overhead costs should also be factored in to the calculations.

Using a Benefit–Cost lens in FGDs yielded useful qualitative insights into the perceived benefits and challenges associated with the uptake of CCA options, in some cases differentiated by social group (e.g., poor, women). The discussions allowed an opportunity to gather information on externalities and characteristics that are not readily quantifiable, such as positive changes in household dynamics and collective action. Within projects, the perceived costs and benefits were similar across study sites, with a few notable exceptions. For GRAD, benefits of community-based land management were more pronounced for Tigray than for SNNPR, where population density is high and there is little communal land. Regional differences are related specifically to the different activities the project is implementing; i.e., rope-and-washer pumps only in SNNPR, crop insurance and herd diversification only in Tigray. Between the two PRIME regions, there were no clear differences in perceived costs. Of note, however, the issue of governance related to natural resource management dominated much of the discussions in the Borena where participants called for strengthening the customary grazing, or *gadaa*, system as a strategy for dealing both with drought and land degradation. At both community and regional levels, participants suggested that implementation of the technical aspects involved with dry/wet season grazing would not be effective without either first or concurrently strengthening governance of this system.

The initial intent of the quantitative research component was to examine data bases from each of the GRAD and PRIME projects to explore in more detail the characteristics of adopters of specific CCA strategies, estimate the possible benefits to households of adopting specific practices, and identify key differences in the characteristics between adopting and nonadopting households that could be used to forecast the total number of likely adopters of individual practices.

Unfortunately, much of this exploratory analysis was not possible. The annual monitoring surveys of the two projects collect very little information about households beyond adoption of specific practices and information needed to measure project indicators. These surveys serve the very specific purpose of collecting information to track project indicators on an annual basis. However, without information about household food security outcomes, economic conditions, or livelihoods, it was not possible to compare profiles of adopters and nonadopters, or to compare information about project participants from the annual surveys with the wider population captured in the baseline surveys.

Despite these limitations, the quantitative data sets provided insights into rates of adoption by project and by research site for some interventions. In some cases, these findings corroborated results from the qualitative analysis, while for certain CCA options qualitative and quantitative results were inconsistent. Unless otherwise specified, percent figures reported below reflect adoption rates for the projects overall.

For GRAD participants, the highest adoption rates for selected CCA options were for watershed improvements (76%), savings (71%), and improved seed varieties (70%). Adoption rates of the CCA options were similar in both Tigray and SNNPR project areas, and compared with the project overall. Adoption of savings practices was notably higher in the Tigray and SNNPR regions, where almost 85% of households reported savings (84% of which initiated savings with support from GRAD). At the community level, FGDs consistently ranked savings as a priority option.

Only around 10% of households have subscribed to weather-indexed insurance. This was the lowest reported adoption rate among the selected CCA options. Similarly, FGDs in Tigray indicated that the benefits of insurance are minimal in that payouts were too small or too late to compensate for crop losses. Of note, insurance is a relatively new strategy introduced in limited project areas. Overall, the majority of households adopted adaptation strategies as a result of project support.

For PRIME, adoption rates of CCA options tend to be somewhat lower than for GRAD and there is less consistency between quantitative and qualitative findings. Nearly one-half (49%) of beneficiary households in Oromia and Somali regions (and 46% overall) adopted savings—the highest rate of adoption among the CCA options included in the study. This is counter to qualitative findings, in which savings consistently ranked low in terms of its perceived effectiveness at helping households deal with drought. The other CCA strategies with the highest adoption rates in the study regions, as well as for the project overall, included gully treatment (38%), improved postharvest technology/storage bags (38%), and managing wet and dry season grazing (37%). Of those who reported adoption of a CCA option, the majority indicated they had done so before 2013, when the program was initiated.

One unexpected finding from the quantitative analysis of PRIME interim monitoring data was that livelihood diversification did not seem to have a positive impact on managing shocks (note, interim monitoring data is not available for GRAD). This is consistent with findings from the qualitative analysis for PRIME, in that alternative/diversified livelihoods consistently ranked low by most groups in the PRIME study area. Findings from recurrent monitoring in PRIME operational areas indicated investments that strengthen social capital have a positive impact on household abilities to manage climate shocks (Box 21.3), consistent with qualitative findings for both projects.

BOX 21.3 Social Capital

Both PRIME and GRAD implement a number of activities that involve group formation for collective action. These include forming savings groups, reforestation, community-based upland management, and management of dry and wet season grazing. Other investments that seem to have a positive effect on managing shocks include improving access to markets, increasing household accumulation of assets and access to financial services. Investments in human capital also play a role in managing shocks (i.e., skills training).

21.4.1.4 Findings From the Analysis: Synthesis

As noted above, the variation in rankings of CCA options both within and between projects points to the importance of identifying a suite of complementary interventions suited to specific operational and environmental contexts. While it would be impractical to derive a simple ranking of individual CCA options for Feed the Future in Ethiopia, we identify three main "clusters" of priority interventions.

- *Access to water*: point source rehabilitation/upgrade, rain-water harvesting, and water conservation. Given the drought-prone PRIME and GRAD project operational areas, interventions to promote rain-water harvesting, soil and moisture conservation, and access to water ranked very high across all research sites. In some areas of SNNPR, where water table levels are conducive to rope-and-washer pump technologies, this option ranked among the top five in PRA and MCA rankings. The benefits of improving access to water are considered "life-saving."
- *Access to income*: savings and livelihood diversification to buffer climate shocks. Whereas water management was highly ranked across all of the research sites, savings and alternative income generation activities were highlighted as important ways to manage a range of shocks across GRAD research sites. Access to income in the form of own savings, together with skills and networks to diversify income, enables households to avoid or reduce negative strategies they have typically used to cope with climate-related shocks. Participants also expressed the value of savings and income-generating activities to improve household decision-making and the division of household labor, and enable self-reliance when a shock occurs.
- *Access to grazing land*: rangeland management and governance systems. In the Borena zone of PRIME, strengthening the customary system for managing rangelands was ranked as a priority strategy to deal with drought (as well as associated land degradation and conflict).

21.4.2 Findings Regarding Implementation of the Decision-Making Tools

Collectively, the selected tools yielded a high level of insights about relevant CCA strategies in a limited amount of time and show potential for informing decision-making around adaptation investments. In part, this is attributed to complementarity among the tools for investigating a specific set of ongoing climate-related interventions. For example, the use of guided exploration of the CCA options in FGDs provided valuable contextual information on various strategies and their adoption. FGDs also identified important strategies that were not necessarily on the initial list of options, such as crop insurance, village settlements, and governance systems. Use of a modified-Delphi approach facilitated consensus-building and verification across a range of stakeholders. The ranking exercises, MCA and CBA, highlighted specific priorities and trade-offs among the suite of interventions, as well as the importance of clustering options. These exercises helped to refine and contain the analysis, thereby enhancing the productivity of research on complex and multivariate topics. Despite the limitations, the quantitative analysis provided a broader scale perspective on the uptake of options and the importance of using CCA interventions to enhance capacities of people to cope with climate shocks. Together, they augmented the effectiveness of the inquiry (i.e., the sum was greater than the parts) and provided insights to help leverage investments in CCA.

21.5 CONCLUDING REMARKS AND RECOMMENDATIONS

21.5.1 Recommendations for CCA Investment

To deal with climate uncertainty and variability, people need multiple options to bolster their resilience to climate shocks—options that can be interchanged or scaled up or down depending on dynamic conditions. They also need access to information to effectively consider these options. The findings of this pilot study suggest it is critical to promote a suite of technically feasible and context-specific interventions that work together to strengthen the ability of households and communities to deal with climate change. CCA activities need to be considered in terms of their effectiveness to manage climate shocks, as well as the ways in which the options can be sequenced or clustered to best enhance resilience to climate shocks and promote livelihoods. To achieve greater impact, it is important to promote targeted CCA investments at multiple levels; i.e., household, community, watershed/zone.

> *Decision analysis promotes contextual thinking—i.e., recognizing the precise nature of the decision problem, defining the larger social and ecological context in which it is embedded, and identifying and evaluating alternatives.*
>
> Johnson et al. (2013:27)

21.6 RECOMMENDATIONS FOR IMPLEMENTATION OF THE DECISION-MAKING TOOLS

Each of the four pilot tools demonstrated value in investigating and prioritizing the CCA activities implemented by GRAD and PRIME. Together, they identified a set of project-level options that demonstrate best fit for context, based on multiple decision-making criteria, costs and benefits, and, ideally, contribution to achieving desired development outcomes. To optimize the effectiveness of the tools, it is important to:

Select complementary qualitative and quantitative methods. Used together, the various qualitative and quantitative methods enabled prioritization of CCA options that was highly contextualized. For example, the ranking exercises facilitated through MCA were informed by and contextualized through FGDs and discussions around costs and benefits. The findings provided insight into uptake and utilization of the interventions, as well as decision-making processes around the adoption of CCA options.

Sample to capture diversity. The use of the tools highlighted variations in preferred options depending on the agro-ecological zone, livelihood system, technical feasibility and resource requirements, and severity of the shock, for example. To ensure that selected CCA options are tailored to the local context, the sample design needs to adequately capture this variation within a project's operational area. Similarly, the use of the range of tools with community and regional level stakeholders made it possible to gather data from multiple perspectives at multiple levels.

Scale up or down based on objectives and resources. It is important to carefully consider the data requirements, data availability, and time and resource requirements, and adapt use of the tools accordingly to best meet the objectives of the study. This recommendation applies to each of the four tools. For the qualitative component, fieldwork can be designed to merge data collection for PRA, MCA, and for the mixed CBA tool. In this way, time requirements can be scaled down to accommodate research budget constraints and minimize demands on participants. Alternatively, each tool can be used independently and/or extended (e.g., to add sites, FGDs, or KIIs). Of course, more data collection translates to more time required for analysis.

For the mixed CBA tool, it is important to determine what level of CBA (e.g., full or scaled down to selected interventions) will be most beneficial to the CCA options assessment and integrate relevant questions into existing project surveys, to the extent possible. To conduct a full CBA for selected interventions, it would be necessary to collect quantitative cost data from households and projects. Questions related to costs and benefits (i.e., cost savings attributed to adoption) could be included in project surveys. Additionally, data related to production and environmental or health externalities could be collected in tandem with project or sectoral specialists. In the Ethiopia pilot study, integrating a scaled-down CBA into the qualitative component was an effective way to frame discussions around the CCA options.

Similarly, for the quantitative analysis, annual surveys provided useful information on adoption rates for some CCA options; however, to facilitate the quantitative analysis component it is important to include in these surveys additional information about household characteristics. This would be extremely valuable as much more extensive comparison and analysis of differences between adopters and nonadopters would then be possible.

21.7 INVESTING IN ENABLING CONDITIONS

This study focused primarily on a set of technical interventions for CCA. However, the research findings suggest that it is equally important to maintain investments in the enabling conditions that contribute to uptake—i.e., to invest in strengthening transformative capacity, as well as the absorptive and adaptive capacities enhanced through a suite of technological interventions such as drought-tolerant seeds or rehabilitation of water points. In the GRAD and PRIME projects, a number of elements emerged that have potential to influence the capacity for adoption and, ultimately, build resilience to climate shocks. Briefly, these include:

- *Promote collective formation.* Building social cohesion enhanced the effectiveness of options ranging from reforestation to savings. VESA and VSLA groups provided platforms for training, skills development, information exchange, and effective follow up among field coordinators.
- *Build on local knowledge, capacities, resources, and services.* Climate-sensitive assessments through Climate Vulnerability and Capacity Assessments and value-chain analysis, e.g., enabled the projects to build on existing human, material, and institutional resources. This approach promoted uptake and linked participants with existing markets and technical support.
- *Integrate CCA interventions with livelihoods programming.* In resource-poor areas affected by a range of climate-related shocks, livelihoods depend on adopting practices that promote CCA. In other words, good climate change programming is good development programming. Uptake and effectiveness is enhanced when CCA options are implemented in an integrated and sequenced way, linked to broader livelihood outcomes, and not as discrete, sector-specific interventions.

This chapter discusses insights gained from an approach to assessing CCA options at the project level. The approach centers on the households, communities, and regional-level stakeholders involved in implementing CCA interventions in drought-affected areas of Ethiopia. Reflecting on the use of the tools to inform investments in CCA, a final recommendation is to apply a similar approach to prioritizing and contextualizing CCA options at a larger (e.g., national) level, with government, donor agencies, and implementing partners. In this way, the tools can be used to align CCA priorities that are effective and appropriate at the local and regional levels with national-level adaptation strategies.

ACKNOWLEDGMENTS

The authors thank the many individuals who contributed to this study and made it possible. We extend our appreciation to Chris Perine and Flora Lindsay-Herrera of Chemonics who provided comments on drafts of this work. Monica Mueller and Maryada Vallet of TANGO International conducted research on GRAD and PRIME CCA strategies and assessment tools in climate change decision science. A. K. M. Towfique Aziz, also of TANGO International, contributed to the quantitative analysis for the study. We also wish to thank the research team from Green Professional Services in Addis Ababa and GRAD and PRIME project staff in Ethiopia for their intellectual and logistic support, and for participating in the piloting of the CCA decision-making tools. Their contributions have been invaluable. We are especially grateful to the numerous GRAD and PRIME project participants and stakeholders who gave their time and shared insights into their experience of climate variability and their ongoing efforts to address its impacts.

REFERENCES

ATLAS, 2016. Climate change risk in Ethiopia: country fact sheet. Climate Change Adaptation, Thought Leadership and Assessments Task Order AID-OAA-I-14-00013, July.

Ayers, J., Anderson, S., Pradhan, S., Rossing, T., 2012. Participatory Monitoring, Evaluation, Reflection and Learning for Community-based Adaptation: A Manual for Local Practitioners. CARE International.

Beaulieu, N., 2010. Evaluation as an integral part of climate change adaptation in Africa. International Symposium of the Secretariat international francophone pour l'evaluation environnementale (SIFEE), Climate Change and Environmental Assessment: Issues and tools for assessing impacts and the development of adaption plans, Niamey, Niger, 26 May 2009.

Béné, C., Frankenberger, T., Nelson, S., 2015. Design, Monitoring and Evaluation of Resilience Interventions: Conceptual and Empirical Considerations. Institute of Development Studies, Brighton, IDS Working Paper No. 459.

Bours, D., McGinn, C., Pringle, P., 2013. Monitoring and Evaluation for Climate Change Adaptation and Resilience: A Synthesis of Tools, Frameworks and Approaches. SEA Change CoP, Phnom Penh and UKCIP, Oxford.

Brooks, N., Anderson, S., Ayers, J., Burton, I., Tellam, I., 2011. Tracking Adaptation and Measuring Development. International Institute for Environment and Development, London, IIED Climate Change Working Paper No. 1.

Constas, M., Frankenberger, T., Hoddinott, J., 2014. Resilience Measurements Principles: Toward an Agenda for Measurement Design. FSIN Technical Working Group. World Food Programme, Rome, Technical Series No. 1.

Conway, D., Schipper, E.L.F., 2010. Adaptation to climate change in Africa: challenges and opportunities identified from Ethiopia. Global Environ. Change 21 (2011), 227−237.

Dazé, A., 2014. Using Analysis of Climate Change Vulnerability and Adaptive Capacity for Program Decision-Making: Lessons from CARE Ethiopia's Experience. CARE Ethiopia, August.

Deressa, T.T., Hassan, R.M., Ringler, C., Alemu, T., Yesuf, M., 2009. Determinants of farmers' choice of adaptation methods to climate change in the Nile Basin of Ethiopia. Global Environ. Change 19 (2009), 248−255.

Dinshaw, A., Fisher, S., McGray, H., Rai, N., Schaar, J., 2014. Monitoring and evaluation of climate change adaptation: methodolgial approaches. OECD Environment Working Papers No. 74.

European Commission [EC], 2017. Climate Action: Adaptation to Climate Change. http://ec.europa.eu/clima/policies/adaptation/index_en.htm

Deutsche Gesellschaft für Internationale Zusammenarbeit (GIZ) GmbH, 2013. Economic approaches for assessing climate change adaptation options under uncertainty: Excel tools for Cost-Benefit and Multi-Criteria Analysis. Bonn and Eschborn, Germany, December.

Golrokhian, A., Browne, K., Hardin, R., Agrawal, A., Askew, K., Beny, L., et al., 2016. A National Adaptation programme of action: Ethiopia's responses to climate change. World Dev. Persp. 1 (2016), 53−57.

Grafakos, S., Olivotto, V., 2012. Choosing the Right Adaptation Assessment Method. Presentation at Resilient Cities 2012, Bonn. Institute for Housing and Urban Development Studies, Erasmus University, Rotterdam.

Helmer, O., 1967. Analysis of the Future: The Delphi method. The Rand Corporation.

Hsu, C., Sanford, B., 2007. The Delphi technique: making sense of consensus. Pract. Assessment Res. Evaluat. 12 (10), 1−8.

International Institute for Sustainable Development [IISD], 2012. CRiSTAL User's Manual Version 5: Community-based Risk Screening Tool— Adaptation and Livelihoods. IISD, Winnipeg, Canada.

Intergovernmental Panel on Climate Change [IPCC], 2007. IPCC Fourth Assessment Report: Climate Change 2007. http://www.ipcc.ch/publications_and_data/ar4/wg2/en/annexessglossary-a-d.html (accessed 01.04.17.).

Johnson, F.A., Williams, B.K., Nichols, J.D., 2013. Resilience thinking and a decision-analytic approach to conservation: strange bedfellows or essential partners? Ecol. Soc. 18 (2), 27.

Kosec, K., Mo, C.H., 2017. Aspirations and the role of social protection: evidence from a natural disaster in rural Pakistan. World. Dev. (97), 49−66.

OECD, 2015. Climate Change Risks and Adaptation: Linking Policy and Economics. OECD Publishing, Paris.

Somda, J., Faye, A., N'Djafa Ouaga, H., 2011. Handbook and User Guide of the Toolkit for Planning, Monitoring and Evaluation of Climate Change Adaptive Capacities. AGRHYMET Regional Center, Niger.

TANGO International, 2015a. Report 1: Adaptation Options. Developing a Method to Analyze and Evaluate Adaptation Options in Ethiopia for the Climate Change Adaptation Assessments, Thought Leadership and Learning Project (ATLAS), USAID. DRAFT. July 31, 2015.

TANGO International, 2015b. Report 2: Summary of Tools. Developing a Method to Analyze and Evaluate Adaptation Options in Ethiopia for the Climate Change Adaptation Assessments, Thought Leadership and Learning Project (ATLAS), USAID. July 31, 2015.

United Nations Framework Convention on Climate Change [UNFCC], 2011. Assessing the costs and benefits of adaptation options: an overview of approaches. The Nairobi Work Programme On Impacts, Vulnerability and Adaptation to Climate Change. Secretariat, Bonn, Germany.

Vermeulen, S., Challinor, A., Thornton, P., Campbell, B., Eriyagama, N., Vervoort, J., et al., 2013. Addressing uncertainty in adaptation planning for agriculture. Proc. Natl. Acad. Sci. 110 (21), 8357−8362.

Watkiss, P., Hunt, A., Blyth, W., Dyszynski, J., 2014. The use of new economic decision support tools for adaptation assessment. A review of methods and applications, towards guidance on applicability. Clim. Change 132, 401−416.

Webb, P., von Braun, J., Yohannes, Y., 1992. Famine in Ethiopia: Policy Implications of Coping Failure at National and Household Levels. Research Report 92. International Food Policy Research Institute, Washington, DC.

World Bank, 2010. Economic Evaluation of Climate Change Adaptation Projects: Approaches for the Agricultural Sector and Beyond. Washington, DC.

FURTHER READING

Boudreau, T., 2013. Reducing the risk of disasters and adapting to climate change: evidence from the consolidated Household Economy Analysis database. Livelihoods at the Limit. Food Economy Group. Save the Children Fund, London.

Vallejo, L., 2017. Insights from national adaptation monitoring and evaluation systems. Climate Change Expert Group. OECD/International Energy Agency, Paris, Paper No. 2017(3).

Chapter 22

Social Capital as a Determinant of Resilience: Implications for Adaptation Policy

Siobhan E. Kerr

University of Maryland College Park, College Park, MD, United States

Chapter Outline

22.1 INTRODUCTION

In recent years, resilience has become a prominent concept in climate and disaster recovery research, adding a substantial new dimension to previous methods for assessing and addressing climate and disaster risk. Twigg argues that the main difference between resilience and more traditional risk analysis approaches to adaptive capacity is that resilience "goes beyond specific behavior, strategies and measures... that are generally understood as capacities" (Twigg et al., 2013). It takes a broader, more pragmatic approach to addressing risk, where "agility and discipline" (Tierney, 2014) collide.

Traditionally, risk analysis work has put its primary focus on protecting and strengthening infrastructure systems so that they can withstand anticipated shocks. By contrast, resilience demands a more holistic approach and stresses that resilient infrastructure must be complemented by resilient communities and resilient systems of governance. In doing so, the resilience approach requires that focus be directed toward more than simply withstanding the anticipated hazard, but also learning and organizing in such a way that the system or community will be better prepared to face a broad spectrum of potential shocks in the future (Godschalk, 2003). This is particularly important because the increasing interconnectedness of our communities and our reliance on countless interdependent infrastructural systems and subsystems creates unprecedented levels of complexity, for which it would be nearly impossible and wholly impractical to identify and address the risks and possible failures of each system component (Linkov et al., 2014).

The complexity and unpredictability of system and community vulnerabilities are even more prominent when climate change is brought to the forefront of the resilience conversation. Although climate change is a scientific certainty, there are still many unknowns, especially with regards to the anticipated time horizons, severity, scale, and nature of impacts (Linkov et al., 2014). These unknowns make it nearly impossible and economically infeasible for a community to prepare for these changes solely by relying on more traditional risk analysis and hazard mitigation techniques. Although these traditional hardening and disaster preparedness approaches continue to have a valuable place in a strong disaster management strategy, they are no longer, and perhaps never were, sufficiently robust on their own (Godschalk, 2003). A resilience approach is what is needed to fill in these gaps.

Resilience. DOI: https://doi.org/10.1016/B978-0-12-811891-7.00022-0

22.2   RISK AND UNEVEN RESILIENCE

There have been many attempts over the years to define resilience, and the precise meaning has evolved considerably since it emerged as a key issue in the social environmental sphere (Twigg et al., 2013). Different definitions choose to include or omit certain elements of resilience, and there is often disagreement over whether resilience is best seen as an outcome (the ability to recover after a shock), or a process (the ongoing act of learning and improving) (Cutter et al., 2008). Despite these differences, there is broad agreement that resilience encompasses the following characteristics: the capacity to absorb; the capacity to adapt; the capacity to recover; and the capacity to organize as a society (Adger et al., 2005). More recent conceptions of resilience emphasize that the goal of these phases is not to simply withstand the hazard, but also to reshape and transform society, thereby minimizing future risk (UN Secretary-General's Initiative Aims to Strengthen Climate Resilience of the World's Most Vulnerable Countries and People, 2015).

These capacities, of course, present unevenly across society. In order to properly understand the causes of uneven resilience, we must understand the nature of risk (Tierney, 2014). Beck's risk society theory suggests that modern society is the product of long-term society-driven change, and that the risks (particularly the environmental risks) to which we are exposed today are not the product of natural inevitability but instead the result of prior decision-making and societal power structures (Beck and Ritter, 1992). This is of great relevance to the concept of resilience and more specifically to the question of what causes certain communities to cope, absorb, and return to their previous baseline more quickly than others after being subjected to a shock.

Wisner et al. (2003) describe the physical shocks as the "triggers" for disasters, and argue that social and historical forces cause risk to build up in such a way that a trigger is able to set it off. This gives rise to the inevitability that vulnerable and marginalized populations are the least resilient, not only because they tend to live in the neighborhoods and buildings that are more exposed, but also because they lack the resources necessary to absorb and cope with the aftermath (Wisner et al., 2003). This increased hardship during the recovery process results in worse long-term recovery outcomes (Phillips and Fordham, 2009).

Generally speaking, low-income communities and individuals struggle more than others during every phase of building resilience, and particularly during the recovery process (Fothergill and Peek, 2004). This is in part due to a basic lack of resources and income, which compounds hardship and stress (Bolin and Stanford, 1998), and a lack of access to trans-local social networks of family and friends who were able to provide material and moral support during times of stress and hardship (Elliott et al., 2010). In addition, low-income residents are found to be less capable of navigating the bureaucratic systems necessary in order to obtain government-issued aid, whereas higher-income residents are more capable of dealing with these sorts of administrative obstacles. As a result, low-income residents are less likely to apply for, and therefore receive, disaster recovery funds (Fothergill and Peek, 2004). Low-income residents are also more likely to experience severe damage to their homes in the aftermath of disasters. This can lead to homelessness and severe shortages of low-income housing in the aftermath of disaster (Greene, 1992).

Other factors that are commonly thought to influence resilience and recovery outcomes include the quality of governance, amount of aid, extent of damage, and population density. Governance has been found to make macrolevel differences in disaster outcomes: an analysis of disaster deaths in 73 countries between 1980 and 2002 found that democracies and other countries with well-functioning institutions experience fewer natural disaster deaths (Kahn, 2005). However, on a more microlevel, this association does not always hold, as nearby neighborhoods under the same governance that have experienced similar levels of damage often do not experience similar rates of recovery. Research has failed to find a causal link between the amount of aid money and the rate of recovery, and research on the relationship between damage and recovery is inconclusive (Aldrich, 2012).

More recently, researchers have begun to point to social capital as a significant, and perhaps the most significant driver of resilience and recovery (Aldrich, 2012; Elliott et al., 2010; Kawamoto and Kim, 2016). Indeed, Aldrich calls social capital the "core engine of recovery" and argues that social capital is an even better predictor of recovery outcomes than socioeconomic status (Adger, 2003). It is critical to understanding differences in vulnerability both within and among communities, and it is central to coping with risk.

22.3   DEFINING SOCIAL CAPITAL

Social capital emerged as a clearly articulated concept in the 1980s. It was first introduced by Pierre Bourdieu, who defines social capital as "the aggregate of the actual or potential resources which are linked to a possession of a durable network of more or less institutionalized relationships of mutual acquaintance or recognition" (Bourdieu, 1985). Breaking this down, he is simply saying that having a social network gives an individual access to more potential and/

or realized resources than they would have in isolation, and that these socially-accessible resources are social capital. Coleman expanded on this definition in his 1988 work, describing social capital as a "variety of entities with two elements in common: they all consist of some aspect of social structures, and they facilitate certain action of actors—whether persons or corporate actors—within the structure" (Coleman, 1988). This extends the definition introduced by Bourdieu by clarifying that social capital need not simply be limited to the exchange or potential exchange of resources, but can take the form of facilitating any action, whether resource-oriented or more abstract.

Putnam (2000) transformed social capital literature by theorizing that it has two subcategories: bonding social capital and bridging social capital, a distinction that he summarizes by saying "bonding social capital constitutes a kind of sociological superglue, whereas bridging social capital provides a sociological WD-40." Bonding social capital is gained from relations in more insular settings, typically among fairly homogenous groups of people. This helps to foster a strong group identity and a great deal of loyalty between group members. However, because of the insular nature of these communities, they tend not to have many social relations outside of the group. Because of this, very little is brought in from the outside, which limits the group's potential pool of resources (Putnam, 2000). This type of social capital is more commonly observed in poorer communities (Elliott et al., 2010) and low functioning states (Adger, 2003).

Bridging social capital tends to be built in more heterogeneous groups that have more outside connections, often to more powerful elements of society such as governance institutions, civil society, and the private sector. This naturally creates a great deal more opportunity for the members of such groups, as they have access to a much broader range of outside resources. This is sometimes referred to as networking social capital in the literature (Adger, 2003). It is associated with wealthier communities (Elliott et al., 2010), well-functioning states, and formal collective action (Adger, 2003). Evidently, these two different types of social capital bring very different kinds of benefits and opportunity.

In some literature, linking social capital is introduced as a third category, and can be understood as a subgroup of the bridging social capital that was introduced above. In such cases, linking social capital specifically refers to informal vertical connections that link individuals of groups of different economic, social, or organizational status and bridging social capital refers to specifically horizontal connections between diverse individuals and groups (Delisle and Turner, 2016). However, for the purposes of this chapter, bridging and linking social capital will be treated as a single category, as they tend to arise in similar situations and function in similar ways.

Although social capital might seem to arise naturally, Portes (1998) stresses that the social networks that foster social capital must not be seen as an inevitability, but rather "constructed through investment strategies oriented to the institutionalization of group relations, usable as a reliable source of other benefits." Though like any other resource, the unequal distribution of social capital in our society is in part attributable to unearned privilege, the social networks that provide returns to their members have to be developed, fostered, and nurtured. Further, it is important to recognize that although social capital is a mechanism through which people can derive gains, there must nonetheless be a transfer of benefits, and such a transfer demands both a recipient and a donor (Portes, 1998).

22.4   SOCIAL CAPITAL AND RESILIENCE

Social capital has significant explanatory power with regards to community-level resilience capacities. The primary way that social capital contributes to resilience is its impacts on recovery; it enables communities to organize in the aftermath of a disaster. This is particularly true in groups exhibiting strong bridging social capital, because they often have networks that reach beyond the disaster-struck area, thereby giving them access to the much needed resources and support necessary for a speedy recovery. This is demonstrated in Consoer's study of the role of social capital in Vermont after Tropical Storm Irene, where the organization of informal "recovery groups" was driven by social capital in some, though not all communities impacted by the storm. These communities enjoyed "proliferating social capital and access to high value resources" (Consoer and Milman, 2016). Although communities that did not experience this group formation eventually caught up to the high-social capital communities' recovery progress, they required increased government efforts in order to close the recovery gap (Box 22.1).

Bonding social capital also plays a role in the recovery process, but it primarily helps residents to cope with the short-term effects of the shock. Communities that are exclusively rich in bonding social capital tend to be quite tightly knit, homogenous, and closed off. As a result, when a disaster hits, these communities will only have networks connecting them to others who are also affected by the disaster. Although these connections are certainly important with regards to dealing with the immediate aftermath of a disaster, they are less useful in mobilizing the resources needed for a successful recovery effort (Aldrich, 2012).

BOX 22.1 Uneven Recovery following the 2004 Indian Ocean Tsunami

A case study of uneven recovery following the 2004 Indian Ocean Earthquake and Tsunami makes a convincing case for both the importance of social capital in resilience recovery, as well as its potentially exclusionary nature (Aldrich, 2012). When considering community-level recovery following the disaster, the fast recovering communities were all observed to have local councils. These councils had long been an important source of bonding social capital, but in the aftermath of the tsunami they transformed into important engines of bridging social capital. They acted as intermediaries between the aid efforts and their communities by communicating community requests, advocating on behalf of the community, and storing and distributing aid, thereby enabling community-level collective action (Aldrich, 2012). The strong bridging social capital facilitated by these community councils improved the recovery outcomes of the communities as a whole, but not everyone benefited equally from their efforts. Prior to the tsunami, the councils had primarily been a source of bonding social capital and despite the development of bridging networks, the exclusionary nature of the bonding social capital persisted. Because the councils took control of aid storage and distribution, they also had the power to exclude certain out-groups in the community from receiving aid for which they qualified, which ultimately reinforced and exacerbated their vulnerability (Aldrich, 2012).

Similarly, social capital has been found to be instrumental in building and strengthening community resilience prior to an adverse event. Given the many capacities associated with a resilient society, this of course manifests in a multitude of ways. A study of flood preparations in China found that social ties were instrumental in encouraging households to invest in risk mitigation efforts. Individuals reported that family members, friends, and neighbors expected them to engage in preparation efforts, which had a substantial influencing effect. Information about risks and resilience strategies were also considered to be more trustworthy when coming from within one's social networks rather than outside agencies (Lo et al., 2015). Similarly, bonding social capital contributes to community resilience by encouraging community members to promote and share the indigenous local knowledge and traditional skills that are instrumental in helping communities cope with climate change (Petzold and Ratter, 2015). However, such measures tend to be short-term solutions, and will be insufficient to deal with the full extent of climate risk (Delisle and Turner, 2016).

The social ties and social trust that are associated with strong social capital promote the collective action that is necessary for a community to come together in order to organize and adapt. Households with higher levels of social capital are more likely to participate in community organizations and NGO initiatives designed to strengthen resilience, although it is unclear whether these bonds make people more willing to take part in these programs or if, more fundamentally, the opportunity to participate is selective and extended based on social connections (Akamani and Hall, 2015). More broadly, however, social ties can be instrumental in making a community aware of climate risks and the potential to build resilience as a collective group. They can also work to combat narratives of hopeless victimhood and empower communities to work together to develop a collective solution (Warrick et al., 2017).

Some literature on social capital is incredibly optimistic about its potential, if properly fostered, to improve social well-being including recovery outcomes without exception (Jordan, 2015). For example, Adger links social capital with health outcomes, stronger governance, and economic growth, going so far as to call collective social capital and social networks a public good (Adger, 2003). However, other authors are more reserved in their analysis, acknowledging that along with its clear benefits as a social transmitter of resources, it also has clear drawbacks. Indeed, Aldrich explicitly specifies social capital ought not be thought of as a public good, because it does not benefit everyone. Rather than a solution in itself, it is simply a tool by which a solution can, in some cases, be facilitated; "a potential source of benefits rather than a benefit in itself" (Aldrich, 2012).

In this vein, some argue that a view of social capital as the social networks that transmit opportunity and resources is an oversimplistic analysis (Barnshaw and Trainor, 2007). The number of social networks to which an individual has access is perhaps less important than the quality of those networks, which is largely determined by the way that cultural and economic power is distributed throughout society. A person who has a very large network that is made up of individuals who lack any significant form of power is likely to be worse off than a person whose network is smaller but filled with elites. In that respect, social capital is not actually capital in itself, but rather the way that social connections can help an individual access other types of capital. If those to whom they are connected do not have influence, wealth, power, or social standing, their ability to transfer resources to others within their network will be severely limited.

Lin (2001) notes that people's social networks tend to be filled with others who share a similar economic and/or cultural status in society. This creates a system where the most privileged in society have access to a network full of similarly powerful people with whom to share resources and opportunities. While the cultural and economically powerful are able to use social capital to secure high-paying jobs and political influence, social networks in more marginalized communities are likely to only have the resources necessary to help each other with more basic day-to-day coping. This serves to consolidate power within the upper-echelons of society, thereby exacerbating preexisting social and economic divides (Lin, 2001).

There are also concerns that strong social capital within a community can lead to a great deal of social control exerted over its members (Portes, 1998). This is often observed in religious communities and small towns, where there are incredibly strong social ties and an acute sense of in-group versus out-group. Bonding social capital in particular can perpetuate narratives within the group, which can ultimately reinforce incorrect information and entrench damaging societal norms (Chamlee-Wright and Storr, 2011; Wolf et al., 2010). In such cases, significant pressure can be placed on members of the in-group to conform with community norms, with which they must comply for fear of exclusion and social isolation if they resist. In a similar way, social capital can manifest in negative ways in groups that are brought together through shared struggles and feelings of rejection and isolation from mainstream society. This can lead to antisocial and even violent behavior, which is made more extreme by the social bonds shared among these groups (Bourgois, 1995).

Taken together, the literature warns that a blind trust in the positive benefits of social capital is unadvisable. When used carelessly as a policy tool,"[i]t's utility and practical application are hampered by a lack of attention to social relations and power inequalities, which risks reinforcing vulnerability" (Jordan, 2015). Like all other forms of capital, its effectiveness will be a function of the distribution of power, privilege, and wealth in society. Although it is certainly a tool that can be incredibly useful if wielded with care, it is crucial for policy makers to anticipate its shortfalls, and plan for equalization efforts to ensure that its benefits are enjoyed more equally across society.

In Elliott's (2010) study of neighborhood resilience and recovery after Hurricane Katrina, social capital manifested in different ways, but produced similar results. He found that inequalities in social capital were magnified as residents prepared for the disaster in the days leading up to the hurricane and coped with the damages in the aftermath. Residents from wealthier neighborhoods were able to access what is referred to as "trans-local ties," which are effectively bridging social capital—connections outside of the city that were able to offer assistance in the immediate aftermath of the hurricane and support during the long-term recovery. The paper hypothesizes that either lower income neighborhoods lack these trans-local connections altogether, their trans-local connections are less equipped to help, or they are unable to access these connections during times of crisis (Elliott et al., 2010).

The important role that trans-local ties play in community resilience is also observed in less developed countries. For example, in Vietnam, "clanic exogamy," or the practice of women moving outside of their family's village when they get married, is instrumental in building resilience. This practice means that familial ties are not concentrated in a single village, but spread out over different nearby valleys that experience different microclimates, and thereby experience routine climatic hardships at different times. As a result, when one village experiences drought or a cold snap, they have family members in different villages upon whom they can rely for small loans or other forms of assistance. However, this assistance tends to be contingent on the perception that the recipient will one day be able to return the favor. In households where the patriarch has a bad reputation in the community, receiving assistance will be a struggle. Widows face similar barriers, even in cases where she is known to be a hard worker and an upstanding citizen (Delisle and Turner, 2016).

In a study of the relationship between social capital and adaptive behaviors among Ethiopian farmers, it was found that individuals exhibiting higher levels of social capital in the form of social trust were more likely to contribute to community-wide adaptation efforts, but were less likely to engage in private household-level initiatives. The authors hypothesize that this social trust and engagement in community-level resilience gave people a sense of security, which discouraged them from engaging in private resilience building. While not entirely a loss, it is important to be mindful of such behaviors when designing and implementing resilience policies in order to ensure that the importance of household-level resilience is not discarded entirely in favor of community-level initiatives (Paul et al., 2016).

In some contexts, social capital has been a driver of exclusively negative outcomes. A study of the way social networks impact individual responses to heat waves found that bonding social capital among the elderly could actually increase their vulnerability to heat waves (Wolf et al., 2010). As discussed earlier, one of the many functions of social capital is the proliferation of narratives, which has a significant impact on people's outlooks, and by extension their outcomes (Chamlee-Wright and Storr, 2011). Interviews with elderly residents of London and Norwich indicated that they did not see heat waves as a legitimate threat to their health and well-being, and felt that they were adequately equipped to cope with soaring temperatures. The authors hypothesize that narratives of resilience and self-reliance are transmitted and reinforced through social networks. This leads to network members being very reluctant to ask for outside help, even when it is desperately needed (Wolf et al., 2010).

Overall, it is clear that social capital can be a powerful engine of resilience both prior to and in the aftermath of a disaster or shock. Well-organized communities are capable of collective action that helps them harness outside resources, information, and knowledge, which is instrumental in putting best practices into action quickly and efficiently. Although social capital arises naturally, it is possible to implement policies that encourage its development. With this in mind, it is important to foster the development of community groups prior to disaster in order to build resilience, and to prioritize keeping communities intact during the recovery process. However, it is unwise to put too much faith in this process. Without oversight and intervention, it is likely that social-capital driven resilience will exclude community out-groups and favor wealthier groups that have better access to trans-local connections, thereby exacerbating the circumstances of society's most vulnerable. In addition, bonding networks can proliferate the spread of inaccurate and unsafe information and coping strategies, causing members to miscalculate risk and choose not to seek help when needed. Efforts must be made to create well-connected bridging networks throughout all facets of society to ensure a more equitable flow of information, resources, and aid.

22.5 MEASURING SOCIAL CAPITAL

In order to link the presence of social capital to concrete outcomes, it is necessary to have a conceptual framework of social capital that can be empirically evaluated. Measuring social capital is an incredibly difficult task, as it requires that the researcher operationalize a very abstract idea: the presence and quality of social networks and social bonds. As with the study of resilience more broadly, taking a quantitative approach to measurement is attractive because it can be compared, indexed, and applied to quantitative models but this necessitates the challenging task of selecting proxies and metrics that can be validated and are not overly correlated with other potential influencers of recovery outcomes, such as income. These issues can be more easily addressed with qualitative approaches to data collection, but these methods have the downsides of lacking scalability, comparability, and straightforward applications to quantitative analyses.

Questions have been included in a number of large-scale surveys, such as the World Values Survey (1981−95), the Social Capital Community Benchmark Survey, the European Social Survey, and the Current Population Survey Civic Engagement Supplement that can be used to approximate social capital at the community level. Such questions typically focus on association to groups and organizations, volunteerism, social trust, and political engagement. An example of operationalizing this sort of survey instrument in order to draw quantitative conclusions about the nature of social capital can be found in Guillen et al.'s 2010 paper. Questions from the European Social Survey, which focused on the amount of formal and informal contact people had with others over a set period of time, were used as a proxy for social participation, and the formal/informal distinction was used to roughly distinguish between bridging and bonding social capital (Guillen et al., 2011).

Despite the appeal of big data approaches to measuring social capital, oftentimes the literature relies on small primary data in order to get a more complete picture of the presence of social capital in a much more localized setting. Some use survey instruments in order to collect data that can be used for quantitative analysis such as a study on earthquake recovery in Japan, which used an online survey to investigate the way that social capital impacted the efficiency of waste management and recovery in the aftermath of an earthquake. The survey asked questions about trust, interactions with neighbors and friends, and social participation (Kawamoto and Kim, 2016). Similarly, a study on the way that social capital impacts the public acceptability of different adaptation policies used a mail survey to ask questions about social trust, institutional trust, networks, and reciprocity. Perhaps this study's most unique contribution was its approach to reciprocity: respondents were asked whether they believed that their neighbors, family, and close friends would help out if the respondent's home was in danger of flooding. This captures the degree to which an individual feels that they can rely on their community, which is central to the concept of social capital (Jones and Clark, 2014).

Qualitative methods such as open-ended interviews and focus groups remain the most common way that researchers study social capital. Although this limits the extent to which quantitative analysis is possible, these exploratory approaches allow for a more robust understanding of the multidimensional nature of social capital, and lend themselves well to the theory-building that is necessary in order to strengthen and validate more quantitative approaches. For example, Wolf et al. used semistructured interviews to explore the way that UK seniors and their social contacts coped with heat wave. Coding software was used to draw common themes and narratives from the interviews, and illustrative quotes were used to emphasize key points (Wolf et al., 2010). Jordan's study of the role of social capital in disaster resilience in Bangladesh also used semistructured interviews accompanied by focus group discussions (Jordan, 2015). Again, no quantitative methods were employed, but instead interpretive analysis and illustrative quotes were used to develop a narrative.

22.6 CONCLUSIONS AND FURTHER STUDY

The identification of social capital as a major influencer of community-level resilience is an important development in the literature of recent years, but it can be difficult to disentangle the effects of social capital from the advantages that are associated with socioeconomic status. Social capital refers to the process by which people can use their social networks and connections to transfer resources and opportunities both within and outside their communities. However, as we have seen, people tend to network with those to whom they are similar. As a result, the wealthy, powerful, and well-educated will have connections with others like them, and the networks of low-income households will be similarly disadvantaged.

Some research has indicated that in the aftermath of shocks, social capital only benefits those who have access to other forms of capital (Elliott et al., 2010; Jordan, 2015). Even in cases when social capital is used to foster collective action that helps communities secure access to resources (Aldrich, 2012), its impacts tend to disproportionately benefit the more powerful community members. This leads to the question of how policy makers can use a community's social capital to their advantage so that it builds resilience in an equitable manner rather than simply amplifying preexisting advantage, thereby deepening socioeconomic divides.

If we are to accept the proposition that social capital offers unique contributions to the broader recovery process, then it is important to develop strategies for taking advantage of the opportunities that it provides. The difficulty of measurement and the relative newness of the concept of social capital in the resilience literature has made it difficult to draw clear conclusions about its utility as a policy tool, but one of the most enduring themes is that social capital in the absence of useful resources, services, and connections to outside agents is virtually useless with regards to building resilience that will be effective in the long term. As a result, it is crucial for government agents and NGOs to develop meaningful ties within the communities they serve. The Indian Ocean Tsunami case indicates that forming community councils can be a useful way of establishing and strengthening such links (Aldrich, 2012). Recent research suggests that social media could also be a promising method building bridging social capital at the community level, although more research is needed into how to most effectively leverage these networks (Houston et al., 2015).

Resilience strategies that are brought into a community from the outside must also be mindful of existing formal and informal social structures, networks, and hierarchies (Petzold and Ratter, 2015). Social capital occurs naturally; particularly in small close-knit communities, and in its raw form it has the potential to be either a help or hindrance to resilience building efforts. Information about risk and resilience strategies is more effective when it is spread within a social network rather than delivered by an outsider, so the simple act of recruiting well-connected members of a community to communicate key information can be helpful (Lo et al., 2015). This awareness of prior social networks is of particular importance during recovery efforts. In times of extreme stress, existing social connections in a community can be central to both physical and psychological coping. Therefore, efforts to keep these networks intact during evacuation and rebuilding can have a profound impact on community well-being (Aldrich, 2012).

Although it can be stated with reasonable confidence that such local-level strategies are effective in increasing resilience at the community level, it is unclear how to ensure that these benefits are spread equitably at the household level (Béné et al., 2016). Approaches to building resilience through social capital must avoid reinforcing and magnifying the vulnerability and destitution of more marginalized segments of society who have in the past been excluded from enjoying social capital's benefits. As discussed, the literature has been quite vague in this regard. It is often suggested that good governance and institutional oversight are instrumental in ensuring a more equitable distribution of benefits throughout society (Aldrich, 2012; Elliott et al., 2010; Paul et al., 2016), but these papers tend to lack the data and specific actions necessary to operationalize such strategies. An approach to recovery that makes the poor worse-off is unjust, regardless of the extent to which it benefits the most powerful (Rawls, 2001), so if social capital continues to be touted as a recovery tool these are questions that must be answered.

REFERENCES

Adger, N., 2003. Social capital, collective action, and adaptation to climate change. Econ. Geogr. 79 (4), 387–404. Available from: https://doi.org/10.1111/j.1944-8287.2003.tb00220.x.

Adger, W.N., Arnell, N.W., Tompkins, E.L., 2005. Successful adaptation to climate change across scales. Glob. Environ. Change 15 (2), 77–86. Available from: https://doi.org/10.1016/j.gloenvcha.2004.12.005.

Akamani, K., Hall, T.E., 2015. Determinants of the process and outcomes of household participation in collaborative forest management in Ghana: a quantitative test of a community resilience model. J. Environ. Manage. 147, 1–11. Available from: https://doi.org/10.1016/j.jenvman.2014.09.007.

Aldrich, D., 2012. Building Resilience: Social Capital in Post-Disaster Recovery. University of Chicago Press, Chicago, IL and London.

Barnshaw, J., Trainor, J., 2007. Race, class and capital amidst the Hurricane Katrina diaspora. In: Brunsma, D.L., Overfelt, D., Picou, J.S. (Eds.), The Sociology of Katrina: Perspectives on a Modern Catastrophe. Rowman & Littlefield, Lanham, MD.

Beck, U., Ritter, M., 1992. Risk Society: Towards a New Modernity. Sage Publications, London.

Béné, C., Al-Hassan, R.M., Amarasinghe, O., Fong, P., Ocran, J., Onumah, E., et al., 2016. Is resilience socially constructed? Empirical evidence from Fiji, Ghana, Sri Lanka, and Vietnam. Glob. Environ. Change 38, 153−170. Available from: https://doi.org/10.1016/j.gloenvcha.2016.03.005.

Bolin, R., Stanford, L., 1998. The Northridge earthquake: community-based approaches to unmet recovery needs. Disasters 22 (1), 21−38.

Bourdieu, P., 1985. The social space and the genesis of groups. Theory Soc. 14 (6), 723−744.

Bourgois, P., 1995. In Search of Respect: Selling Crack in El Barrio. Cambridge University Press, New York, NY.

Chamlee-Wright, E., Storr, V.H., 2011. Social capital as collective narratives and post-disaster community recovery. Sociol. Rev. 59 (2), 266−282. Available from: https://doi.org/10.1111/j.1467-954X.2011.02008.x.

Coleman, J.S., 1988. Social capital in the creation of human capital. Am. J. Sociol. 94, S95−S120. Available from: https://doi.org/10.1086/228943.

Consoer, M., Milman, A., 2016. The dynamic process of social capital during recovery from Tropical Storm Irene in Vermont. Nat. Hazards 84 (1), 155−174. Available from: https://doi.org/10.1007/s11069-016-2412-z.

Cutter, S.L., Barnes, L., Berry, M., Burton, C., Evans, E., Tate, E., et al., 2008. A place-based model for understanding community resilience to natural disasters. Glob. Environ. Change 18 (4), 598−606. Available from: https://doi.org/10.1016/j.gloenvcha.2008.07.013.

Delisle, S., Turner, S., 2016. 'The weather is like the game we play': coping and adaptation strategies for extreme weather events among ethnic minority groups in upland northern Vietnam. Asia. Pac. Viewp. 57 (3, SI), 351−364. Available from: https://doi.org/10.1111/apv.12131.

Elliott, J.R., Haney, T.J., Sams-Abiodun, P., 2010. Limits to social capital: comparing network assistance in two New Orleans neighborhoods devastated by hurricane Katrina. Sociol. Q. 51, 624−648.

Fothergill, A., Peek, L.A., 2004. Poverty and disasters in the United States: a review of recent sociological findings. Nat. Hazards 32 (1), 89−110. Available from: https://doi.org/10.1023/B:NHAZ.0000026792.76181.d9.

Godschalk, D.R., August 2003. Urban hazard mitigation: creating resilient cities. Nat. Hazards Rev. 136−143. Available from: https://doi.org/10.1061/(ASCE)1527-6988(2003)4:3(136).

Greene, M., 1992. Housing Recovery and Reconstruction: Lessons from Recent Urban Earthquakes.

Guillen, L., Coromina, L., Saris, W.E., 2011. Measurement of social participation and its place in social capital theory. Soc. Indic. Res. 100 (2), 331−350. Available from: https://doi.org/10.1007/s11205-010-9631-6.

Houston, J.B., Hawthorne, J., Perreault, M.F., Park, E.H., Goldstein Hode, M., Halliwell, M.R., et al., 2015. Social media and disasters: a functional framework for social media use in disaster planning, response, and research. Disasters 39 (1), 1−22. Available from: https://doi.org/10.1111/disa.12092.

Jones, N., Clark, J.R.A., 2014. Social capital and the public acceptability of climate change adaptation policies: a case study in Romney Marsh, UK. Clim. Change 123 (2), 133−145. Available from: https://doi.org/10.1007/s10584-013-1049-0.

Jordan, J.C., 2015. Swimming alone? The role of social capital in enhancing local resilience to climate stress: a case study from Bangladesh. Clim. Dev. 7 (2), 1−14. Available from: https://doi.org/10.1080/17565529.2014.934771.

Kahn, M., 2005. The death toll from natural disasters: the role of income, geography, and institutions. Rev. Econ. Stat. 87 (2), 271−284. Available from: https://doi.org/10.1162/0034653053970339.

Kawamoto, K., Kim, K., 2016. Social capital and efficiency of earthquake waste management in Japan. Int. J. Disaster Risk Reduct. 18, 256−266. Available from: https://doi.org/10.1016/j.ijdrr.2015.10.003.

Lin, N., 2001. Social Capital: A Theory of Social Structure and Action. Cambridge University Press, London and New York, NY.

Linkov, I., Bridges, T., Creutzig, F., Decker, J., Fox-Lent, C., Kröger, W., et al., 2014. Changing the resilience paradigm. Nat. Clim. Change 4 (6), 407−409. Available from: https://doi.org/10.1038/nclimate2227.

Lo, A.Y., Xu, B., Chan, F.K.S., Su, R., 2015. Social capital and community preparation for urban flooding in China. Appl. Geogr. 64, 1−11. Available from: https://doi.org/10.1016/j.apgeog.2015.08.003.

Paul, C.J., Weinthal, E.S., Bellemare, M.F., Jeuland, M.A., 2016. Social capital, trust, and adaptation to climate change: evidence from rural Ethiopia. Glob. Environ. Change 36, 124−138. Available from: https://doi.org/10.1016/j.gloenvcha.2015.12.003.

Petzold, J., Ratter, B.M.W., 2015. Climate change adaptation under a social capital approach-an analytical framework for small islands. Ocean Coast. Manage. 112, 36−43. Available from: https://doi.org/10.1016/j.ocecoaman.2015.05.003.

Phillips, B., Fordham, M., 2009. Understanding social vulnerability. In: Phillips, B.D., Thomas, D.S., Fothergill, A., Blinn-Pike, L. (Eds.), Social Vulnerability to Disasters. CRC Press, Boca Raton, FL.

Portes, A., 1998. Social capital: its origins and applications in modern sociology. Annu. Rev. Sociol. 24 (1), 1−24. Available from: https://doi.org/10.1146/annurev.soc.24.1.1.

Putnam, R.D., 2000. Bowling Alone: The Collapse and Revival of American Community. Simon & Schuster, New York, NY.

Rawls, J., 2001. In: Kelly, E. (Ed.), Justice as Fairness: A Restatement. President and Fellows of Harvard College, Cambridge.

Tierney, K., 2014. The Social Roots of Risk: Producing Disasters, Promoting Resilience. Stanford University Press, Stanford., http://www.sup.org/books/title/?id = 18573.

Twigg, J., Tanenbaum, J.J.G., Williams, A.M., Desjardins, A., Tanenbaum, K., Peres, E., et al., 2013. An integrated conceptual framework for long-term social-ecological research. Int. J. Prod. Res. 7. Available from: https://doi.org/10.1890/100068.

UN Secretary-General's Initiative Aims to Strengthen Climate Resilience of the World's Most Vulnerable Countries and People, 2015.

Warrick, O., Aalbersberg, W., Dumaru, P., McNaught, R., Teperman, K., 2017. The 'Pacific Adaptive Capacity Analysis Framework': guiding the assessment of adaptive capacity in Pacific island communities. Reg. Environ. Change 17 (4), 1039–1051. Available from: https://doi.org/10.1007/s10113-016-1036-x.

Wisner, B., Cannon, T., Davis, I., 2003. At Risk: Natural Hazards, People's Vulnerability and Disasters, second ed. Routledge, London.

Wolf, J., Adger, W.N., Lorenzoni, I., Abrahamson, V., Raine, R., 2010. Social capital, individual responses to heat waves and climate change adaptation: an empirical study of two UK cities. Glob. Environ. Change 20 (1), 44–52. Available from: https://doi.org/10.1016/j.gloenvcha.2009.09.004.

Emerging Issues

Chapter 23

Climate-Resilient Development in Fragile Contexts

Sarah Henly-Shepard[1], Zinta Zommers[2], Eliot Levine[3] and Daniel Abrahams[4]

[1]Mercy Corps, Washington, DC, United States, [2]Food and Agriculture Organization of the United Nations, Freetown, Sierra Leone, [3]Mercy Corps, Portland, Oregon, United States, [4]University of South Carolina, Columbia, SC, United States

Chapter Outline

23.1 INTRODUCTION

Fragility is a major issue on the global agenda (OECD, 2016). Over 1.6 billion people, 22% of the global population, live in places that experience complex crises (Mercy Corps, 2016a; OECD, 2016). While the number of people living in extreme poverty has decreased, the population living in extreme poverty in fragile states is expected to increase from 480 million in 2015 to 542 million in 2035 (OECD, 2016). Sociopolitical fragility—defined as the lack of stable, accountable and transparent governance across economic, political, security, individual, and environmental dimensions (OECD, 2016)—increases vulnerability to climate change and variability (Mercy Corps, 2016a) (Fig. 23.2). In turn, climate change and variability exacerbates underlying drivers of risk by undermining development gains and increasing the frequency and intensity of humanitarian crises, creating negative feedback loops (Fig. 23.2). This is particularly true for those with the lowest-incomes who rely heavily on natural resource-based livelihoods (IPCC, 2014). Understanding and operationalizing humanitarian assistance and climate-resilient development in a fragile context is still relatively underexplored, but ever more necessary as an increasing number of people are living in fragile contexts (OECD, 2016).

A resilience approach may help address root causes of fragility and identify viable response options. A resilience approach is multidimensional and helps identify capacities that enable households and communities to effectively function in the face of shocks and stressors—climate related or otherwise (Box 23.1). A climate-resilient, systems-driven approach integrating these dynamics, offers a bridging solution between humanitarian and development spaces. International nongovernmental organizations (INGOs), research institutions, and civil society organizations have made considerable progress in building an evidence base for this area. This chapter reviews the current state of the evidence of climate-resilient development in fragile contexts and identifies gaps and challenges. We utilize two case studies to highlight key lessons in the proposed conceptual framework for climate-resilient development in fragile states (Fig. 23.2). In Karamoja, Uganda, the case study illustrates the challenges inherent with climate-resilient programming in fragile contexts, that seeks to balance short-term response (e.g., crisis modifiers) with longer-term adaptation and recovery. In Ethiopia, the case study illustrates how climate-resilient development can decrease fragility by increasing

Resilience. DOI: https://doi.org/10.1016/B978-0-12-811891-7.00023-2

BOX 23.1 Resilience Framing

Resilience can be understood and defined through a line of inquiry exploring a series of analytical questions. These questions include:

1. Resilience of what (systems);
2. Resilience of who (vulnerable populations);
3. Resilience to what (shocks and stresses);
4. Resilience through what (absorptive, adaptive, and transformative capacity-building); and,
5. Resilience to what end (in order to achieve development and well-being outcomes) (Mercy Corps, 2016a, 2016b).

In addition, **resilience capacities in response to shocks and stresses can be understood through** the following dynamics:

1. Malleability/flexibility, or the speed and degree at which the system can change back to the previous state, or to a positive alternate state;
2. Diversity of stakeholders or components, which help diversify risks;
3. Anticipation and adaptive learning in the face of change;
4. Social action and cohesion within and across communities and institutions;
5. Self-reliance and resourcefulness using local resources and knowledge.
   (USAID, 2012, pp 44−56; UNICEF, 2011; Henly-Shepard, 2014).

trust between communities and government agencies, while simultaneously building capacity of communities to adapt to climate change and variability. The case studies are followed by a critical discussion of the scientific and practical challenges and opportunities to bridge research, policy, and practice of climate-resilient development in fragile contexts.

23.2 STATE OF THE EVIDENCE

23.2.1 Climate−Fragility Nexus

Climate change and fragility are deeply linked. As of 2009, more than half of the populations experiencing natural hazard shocks and stresses were living in fragile, conflict-affected states (Kellet and Sparks, 2012 from Levine et al., 2016b). Climate variability and extreme events negatively impact the achievement of humanitarian and development outcomes and in many cases drive conflict and poverty (Fig. 23.2). In fact, climate change has been called one of the greatest threats to global health, livelihoods, food and water security, human safety and well-being (Levine et al., 2016a; Tanner et al., 2015). Evidence links climate change indirectly to increasing risks of violent conflict by exacerbating and accelerating the drivers and root causes of these conflicts (IPCC, 2014). A meta-analysis of 60 of the strongest quantitative studies drawing the connection between climate change and violence and conflict found that "for each one standard deviation change in climate toward warmer temperatures or more extreme rainfall, median estimates indicate that the frequency of interpersonal violence rises 4% and the frequency of intergroup conflict rises 14%" (Hsiang et al., 2013) though these findings have been widely challenged.

Fragility, like climate change or resource degradation, can also be a transboundary issue. Fragility extends across borders and systems, creating contexts which drive millions of internally displaced persons and refugees within and across states. From a conflict analysis lens, resilience is often interpreted as the ability of society to support nonviolent resolution to grievances (USAID, 2012). In fragile socioecological systems with poor governance, inequity, and low development, when there are multiple sequential or compounding shocks and stresses, societies experience deeper states of fragility, degrading resilience and inciting further root causes of complex crises, fragility, and conflict. Such negative feedback loops can be self-reinforcing (Berkes et al., 2003), as the effects between climate change, conflict, and fragility drive and compound one another, undermining food and nutritional security, household health and well-being, livelihood security, social stability, and peace (Levine et al., 2016b).

In fragile contexts, communities are unable to access basic development services nor the sufficient coping and adaptive capacity support they need from a weak or underresourced government. Where conflict and violent extremism are present, the systems are often at (or past) negative tipping points of ecological degradation, social unrest, livelihood and food insecurity (Fig. 23.2). According to the OECD, states' fragility is a result of how they measure across five dimensions: (1) violence; (2) access to justice; (3) accountable and inclusive institutions; (4) economic inclusion and stability; and (5) capacities to prevent and adapt to social, economic, and environmental shocks and disasters (OECD, 2016). These same dimensions that contribute to fragility also contribute to the capacity of a community to cope and adapt to the effects of climate change.

23.2.2 Challenges to Breaking Down the Humanitarian Aid/Development Wall

The links between climate change and humanitarian aid and peace-building are acknowledged across policy and scientific communities and mentioned in international policy frameworks (e.g., the Sendai Framework for Disaster Risk Reduction (DRR), the Sustainable Development Goals, and the Paris Agreement). However, linking these frameworks to specific human rights mechanisms and integrating localized, appropriate programming, has proven difficult. According to an OECD study of fragile states "policy makers and practitioners know relatively little about how to reduce fragility and increase resilience (and) policy commitments to investing in fragile, at-risk and crisis-affected contexts need to be matched with investments in the ability to spend that money in an effective way" (OECD, 2016). Many of these frameworks and associated targets assume a degree of good governance and baseline development, which is often not the case in fragile states and places with low levels of development.

Humanitarian aid and disaster response has traditionally operated within acute, urgent timescales to meet immediate life-saving and sustaining needs (Abrahams, 2014). Transitioning societies from humanitarian aid-dependency to early recovery and sustainable development requires engaging from the beginning of a crisis and remaining for an extended and committed time period, in order to understand systems and cultivate socioecological systems transformation to reduce underlying drivers of fragility. Long-term presence of humanitarian actors and concern of donor fatigue has stimulated demand for generation of evidence of value-for-money and proof of sustainable impacts and reduction of chronic aid-dependency. This is a positive trend. The impact of development investments should be tracked, evaluated, and adjusted as needed. However, engaging in measurement processes in fragile states has inherent challenges related to security, safety, and access, which challenge operations, data collection, and validity.

Further, challenges to program design and implementation are posed by the rate at which ecosystems are degrading and the scale of impacts across landscapes. As climate impacts increase fragility within and across borders, humanitarian and development actors must quickly develop strategies and approaches tailored to helping countries and the communities within them to prepare for, manage, and recover from climate shocks and stresses.

Implementing agencies (such as local and international NGOs) operating in fragile contexts are under pressure to implement increasingly complex programs, and prioritize the development of evidence that shows programs are building resilience. These institutions are subjected to a complex funding landscape, working across a range of donor institutions (with various priorities and expectations), and most often juggling a portfolio of overlapping and distinct programs, many of which lack the timelines and resources necessary to meet expected resilience gains. Constant turnover in political regimes and the associated funding and policies that support humanitarian, development, and climate adaptation programs create further complications. Limited and short-term funding can increase competition and limit transparent partnerships. This challenges coordination, resource and information sharing within and across the humanitarian-development divide, and can limit the extent to which climate change considerations are incorporated into the work.

23.2.3 The Need for New Frameworks

A recent review of 35 frameworks, indices, and tools related to resilience, risk, and fragility, concluded that no framework covers risk, fragility, and resilience in an encompassing way, and most have not been empirically tested (UNU, 2016; Levine et al., 2016a). This review also showed that fragility frameworks are often applied to the state-level with focus on official institutions, without adequate attention to local governance, informal institutions, and civil society structures, which have particularly relevant roles in fragile contexts (UNU, 2016). According to most frameworks, fragility manifests within a state as an "authority failure" (where the state "lacks the authority to protect its citizens from violence of various kinds"), a "service failure" (where the state "fails to ensure that all citizens have access to basic services"), and "legitimacy failure" (where the state "lacks legitimacy, enjoys limited support among the people, and is typically not democratic") (UNU, 2016). In this conceptualization, there is little emphasis on interactions between political and social, economic, or ecological systemic risks. Transboundary or landscape-scale systems risks are not adequately considered and as such could produce serious unintended negative consequences. Fragility is associated with constant flux and there is little agreement on the complexity of contributing factors, let alone what should be done. However, it is clear that fragile states cannot apply most of the conventional development tools used or demanded by the international community.

There is an urgent need for frameworks that reconceptualize resilience and fragility in an integrated manner to help practitioners, researchers, and donors better align, assess, and address political, economic, social, and environmental risks. For example, the Sustainable Livelihoods Framework and approach, although highly successful in influencing the development community, was not designed for disasters or conflict (UNU, 2016). As such, this framework alienates the humanitarian community from engaging in livelihood programming. The Food and Agriculture Organization of the United Nations later developed the Sustainable Livelihoods and Emergencies Approach (SLA) to address vulnerability, poverty, and recurrent shocks and stresses, through SLA assessments and analyses within development and humanitarian contexts across the emergency response cycle (FAO, accessed 2017; Frankenberger, 2012). However, this framework fails to include a climate-inclusive approach despite the climate sensitivity of many livelihoods.

A conceptual framework for resilience was later developed by integrating a climate change lens to a coalescence of multiple adapted livelihoods and DRR frameworks (Frankenberger, 2012). This resilience framework illustrates how long-term contextual drivers affect root causes of vulnerability, and the role of capacities to influence people's resilience trajectories relevant to particularly development outcomes like food, livelihood and environmental security (Frankenberger, 2012). In addition, much progress has been made in utilizing livelihood diversification and market systems development as a gateway to socioeconomic resilience-building and reducing fragility (Levine et al., 2016a), however this approach has yet to be applied comprehensively in fragile states, warranting more evidence and programming.

The interweaving—and reinforcing—of the relationships between climate change, risk and fragility, necessitates an evidence base that speaks to unified humanitarian and development solutions. An adaptive, flexible climate-resilient development approach facilitates this paradigm shift. The following case studies present lessons learned and opportunities for innovation within two climate-resilient development experiences in fragile contexts—Karamoja, Uganda, and the pastoral arid lands in Ethiopia. We then propose a new conceptual framework illustrating climate-resilient development within the climate—fragility nexus (Fig. 23.2).

23.3 ON THE FRONTLINES

23.3.1 Overview

Mercy Corps, in close collaboration with its partners and donors, works on the frontlines of complex crises. In this context, the organization has worked to develop an evidence base that integrates climate-resilient development solutions in humanitarian and development programs (Mercy Corps, 2016a). These solutions are generated through adaptive program design, implementation, learning and knowledge management, including resilience monitoring and evaluation components, to enlighten successes and challenges. Below are two case studies that typify Mercy Corps' work to integrate climate change, resilience, and complex crises. The first case study examines programming in Karamoja, Uganda, and illustrates the inherent challenges in balancing immediate needs and coping capacities with longer-term adaptation strategies. The second case study examines efforts in Ethiopia to build resilience capacities in pastoralist areas. Core findings include the critical need for improved technology and access to climate information. Climate information services (CIS) can inform climate-smart livelihood decision-making and risk diversification, as well as policy formulation for good governance of natural resource management planning. Improved access to CIS supports risk management, and when coupled with progressive Participatory Scenario Planning (PSP) it can enable programs to strengthen resilience of vulnerable households to climate and other shocks and stresses. Both case studies offer distinct insight in engaging climate-resilient development in fragile contexts, proposing adaptive approaches and methodologies to address underlying drivers of risk, reducing dependency on humanitarian aid, and increasing efficacy and sustainability of achieving development and wellbeing outcomes.

23.3.2 Case Study 1: The Climate—Conflict Nexus: Lessons from Karamoja

By most metrics, the Karamoja region is the least developed in Uganda (Mercy Corps, 2016b; USAID, 2017a). The population is just over 1,000,000 people; roughly 700,000 of whom live below the national absolute poverty line (Nakalembe et al., 2017). By comparison, across the rest of country, 19.7% of the population lives below the absolute poverty line (Nakalembe et al., 2017; UBOS, 2013). Other development indicators point to comparable challenges: food security, infant mortality, and literacy lag significantly behind the rest of the country (USAID, 2017a). The socioeconomic situation of the region, while improving, is marked by fragility and limited adaptive capacity (Mercy Corps, 2016b; USAID, 2017a). This fragility is linked through a wider political economy to its relatively recent history of intertribal conflict (Eaton, 2008a,b; Stark, 2011). At its surface, these conflicts were driven by cattle raids, a commercialization of cattle raiding, an influx of arms, and to a degree, competition over resources.[1] Frameworks like the Relief-Development Continuum fall short of understanding relief and early recovery in a context of long-term stresses like conflict and violent extremism.

---

1. A full review of these factors is outside the scope of the paper, for more comprehensive review of cattle raiding see in the region see Eaton (2008a, b), Stark (2011), Howe et al. (2015).

The climate in Karamoja is semiarid. The spatial and temporal nature of the climate is an important driver of a socioeconomic system traditionally centered on pastoralism (Stark, 2011). Rainfall is unimodal with rains from March to August; it is also variable both temporally and in terms in geographic distribution. Total rain ranges from 350 to 1500 mm per year (Nakalembe et al., 2017), with most regions receiving 500–1000 mm per year (USAID, 2017a). The impacts of climate change and variability on the region are multifaceted. Despite widespread perceptions that climate change is shortening the dry season, it is unclear that total rainfall has changed (Mercy Corps, 2016b). However, by most accounts, the temporal and spatial patterns, as well as intensity of rainfall are shifting and becoming less predictable—a concern for most Karamajong who are reliant upon primary production for their livelihoods (USAID, 2017a).

As in all places, the political ecology of the climate–conflict nexus in Karamoja is multidirectional and complex (Abrahams and Carr, 2017). Competition over resources, most notably water and pasture, do exist and can interact with other factors that affect conflict, such as land disputes and historic conflict between tribal and political rivals (Stark, 2011). That is, the effects of climate change and variability do not drive conflict, but instead follows the discourse of climate change as a "threat multiplier" whereby climactic events can exacerbate or hasten conflict, but will not cause it (Abrahams and Carr, 2017; CNA 2007).

Karamoja is, in many ways, the ideal enabling environment for NGOs looking to address the links between climate change, security, and conflict. Despite a relatively recent history of conflict, a disarmament of the people of Karamoja, coupled with a presence of the Uganda People's Defence Forces in areas prone to cattle raiding and conflict, has kept armed, violent conflict to a minimum. Thus, at present, the area is relatively stable. Coupled with rapidly improving infrastructure, the current security situation allows for easy access by humanitarian and development organizations. Further demonstrating the role conflict can play in increasing vulnerability to climate shocks and stressors, years of conflict in the region have increased the fragility to climactic shocks, decreased the adaptive capacity of communities, and increased vulnerability to hydrometeorological changes in the area (Rüttinger et al., 2015; USAID, 2017a). Moreover, many government officials in the area are deeply concerned about climate–conflict connections, and therefore are willing to offer support for these issues. Yet there are inherent challenges to developing and implementing programming that seeks to address the climate–conflict nexus. Below we outline two programmatic barriers facing implementation of ongoing programming in the region that are likely to be able to be generalized beyond the specific context.

Our examples are grounded in two programs being implemented by Mercy Corps in Karamoja: Building Resilience Against Climate Extremes and Disasters (BRACED) (Mercy Corps, 2017a) and the third iteration of Peace (PEACE III) in East and Central Africa (USAID, 2017b).[2] BRACED is a program with a broad remit of resilience and climate change adaptation (CCA) programming that explicitly addresses markets and livelihoods, agriculture, gender, natural resource management, and governance. PEACE III is a cross-border peacebuilding program that seeks to mitigate conflict and increase cooperation. A component of PEACE III is integrating climactic shocks that risk exacerbating or triggering conflict between different pastoralist groups and/or pastoralists and sedentary agriculturalists into broader peacebuilding efforts. Therefore, the two programs are complementary: BRACED is integrating programming related to climate change and variability in the context of conflict and postconflict dynamics whereas PEACE III is a cross-border peacebuilding program seeking to integrate concerns related to climate change and variability (Mercy Corps, 2017b; USAID, 2017b).

The ability to build and utilize adaptive capacities to employ longer-term land use, livelihood, mobility, financial, and other decisions is confronted with the challenges of more immediate needs of the community, and the associated coping strategies. This is an acute challenge in areas experiencing, or that have recently experienced, a complex crisis (Abrahams, 2014). This challenge was evident at multiple stages of the implementation of BRACED. For example, while implementing a program focused on manipulating plots of land to retain water and increase agricultural output in a community that had traditionally relied upon pastoralism but is now shifting to agriculture, the program officer responsible for the work described difficulties in recruiting community members to assist in the project. He attributed this to the challenges facing the community during the end of the dry season, namely lower agricultural production and acute food insecurity. Even though this programming was explicitly designed to mitigate the impacts of climatic variability on food insecurity, that same food insecurity and its inherent immediacy created a programmatic hurdle. The inherent challenge of misaligned temporal frames across the multiple stages of development represents an easily overlooked challenge that could push programming away from core goals. There is a need for program adjustments to address food insecurity and short-term needs (e.g., via crisis modifiers like cash transfers, food aid), in parallel with global adaptation programming underpinned by core development (Fig. 23.2).

---

2. Both PEACE III and BRACED have programming that exists outside the region, however, for the sake of these examples we only examine the programming taking place in Karamoja.

There is a limited understanding of the climate−conflict nexus beyond its potential role as a threat multiplier. This lack of understanding, coupled with the inherent complexity of the climate−conflict nexus and a lack of personnel with expertise in both the biophysical impacts of climate change and conflict response and mitigation, can create tangible challenges for program design and implementation. This was evident at a training put on by PEACE III. The purpose of the training was to educate representatives of local and central government agencies from northwest Kenya and northeast Uganda, as well as representatives from select NGOs, on issues of climate change and conflict and to develop locally driven strategies to address the link. During the training the facilitator separated the attendees into their respective geographic areas and asked each group to develop a plan to address the climate−conflict nexus for the geographic region each group represented. All groups, despite working independently, came up with plans to try and reduce charcoal production and therefore timber extraction largely through community sensitization. It was thought this would limit biophysical changes that are thought to drive conflict. Though charcoal extraction can be problematic, the extent of it is unlikely to affect the local climate or hydrological cycle (Chidumayo and Gumbo, 2013). Likewise, the scale of carbon emissions from charcoal extraction is miniscule when compared with global commerce, and there is no evidence charcoal is driving conflict in the region. This misunderstanding does not represent a failure of the group—nor of the training. The inherent complexity of the issue and the nascent state of understanding is embedded in the vague threat multiplier discourse (Abrahams & Carr, 2017). Understanding what precisely is meant by climate change and how that affects conflict can vary by geography. This example, demonstrates a fundamental challenge: the myriad components that link climate change to conflict, and the spatially and temporally unconfined nature of the nexus, can complicate program design.

Evident in these two examples are some of the systematic challenges facing donors, government agencies, and NGOs seeking to understand and address the joint challenges of climate change and conflict. The climate impacts that drive or influence conflict may happen in a place and/or time disconnected from where the conflict that it impacts take place. Conceptualizing climate change impacts on conflict and security is difficult; knowing what to target, as well as when, where, and how is very difficult. To better understand the complex systems and interconnected drivers of instability and how they threaten progress in Karamoja, Mercy Corps employed a Strategic Resilience Assessment (Mercy Corps, 2016b). This assessment enabled practitioners to use resilience systems thinking, overlaid with the shocks, stresses and drivers like conflict and climate that affect vulnerable groups differently, and identify what capacities are needed to enable these groups to cope, adapt and recover without backsliding. This resilience approach, engaged through a Strategic Resilience Assessment methodology, supported practitioners and partners to design a contextualized resilience theory of change, including targeted resilience-building strategies and interventions to support communities in achieving long-term well-being outcomes and transformational change amidst conflict and climate impacts (Mercy Corps, 2016b).

These examples raise important questions for those seeking to address the nexus with development programming, including (1) how does scale, temporal and spatial, affect the conflict and therefore the intervention; (2) does the link between climate change, conflict, and security offer any viable intervention points? (3) does it make more sense to frame the nexus as an issue of security and peacebuilding as compared to a threat multiplier in need of risk mitigation? and (4) do our approaches, methodologies, and partnerships enable adequate understanding of systems thinking and resilience-building for vulnerable populations? Explicit examination of these questions across the multiple stages of program design, implementation, and evaluation represents a potentially powerful, transparent and robust model on the development and implementation of programming adaptive to the climate−conflict nexus.

23.3.3   Case Study 2: Climate Change and Famine in Ethiopia

By many measures, Ethiopia is highly vulnerable to climate risks (GAIN, 2017; USAID, 2012). Increasingly frequent and intense weather events, such as drought, increasing temperatures, and increasing variability in important seasonal rainfall periods, are causing significant challenges, threatening a decade of rapid progress captured by key developmental indicators (Stark, 2011). However, climate factors do not operate in a vacuum, but are instead amplifying additional development challenges which are driving an increasingly fragile context through continuous feedback loops (Fig. 23.2).

The Government of Ethiopia (GOE) has made strong investments in both infrastructure and agriculture over the past decade. However, much of this growth is the result of strong public investment, restricting and often completely closing out private sector participation. As such, many services which the GOE are not prioritizing go unfulfilled. Ultimately, this limits poor Ethiopians from finding less climate-sensitive off-farm jobs, and getting access to financial and veterinary services, important for dealing with drought, and the access to climate information (e.g., climate early warning information) needed to lead prosperous, climate-resilient livelihoods (Mercy Corps, 2017).

Additionally, government initiated land-use planning and federal land policies have limited access to customary grazing lands, eroding the standing of traditional natural resource structures and processes. These policies have led to the redistribution of land, nullifying traditional land and resource access rights, while formalizing and reducing flexibility of boundaries (Mercy Corps, 2017). This has a compounding impact on mobility—an essential capacity of a pastoralist system in the face of an increasing population competing for a shrinking pasture and water resource base.

To address these issues Mercy Corps has implemented the Pastoralist Areas Resilience Improvement through Market Expansion (PRIME) program. Launched in 2012 in the Afar, Oromia, and Somali regions, PRIME is a 5-year multiagency program that focuses on supporting pastoralists via expansion of markets and long-term behavior change, aiming to increase the incomes of 250,000 households. As part of its approach, PRIME integrates strategies to support communities to become more resilient to climate change and its effects on society at various levels (Levine et al., 2016). As part of the PRIME program, Mercy Corps and its partner CARE, are working to restructure the means through which communities receive seasonal weather information, and in doing so increase cooperation, communication, and learning between the GOE and the communities it serves.

As part of its larger public safety net and development strategy, the GOE has invested in DRR under the Ministry of Agriculture's Disaster Risk Management and Food Security Sector (DRMFSS). Among its roles, the DRMFSS supports general management and overall coordination of early-warning disaster risk monitoring, response, and local institutionalization. The DRMFSS conducts continuous monitoring of seasonal weather risks and related impacts on livelihoods, aiding government and nonstate actors in making evidence-based DRR decisions. However, communities have limited opportunity to engage in the risk assessment process, by reporting on their nutrition and food security status. For the duration of the assessment process, facilitators fail to: (1) share formative summary findings in a timely, relevant, or user-friendly format, or (2) include communities in DRR decision-making planning processes. Instead, important decisions are discussed and made at higher levels. These gaps only exacerbate the increasingly fragile context across Ethiopia, by decreasing the inclusivity of communities in key processes which determine risk reduction investments, and furthering a lack of critical information which could inform communities' ability to more proactively manage scarce land and water resources in the face of seasonal risks.

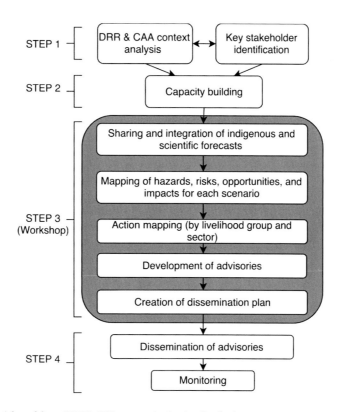

FIGURE 23.1 PSP process map. *Adapted from CARE's PSP process by Jessica Omukuti.*

PSP is one of a suite of progressive climate-resilient development strategies, implemented by PRIME, to address climate risks and build resilience capacities (Fig. 23.1). PRIME fosters a coordinated approach, which leverages a variety of stakeholders to explore potential climate and weather changes and their impacts. The process then uses that information to develop locally relevant action plans that support livelihoods, social and ecosystem adaptation.

Central to PSP is the creation of inclusive and respectful spaces for sharing and integrating indigenous and scientific knowledge and seasonal weather forecasts, so that all information may be interpreted and contextualized to inform local adaptation strategies. Within these spaces, stakeholders—households, private sector actors, and NGOs—collectively share, analyze, and interpret these seasonal weather forecasts to develop contingency plans aimed at supporting households, communities, and businesses in responding to different levels of risk and uncertainty.

Adopted from CARE's Adaptation Learning Program (CARE, 2014), the PSP process begins with a context analysis—providing an overview of existing DRR, CCA information, and early warning structures in target areas—and an exercise mapping key existing and potential DRR and CCA stakeholders (e.g., households, local NGOs, private sector actors). From there, the PRIME team trains all stakeholders on how to participate in the PSP process, and holds a multiday workshop where scientific and indigenous forecasts are shared, hazards and risks are mapped, and strategies are codeveloped. These strategies are then integrated into advisories which are disseminated through a variety of formal and informal channels.

While more formal evaluation processes are ongoing, early observations indicate that the PSP processes are allowing communities to plan more effectively for seasonal rainfall variability. This appears to reduce both their vulnerability to, and the impacts of, droughts and floods on communities. Observations suggest several additional positive effects of PSP processes on aspects of DRR and CCA planning, including: more comprehensive risk assessment and integrated (GOE-community) adaptation planning; increased inclusivity and transparency in DRR planning; increased government investment in responsiveness to community adaptation needs; and, greater social capital between communities and households.

The benefits of increasing access to seasonal weather and risk information also support other aspects of the PRIME program which seek to decrease fragility. One effort focuses on reducing risk to climate shocks and stresses by increasing engagement in climate-resilient livelihoods. PRIME works with women's cooperatives to train them on camel milk collection and transportation for local producers, while also investing funds into local camel milk production facilities. Relying on camels reduces climate-sensitivity due to their more diverse diet (leaves and grasses) than that of cattle (Levine et al., 2016).

An additional component of PRIME is to improve the management of the increasingly scarce rangeland resources, necessary to support pastoralist livelihoods. To do this the PRIME team has worked collaboratively to map the extent and boundaries of rangeland and reengage traditional rangeland councils to more sustainably, adaptively, and cooperatively manage them. Revitalization activities include the facilitation of more regular council meetings, organizing community rehabilitation activities, and establishing community dry season grazing reserves (Levine et al., 2016).

These strategies appear to be working as well. According to an evaluation of the PRIME program which looked at community resilience in the face of the 2016 El Nino drought, participating communities fared better than other households in their ability to maintain sufficient food, avoid impoverishment, maintain productive household assetsand animal health (Sagara and Hudner, 2017).

While this work is still ongoing, and the scale and impact of its strategies continue to be evaluated, there are important lessons and considerations which future programming should consider when addressing the links between climate and fragility: (1) development strategies should seek to increase *Bonding* and *Bridging* social capital in order to increase cooperation and management of critical resources in the face climate shocks (Mercy Corps, 2017c); (2) enhanced *Linking* social capital should be prioritized to increase transparency and cooperation between governing institutions and communities (Mercy Corps, 2017c); (3) increasing access to user-driven information can help inform and support willingness to engage in proactive risk reduction activities, help ensure a more unified understanding of risk, and support a more iterative process of adaptation over time. While efforts to embed this learning in development agendas will be challenging, especially in already fragile contexts, it is essential that donors increase flexible funding and longer-term trajectories practitioners develop innovative strategies.

23.4 CONCLUSIONS AND RECOMMENDATIONS

This chapter posits that a climate-resilient development approach can help bridge the humanitarian aid and development divide in fragile states, helping anticipate, prevent, mitigate, and reduce complex crises and positively transform systems. The two case studies presented here ground this position in three ways. First, the factors that define resilience are also those which determine a community's ability to prepare for, manage, and recover more quickly and effectively from climatic (and nonclimatic) shocks and stresses. These factors or capacities should therefore be strengthened. Second, engaging without a climate-resilient approach to humanitarian and development work implies programming with a climate-blind approach. Insufficient planning for the effects of climate change can multiply drivers of fragility in the long-term and potentially move a context from a relatively low level of fragility to a complex crisis. This could imply significant programmatic setbacks if beneficiaries backslide on the well-being outcomes gained, due to climate shocks and stresses degrading their coping and adaptive capacities and resources, and pushing them further into a negative alternate state of increased fragility (Fig. 23.2). Third, understanding of the root causes and drivers of risk and fragility, with a climate resilience approach to analyzing systems, enables entry points and intervention targets for transformation of governance, social, and economic systems to enable true climate-resilient development (Fig. 23.2).

Integration of climate change adaptive measures within resilience frameworks for humanitarian aid and development, through no-regrets adaptation strategies that are mutually beneficial for sustainability and ecosystem, are critical to address these lacunae. At the same time, CCA measures should help build resilience by directing funding and attention to reducing vulnerability, not just to climate change but also to environmental degradation, poverty, and conflict (Brown et al., 2007). A new conceptual framework is offered to illustrate these dynamics, and address the gaps identified and discussed above (Fig. 23.2).

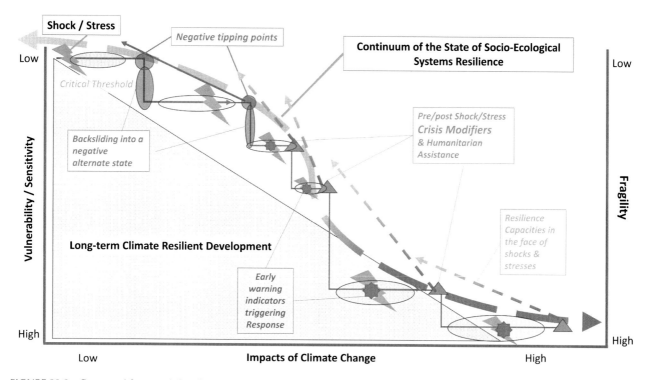

FIGURE 23.2 Conceptual framework for climate-resilient development in fragile contexts.

23.4.1 Bridging Solutions

Shift in time, and space. Resilience dividends and changing root causes of vulnerability within 1−5-year time frames in fragile contexts is unrealistic. Funders and policy-makers must acknowledge and adapt to this if true gains on value-for-money and supporting development of vulnerable populations in complex crises is a priority. Indeed, effective resilience to climate change involves a variety of actors from local civil society to economic development ministries to the security sector over extended timeframes. There are a number of entry points beyond the international climate change process to ensure that all relevant policy is designed in a climate-compatible and conflict-sensitive way (IISD, 2015). Evidence on effective climate resilience-building measures must be utilized and packaged well, in order to open such pathways for policy and institutional funding shifts.

We're all in the same sandbox. Epistemological divides, different world views and priorities, practical and policy differences often put humanitarian aid workers, development actors, and CCA practitioners on opposite sides of the playground. But these systemic challenges are intertwined and interdependent, across time, space, and the social constructs of addressing them. The long-term challenge of addressing root causes of complex crises, whilst supporting immediate lifesaving safety nets for populations within them, coupled with the uncertainty and unprecedented challenges that climate change fuels, is the shaky platform upon which humanitarian and development policies, programs, and actors must cocreate and coinvest.

Identify and address root causes and drivers of risk and fragility. Development and humanitarian response policies and programs are thus challenged to adequately identify and address underlying causes of vulnerability and exposure to conflict and disaster, whilst building resilience capacities of people to cope, adapt, and transform. Given the paucity of evidence, ongoing tracking of return on investment is needed to track gains in building resilience whilst reducing root causes of systemic vulnerability and impacts of shocks and stresses. Core to this success, governance systems-strengthening is needed to address drivers of inequity, human rights abuses, violence and conflict, and corruption, and enable transformation of sociocultural and gender norms that can drive risk.

We must never stop adapting. By recognizing and deconstructing our inherent biases and world views, and understanding the constant flux of climate and fragility dynamics across the world, humanitarian and development actors, funders, and policy-makers must partner closely with impacted communities to regularly adapt our approaches, methods and learning. This includes flexible, long-term funding and policy frameworks integrated undivided across humanitarian, development, and climate adaptation. Donor and though partner support for transparent, adaptive management processes embedded within learning and knowledge catalyzing platforms will be critical to more effectively sharing lessons and enhancing partnership and solidarity, and ultimately improving security and well-being outcomes.

Increase access to information, early warning and resilience analysis. A long-term climate-resilience approach engaging systems thinking, driven by comprehensive research and learning agendas to generate evidence guiding programs and policies, is a necessary lens to build long-term, equitable climate-resilient development and humanitarian efforts (Levine et al., 2016a; Mercy Corps, 2016a). This responds to the demand to increase our understanding and analysis capacities to measure and mitigate humanitarian and development program impacts on the environment, beyond limited environmental impact assessments. This is needed to transparently support cross-system resilience and sustainability. Scaling out such analyses at the household, community, and landscape scales through community engagement and mobilization processes, is critical to increase sustainability of evidence, and informing adaptive programming. We must build the evidence base inter- and intrasocioecological systems resilience metrics to understand programmatically within and across systems, and increase transparency and robustness of landscape-scale systems-resilience. As climate and conflict drive displacement, resilience-building efforts must explore supporting flexible, experimental monitoring and evaluation frameworks. This will enable adaptation Utilizing geospatial and mobile technologies can complement local data collection and program adaptation. Such adaptation includes establishing critical thresholds and benchmarks within programs, to trigger early warning indicators that solicit crisis modifier interventions (e.g., food aid, cash transfers, social protection measures) for avoidance and mitigation of backsliding for high-risk populations in fragile contexts (Fig. 23.2, Box 23.2).

Humanity, ecosystems, and stability are under grave threat around the world. There is a clear call to action to challenge and improve our understanding and practices for effectively reducing risk drivers through monitoring and engaging in proactive response to early warning triggers for conflict, climate, ecological degradation, food, nutritional, water, and human security. Integrated undivided funding and frameworks, information and evidence will enable improved protection and will empower high-risk communities and systems to move from a state of extreme vulnerability and fragility to one of increased resilience and stability. Evidence-generation and genuine partnership will be the backbone of our success, or failure, in responding to this urgent state of the world's most fragile contexts and challenging problems.

BOX 23.2 High-level Recommendations

1. **Long-term climate resilience-building development strategies addressing root causes and drivers** of vulnerability and exposure (Fig. 23.1);
2. **Governance systems-strengthening** to address drivers of inequity, human rights abuses, violence, conflict, and corruption, and enable transformation of cultural and gender norms;
3. **Access to, and understanding of, climate and environmental information** is critical to improve decision-making; and
4. **Understanding and measuring the resilience-state of all systems** (and their interactions) is critical to:
   - **Establishing critical thresholds and benchmarks** to trigger **early warning indicators and solicit crisis modifier interventions** to support **avoidance and mitigation of backsliding** (Fig. 23.2);
   - Improve insight into effective program strategies to **foster appropriate climate resilience absorptive, adaptive, and transformative capacities.**

ACKNOWLEDGEMENTS

The authors extend their sincere thanks to David Nicholson of Mercy Corps for reviewing this chapter, to Jessica Omukuti for her investigation and documentation of the PSP project in Ethiopia, to all of the Mercy Corps staff that worked on the projects listed in the case studies, to USAID and DFID for funding in support of the projects described in the case studies, and to UN Environment for funding presentation of this work at the Good Hope for Sciences Conference in Cape Town, August 2017.

REFERENCES

Abrahams, D., 2014. The barriers to environmental sustainability in post-disaster settings: a case study of transitional shelter implementation in Haiti. Disasters 38 (S1). Available from: https://doi.org/10.1111/disa.12054.

Abrahams, D., Carr, E.R., 2017. Understanding the connections between climate change and conflict: contributions from Geography and Political Ecology Climate Change Reports, Forthcoming.

Berkes, F., Colding, J., Folke, C. (Eds.), 2003. Navigating Social-Ecological Systems: Building Resilience for Complexity and Change. Cambridge University Press, United Kingdom.

Brown, O., Hammill, A., Mcleman, R., 2007. Climate change as the 'new' security threat: implications for Africa. Int. Affairs 83 (6), 1141–1154.

CARE, 2014. Community Based Adaptation: An empowering approach for climate resilient development and risk reduction Adaptation Learning Programme for Africa. http://careclimatechange.org/wp-content/uploads/2014/08/CBA_Brief_ALP_English.pdf (accessed May 2017.).

Chidumayo, E.N., Gumbo, D.J., 2013. The environmental impacts of charcoal production in tropical ecosystems of the world: a synthesis. Energy Sustain. Develop. . Available from: https://doi.org/10.1016/j.esd.2012.07.004.

CNA. 2007. National security and the threat of climate change. Alexendria, VA. https://www.npr.org/documents/2007/apr/security_climate.pdf.

Eaton, D., 2008a. The business of peace: raiding and peace work along the Kenya–Uganda border (Part I). Afr. Affairs 107 (426), 89–110.

Eaton, D., 2008b. The business of peace: raiding and peace work along the Kenya–Uganda border (Part II). Afr. Affairs 107 (427), 243–259.

FAO, 2017 access. Sustainable Livelihoods and Emergencies. http://www.fao-ilo.org/fileadmin/user_upload/fao_ilo/pdf/Sustainable_Livelihoods_and_Emergencies.pdf (accessed August 2017.).

Frankenberger, T.R., 2012. Enhancing Resilience to Food Insecurity amid Protracted Crisis. Prepared by: Timothy R. Frankenberger, President of TANGO International Inc.; Tom Spangler, TANGO International Inc.; Suzanne Nelson, TANGO International, Inc.; Mark Langworthy, TANGO International Inc. August 17, 2012 http://www.fao.org/fileadmin/templates/cfs_high_level_forum/documents/Enhancing_Resilience_FoodInsecurity-TANGO.pdf.

Henly-Shepard, S., Anderson, C., Burnett, K., Cox, L.J., Kittinger, J.N., Ka'aumoana, M., 2014. Quantifying household social resilience: a place-based approach in a rapidly transforming community. Nat. Hazards 75 (1), 343–363. January 2015. http://link.springer.com/article/10.1007%2Fs11069-014-1328-8# https://doi.org/10.1007/s11069-014-1328-8.

Howe, K., Stites, E., Akabwai, D., 2015. "We Now Have Relative Peace"; Changing Conflict Dynamics in Northern Karamoja, Uganda.

Hsiang, S.M., Burke, M., Miguel, E., 2013. Quantifying the influence of climate on human conflict. Science 341 (6151), 1235367. http://science.sciencemag.org/content/341/6151/1235367 Accessed May 2017.

IISD, 2015. Promoting Climate-Resilient Peacebuilding in Fragile States. March 2015. Written by A. Crawford, A. Dazé, A. Hammill, J-E. Parry and A.N. Zamudio. Support by Ministry of Foreign Affairs of Denmark.

IPCC, 2014. Summary for policymakers, Impacts, Adaptation and Vulnerability. https://www.ipcc.ch/pdf/assessment-report/ar5/wg2/ar5_wgII_spm_en.pdf 2014. Accessed April 2017.

Levine, E., McMahon, K., Nicholson, D., Wolfe, R., 2016a. Climate Change and Complex Crises. A Discussion Paper. Mercy Corps, July 2016.

Levine, E., Murphy, A., Nicholson, D., 2016b. Mercy Corps. Climate Change and Development: Experiences from Ethiopia and the Mercy Corps' PRIME program. Supported by the United States Agency for International Development (USAID). https://www.mercycorps.org/sites/default/files/ Mercy_Corps_CRD_Narrative_Case_study_Ethiopia-PRIME.pdf (accessed February 2017.).

Mercy Corps, 2016a. Mercy Corps Resilience Approach. https://www.mercycorps.org/sites/default/files/Resilience_Approach_Booklet_ English_121416.pdf (accessed May 2017.).

Mercy Corps, 2016b. Building an Empowered Karamoja STRESS Summary and Capacities. Mercy Corps Strategic Resilience Assessment Findings. August 2016 https://www.mercycorps.org/research/building-empowered-karamoja-stress-summary-and-capacities.

Mercy Corps, 2017a. A Strategic Resilience Assessment of Ethiopia. Internal Resource.

Mercy Corps, 2017b. BRACED http://www.braced.org/fr/about/about-the-projects/project/?id = b34bef40-1170-4503-b51b-1bb2c1fd179d (accessed May 2017.).

Mercy Corps, 2017c. Social Capital and Good Governance, A Governance in Action Research Brief. APRIL 2017. https://www.mercycorps.org/sites/ default/files/Social-Capital-Good-Governance-Mercy-Corps-2017.pdf (accessed August 2017.).

Nakalembe, C., Dempewolf, J., Justice, C., 2017. Agricultural land use change in Karamoja Region, Uganda. Land Use Policy 62, 2−12. Available from: https://doi.org/10.1016/j.landusepol.2016.11.029.

Notre Dame Global Adaptation Initiative (GAIN). http://index.gain.org/about. Accessed: September 2017.

OECD, 2016. States of Fragility 2016: Understanding Violence. Organization for Economic Cooperation and Development OECD Publishing, Paris. Available from: https://doi.org/10.1787/9789264267213-en.

Rüttinger, L., Smith, D., Stang, G., Tänzler, D., Vivekananda, J., Brown, O., Pohl, B., 2015. A New Climate for Peace. Berlin.

Sagara, B., Hudner, D., 2017. Enhancing Resilience to Severe Drought: What Works? Evidence from Mercy Corps' PRIME Program in the Somali Region of Ethiopia. Mercy Corps, Portland, OR.

Stark, J., 2011. Climate Change and Conflict in Uganda: The Cattle Corridor and Karamoja. United States Agency for International Development, Washington, DC.

Tanner, T., Lewis, D., Wrathall, D., Cradock-Henry, N., Huq, S., Lawless, C., et al., 2015. Livelihood resilience: preparing for sustainable transformations in the face of climate change. Nat. Clim. Chang. 1, January 2015. Published Online: 18 December 2014. https://doi.org/10.1038/ NCLIMATE2431.

UBOS, 2013. Uganda National Panel Survey 2010/2011 Wave II Report. Uganda Bureau of Statistics, Kampala, Uganda, Technical Report, June.

UNICEF, 2011. Humanitarian Action for Children: Building Resilience. UNICEF, New York.

UNU, 2016. United Nations University Centre for Policy Research. Background Paper. October 2016. Fragility, Risk, and Resilience: A Review of Existing Frameworks. Louise Bosetti Policy Officer. United Nations University, Tokyo, Japan.

USAID, 2012. USAID Conflict Assessment Framework version 2.0. United States Agency for International Development.

USAID, 2017a. Climate Risk Screening for Food Security: Karamoja Region, Uganda. Washington, DC.

USAID, 2017b. PEACE III Fact Sheet. https://www.usaid.gov/sites/default/files/documents/1860/PEACE_III_Fact_Sheet_August_2017.pdf (accessed August 2017.).

FURTHER READING

Bosetti, L., Ivanovic, A., Munshey, M., 2016. Fragility, Risk, and Resilience: A Review of Existing Frameworks. United Nations University Centre for Policy Research Background Paper.

CRED, 2015. The Human Cost of Natural Disasters 2015: A Global Perspective. Center for Research on Epidemiology of Disasters, CRED.

DARA, 2010. Climate Vulnerability Monitor 2010—The State of the Climate Crisis. http://daraint.org/wp-content/uploads/2010/12/CVM_Complete- 1-August-2011.pdf (Accessed June 2017.).

Maplecroft, 2016. Climate Change Vulnerability Index 2017. http://reliefweb.int/sites/reliefweb.int/files/resources/verisk%20index.pdf (accessed September 2017.).

Stark, J., Terasawa, K., Ejigu, M., 2011. Climate Change and Conflict in Pastoralist Regions of Ethiopia: Mounting Challenges, Emerging Responses. CMM Discussion Paper. United States Agency for International Development (USAID).

Stein, S., Walch, C., 2017. The Sendai Framework for Risk Reduction as a Tool for Conflict Prevention. Source: Social Science Research Council, https://environmentalpeacebuilding.org/library/show/libraryitem-3000 (accessed May 2017.).

Chapter 24

Ecological, Agricultural, and Health Impacts of Solar Geoengineering

Christopher H. Trisos[1], Corey Gabriel[2,3], Alan Robock[2] and Lili Xia[2]

[1]National Socio-Environmental Synthesis Center (SESYNC), University of Maryland, Annapolis, MD, United States, [2]Department of Environmental Sciences, Rutgers University, New Brunswick, NJ, United States, [3]Scripps Institution of Oceanography, La Jolla, CA, United States

Chapter Outline

Proposals to engineer the climate are not new (Budyko, 1974; Latham, 1990), but received little attention until Nobel Laureate Paul Crutzen's 2006 essay on the potential to cool the Earth by injecting sunlight-reflecting aerosols into the stratosphere (Crutzen, 2006)—breaking the taboo on geoengineering research. Since Crutzen's essay, slow and stalling progress on emissions reductions (Peters et al., 2013), increasing attribution of negative impacts in social and environmental systems to climate change (Carleton and Hsiang, 2016), and the ambitious goal in the Paris Climate Agreement to hold global warming to below 2°C (Rogelj et al., 2016) have all heightened attention on solar geoengineering as a potential tool to reduce the impacts of climate change.

The term "geoengineering" (also called "climate engineering" or "climate intervention") refers to suggestions to artificially enhance Earth's albedo, called albedo modification or solar radiation management (SRM), and to suggestions to remove carbon dioxide from the atmosphere to reduce the greenhouse effect, called carbon dioxide removal (CDR). They are quite distinct in technology, risks, costs, and benefits. Here we only address SRM. The aim of proposed SRM techniques is to increase the reflection of sunlight back to space to cool the climate. SRM is not typically considered climate change adaptation because the aim of proposed schemes is to reduce climate impacts on human and natural systems by reducing the amount of global warming rather than human and natural systems adapting to cope with higher temperatures. Thus, SRM might be used in combination with, or to buy time for, further mitigation, adaptation, and CDR efforts. For example, if climate sensitivity to carbon dioxide increasing is at the mid–high end of estimates then using some moderate and temporary amount of SRM in combination with aggressive emissions mitigation may keep global mean temperatures below 2°C and prevent climate impacts from more extreme warming (Smith and Rasch, 2013; Keith and MacMartin, 2015). The consideration of SRM technologies therefore requires the development of strong governance mechanisms and must be considered in close coordination with mitigation and adaptation efforts. Indeed, to deploy SRM in isolation would be reckless (Box 24.1).

Climate model simulations have shown that SRM techniques may be able to offset a substantial fraction of the change in temperature and precipitation from increased atmospheric greenhouse gas concentrations (Kravitz et al., 2013).

Resilience. DOI: https://doi.org/10.1016/B978-0-12-811891-7.00024-4

BOX 24.1 Potential Responses to Global Warming

Global warming is a real threat to humanity (IPCC, 2013), and there are a number of possible societal responses:

1. **Do nothing**, and hope that the problem is not so bad, or future technology will address most of the impacts. This has been the overwhelming global response so far.
2. **Mitigation**. Reduce the anthropogenic emissions of greenhouse gases and aerosols that are causing global warming. This is the most important response, and some steps are being taken, as outlined in the 2015 Paris Agreement at the 21st Conference of the Parties of the United Nations Framework Convention on Climate Change. But additional mitigation beyond the Paris pledges will be needed to prevent dangerous anthropogenic climate change.
3. **Adaptation**. Reducing the impacts of global warming by such actions as retreating from flooding sea coasts and new farming practices for different climates are already starting, but will not be enough to prevent the worst climate impacts.
4. **Geoengineering** (also called "climate engineering" or "climate intervention"). Geoengineering is defined as "deliberate large-scale manipulation of the planetary environment to counteract anthropogenic climate change" (Shepherd et al., 2009). The term is applied to two quite distinct ideas, removing the primary cause of global warming, CO_2, directly from the air (carbon dioxide removal or CDR), and reflecting sunlight to cool Earth (solar radiation management or SRM). CDR is probably a good idea, if it could be done on a large scale and inexpensively (National Research Council, 2015a), but it would take effect slowly. SRM, so far an unproven technology, might cool Earth quickly (Fig. 24.1), but would come with many potential risks and concerns (National Research Council, 2015b; Robock, 2016; and see Box 24.2). For example, the trade-offs in SRM effects on temperature and precipitation are such that no currently feasible SRM technique is able to return both temperature and precipitation to a preindustrial state (Kravitz et al., 2013).

BOX 24.2 Solar Radiation Management Schemes, and Their Potential Benefits and Risks

Although reflecting sunlight using space-based satellites, by brightening the ocean or land surface, or by brightening clouds over the oceans has been studied, the most serious proposal for SRM is creating a cloud in the stratosphere to reflect sunlight back to space, mimicking the effects of large volcanic eruptions (National Research Council, 2015b). The benefits, risks, and concerns of such a scheme are summarized in Table 24.1. See sections of this chapter for further details.

However, simulations have also shown trade-offs in SRM effects on temperature and precipitation so that no SRM technique is able to completely reverse anthropogenic climate change and return the climate to a preindustrial state (Kravitz et al., 2013). Furthermore, the risks associated with stratospheric SRM may outweigh the potential benefits (Robock, 2008, 2016), and more research is needed so that if policy-makers are tempted to implement SRM they will be making an informed decision.

Research to date has focused almost exclusively on climate and physical environment responses to SRM such as precipitation, temperature, and sea ice extent (Irvine et al., 2017). However, any decision to deploy SRM would rest fundamentally on the technology's impacts on natural and human systems (e.g., health, agriculture and ecosystems). Thus, SRM research is at a critical juncture where evaluation of SRM from the climate impacts community is required urgently in order to advance any contribution of SRM to reducing climate risks. In this chapter, we summarize existing knowledge of SRM impacts on climate, agriculture, ecosystems, and human health, and highlight priorities for future research. We focus our summary on the two most widely discussed and, at this time, most plausible SRM techniques: stratospheric aerosol injection (SAI) and marine cloud brightening (MCB).

24.1 SOLAR RADIATION MANAGEMENT TECHNIQUES

Large tropical volcanic eruptions clearly demonstrate the potential efficacy of SAI. For example, the Mt Pinatubo eruption in 1991 injected 15−20 million tons of sulfur dioxide gas into the stratosphere, which converted to a global cloud of sulfuric acid droplets, reflecting incoming sunlight back to space and causing a reduction in mean global surface air temperature of 0.3−0.5°C for 2 years (Soden et al., 2002). However, unlike the aerosols that naturally fall out of the stratosphere in the years after a volcanic eruption, SAI involves the continued injection of either aerosol particles (e.g., calcite) or their precursors (e.g., sulfur dioxide) into the lower stratosphere to sustain global cooling (Fig. 24.1; Irvine

TABLE 24.1 Risks or Concerns and Benefits of Stratospheric Geoengineering, Adapted from Robock (2008, 2014, 2016). See Relevant Chapter Sections for Further Details

Potential Benefits	Potential Risks or Concerns
Physical and biological climate system	***Physical and biological climate system***
1. Reduce surface air temperatures, which could reduce or reverse negative impacts of global warming, including floods, droughts, stronger storms, sea ice melting, and sea level rise.	1. Drought in Africa and Asia
	2. Decreased primary productivity
	3. Unexpected shifts in species regional distributions
	4. Ozone depletion
2. Increased primary productivity (land and oceans)	5. Continued ocean acidification
3. Increased terrestrial CO_2 sink	6. May not stop ice sheets melting
4. Reduced heat stress for coral reefs and other heat-sensitive ecosystems.	7. Impacts tropospheric chemistry
	8. Rapid warming if geoengineering stopped suddenly
Human impacts	***Human impacts***
5. Beautiful red and yellow sunsets	9. Less solar electricity generation
6. Increased crop yields relative to global warming impacts	10. Decreased crop yields
	11. More sunburn
7. Reduced heat-related mortality	12. Degrade passive solar heating
Governance	13. Effects on airplanes flying in stratosphere
8. Prospect of geoengineering being implemented could increase drive for mitigation efforts	14. Effects on electrical properties of atmosphere
	15. Affect satellite remote sensing
Unknowns	16. Degrade terrestrial optical astronomy
9. Unexpected or surprise benefits	17. White appearance of the sky
	18. Affect stargazing
	Governance
	19. Cannot stop geoengineering effects quickly
	20. Potential for commercial control of technology
	21. Who sets the thermostat?
	22. Societal disruption, and conflict among countries over optimal climate
	23. Conflicts with current international treaties
	24. Moral hazard – the prospect of geoengineering working could reduce effort for mitigation efforts
	Ethics
	25. Military use of geoengineering technology
	26. Moral authority – do we have the right to do this?
	Unknowns
	27. Human error during implementation
	28. Unexpected or surprise consequences

et al., 2016). In this chapter, we focus primarily on the consequences of the most widely simulated scenario—injection of sulfur dioxide (SO_2) that reacts with water to form a layer of sulfuric acid droplets. Depending on their size, the resultant aerosol particles would have a lifetime of approximately 1−3 years. Injecting the aerosols into the equatorial stratosphere at ∼20 km altitude would make the most effective use of stratospheric currents to spread the aerosol layer globally and achieve as even as possible a distribution of radiative forcing (Irvine et al., 2016). Initial estimates indicated that an injection of 3−5 million tons of SO_2 per year would be sufficient to offset warming from a doubling of preindustrial CO_2 concentration, but accounting for the growth of aerosol particles showed that emissions of 90 million tons of SO_2 per year would be required to offset business as usual greenhouse gas emissions by the end of the current century (Heckendorn et al., 2009; Niemeier and Timmreck, 2015). This would be the equivalent of 5−7 Mt Pinatubo eruptions per year. High-altitude aircraft are the most feasible option to deliver the aerosols at an estimated cost of USD 1−10 billion per million tons of material per year (Robock et al., 2009), which is a relatively inexpensive deployment, at least in terms of direct economic cost and orders of magnitude less than the estimated cost of decarbonizing the world's economy.

Placing an array of mirrors in space to block a small percentage of incoming sunlight has been proposed (Angel, 2006). Although space mirrors remain infeasible at present, simply turning down incoming solar radiation in climate

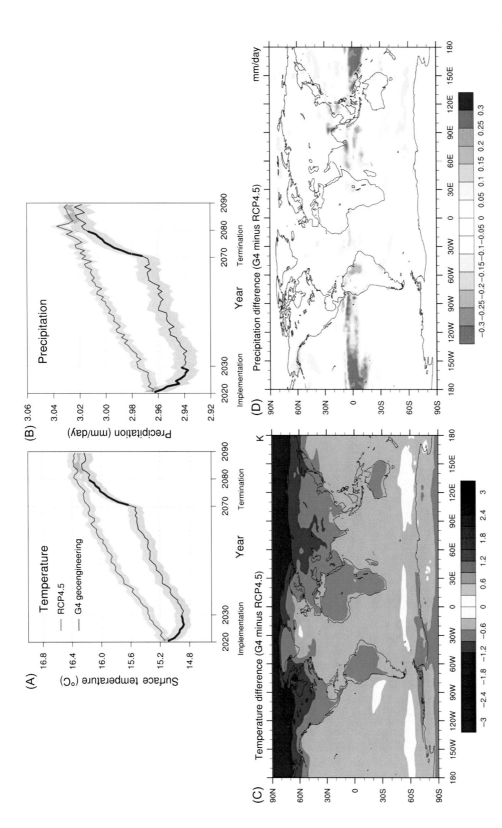

FIGURE 24.1 Comparison of global mean temperature (A) and precipitation (B) trends for RCP4.5—a scenario with moderate greenhouse gas mitigation equal to RCP4.5, but with stratospheric sulfur dioxide injections of 5 million tons per year beginning in 2020 and ending abruptly in 2070. The increasing temperatures in G4 after 2030 result from the continuation of greenhouse gas emissions increasing radiative forcing. Absent more aggressive mitigation, maintaining lower temperatures would require increasing aerosol injections. The difference between G4 and RCP4.5 scenarios for mean temperature (C) and precipitation (D) for the period 2030–2069, showing cooling and reduced precipitation for G4. Solid lines in A and B show global means and shaded regions show one standard deviation across 12 ensemble members from four climate models (see Trisos et al., 2018). Bold lines in A, B show the first 10 years of geoengineering ("Implementation") and the 10 years after geoengineering is stopped ("Termination").

simulation models (called "sunshade geoengineering") provides a useful proxy for understanding SAI. For example, the G1 scenario of the Geoengineering Model Intercomparison Project (GeoMIP) used models that turned down insolation by up to 5% instead of simulating the more realistic but computationally intensive creation of a stratospheric aerosol layer (Kravitz et al., 2013). We discuss sunshade geoengineering model results relevant to SAI and note the significant differences in the resulting climate responses of precipitation (Niemeier et al., 2013), solar radiation partitioning (Xia et al., 2016), and atmospheric chemistry (Nowack et al., 2016; Xia et al., 2017).

In contrast to SAI, marine cloud brightening (MCB) is a more geographically-specific approach. The proposed MCB scheme would inject salt from evaporated sea water into the marine boundary layer to directly scatter light and to increase the reflectivity (albedo), and potentially increase the persistence, of low-lying maritime clouds that reflect incoming solar radiation (Latham et al., 2008). MCB has an analogue in ship tracks, highly reflective marine clouds produced by particulate matter in ship exhausts that act as effective cloud condensation nuclei, producing a cooling effect on the order of -0.1 Wm^{-2} (Schreier et al., 2006). Calculations suggest an increase of approximately 0.06 in cloud-top albedo would generate cooling sufficient to offset a doubling of CO_2 concentration (Latham et al., 2008). Salter et al. (2008) proposed MCB deployment could be achieved with ships sailing perpendicular to the prevailing wind so that the injected plume of sea salt did not trail the ship. However, aligning ships at all times perpendicular to an always changing wind would require excellent weather forecasting, as well as rapid communication of forecasts to the ships to automatically adjust route in response to anticipated wind changes. Climate models that apply MCB globally or everywhere within 30° of the equator simulate additional particles in regions that will not form stable low-lying marine clouds, and have found that the increased direct reflection from those particles may be more effective than cloud brightening (Ahlm et al., 2017). Partanen et al. (2012) estimated marine boundary layer conditions to be suitable for MCB across only 3.3% of the global ocean area in three distinct regions that, to achieve a substantial global mean temperature reduction, would be exposed to very significant light limitation and local cooling: the west coast of the United States, the west coast of Africa in the Southern Hemisphere Tropics and subtropics, and the west coast of South America. These regions are also those with upwelling of nutrient-rich waters that make for rich fishing grounds, and are close to large population centers.

24.2 SOLAR RADIATION MANAGEMENT SCENARIO DEVELOPMENT FOR IMPACT ASSESSMENT

SRM poses a significant challenge to researchers assessing climate impacts on human and natural systems because of the flexibility to design SRM deployment to achieve a wide variety of specific climate outcomes; for example, either restoring Arctic sea ice cover or preindustrial precipitation (Irvine et al., 2017). This flexibility in how SRM is implemented (e.g., restricting SAI to a single hemisphere) thus presents both the climate simulation and impact assessment communities with a potentially overwhelming number of scenarios for analysis. To date, SRM research has focused on a prescribed set of idealized SRM scenarios developed primarily to understand climate responses to SRM—for example, SRM used to completely offset all warming from $4 \times CO_2$—rather than address specific climate policy goals (Keith and MacMartin, 2015). From a policy perspective the more relevant research focus is instead likely to be an engineering and design perspective that asks: What SRM strategy will achieve a particular set of climate, and more importantly, human and natural system outcomes?

One recent suggestion has been to focus impact assessment efforts on a set of more policy-relevant scenarios; for example, the temporary use of SRM to only partially offset warming and reduce climate impacts as mitigation proceeds (Keith and MacMartin, 2015; Kravitz et al., 2015). However, this smaller number of scenarios will still be insufficient to capture the variety of SRM deployment choices and thus address the wide range of questions about SRM impacts on natural and human systems. A partial solution could be to develop a smaller set of policy-relevant reference scenarios that have been assessed using complex Earth System and climate impacts models, and then use these to develop computationally efficient emulators for the climate effects of a wider range of SRM scenarios (Irvine et al., 2017). These simplified methods for climate emulation would include considerable uncertainties in climate responses to greenhouse gas concentrations and SRM that would limit the confidence in their projections. Thus, while a full scoping of the natural and human impacts of SRM will remain computationally limited, the integration of climate emulators with impact assessment could allow for some comparisons and improved understanding of the trade-offs among SRM options beyond a small and idealized set of scenarios (Irvine et al., 2017).

24.3 CLIMATE RESPONSES TO SOLAR RADIATION MANAGEMENT

Stratospheric aerosol injections, based on a volcanic analogue, operate as follows: solar radiation is scattered back to space and the surface cools (Robock, 2000). By studying observations and using volcanism as an analogue for SAI, Trenberth and Dai (2007) pointed out the possibility that drought, particularly in the tropics, could result from SAI. Many of the larger eruptions in the past millennium have also forced El Niño/Southern Oscillation (ENSO) variability (Emile-Geay et al., 2007; Maher et al., 2015; Khodri et al., 2017).

A "rich get richer, poor get poorer" paradigm states that, with global warming, increased moisture convergence in areas that already get a lot of precipitation will result in the "wet getting wetter," while increased moisture divergence in dry areas will result in the "dry getting drier" (Held and Soden, 2006). However, this paradigm does not hold up in an SRM world, where the response is very different from that under global warming. Tilmes et al. (2013) analyzed the hydrological cycle under geoengineering regimes where the solar constant was reduced to achieve preindustrial temperatures in a high CO_2 world and compared that to year 1850 preindustrial conditions. They found a strong reduction in global monsoon rainfall, including in the Asian and West African monsoon regions (see also Fig. 24.1). This illustrates the trade-offs in SRM effects on temperature and precipitation so that no currently feasible SRM technique is able to return the climate to a preindustrial state (Kravitz et al., 2013). Modeling of SAI that achieves $\sim 0.9°C$ of cooling over 40 years when compared against a concurrent simulation of Representative Concentration Pathway (RCP) 6.0 leads to a 3% reduction in precipitation (Xia et al., 2016).

If SAI were ever terminated abruptly, since the sulfate layer that reflects incoming solar radiation and reduces global mean temperature could only persist, without replenishment, for a couple of years, the climate would return rapidly to what it would have been had geoengineering not been imposed (Fig. 24.1). In geoengineering experiments that simulate SAI sufficient to generate a temperature reduction of $1-1.5°C$ when compared to a scenario with a 1% increase in CO_2 per year, that same amount of warming would occur over a 5 to 10-year period following termination (Jones et al., 2013). This rate of warming is approximately an order of magnitude faster than what would occur with global warming. The consequence of rapid termination would be to compress the equivalent temperature change experienced during multiple decades of global warming into less than one decade. Such rapid termination of geoengineering would not be a desirable policy response, but in the absence of international comity about how to respond to global warming, a climate policy response that includes rapid termination is possible.

There is no single aerosol optical depth, or latitudinal distribution of aerosol optical depth for sulfuric acid aerosols that could achieve a spatially uniform cooling that would be equally desirable everywhere (Ricke et al., 2013). Particular regions have distinct preferences for what constitutes an optimal climate, let alone what constitutes an optimal response to climate change. Thus, with currently feasible SRM technologies, it is unlikely that any single stratospheric geoengineering regime could be implemented that would be simultaneously optimal for all regions. Even if the average global temperature could be reduced with stratospheric geoengineering, there would still be important regional differences (e.g., Kravitz et al., 2013). For example, the northward shift in the jetstream in the Northern Hemisphere that has been shown to occur in observations of winters after volcanism and in model simulations of stratospheric geoengineering could lead to precipitation decreases in areas that receive a large percentage of their rainfall from midlatitude storms. Many of these incentives are alien to the mitigation discussion, which assumes a global collective interest. Therefore, unlike with mitigation, because different geoengineering approaches may have disparate regional impacts, powerful alliances could form to elevate specific climate engineering interests of benefit to some nations and at the expense of others (Ricke et al., 2013).

Geographical limitations on where effective MCB is possible may be an important constraint on its potential efficacy to reduce global mean temperature and avoid temperature-dependent effects of global warming. Jones et al. (2009, 2012) applied a global mean radiative forcing of $\sim 1 \ Wm^{-2}$ from MCB in the North Pacific, South Pacific, and South Atlantic. This led to a decrease in global mean temperature of $\sim 0.5°C$ relative to a moderate emissions mitigation scenario (RCP4.5). However, local climate change brought about by MCB led to substantial increases in precipitation in India and declines in northeast South America. How political coalitions would develop in an MCB-world versus an SAI-world is not known, but future research on MCB may emphasize the potential to use the technology as a regional approach to either supplement SAI, or deal with specific regional climate impacts, such as rapid warming in areas near coral reefs (Latham et al., 2013).

24.4 AGRICULTURAL IMPACTS

Changes in temperature, precipitation, solar radiation, surface air quality, and CO_2 concentration all impact agricultural productivity. SRM would impact all of these factors affecting agriculture. As changes in these factors become more severe the impacts on agricultural productivity will be exacerbated and will vary by crop and by region, generating potential winners and losers with respect to agriculture for a given geoengineering deployment. Here, we summarize knowledge on the impact of each factor on crop yields and the interactions among them.

24.4.1 Temperature

The length of the growing season is defined as the longest continuous period of time in a year that soil temperature and moisture conditions can support plant growth. Increases in the number and the magnitude of extreme daily maximum temperature events would limit the growing season and negatively affect agricultural production, often causing steep declines in yield or even crop failure when the temperature threshold (e.g., >30°C) of a particular cultivar is crossed (Carleton and Hsiang, 2016). Surprisingly, so far, effective adaptation to climate in agriculture has been modest, even when warming effects are gradual (Carleton and Hsiang, 2016). Thus, SRM might provide a potentially useful tool to reduce temperature impacts on agriculture. Indeed, with no changes in agriculture practices, models suggest cooling from SRM would benefit crop production across the tropics as crops are released from heat stress while it would damage crops in high latitudes as SRM would bring temperature below the optimal level for crop growth (Xia et al., 2014).

24.4.2 Precipitation

Evapotranspiration is the removal of water from soil through evaporation from the soil surface and transpiration from plants. Precipitation affects crop production coupled with temperature by determining the evapotranspiration rate of crops. There is the potential for large regional differences in precipitation-mediated agricultural impacts from SAI. In particular, a reduction in monsoon rainfall is a potential consequence of SAI (Robock et al., 2008; Tilmes et al., 2013) exposing countries such as India—where agriculture productivity is largely governed by the monsoon circulation—to potential negative agricultural impacts. However, as the cooling effect of SAI also reduces surface evaporation (Tilmes et al., 2013), reduced precipitation may not necessarily result in decreasing soil moisture—the available water for plant growth. Further studies are needed to understand how precipitation change from geoengineering would affect agriculture.

24.4.3 Solar Radiation

Solar radiation reaching Earth's surface is the primary driver of plant photosynthesis. Plant photosynthesis tends to increase nonlinearly with incident photosynthetically active radiation (PAR), and saturates at light levels that are often exceeded on bright days during the growing season. Once a saturation level is reached, the photosynthesis process stops, as does crop growth. In contrast, under cloudy skies or those with light-scattering aerosols, incoming radiation is more diffuse, producing a more uniform irradiance of the plant canopy with a smaller fraction of the canopy likely to be light-saturated (Mercado et al., 2009). As a result, such diffuse radiation results in higher light use efficiencies by plant canopies and has much less tendency to cause canopy photosynthetic saturation (Roderick et al., 2001). In addition, diffuse radiation leads to plant radiation use efficiencies at least twice those for direct sunbeam radiation (Gu et al., 2002). Hence, the net effect on photosynthesis of radiation changes associated with an increase in clouds or scattering aerosols depends on the balance between a reduction in the total PAR (which tends to reduce photosynthesis) and an increase in the diffuse fraction of the PAR (which tends to increase photosynthesis). SAI, which would scatter incoming sunlight, would decrease total solar radiation but increase diffuse radiation reaching the surface. This diffuse radiation enhancement would promote terrestrial plant photosynthesis and may benefit agriculture (Xia et al., 2016).

24.4.4 Surface Ozone Concentration

Surface ozone adversely affects agriculture (e.g., Mauzerall and Wang, 2001; Ainsworth et al., 2012). The causes of this reduction in agricultural productivity include reduced plant net photosynthesis, increased susceptibility of crops to disease, and reduced growth rate (Mauzerall and Wang, 2001). Avnery et al. (2011) concluded that in the year 2000 ozone-induced global crop yield reductions ranged from 8.5−14% for soybeans, 3.9−15% for wheat, and 2.2−5.5% for maize in comparison to the theoretical yield without ozone damage. Artificially reducing solar insolation with sulfate SAI would cause changes in atmospheric chemistry and dynamics that would impact surface ozone concentration, with strong regional differences. For example, if SRM started at 2020 and continued for 50 years, sunshade geoengineering would increase surface ozone concentration whereas sulfate SAI would reduce surface ozone concentration (Xia et al., 2017), and may thus benefit agriculture.

24.4.5 Ultraviolet (UV) Radiation

UV irradiance at Earth's surface is dependent on latitude, altitude, cloud coverage, and season. Under sulfate SAI, both the optical properties of the aerosols themselves and the depletion of the stratospheric ozone layer—which protects the Earth's surface from UV radiation from the Sun—would impact surface UV radiation exposure (Heckendorn et al., 2009; Tilmes et al., 2012). The amount of sulfate injection and its distribution determine stratospheric ozone depletion and changes in global cloud coverage. Simulating SAI of 2 million tons of sulfur per year, Tilmes et al. (2012) predicted a UV radiation increase of 6% and 12% in northern and southern high latitudes, respectively. Heckendorn et al. (2009) found that an injection of 5 million tons per year would cause a stratospheric ozone depletion of 7 and 11% in northern and southern high latitudes which would result in surface UV increases of 21 and 33%, respectively.

Since the ozone hole was discovered there have been extensive studies on the potential damage to plants from UV (reviewed by Ballaré et al., 2011). Field experiments manipulating UV showed that with a 20−30% UV increase due to a 10% ozone depletion, the reduction in plant growth rate was 6% or less (Allen et al., 1998; Ballaré et al., 2011). Whether increased UV radiation damages or benefits crop production is debated (e.g., Wargent and Jordan, 2013; Williamson et al., 2014). Increased UV inhibits leaf expansion (Searles et al., 2001) with smaller and fewer leaves slowing plant growth, but plants may also adapt to higher UV environments by producing pigmentation compounds that reduce UV penetration and protect plant photosynthesis (Bassman, 2004). Further studies are needed on whether UV radiation changes from SAI would impact crop yields. However, the above-mentioned increases in surface UV predicted for sulfate SAI are mild and restricted to high latitude regions such that the impact on agriculture may be relatively small, with the potential exception of regions already exposed to the ozone hole. Proposals to use alternative aerosol particles such as calcite for SAI may instead increase stratospheric ozone (Keith et al., 2016) and reduce UV surface radiation, potentially increasing plant growth.

24.4.6 Combined Effects on Agriculture

Very few studies have examined SRM impacts on agriculture in any detail. Pongratz et al. (2012) built a statistical model using observations of temperature, precipitation, CO_2 concentration, and crop yields, and used climate model output to study agriculture impacts of sulfate injection geoengineering. Global rice, maize, and wheat yields were predicted to increase due to the combination of CO_2 fertilization and reduced heat stress in sulfate injection geoengineering compared to a doubling of CO_2 without geoengineering. However, possible rice yield losses were predicted to occur in the middle latitudes of the Northern Hemisphere.

Xia et al. (2014) used a process-based crop model, the Decision Support System for Agrotechnology Transfer (DSSAT), to simulate crop responses to sunshade geoengineering in China. The crop model used the output from 10 global climate models that simulated the GeoMIP G2 scenario—starting in 2020, a reduction in solar radiation to balance a 1% per year increase in CO_2 concentration (1pctCO2) for 50 years. Without changing land management practices, compared to CO_2 fertilization and climate change from a 1% per year increase in CO_2, the effect of SRM with CO_2 fertilization was predicted to have little impact on rice production in China (-3.0 ± 4.0 million tons per year). This is because the CO_2 fertilization effect compensates for the negative effects on rice production of changes in precipitation, temperature, and sunlight from SRM. In contrast, SRM was predicted to increase Chinese maize production by 18.1 ± 6.0 Mt/yr ($13.9 \pm 5.9\%$) relative to 1pctCO2 because lower temperatures from SRM reduced heat stress in more heat-sensitive maize seeds. The termination of SRM showed negligible impacts on rice production but led to a 19.6 Mt/yr (11.9%) reduction of maize production.

A recent study on groundnuts in India found that, relative to climate change from a moderate emissions scenario (RCP4.5), groundnut yields decreased up to 20% for a scenario with SAI sufficient to offset all temperature increase from RCP4.5 (Yang et al., 2016). Yield reductions were mainly a result of enhanced water stress from reduced summer monsoon rainfall. When SAI was terminated, the groundnut yield tended back to the level of the RCP4.5 scenario. Yang et al. considered changes in temperature, precipitation, solar radiation, and CO_2. However, since the comparison was between scenarios which have the same CO_2 concentration, the CO_2 fertilization effect was canceled out.

A single study has simulated crop responses to MCB geoengineering. Parkes et al. (2015) focused on spring wheat in northeastern China and groundnut in West Africa under global warming from a 1% CO_2 increase per year (capped at 560 ppm), and compared this with a MCB scenario for three ocean regions (North Pacific, South Pacific, and South

Atlantic) totaling 5% of the global ocean surface. Relative to the global warming scenario, MCB was predicted to increase yields for both wheat and groundnuts and reduce crop failure rates as a result of cooling and, in the West African case, enhanced precipitation.

To date, agriculture impact studies have considered climate changes of temperature, precipitation, and total solar radiation. However, other climate factors discussed at the beginning of this section, including UV, partitioning of solar radiation (diffuse and direct), and ozone are important to agriculture. Crop model improvements are needed to include these processes, and to further understanding of how solar geoengineering impacts on agriculture will vary by crop and by region. In addition, climate changes from solar geoengineering might influence the ecology of pest and pathogen species, as well as the crop transport and storage chain. Further studies that include these ecosystem and economic impacts are needed to fully understand how geoengineering might influence agriculture and food availability.

24.5 ECOLOGICAL IMPACTS

Assessments of the ecological impacts of SRM are extremely limited, especially regarding impacts on biodiversity and ecosystems. The creation of a high CO_2 and low temperature climate that could result from SRM is unprecedented in recent Earth history. This lack of historical analogues limits inference of ecological impacts and raises novel questions about how organisms and ecosystems may respond (McCormack et al., 2016).

24.5.1 Productivity on Land and in the Oceans

Increased CO_2 concentrations increase plant photosynthesis (Allen et al., 1987) and reduce transpiration, improving plant water-use efficiency (reviewed by Leakey et al., 2009). These direct effects of CO_2 are a major driver of vegetation change, increasing plant growth across the tropics, especially in arid regions. Because SRM does not reduce CO_2 concentrations this CO_2 fertilization effect is common to scenarios with and without SRM. However, changes in other climate variables due to SRM have potentially significant impacts on plant productivity. Compared to global warming without SRM, sunshade geoengineering is predicted to increase net primary productivity (NPP)—a measure of the total carbon flux from the atmosphere to plants—in tropical regions by reducing heat stress on plants, but decrease NPP at high latitudes as reduced temperature increases from SRM prevent the increased plant growth forecast to occur in these cold regions with global warming (Glienke et al., 2015). However, this result varies among models mainly due to model inclusion or exclusion of a nitrogen cycle (Jones et al., 2013; Glienke et al., 2015). In addition to these regional effects of cooling on vegetation growth, SAI may increase plant photosynthetic rates due to the increase in diffuse radiation from an aerosol layer producing a more uniform irradiance of the plant canopy (Xia et al., 2016). Theoretically, under SAI, the combination of enhanced diffuse radiation and cooling would increase plant photosynthesis, and the cooling would suppress plant and soil respiration, as shown by a modeling study of the effects of the Mt. Pinatubo eruption in 1991 (Mercado et al., 2009). Those effects from SAI are expected to increase the land carbon sink substantially, moderating CO_2 increases. However, these benefits may be overstated as SAI simulations have not yet included nitrogen and phosphorous nutrient limitations on plant growth explicitly, and the increase in photosynthesis due to more diffuse light may be balanced by a decrease in photosynthesis from the lower direct sunlight due to SAI (Kalidindi et al., 2014).

Similar to SAI, relative to no geoengineering, MCB is predicted to increase terrestrial NPP in the tropics and suppress NPP at high latitudes (Jones et al., 2012). However, an important difference is the potential for stronger differences in NPP responses among tropical regions, depending on which ocean regions are selected for MCB. In particular, climate simulations show MCB in the South Atlantic could reduce precipitation in the Amazon or northeastern South America substantially with corresponding negative impacts on plant productivity (Jones et al., 2009; Jones et al., 2012).

Only two studies have assessed potential impacts of SRM on marine ecosystem productivity, none for SAI. Using an Earth system model, Partanen et al. (2016) found that MCB decreased global ocean NPP slightly ($\sim 1\%$) compared to global warming without MCB. However, there were major regional differences in NPP due to reduced light availability in regions where MCB was deployed, especially off the coast of Peru where phytoplankton growth is light-limited in the model. In contrast, Hardman-Mountford et al. (2013) suggest a 90% reduction of light from MCB would redistribute but not decrease NPP within the water column. However, their model used a one-dimensional water column (i.e., depth cross-section) for a single region, and although it depicts the marine ecosystem in more detail than Partanen et al. (2016) it does not include movement of nutrients across regions as in an Earth system model. A clear next step is to repeat the Hardman-Mountford et al. (2013) study in a water column more typical of water columns in candidate MCB regions.

24.5.2 Biodiversity and Ecosystem Impacts

To avoid extinction from climate change, species can either respond by adapting to new conditions within their current geographic ranges or by moving to track their climate conditions across space (e.g., geographic range shifts) or time (e.g., earlier spring emergence). Only a single study has assessed potential SRM impacts on global biodiversity. By calculating climate velocities—the speeds and directions that species would need to move to track climate changes (i.e., to stay in their climate niche)—Trisos et al. (2018) estimated that were SRM ever terminated abruptly the movement speeds required to keep pace with climate change would likely exceed dispersal capacities for many species, increasing local extinction risk in marine and terrestrial biodiversity hot spots compared to global warming without SRM.

In the ocean, studies have assessed tropical coral reef responses to SRM. Anomalously high ocean temperatures induce coral bleaching when corals eject their photosynthetic symbionts, often resulting in colony death. By reducing sea surface temperatures, both SAI and MCB could reduce the likelihood of high temperature events, enhancing coral reef survival and the extent of suitable habitat compared to global warming without SRM (Latham et al., 2013; Kwiatkowski et al., 2015). Compared to a reference global warming scenario without SRM, ocean acidification would increase slightly under SRM due to the increased solubility of CO_2 in cooler water (Keller et al., 2014). Although corals are sensitive to increasing ocean acidification this was of secondary importance to heat stress in the maintenance of suitable coral habitat (Couce et al., 2013; Kwiatkowski et al., 2015).

On land, positive effects of SRM on plant productivity have the potential to drive significant change in grassland and savannah biomes. Elevated CO_2 concentration has been suggested as a major driver of tree and shrub invasion into tropical grassy ecosystems (Bond and Midgley, 2000, 2012), with negative impacts for grassland biodiversity, ecotourism, and livestock grazing. Additional cooling from SRM could further advantage C3 photosynthetic trees over C4 photosynthetic grasses, speeding up the woody encroachment of grassland biomes and the transformation of savannahs to closed forests. Enhanced woody thickening of these ecosystems could reduce wildfire frequencies relative to global warming without geoengineering and alter the global carbon cycle. Earth system models that include dynamic vegetation and wildfire dynamics are needed to further understand the potential impacts of SRM on terrestrial tropical biomes.

In the case of SAI, deposition of sulfate aerosols will in general increase the acidity of precipitation which is known to damage ecosystems when the sulfate is sufficiently concentrated. However, for an injection of 5 million tons of SO_2 per year, Kravitz et al. (2009) found that only ecosystems already close to thresholds for acid rain deposition would be susceptible to damage.

24.6 HEALTH IMPACTS

There have been no quantitative studies published of SRM impacts on human health. Sunshade geoengineering could increase surface ozone (Nowack et al., 2016; Xia et al., 2017) which, in addition to reducing crop yields (Mauzerall and Wang, 2001), would have substantial negative impacts on human health (Silva et al., 2013). In contrast, simulations of atmospheric chemistry suggest SAI would decrease surface ozone (Xia et al., 2017), but with an increase of UV radiation, which increases the risk of skin cancer (Tilmes et al., 2012). Aerosols from SAI would be deposited at the Earth's surface posing a risk of chronic health effects from prolonged exposure. Most aerosols proposed for SAI have known or suspected negative effects on respiratory and cardiovascular health with other health effects such as metabolic abnormalities depending on the aerosol (Effiong and Neitzel, 2016). A preliminary study by Eastham (2015) suggested that for sulfate SAI almost all of the descending aerosols would be removed by wet deposition so that the direct contribution of sulfate aerosols to aerosol particulate matter at the surface would be very low. The same analysis suggested that, per degree of cooling, sulfate SAI could result in an additional 26,000 premature deaths per year. However, these direct impacts from SAI on human health are likely minor compared to the direct and indirect effects from changes in crop yield, heat-related mortality, drought, or flood exposure under geoengineered climates. The inclusion of these impacts into integrated assessments is required to generate more complete forecasts of solar geoengineering impacts on human health relative to other global warming scenarios.

24.7 CONCLUSION

Little is known about the potential environmental and social impacts of SRM. The increasing attention on SRM as a potential tool to offset climate impacts, buying time for additional mitigation, CDR, and adaptation efforts, demands development of more policy-relevant SRM scenarios and their inclusion into the mainstream of climate impacts assessment. Ultimately decisions on whether and how to deploy SRM will be based primarily on the potential for SRM to reduce environmental and social impacts. Estimates of these impacts must be up to the task.

ACKNOWLEDGMENTS

Trisos is supported by U.S. National Science Foundation grant DBI-1052875. Gabriel, Robock, and Xia are supported by U.S. National Science Foundation grants GEO-1240507 and AGS-1617844.

REFERENCES

Ahlm, L., Jones, A., Stjern, C.W., Muri, H., Kravitz, B., Kristjánsson, J.E., 2017. Marine cloud brightening—as effective without clouds. Atmos. Chem. Phys 17, 13071—13087. Available from: https://doi.org/10.5194/acp-17-13071-2017. https://doi.org/10.5194/acp-2017-484.

Ainsworth, E.A., Yendrek, C.R., Sitch, S., Collins, W.J., Emberson, L.D., 2012. The effects of tropospheric ozone on net primary productivity and implications for climate change. Annu. Rev. Plant. Biol. 63, 637—661.

Allen, D.J., Nogués, S., Baker, N.R., 1998. Ozone depletion and increased UV-B radiation: is there a real threat to photosynthesis? J. Exp. Bot. 49, 1775—1788.

Allen Jr, L.H., Boote, K.J., Jones, J.W., Jones, P.H., Valle, R.R., Acock, B., et al., 1987. Response of vegetation to rising carbon dioxide: Photosynthesis, biomass, and seed yield of soybean. Global. Biogeochem. Cycles. 1, 1—14.

Angel, R., 2006. Feasibility of cooling the Earth with a cloud of small spacecraft near the inner Lagrange point (L1). Proc. Nat. Acad. Sci. 103, 17184—17189.

Avnery, S., Mauzerall, D.L., Liu, J.F., Horowitz, L.W., 2011. Global crop yield reductions due to surface ozone exposure: 1. Year 2000 crop production losses and economic damage. Atmos. Environ. 45, 2284—2296.

Ballaré, C.L., Caldwell, M.M., Flint, S.D., Robinson, S.A., Bornman, J.F., 2011. Effects of solar ultraviolet radiation on terrestrial ecosystems: Patterns, mechanisms, and interactions with climate change. Photochem. Photobiol. Sci. 10, 226—241.

Bassman, J.H., 2004. Ecosystem consequences of enhanced solar ultraviolet radiation: secondary plant metabolites as mediators of multiple trophic interactions in terrestrial plant communities. Photochemistry Photobiology 79, 382—398.

Bond, W.J., Midgley, G.F., 2000. A proposed CO2-controlled mechanism of woody plant invasion in grasslands and savannas. Global Change Biology 6, 865—869.

Bond, W.J., Midgley, G.F., 2012. Carbon dioxide and the uneasy interactions of trees and savannah grasses. Phil. Trans. R. Soc. B 367, 601—612.

Budyko, M.I., 1974. Climate and Life. Academic Press, New York, New York.

Carleton, T.A., Hsiang, S.M., 2016. Social and economic impacts of climate. Science 353, 1112.

Couce, E., Irvine, P.J., Gregorie, L.J., Ridgwell, A., Hendy, E.J., 2013. Tropical coral reef habitat in a geoengineered, high-CO_2 world. Geophys. Res. Lett. 40, 1799—1805.

Crutzen, P.J., 2006. Albedo enhancement by stratospheric sulfur injections: A contribution to resolve a policy dilemma? Clim. Change. 77, 211—220.

Eastham, S.D., 2015. Human health impacts of high altitude emissions. Mass. Inst. Technol., Cambridge, Mass.

Effiong, U., Neitzel, R.L., 2016. Assessing the direct occupational and public health impacts of solar radiation management with stratospheric aerosols. Environ. Health 15, 1—9.

Emile-Geay, J., Seager, R., Cane, M.A., Cook, E.R., Haug, G.H., 2007. Volcanoes and ENSO over the past millennium. J. Climate 21, 3134—3148.

Glienke, S., Irvine, P.J., Lawrence, M.J., 2015. The impact of geoengineering on vegetation in experiment G1 of the GeoMIP. J. Geophys. Res. Atmos. 120, 10196—10213.

Gu, L.H., Baldocchi, D., Verma, S.B., Black, T.A., Vesala, T., Falge, E.M., et al., 2002. Advantages of diffuse radiation for terrestrial ecosystem productivity. J. Geophys. Res. Atmos. 107 (D6). Available from: http://dx.doi.org/10.1029/2001JD001242.

Hardman-Mountford, N.J., Polimene, L., Hirata, T., Brewin, R.J., Aiken, J., 2013. Impacts of light shading and nutrient enrichment geo-engineering approaches on the productivity of a stratified, oligotrophic ocean ecosystem. Journal of The Royal Society Interface 10, 20130701. http://dx.doi.org/10.1098/rsif.2013.0701.

Heckendorn, P., Weisenstein, D., Fueglistaler, S., Luo, B.P., Rozanov, E., Schraner, M., et al., 2009. The impact of geoengineering aerosols on stratospheric temperature and ozone. Environ. Res. Lett. 4, 045108.

Held, I.M., Soden, B.J., 2006. Robust responses of the hydrological cycle to global warming. J. Climate 19, 5686—5699.

IPCC, 2013. Summary for Policymakers. In: Stocker, T.F., Qin, D., Plattner, G.-K., Tignor, M., Allen, S.K., Boschung, J., et al.,Climate Change 2013: The Physical Science Basis. Contribution of Working Group I to the Fifth Assessment Report of the Intergovernmental Panel on Climate Change. Cambridge University Press, Cambridge, UK and New York, NY, USA.

Irvine, P.J., Kravitz, B., Lawrence, M.G., Muri, H., 2016. An overview of the Earth system science of solar geoengineering. Wiley Interdisciplinary Reviews: Climate Change 7, 815—833.

Irvine, P.J., Kravitz, B., Lawrence, M.G., Gerten, D., Caminade, C., Gosling, S.N., et al., 2017. Towards a comprehensive climate impacts assessment of solar geoengineering. Earth's Future 5, 93—106.

Jones, A., Haywood, J.M., 2012. Sea-spray geoengineering in the HadGEM2-ES earth-system model: radiative impact and climate response. Atmospheric Chemistry and Physics 12, 10887—10898.

Jones, A., Haywood, J., Boucher, O., 2009. Climate impacts of geoengineering marine stratocumulus clouds. Journal of Geophysical Research: Atmospheres 114, D10.

Jones, A., Haywood, J.M., Alterskjær, K., Boucher, O., Cole, J.N., Curry, C.L., et al., 2013. The impact of abrupt suspension of solar radiation management (termination effect) in experiment G2 of the Geoengineering Model Intercomparison Project (GeoMIP). J. Geophys. Res. Atmos. 118, 9743−9752.

Kalidindi, S., Bala, G., Modak, A., Caldeira, K., 2014. Modeling of solar radiation management: a comparison of simulations using reduced solar constant and strato-spheric sulphate aerosols. Clim. Dyn. 44, 2909−2925.

Keith, D.W., MacMartin, D.G., 2015. A temporary, moderate and responsive scenario for solar geoengineering. Nature Climate Change 5, 201−206.

Keith, D.W., Weisenstein, D.K., Dykema, J.A., Keutsch, F.N., 2016. Stratospheric solar geoengineering without ozone loss. Proc. Nat. Acad. Sci. 113, 14,910−14,914.

Keller, D.P., Feng, E.Y., Oschlies, A., 2014. Potential climate engineering effectiveness and side effects during a high carbon dioxide-emission scenario. Nat. Commun. 5, 3304.

Khodri, M., Izumo, T., Vialard, J., Janicot, S., Cassou, C., Lengaigne, M., et al., 2017. Tropical explosive volcanic eruptions can trigger El Niño by cooling tropical Africa. Nature Communications 8, 778. Available from: https://doi.org/10.1038/s41467-017-00755-6 .

Kravitz, B., Robock, A., Oman, L., Stenchikov, G., Marquardt, A.B., 2009. Sulfuric acid deposition from stratospheric geoengineering with sulfate aerosols. J. Geophys. Res. Atmos. 114, D10107. Available from: https//doi.org/10.1029/2008JD011652.

Kravitz, B., Caldeira, K., Boucher, O., Robock, A., Rasch, P.J., Alterskjær, K., et al., 2013. Climate model response from the geoengineering model intercomparison project (GeoMIP). J. Geophys. Res. Atmos 118, 8320−8332.

Kravitz, B., Robock, A., Tilmes, S., Boucher, O., English, J.M., Irvine, P.J., et al., 2015. The geoengineering model intercomparison project phase 6 (GeoMIP6): Simulation design and preliminary results. Geoscientific Model Development 8, 3379−3392.

Kwiatkowski, L., Cox, P., Halloran, P.R., Mumby, P.J., Wiltshire, A.J., 2015. Coral bleaching under unconventional scenarios of climate warming and ocean acidification. Nat. Clim. Change 5, 777−781.

Latham, J., 1990. Control of global warming? Nature 347, 339−340.

Latham, J., Rasch, P., Chen, C.C., Kettles, L., Gadian, A., Gettelman, A., et al., 2008. Global temperature stabilization via controlled albedo enhancement of low-level maritime clouds. Phil. T. Roy. Soc. A 366, 3969−3987.

Latham, J., Kleypas, J., Hauser, R., Parkes, B., Gadian, A., 2013. Can marine cloud brightening reduce coral bleaching? Atmos. Sci. Lett. 14, 214−219.

Leakey, A.D.B., Ainsworth, E.A., Bernacchi, C.A., Rogers, A., Long, S.P., Ort, D.R., 2009. Elevated CO_2 effects on plant carbon, nitrogen, and water relations: six important lessons from FACE. J. Experimental Botany 60, 2859−2876.

Maher, N., McGregor, S., England, M.H., Gupta, A.S., 2015. Effects of volcanism on tropical variability. Geophys. Res. Lett. 42, 6024−6033.

Mauzerall, D.L., Wang, X.P., 2001. Protecting agricultural crops from the effects of tropospheric ozone exposure: Reconciling science and standard setting in the United States, Europe, and Asia. Annual Review of Energy and the Environment 26, 237−268.

McCormack, C.G., Born, W., Irvine, P.J., Achterberg, E.P., Amano, T., Ardron, J., et al., 2016. Key impacts of climate engineering on biodiversity and ecosystems, with priorities for future research. Journal of Integrative Environmental Sciences 13 (2−4), 103−128.

Mercado, L.M., Bellouin, N., Sitch, S., Boucher, O., Huntingford, C., Wild, M., et al., 2009. Impact of changes in diffuse radiation on the global land carbon sink. Nature 458 (7241), 1014−U1087.

National Research Council, 2015a. Climate Intervention: Carbon Dioxide Removal and Reliable Sequestration. The National Academies Press, Washington, DC. Available from: https://doi.org/10.17226/18805.

National Research Council, 2015b. Climate Intervention: Reflecting Sunlight to Cool Earth. The National Academies Press, Washington, DC. Available from: https://doi.org/10.17226/18988.

Niemeier, U., Timmreck, C., 2015. What is the limit of climate engineering by stratospheric injection of SO_2. Atmos. Chem. Phys. 15, 9129−9141.

Niemeier, U., Schmidt, H., Alterskjær, K., Kristjánsson, J.E., 2013. Solar irradiance reduction via climate engineering: Impact of different techniques on the energy balance and the hydrological cycle. J. Geophys. Res. Atmos. 118, 11,905−11,917.

Nowack, P.J., Abraham, N.L., Braesicke, P., Pyle, J.A., 2016. Stratospheric ozone changes under solar geoengineering: implications for UV exposure and air quality. Atmos. Chem. Phys. 16, 4191−4203.

Parkes, B., Challinor, A., Nicklin, K., 2015. Crop failure rates in a geoengineered climate: impact of climate change and marine cloud brightening. Environ. Res. Lett. 10, 084003.

Partanen, A.I., Kokkola, H., Romakkaniemi, S., Kerminen, V.M., Lehtinen, K.E., Bergman, T., et al., 2012. Direct and indirect effects of sea spray geoengineering and the role of injected particle size. J. Geophys. Res. Atmos. 117 (D2).

Partanen, A.I., Keller, D.P., Korhonen, H., Matthews, H.D., 2016. Impacts of sea spray geoengineering on ocean biogeochemistry. Geophys. Res. Lett. 43, 7600−7608.

Peters, G.P., Andrew, R.M., Boden, T., Canadell, J.G., Ciais, P., Le Quéré, C., et al., 2013. The challenge to keep global warming below 2°C. Nature Climate Change 3, 4−6.

Pongratz, J., Lobell, D.B., Cao, L., Caldeira, K., 2012. Crop yields in a geoengineered climate. Nature Clim. Change 2, 101−105.

Ricke, K.L., Moreno-Cruz, J.B., Caldeira, K., 2013. Strategic incentives for climate geoengineering coalitions to exclude broad participation. Environmental Research Letters 8, 014021.

Robock, A., 2000. Volcanic eruptions and climate. Rev. Geophys. 38, 191−219.

Robock, A., 2008. 20 reasons why geoengineering may be a bad idea. Bull. Atomic Scientists 64, 14−18.

Robock, A., 2014. Stratospheric aerosol geoengineering. Issues Env. Sci. Tech. 38, 162−185 (special issue "Geoengineering of the Climate System").

Robock, A., 2016. Albedo enhancement by stratospheric sulfur injection: More research needed. Earth's Future 4. Available from: https://doi.org/10.1002/2016EF000407.

Robock, A., Oman, L., Stenchikov, G., 2008. Regional climate responses to geoengineering with tropical and Arctic SO_2 injections. J. Geophys. Res. 113, D16101. Available from: https://doi.org/10.1029/2008JD010050.

Robock, A., Marquardt, A.B., Kravitz, B., Stenchikov, G., 2009. The benefits, risks, and costs of stratospheric geoengineering. Geophys. Res. Lett. 36, L19703. Available from: https://doi.org/10.1029/2009GL039209.

Roderick, M., Farquhar, G.D., Berry, S.L., Noble, I.R., 2001. On the direct effect of clouds and atmospheric particles on the productivity and structure of vegetation. Oecologia. 129, 21−30.

Rogelj, J., Den Elzen, M., Höhne, N., Fransen, T., Fekete, H., Winkler, H., et al., 2016. Paris Agreement climate proposals need a boost to keep warming well below 2 C. Nature 534, 631−639.

Salter, S., Sortino, G., Latham, J., 2008. Sea-going hardware for the cloud albedo method of reversing global warming. Philos. T. Roy. Soc. A 366, 3989−4006.

Schreier, M., Kokhanovsky, A.A., Eyring, V., Bugiaro, L., Mannstein, H., Mayer, B., et al., 2006. Impact of ship emissions on microphysical, optical and radiative properties of marine stratus: A case study. Atmos. Chem. Phys. 6, 4925−4942.

Searles, P.S., Flint, S.D., Caldwell, M.M., 2001. A meta-analysis of plant field studies simulating stratospheric ozone depletion. Oecologia. 127, 1−10.

Shepherd, J.G.S., et al., 2009. Geoengineering the climate: Science, governance and uncertainty. The Royal Society), London RS Policy Document 10/09. Available from: https://royalsociety.org/~/media/Royal_Society_Content/policy/publications/2009/8693.pdf.

Silva, R.A., West, J.J., Zhang, Y., Anenberg, S.C., Lamarque, J.-F., Shindell, D.T., et al., 2013. Global premature mortality due to anthropogenic outdoor air pollution and the contribution of past climate change. Environ. Res. Lett. 8, 034005.

Smith, S.J., Rasch, P.J., 2013. The long-term policy context for solar radiation management. Clim. Change. 121, 487−497.

Soden, B.J., Wetherald, R.T., Stenchikov, G.L., Robock, A., 2002. Global cooling following the eruption of Mt. Pinatubo: A test of climate feedback by water vapor. Science 296, 727−730.

Trenberth, K.E., Dai, A., 2007. Effects of Mount Pinatubo volcanic eruption on the hydrological cycle as an analog of geoengineeringGeophys. Res. Lett 34, L15702. Available from: https://doi.org/10.1029/2007GL030524.

Tilmes, S., Kinnison, D.E., Garcia, R.R., Salawitch, R., Canty, T., Lee-Taylor, J., et al., 2012. Impact of very short-lived halogens on stratospheric ozone abundance and UV radiation in a geo-engineered atmosphere. Atmos. Chem. Phys. 12, 10,945−10,955.

Tilmes, S., Fasullo, J., Lamarque, J.-F., Marsh, D.R., Mills, M., Alterskjaer, K., et al., 2013. The hydrological impact of geoengineering in the Geoengineering Model Intercomparison Project (GeoMIP). J. Geophys. Res. Atmos. 118, 11036−11058. Available from: https://doi.org/10.1002/jgrd.50868.

Trisos, C.H., Amatulli, G., Gurevitch, J., Robock, A., Xia, L., and Zambri, B., 2018. Potentially dangerous consequences for biodiversity of solar geoengineering implementation and termination. Nature Ecology & Evolution 1. Available from: http://dx.doi.org/10.1038/s41559-017-0431-0.

Wargent, J.J., Jordan, B.R., 2013. From ozone depletion to agriculture: understanding the role of UV radiation in sustainable crop production. New Phytologist 197, 1058−1076.

Williamson, C.E., Zepp, R.G., Lucas, R.M., Madronich, S., Austin, A.T., Ballaré, C.L., et al., 2014. Solar ultraviolet radiation in a changing climate. Nature Climate Change 4, 434−441.

Xia, L., Robock, A., Cole, J.N.S., Ji, D., Moore, J.C., Jones, A., et al., 2014. Solar radiation management impacts on agriculture in China: A case study in the Geoengineering Model Intercomparison Project (GeoMIP). J. Geophys. Res. Atmos. 119, 8695−8711. Available from: https://doi.org/10.1002/2013JD020630.

Xia, L., Robock, A., Tilmes, S., Neely III, R.R., 2016. Stratospheric sulfate geoengineering could enhance the terrestrial photosynthesis rate. Atmos. Chem. Phys. 16, 1479−1489. Available from: https://doi.org/10.5194/acp-16-1479-2016.

Xia, L., Nowack, P.J., Tilmes, S., Robock, A., 2017. Impacts of stratospheric sulfate geoengineering on tropospheric ozone. Atmos. Chem. Phys. 17, 11913−11928. Available from: https://doi.org/10.5194/acp-17-11913-2017.

Yang, H., Dobbie, S., Ramirez-Villegas, J., Feng, K., Challinor, A.J., Chen, B., et al., 2016. Potential negative consequences of geoengineering on crop production: A study of Indian groundnut. Geophys. Res. Lett. 43, 11,786−11,795. Available from: https://doi.org/10.1002/2016GL071209.

Chapter 25

The Progression of Climate Change, Human Rights, and Human Mobility in the Context of Transformative Resilience—A Perspective Over the Pacific

Cosmin Corendea[1] and Tanvi Mani[2]

[1]United Nations University – Institute for Environment and Human Security, Bonn, Germany, [2]The Graduate Institute of International and Development Studies, Geneva, Switzerland

Chapter Outline

25.1 INTRODUCTION

The Paris Agreement has within Article 8 recognized the immediate need to avert, minimize, and address the loss and damage associated with the debilitating impacts of climate change. This includes extreme weather conditions as well as slow onset events. There has also been an understanding of the need to enable and encourage sustainable development in order to reduce the risk of climate related loss and damage. The particular areas of cooperation identified include the establishment of early warning systems as well as adequate networks of preparedness in case of emergencies, events that involve permanent loss and damage and slow onset events. There is a particular need for comprehensive risk assessment and management in the form of risk insurance facilities, pooling of climate risks and alternative insurance solutions. (Article 8, Paris Agreement). The noneconomic losses sustained by communities as well as the resilience of their livelihoods and ecosystems must also be taken into consideration. This is especially important in the context of Pacific Island Countries.

Approximately 95% of the land in the Pacific Island Countries falls within a form of customary ownership wherein the land belongs to a group linked together through kinship. The land cannot be bought or sold and is passed on across generations. It therefore is integral to the core identity of the Pacific Islanders and cannot be separated from the people. The people, the land, and the culture are mutually coexistent and support the continued sustenance of one another. This connect the people have to the land is deeply emotional and surpasses traditional structures of economic valuation (Barnett, 2008). Forced community relocation has the potential effect of disrupting this bond as a result of which the relationship between the land and its people may itself be challenged and the cultural identity of individuals as well as communities could face significant endangerment (Barnett, 2008).

Climate change has the potential to impact land security or the physical presence of habitable land, livelihood security that encompasses the productivity of the subsistence and commercial activities undertaken on the land, and habitat security, which enables the sustenance of a secure and vigorous environment (Campbell and Hamilton, 2014). The

Resilience. DOI: https://doi.org/10.1016/B978-0-12-811891-7.00025-6

increase in temperature within both the air and water would result in an increase in the severity of hurricanes and cyclones, according to scientific records (IPCC, 2007). Since the 1970s the number of sudden onset natural disasters has tripled and almost 90% of the natural disasters recorded today are climate related (EM-DAT: The OFDA/CRED International Disaster Database). Approximately 6% of the population within Pacific Island countries were affected by sudden onset disasters in the period between 2000−2011. Of these, 87% were climatological and hydrometeorological disasters. For example, the economic losses suffered by Samoa during the period within which it faced the maximum number of disasters was approximately 46% of its GDP; the corresponding figures for Vanuatu and Tonga were 30s and 14%, respectively (Bettencourt et al., 2006). The impact of these natural disasters has been exacerbated by the increased urbanization associated with development within these regions, deforestation, and the destruction of the mangroves, which had served as an effective buffer against flooding and tsunamis (Julca and Paddison, 2010). The disasters have also had the effect of intensifying existing vulnerabilities within the population. The poor, for example, face a disproportional impact of flooding, earthquakes, and landslides, which completely debilitates their makeshift, poorly constructed housing (OHCHR, Regional Office for the Pacific, 2011). In other words, adverse impacts of climate change have resulted in the inability of the land to support the livelihoods of its occupants and in extreme cases, the loss of capacity to sustain life itself causing the forced displacement of the original inhabitants of the land (Campbell and Hamilton, 2014).

The consequent migration of vulnerable populations has resulted in significant loss and damage at an individual as well as a community level. This includes but is not limited to quantifiable economic losses such as those associated with relocation and resettlement, for example, housing and infrastructure. But more importantly, it would also entail the potential loss of identity associated with reintegration into significantly different communities, which may even be hostile to such migration. As a result, seeking employment and higher education within host communities would also be difficult. Added to this, the social, psychological, cultural, and emotional loss suffered by these communities is significantly debilitating and yet often remains sidelined within the climate change loss and damage articulation.

This chapter seeks to understand resilience in the context of migration and attempts to formulate mechanisms, posited in international law, rights-based approaches, and regional governance, which may be able to enhance the resilience of climate migrants. It assesses the need for a bottom-up approach with adequate stakeholder engagement and an appreciation of the culture and identity of the migrants so as to ensure that the policies derived are not devoid of a relevant sociocultural context and are therefore more effective.

25.2 CONTEXTUALIZING RESILIENCE WITHIN THE MIGRATION NARRATIVE IN THE PACIFIC

Resilience has been defined as "the ability of a system and its component parts to anticipate, absorb, accommodate or recover from the effects of a hazardous event in a timely and efficient manner through the preservation, restoration and improvement of its essential structure and basic functions" (IPCC, 2012). The resilience of a social-ecological system is determined by its capacity to anticipate shocks and absorb them so as to maintain and recover the functional relationships that aid in its survival. Social systems have inherent safeguards to maintain these functional relationships such as laws, rules, traditions, and customs. These safeguards maintain the stability and security of a social system, while also providing a predictable environment in which the members of the society can interact with each other in a beneficial, productive manner. The ability of a social structure to effectively address shocks and environmental stresses is derived from the inherent capacity, whether adaptive, absorptive, anticipatory, or transformative, of the social system itself. A system possessing these inherent capacities is less likely to be undermined by shocks and stresses. Thus, the well-being of its members can be secured, both internally and through external interventions that strengthen this internal capacity, in the face of extreme climate change (Bahadur et al., 2015).

A school of thought that is gaining significant traction is that of transformative resilience. This form of resilience calls for an enhanced focus on the capacity of social-ecological systems to renew themselves and develop in a manner that utilizes environmental change as a means to innovate, evolve, and create new pathways that would further improve their ability to adapt effectively (Pelling, 2011). Resilience may therefore also be alternatively defined as "The capacity of people across generations to sustain and improve their livelihood opportunities and well-being despite environmental, economic, social, and political disturbances" (Tanner et al., 2015). The migration of displaced people to environments that offer them stability and security or enhances their capacity to respond more effectively to climate shocks strengthens the transformative resilience of both the migrating communities as well as the host communities.

This argument relies on the notion that the scope of climate risks and the scale on which they manifest requires a transformational shift in the management of risk—a shift that goes beyond merely adapting to the changing environment. Transformative resilience catalyzes the response to climate change in a manner that allows for communities to not just adjust to changes in their environment but also learn from such change and utilize both shared experiences and collaborative mechanisms to reorganize and reinvent themselves. This would enhance their resilience and aid in the development of unique ways to cope better with such dynamic changes in the near future (Wilkinson and Peters, 2015). This is predicated on the recognition that communities themselves are dynamic systems that benefit from social, cultural, and political diversity of their governing institutions, local associations, and coalitions.

The resilience of a social system to climate change relies significantly on the sustained integration of socioculturally and economically diverse influences. This diversity helps strengthen the collaborative networks within the community and enhances its capacity to sustain environmental and economic shocks (Pelling, 2011). Polycentric, diversified, and multilayered institutions assist in the creation of more robust communities, which are then able to draw on diverse resources to enhance their resilience (Pelling, 2011). Transformative resilience therefore offers tangible benefits to both migrating communities as well as host communities by enabling adaptation to irreversible changes through a core restructuring process. This process uses diverse experiences to create a new reservoir of ideas and deliberations. It also forms more widespread linkages to derive comprehensive solutions thereby making the transition within the newly transformed environment, more tenable.

Within the Pacific island region, the resilience of the inhabitants is inherently intertwined with their interaction with the island resources for subsistence purposes. Despite the rampant urbanization, the land still serves as the most important source of subsistence for the islanders. Within most outer islands in the archipelagic countries, the out-migration over the past 50 years has left vast swathes of empty lands (Connell, 2010). This has resulted in an increase in the diversity of communities living in populated areas. An especially significant aspect of the resilience of the Pacific islanders is the close interlinking with the cultural practices and the innate characteristics of the communities themselves. For example, the emphasis on communal living and an interconnected network of support is manifested in the nature of the response of the people to natural disasters and unprecedented changes in the environment (Rasmussen et al., 2009).

The long-standing tradition of community decision-making has also been an integral part of the resilience narrative within the region. As a result of colonization and the subsequent establishment of a central government, this community decision-making weakened to some extent. However the failure of the central government to effectively govern the peripheral parts have strengthened the determination of the communities, leaving them with a significant amount of power in defining their own responses to climate adversity and allowing them to play an important role in the formulation and sustainment of strategies for their development. It is also important to consider the awareness of the Pacific islanders who have an inherent understanding of the vagaries of the environment as it forms an essential component of their subsistence. They are therefore more receptive to information regarding the changes in the environment and mechanisms through which they can respond to it in the most effective manner (Bedford and Ho, 2003). Moreover, Pacific Islanders are literate and well educated. As a result of this, they have a heightened ability to comprehend, in an effective manner, the need to adapt to climate change and may be receptive to appropriate strategies that would aid in their adaptation (Smith and Long, 2000).

External Aid—A Counter Perspective

An argument that is increasingly gaining traction is that the degree of dependence of the Pacific Island nations on external aid is too significant to enable the recipient nations to determine their own course of action for themselves. In the context of climate change-induced migration, this is especially problematic as it might result in the adoption of specific interventions or environmental policies and legislations of the donor country without any due diligence as to its appropriateness within the recipient country. The development of resilience strategies by donor countries may even inhibit in-country community involvement and alienate the stakeholders most directly affected by the changing environment. What might work in a developed country context may not be viable in the context of the Pacific Island countries. Formal legislation can neither be effectively enforced nor disseminated as it is the traditional upholders of customary law such as the chiefs who hold the most sway in most parts of the island nations. There is therefore a need to integrate formal legislation and international law with local customary practices and traditional rules so as to enable an integrated framework that is both viable and sustainable in the long run and can effectively meet the demands of a dynamically diverse, migratory population.

25.3 UNDERSTANDING HUMAN MOBILITY IN THE PACIFIC CONTEXT

Migration is an integral part of the culture of the Pacific Islanders. It was used as an adaptation technique to spread environmental and economic risks over a larger area thereby reducing the exposure of vulnerable communities to these risks. Mobility and the subsequent expansion of social linkages have always been ingrained as part of the Pacific history and tradition (Hau'ofa, 1993). More recently, this mobility has become a strategic mechanism of responding to climate risks and has been subsumed in the everyday life of islanders. Historically, internal and external migration was caused not only by climate change but also by economic, social, and cultural factors. However, climate change has increasingly become an important driver of migration in the region. The reasons for this include the threat to land security caused by coastal inundation or erosion rendering habitation impossible, the threat to subsistence and cash-based livelihoods due to limited employment opportunities and depleting food security, and the deterioration of living conditions due to the loss of water supply, increase in diseases, and greater exposure to extreme events.

Migration presents an opportunity to improve the adaptive capacity of the communities by reducing the risks associated with climate change and enhancing resilience. The movement of populations across the islands can result in the mobility of social capital and aid in the transfer of ideas, knowledge, and useful skills; some of which can also be used to engineer more suitable and effective adaptive responses (Locke, 2009). Further, it would also reduce the demands on the natural resources in the source regions, which are already at risk due to the adverse climate. In spite of this, there is a political and psychological resistance to address the migrants as "climate refugees" as the islanders believe that this would leave them without any rightful claims to the land and threaten their core identity, which is intrinsically linked to the land and the strong ties of community associated with this land (Barnett and O'Neill, 2012).

In order to facilitate migration as a sustainable adaptive strategy, policy interventions must take into account the maintenance of culture and community while moving across borders; thereby aligning the response to climate adversity with an understanding of the needs and obligations of the stakeholders most affected by its impacts. For example, the concept of migration with dignity has been firmly entrenched in the strategic framework that Kiribati has planned to adopt in response to climate change. It has planned for a schematic international migration system comprising of creating opportunities for voluntary migration and supporting the establishment of expatriate communities in I-Kiribati. Further, it has also planned to assist in the education and qualification of Kiribati people so they may be more skilled as migrants and would be able to integrate seamlessly within the host communities (Office of the Republic of Kiribati).

Planned relocation on a large scale would also have to take into account risks such as unemployment, homelessness, community alienation, food insecurity, and possible landlessness. In the Pacific region specifically, most of the risks associated with migration are those related to the relationship the people have with their land. This includes the loss of identity, unstable family linkages, conflicting cultures, and issues of governance arising from the possible resettlement of the migrant community in what were originally customary lands of the host community (Boege, 2011). Although planned relocation would be ideal, it would require an assurance that the culture and identity of the migrants would be maintained, in order to reduce their resistance to such relocation (Kuruppu and Liverman, 2011). The planning of an overarching migration policy should therefore be participatory in nature and involve Pacific communities in the strategic formulation of mechanisms to maintain their culture and identity. It should be implemented in consonance with an adaptation policy so as to facilitate a greater level of resilience to climate adversity and must be a staged, long-term process. Finally, engaging with the islanders and creating awareness about the possibility of migration in the future would give them the opportunity to prepare and respond more favorably to the idea.

Migration with Dignity

Kiribati's migration with dignity policy attempts to construct a relocation strategy that is both widespread and sustainable in the long-term. This policy foresees the creation of opportunities for migrants in the near future through fostering ties with expatriate communities in various receiving countries such as New Zealand and Australia. This would allow migrants to receive the benefits of community living, enable the preservation of their culture and identity, and enhance their access to opportunities to send remittances back. The government has incurred significant costs to improve both educational and vocational qualifications within Kiribati so as to ease the integration of migrants within the regions they migrate to. This will build on existing cross-border labor arrangements and safeguard livelihoods. However, the shortcoming of this policy is that it caters only to voluntary migrants who already have preexisting skills and are literate; leading to the alienation of those with primarily subsistence-based livelihoods. It does not ensure a protective migration mechanism in an equitable manner. Its success therefore rests on sustained positive outcomes in both the sending and receiving countries.

25.4 CASE STUDIES THAT SUPPORT A REGIONAL APPROACH

The following case studies are the greatest proponent of the regional approach as they encapsulate the need to focus on a bottom-up integration of climate migrants in a manner that takes into consideration the various challenges faced by them and the best mechanism to address these challenges at a grassroots level. These studies exemplify the nature of migration in the Pacific and perpetuate the notion that relocation, whether forced or voluntary, is almost always a last resort—one that is undertaken with great personal loss and in the face of immense risk. While there do exist alternatives to migration in the form of adaptation and early action, in most parts these measures are ineffective against the barrage of climate risks, forcing large communities to migrate to more stable environments. The case studies, as elaborated below, provide an insight into the nature of migration in the Pacific and are an extremely useful resource base for the derivation of migration policies in the future, that draw their essence from the lessons learned within these instances of historical voluntary and forced migration.

25.4.1 Fiji

Fiji has experienced unpredictable climate patterns for the last 30 years. The sea level has risen from 1 to 1.5 mm per year prior to 1990 to 3 mm in 2015. The 5th IPCC Assessment Report concluded that at this rate by 2100 the sea level could rise by 50–120 cm leaving large parts of the island uninhabitable, through either coastal erosion, flooding, or saltwater intrusion into the soil and ground reserves. Coral bleaching, changing patterns of rainfall, and severe cyclones also pose significant climate risks to the island.

The National Climate Change Policy was established in 2008 to identify the roles and responsibilities of the ministries as well as formulate strategies to address the debilitating impact of climate change in the short and long term. Although the 2012 national climate change summit resulted in increased awareness and an understanding of the need to enhance community resilience, it was found that there still exist significant gaps in assessing the varying degrees of vulnerability of Fiji's communities. This resulted in the government undertaking a nationwide community risk mapping under the leadership of the Climate Change division (Watch, 2015).

The climate risk assessment has enabled the identification of numerous villages that had been affected by rising sea levels. Forty-five of these villages have been assigned with extensive plans of relocation in the next 5 to 10 years. However the mobilization of the resources required for such relocation is still the subject of much ambiguity (Watch, 2015).

According to a study conducted by A Ravuvu, "The people of the Nakorosule village in Fiji have not been able to live without the physical embodiment of their land, on which the survival of both individuals as well as the entire community depends...Land is therefore an extension of the self and the people are an extension of the land" (Ravuvu, 1988).

A significant precedent for a successful relocation is the migration of the people of the Vunidogoloa Village in the Cakaudrove Province, situated within Fiji's second largest island Vanua Levu. The relocation was undertaken as a response to severe climate change-induced flooding and was planned as a final alternative to protect the inhabitants from further danger to their lives and offer them the prospect of earning a fulfilling livelihood elsewhere as well. The relocation was therefore planned significantly in advance. The foresight, long-term plan, and systematic consultation with the community at the grassroots level made the entire relocation process more tenable, thus contributing significantly to its success—lessons that can be incorporated within future relocation strategies across the Pacific as well. The consultations with the government regarding the prospect of relocation had begun in 2006, when the village had begun to flood severely with every subsequent high tide. The cost of such relocation was over USD 800,000 including the construction of 30 new homes, fish ponds, copra drier, and farms. The communities involved did contribute to the cost of relocation along with other partners. However, there were still costs that had to be upheld by the government such as schools, waste water, sewage and power facilities. It was ultimately a cost intensive and complex process, one that also involved significant emotional turmoil for the uprooted families (IOM, EDGE, Sciences Po, 2015).

Experts who have studied this relocation have identified factors such as the availability of land and sufficient resources, the mutual trust the villagers have for each other, a strong leadership combined with a participatory process, the support of the government, and improved living conditions within the new site as integral to the success of the relocation (Bread for the World/PCC, 2015). These findings are extremely relevant while deriving important lessons that can be incorporated in the strategic policy formulation for undertaking specific relocations, as a response to climate change-induced disasters, in the future. It is therefore essential that a determination of such factors be made prior to undertaking such relocations and their viability be assessed on a metric of these identified factors.

25.4.1.1 Historical Relocation Case Studies

25.4.1.1.1 Banabans in Rabi

Background: In December 1945 the people of the island of Banaba (also known as Ocean Island) in present-day Kiribati were moved to Rabi Island in Fiji to make way for mining by the British Phosphate Commission, a joint British, Australian, and New Zealand enterprise. The Banabans say they had no choice in moving and it was a case of forced relocation; unlike the Vaitupuans in Kioa, who voluntarily took the decision to migrate themselves (McAdam, 2014).

Reason for forced relocation: Ever since the early 1900s, when phosphate deposits were discovered on Ocean Island, the Banabans had been regarded as an "awkward obstacle" to phosphate mining operations (jointly carried out by the UK, Australia, and New Zealand) (Deb, 1975). In 1942, Rabi Island in Fiji was purchased on the Banabans' behalf as a "second home"— essentially as an insurance policy against the time when Ocean Island might be rendered uninhabitable on account of the mining. Later that year, Ocean Island was occupied by Japan, who dispersed most of the Banabans across the colony. At the end of the Second World War, the British colonial authorities considered it expedient to move the Banabans straight to Rabi rather than back to Ocean Island. To this day, the Banabans claim that they were misled about the conditions in Rabi and the nature of the move, and that this was an unjust, forced relocation (McAdam, 2014).

Response and integration: Banabans feel an affinity with Fiji, however, and a sense of gratitude for its hospitality in accommodating them on Rabi. A common refrain among those interviewed was, "I am Banaban. I am only Fijian because of my passport and because I live in Fiji." In his submission to the Fiji Constitutional Commission, the former Rabi representative in the parliament of Kiribati, David Christopher (who was later an MP in the Fijian government), put it this way: The word Fijian is associated with race, with the race of the indigenous community of Fiji. Banaba is an island in the central Pacific Ocean. The indigenous community on Banaba is called the Banabans. I am a descendant of the indigenous community on Banaba and I call myself a Banaban. I find it difficult and most uncomfortable to call myself a Fijian as I was not a Fijian and will never be a Fijian (McAdam, 2013a).

25.4.1.1.2 Vaitupuans in Kioa

Background: In October 1947, part of the population of Vaitupu, an island of present-day Tuvalu, bought and settled in a Fijian island, Kioa, just kilometers from Rabi. The Vaitupuans voluntarily purchased the island of Kioa in Fiji as a safeguard against future overpopulation. They were not motivated by imminent land scarcity or extreme environmental conditions, nor coerced by the authorities (McAdam, 2013a).

Effect of relocation: Their choice to relocate has led to a very different, and much more positive, self-story, in which they cast themselves as pioneering "settlers" rather than forced migrants (McAdam, 2014). As Teresia Teaiwa has observed that "while Rabi Island's settlement (...) has become something of a historical controversy, Kioa was clearly settled by the choice of islanders by Vaitupu, and without as much drama" (Teaiwa, 1997). Today, both communities have very similar living conditions and both groups acknowledge that they are better off in material terms than if they had remained at home (Kälin, 2013). However, for the Banabans, socioeconomic gains are overshadowed by a sense of injustice and disempowerment about the initial move (Connell, 2012). This element of coercion is key to their ongoing perceptions about the success or otherwise of relocation.

Response and integration: "The Kioans see their story as one of community survival. On Kioa Day, the floral wreaths that people wear on their heads represent unity—no matter how many different flowers are bound together, they are ultimately one, just like the people of Kioa. The people say that they have two homes—Kioa, the homeland, and Vaitupu, the motherland. Many people identify as Fijian and Vaitupuan. Many still have land and relatives back on Vaitupu and have visited several times. The council helps to arrange annual visits to Vaitupu for the young people. Since Vaitupu continues to sustain a large community, there is perhaps more openness among the Kioans towards adopting two identities. While the Banabans also say they have two homes, the relationship to Banaba Island is still fraught. They don't see themselves as I-Kiribati and yet, despite being Fijian citizens, most of them don't regard themselves as Fijian either" (McAdam, 2013a).

25.4.1.1.3 Contract for the Sale of Land to Facilitate Relocation

While the international relocation of whole Pacific Island communities has been mooted from time to time, there are no moves afoot to facilitate this (McAdam, 2013a). There is acknowledgement by Pacific leaders that the likelihood of any state ceding land to it is remote, and while Kiribati has purchased a tract of fertile church-owned land in Fiji, this is not (contrary to media reports) to secure a new homeland for the people of Kiribati, but rather to provide food security and possible employment opportunities for its citizens (McAdam, 2013b).

25.4.2 Vanuatu

The loss and damage caused by climate change in Vanuatu can be characterized both in terms of the loss of habitable land as well as the loss of livelihood. This is especially relevant in the context of the fisheries sector within Vanuatu where such a loss has posed a risk to the overall economy, food security, and the livelihood of the people of Vanuatu. The major events contributing to such loss and damage include ocean acidification, siltation, and the rise in sea surface temperature. Cyclones, droughts, and the rising sea level have also adversely impacted the agricultural sector, upon which a majority of the ni-Vanuatu depend for their basic sustenance and livelihood. Erosion, flooding, saltwater intrusion, and volcanic ash have resulted in the positing of an extremely risky narrative of economic duress for the people of Vanuatu. As a consequence, the likelihood of displacement as a result of climate change is becoming an exceedingly real possibility.

Climate-induced migration is however a concept that has not found its rooting within the legal, political, or social articulations of Vanuatu. It does not fall within the functions and responsibilities of any of the primordial sectors. However, internal displacement and population mobility caused by natural disasters is a reality that is presently affecting thousands of its people, who are often left incapacitated due to the lack of appropriate and timely responses from the government. Moreover, the existing mechanisms of enhancing the resilience of communities to climate change are inadequate.

Communities near coastal areas and river banks have been asked to relocate to higher ground. However, there are no projects or programs on migration or displacement currently in place. Neither are there any legislation, plans, policies, tools, or methodologies which may be able to guide or inform the nature of such migration and enable its effectuation in the least disruptive manner. Without any such concrete policy to fall back on, the nature of such migration would be chaotic and unregulated resulting in additional losses not just to the communities but to the government as well.

"Land to the ni-Vanuatu is what a mother is to a baby, it is with the land that he defines his identity and it with the land that he maintains his spiritual strength" (Regenvanu, 1987). Noneconomic loss was a concept that was explored for the first time in Vanuatu in the assessment of the devastation left behind by Cyclone Pam. The losses identified included environmental impacts and the loss of traditional knowledge and focused on traditional systems of coping after a disaster. There are currently no projects in place in Vanuatu that focus directly on noneconomic losses. However, there are adaptation initiatives that have been undertaken to address such losses through a lens of traditional knowledge and customary law.

One such project running from January 201 to January 2018 focuses on the value that customary law can contribute to future environmental legislation in the context of global climate change adaptation. This is a continuation of the 2010 Tabwemasana Project where the research was focused on the sustainability of local communities (Vanuatu Project Portal 2016). Similarly, the Traditional Knowledge and Climate Change Disaster Risk Reduction Guidelines Project running from April 2013 to April 2015 involved the development of guidelines and amendments to legislations addressing climate change and disaster risk reduction through the effective use of traditional knowledge.

25.5 CONCLUSION

The Pacific Islanders strongly believe that blood and mud mix together to create identity. Most Pacific Islanders therefore remain wary of community relocations because they see it as an enduring separation of land and identity (Nansen Initiative, 2013). They hold a lasting fear that it may have a significantly negative impact on the strong ideals of nationhood, control over land and sea, sovereignty, culture, and livelihoods, that they hold important and inalienable (Nansen Initiative, 2013). Thus, in instances of cross-border relocations the issues that rise to the forefront include the nature of the status of the community once it moves, for instance the question of whether it retains its original statehood or does it acquire a self-governing status within the new territory (Rayfuse, 2011). In addition to this there has emerged an imminent need to renegotiate immigration and citizenship rights of the relocated communities.

The Nansen Initiative's Pacific Consultations provided recommendations to facilitate the planned relocation to another country. These included the need to "i) define the legal status of the relocated community within the new state, ii) assist communities in integrating through the adaptation to local customs, practices and laws, iii) include consultations with potential host communities, and iv) formulate measures that would facilitate the maintenance of cultural ties within the migrant community, such as allowing dual citizenship" (Nansen Initiative, 2013).

The inclusion of such measures however provides no absolute guarantee that the migrant community will necessarily integrate with ease. As observed with the instance of the relocation of the Banabans in Fiji, the institution of nuanced constitutional protections such as the rights to land and nationality are not adequate to overcome sentiments of disenfranchisement brought on by forced or induced relocation (McAdam, 2014). This is because this form of permanent relocation does foster deep seated spiritual anxieties and emotional resentment kindled by the loss of identity, culture, and social coherence. These might also manifest in legal dimensions such as the right to self-determination, citizenship, and social and cultural rights.

It is more and more evident that climate change, human rights, and human mobility do represent a nexus that will increase in the coming years. Research shows that addressing climate change without referring to human rights or human mobility (where human mobility is an inclusive concept of migration, displacement, and relocation) will create further gaps in regulating, implementing, and communicating international norms into regional or domestic systems of law (McAdam, 2014). Regardless in which fora the discussion takes place (the climate process conducted by UNFCCC, the migration one governed by various actors such as the UNHCR, IOM, or PDD, or the Human Rights Council) the linkage between the three will continue to progress as countries and regions will try to address the impacts of climate change, facilitate human mobility, and implement human rights values.

It is also clear from the research that the states do not own an individual capacity to develop mobility policies due to different factors such as the lack of financial or human resources. While regulating the internal migration there is also the cross-border movement, which in most cases is not governed when addressing climate change. The states' practices and measures in regards to climate change, human rights, and human mobility should reflect a bottom-up approach, where both levels of governance (Kastom and the domestic system) are complementary and harmonized, strengthened by robust communication between the community leaders and policy makers (founded on people's needs) and by increasing capacity-building efforts to eliminate gaps of understanding and perception of different norms and standards. On the other hand, an international response to domestic or regional requirements should not represent an ultimate solution for the dynamics of people's life, but a complementary action that may add value to the region by diversifying options both people and states could consider in the future.

The immediate answer for both short- and long-term consideration should represent the regional approach, where states galvanized by common or similar values and traditions, solidarity, and historical regional cooperation could address their needs in regards to climate change and human mobility, complemented also by a regional recognition of human rights norms indispensable for multidimensional protection. A regional framework on climate change and human mobility, regardless of its legal form, could only help building local and national resilience and offer alternatives to both states and communities.

These preemptive measures could only capacitate states' reaction when importing and implementing international law and could also establish minimum standards of protection for the people affected by climate change, whose rights are breached and who decide to migrate, as an adaptation strategy.

REFERENCES

Bahadur, A.V. Peter, K Wilkinson, E Pichon, F Gray, K. Tanner, T. The 3As 2015: Tracking Resilience Across Braced. Braced Working Paper. Available at: https://www.odi.org/sites/odi.org.uk/files/odi assets/publications-opinion-files/9812.pdf [03.05.2016].

Barnett, J., 2008. The Effect of Aid On Capacity To Adapt To Climate Change: Insights From Niue. Political Science 60 (1), 31–45.

Bedford, R., Ho, E., 2003. Labour force participation as a measure of progress amongst New Zealand's Pacific population: a cohort approach. In: Gao, J., Le Heron, R., Logie, J. (Eds.), *Windows on a Changing World: Proceedings of the 22nd New Zealand Geographical Society Conference, Auckland, 6-11 July*. GeographicalSociety, Auckland: New Zealand, pp. 229–232.

Bettencourt, Sofia, Croad, Richard, Freeman, Paul, Hay, John, Jones, Roger, King, Peter, et al., 2006. Not if but when: Adapting to Natural Disasters in the Pacific Islands Region: A Policy Note. The World Bank, p. viii.

Boege, V., 2011. Peacebuilding and State formation in post-conflict Bougainville. Peace Review 21 (1), 29–37.

Bread for the World/PCC 2015: Connecting the Dots: Learnings from Community-Based Adaptation and Relocation for Policy Influencing. A Workshop Report. Berlin & Suva.

Campbell John R., Hamilton 2014 '*The implications of climate change for the loss and damage caused by disruption of the essential link between people and their land.*' Submission to the Executive Committee of the Warsaw International Mechanism for Loss and Damage associated with Climate Change Impacts of the United Nations Framework Convention on Climate Change. Available at https://unfccc.int/files/adaptation/groups_committees/loss_and_damage_executive_committee/application/pdf/l_d_submission_j__campbell.pdf [Last accessed on 13.10.2016].

Connell, J., 2010. Pacific islands in the global economy: Paradoxes of migration and culture. Singapore Journal of Tropical Geography 31, 115–129.

Connell, John, 2012. Population Resettlement in the Pacific: Lessons from a Hazardous History? Aust Geographer 127. 13.

Corendea, C., 2016. Hybrid legal approaches towards climate change: concepts, mechanisms and implementation. Annu. Surv. Int. Comp. Law 21, 1. Article 5. <http://digitalcommons.law.ggu.edu/annlsurvey/vol21/iss1/5> (accessed 20.12.17.).

Corendea, C., 2017. Migration and human rights in the wake of climate change, a policy perspective over the Pacific. UNU-EHS Publication Series, Policy Report. United Nations University, pp. 11−44, < http://collections.unu.edu/view/UNU:6305 > (accessed 08.01.18.).

EM-DAT: The OFDA/CRED International Disaster Database, www.emdat.be - Université Catholique de Louvain - Brussels − Belgium, also see www.unisdr.org.

Hau'ofa, E., 1993. Our Sea of Islands. School of Social and Economic Development, The University of the South Pacific, Suva.

HC Deb 18 December 1975, vol 902, col 1857 (Sir Bernard Braine), referring (at col 1856) to notes of a meeting held in October 1945 between the British colonial authority and representatives of the British Phosphate Commission.

Intergovernmental Panel on Climate Change, *IPCC Fourth Assessment Report: Climate Change 2007, Working Group I: The Physical Science Basis, Executive Summary, Direct Observations of Recent Climate Change*, 2007, http://www.ipcc.ch/publications_and_data/ar4/wg1/en/spmsspm-direct-observations.html.

IPCC 2012: Special report: Managing the risks of extreme events and disasters to advance climate change adaptation (SREX). Available at: http://www.ipcc-wg2.gov/SREX/ [04.05.2017].

Julca, Alex, Paddison, Oliver, 2010. Vulnerabilities and migration in Small Island Developing States in the context of climate change. Natural Hazards 717−728.

Kälin Walter, 'Changing Climates, Moving People: Distinguishing Voluntary and Forced Movements of People' in Koko Warner and others (eds), Changing Climates, Moving People: Framing Migration, Displacement and Planned Relocation (2013) UN University Policy Brief No 8, 38, 40 <http://collections.unu.edu/eserv/UNU:1837/pdf11213.pdf> accessed 24 July 2015.

Kuruppu, N., Liverman, D., 2011. Mental preparation for climate adaptation: the role of cognition and culture in enhancing adaptive capacity of water management in Kiribati. Global Environmental Change 21, 657−669.

Locke, J.T., 2009. Climate change-induced migration in the Pacific Region: sudden crisis and long-term developments. The Geographical Journal 175, 171−180.

McAdam Jane 'Inside Story: Caught between homelands' 2013a; available at https://unfccc.int/files/adaptation/groups_committees/loss_and_damage_executive_committee/application/pdf/http___insidestory.org.pdf.

McAdam Jane, Interview with Anote Tong, President of Kiribati (Kiribati, 11 September 2013b).

McAdam, Jane, 2014. Historical Cross-Border Relocations in the Pacific: Lessons for Planned Relocations in the Context of Climate Change. J Pac Hist 301.

Nansen Initiative on Disaster-Induced Cross-Border Displacement, 'Human Mobility, Natural Disasters and Climate Change in the Pacific' (Report from the Nansen Initiative Regional Consultation, Rarotonga, 21−24 May 2013) (Nansen Initiative Pacific Report) <http://www.nanseninitiative. org/pacific-consultations- intergovernmental/> accessed 3 June 2015. 10−11, 17.

ODI (Wilkinson, E. Peters, K.)(Eds.) 2015: Climate extremes and resilient poverty reduction: development designed with uncertainty in mind. Overseas Development Institute, London. Available at: https://www.odi.org/sites/odi.org.uk/files/odi-assets/publications-opinion-files/10130.pdf [02.05.2017].

Office of the Republic of Kiribati; Available at www.climate.gov.ki/category/action/relocation [Last visited 1.7.2017].

OHCHR, Regional Office for the Pacific, *Protecting the Human Rights of Internally Displaced Persons in Natural Disasters*, 2011, http://pacific. ohchr.org/docs/IDP_report.pdf.

Pelling, M., 2011. Adaptation to Climate Change: From Resilience to Transformation. Routledge, Abingdon.

Rasmussen, K., May, W., Birk, T., Mataki, M., Mertz, O., Yee, D., 2009. Climate change on three Polynesian outliers in the Solomon Islands: impacts, vulnerability and adaptation. Geografisk Tidsskrift − Danish Journal of Geography 109, 1−13.

Ravuvu, A. 1988 Development or Dependence: The Pattern of Change in a Fijian Village. Suva, University of the South Pacific. Secretariat of the Pacific Regional Environment Programme,'*Upheaval of Fiji communities as climate change impacts are felt*' available at http://www.sprep.org/ climate-change/upheaval-of-fiji-communities-as-climate-change-impacts-are-felt [accessed on 13.10.2016].

Regenvanu, S. 1987 First Minister of Lands, quoted in HN van Trease, The Politics of Land in Vanuatu: From Colony to Independence Suva, University of the South Pacific, 1987.

Smith, L.E., Long, J.R., 2000. Literacy, writing systems, and development in the Pacific. Studies in the Linguistic Sciences 30, 169−181.

Tanner, T., et al., 2015. Livelihood resilience in the face of climate changeIn: Nature Climate Change 5, 23−26.

Teaiwa, Teresia K., 1997. Rabi and Kioa: Peripheral Minority Communities in Fiji. In: Lal, Brij V., Vakatora, Tomasi R. (Eds.), Fiji in Transition: Research Papers of the Fiji Constitution Review Commission. University of the South Pacific, p. 132.

Vanuatu Project Portal. National Advisory Board available at http://www.nab.vu/projectsearch2 [Last accessed on 13.10.2016].

Watch German, 2015 Climate related loss and damage: Finding a just solution to political challenges. Available at http://climate-neld.com/wp content/ uploads/2015/11/Profil19_E_LossAndDamage.compressed.pdf[Last Accessed on 13.10.2016].

FURTHER READING

Barnett, Jon, Chamberlain, Natasha, 2010. Migration as Climate Change Adaptation: Implications for the Pacific. In: Burson, Bruce (Ed.), Climate Change and Migration: South Pacific Perspectives. Institute of Policy Studies, Victoria University, p. 51.

Barnett, J., O'Neill, S., 2012. Islands, resettlement and adaptation. Nature Climate Change 2 (1), 8−10. McNamara, K.E., Gibson, C., 2012. "We do not want to leave our land": Pacific ambassadors at the United Nations resist the category of "climate refugees". Geoforum 40 (3), 475−483.

Barnett Jon and Michael Webber, 'Migration as Adaptation: Opportunities and Limits' in Jane McAdam (ed) *Climate Change and Displacement: Multidisciplinary Perspectives* (2010) 41.

Biermann and Boas, above n 35, 75−6; see also Docherty, Bonnie, Giannini, Tyler, 2009a. Confronting a Rising Tide: A Proposal for a Convention on Climate Change Refugees. Harvard Environmental Law Review 349.

Biermann Frank and Ingrid Boas, 'Preparing for a Warmer World: Towards a Global Governance System to Protect Climate Refugees' (2010) 10 Global Environmental Politics 60, 61; David Hodgkinson et al, '"The Hour When the Ship Comes in": A Convention for Persons Displaced by Climate Change' (2009b), 13.

Cambodia: IOM, 'Migration Initiatives Appeal 2010', Indonesia: IOM, 'Migration Initiatives Appeal 2010'.

Climate Change Part of Refugee Ruling, Radio New Zealand (Aug. 4, 2014), http://www.radionz.co.nz/news/national/251293/climate-change-part-of-refugee-ruling.

Conference of the Parties, United Nations Framework Convention on Climate Change, Report of the Conference of the Parties on its Sixteenth Session, Held in Cancun from 29 November to 10 December 2010 — Addendum — Part Two: Action Taken by the Conference of the Parties at its Sixteenth Session — Decision 1/CP.16: The Cancun Agreements: Outcome of the Work of the Ad Hoc Working Group on Long-term Cooperative Action under the Convention, UN Doc FCCC/CP/2010/7/Add.1, §14(f).

Constitution of the International Organization for Migration 1953 art 1.

Convention Relating to the Status of Refugees, opened for signature 28 July 1951, 189 UNTS 150 (entered into force 22 April 1954). 3 Ibid art 1 (A)(2).

Corendea, C., 2016. Legal Protection of the Sinking Islands Refugees. Vandeplas Publishing, pp. 1−47.

Crépeau François, Special Rapporteur, Report on the Human Rights of Migrants, UN Document A/67/299 (13 August 2012).

CRIDEAU, 'Draft Convention on the International Status of Environmentally-Displaced Persons' (2008) 39 Revue de Droit de l'Université de Sherbrooke 451, 461−2; CIDCE, 'Draft Convention on the International Status of Environmentally Displaced Persons (Second Version — May 2010)' (2010).

Duncan, R., 2008. Cultural and economic tensions in Pacific Islands' futures. International Journal of Social Economics 35, 919−929.

FAO, 2008. Climate Change and Food Security in Pacific Island Countries. FAO (Food and Agriculture Organization of the United Nations), Rome, Italy.

Indonesia: IOM, Migration Initiatives Appeal 2010 107 <http://www.iom.int/jahia/webdav/site/myjahiasite/shared/shared/mainsite/published_docs/books/Migration-Initiatives-Appeal.pdf>. Timor-Leste: IOM, in 'Migration Initiatives Appeal 2010', ibid 99.

Intergovernmental Panel on Climate Change [Ipcc], Fifth Assessment Report (2009), https://www.ipcc.ch/report/ar5/.

IOM, *Disaster Risk Reduction, Climate Change Adaptation and Environmental Migration: A Policy Perspective* (2010), 3 http://publications.iom.int/bookstore/free/DDR_CCA_report.pdf.

IOM, EDGE, Sciences Po, 'The State of Environmental Migration 2015 − A Review of 2014 Edited by François Gemenne Caroline Zickgraf Dina Ionesco'. Available from: < https://www.sciencespo.fr/psia/sites/sciencespo.fr.psia/files/StateofEnvironmentalMigration2015.pdf > (accessed 12.2.18).

Mcadam, Jane, 2011a. Swimming against the Tide: Why a Climate Change Displacement Treaty is Not the Answer. International Journal of Refugee Law 2. 2.

McAdam Jane, Climate Change, Forced Migration, and International Law (n 4) 119−60; Rosemary Rayfuse, 'International Law and Disappearing States: Maritime Zones and the Criteria for Statehood' (2011b) 41 Enviro Policy & L 281.

McLeman, R., Smit, B., 2006. Migration as an Adaptation to Climate Change. Climatic Change 31. ADB, *Facing the Challenge of Environmental Migration in Asia and the Pacific* (2011).

Mertz, O., Halnæs, K., Olesen, J.E., Rasmussen, K., 2009. Adaptation to climate change in developing countries. Environmental Management 43, 743−752.

Miller, Bradley W., 2009. Beguiled by Metaphors: The 'Living Tree' and Originalist Constitutional Interpretation in Canada. CAN. J. L. & JURISPRUDENCE 331.

Nepal IOM, 'Migration Initiatives Appeal 2010', Bangladesh: IOM, 'Migration Initiatives Appeal 2010', Tajikistan, Kyrgyzstan and Turkmenistan IOM, 'Migration Initiatives Appeal 2010'.

Nicholls, R.J., Marinova, N., Lowe, J.A., Brown, S., Vellinga, P., de Gusmão, D., et al., 2011. Sea-level rise and its possible impacts given a 'beyond 4°C world' in the twenty-first century. Philosophical Transactions of the Royal Society of London A 369, 161−181.

Novaczek, I., Mitchell, J., Veitayaki, J. (Eds.), 2005. Pacific Voices: equity and sustainability in Pacific Island fisheries. Institute of Pacific Studies, The University of the South Pacific, Suva.

Nunn, P.D., 2009b. Responding to the challenges of climate change in the Pacific Islands: management and technological imperatives. Climate Research 40, 211−231.

Rayfuse, R., 2011. International law and disappearing states: maritime zones and the criteria for statehood. Environ. Policy Law 41, 281.

Renaud, F.G., et al., 2011. A Decision Framework for Environmentally Induced Migration. International Migration e5. e12-e14.

Seacrest, S., Kuzelka, R., Leonard, R., 2000. Global climate change and public perception: the challenge of translation. J Am Water Resour As 36, 253−263.

The Moana Declaration, 2009, Available at http://www.oikoumene.org/en/resources/documents/wcc-programmes/justice-diakonia-and-responsibility-for-creation/climate-change-water/pacific-church-leaders-statement [Last accessed 1.7.2017].

The Paris outcome on loss and damage (Article 8 of the Paris Agreement and Decision 1/CP.21 Paragraphs 48−52 (FCCC/CP/2015/L.9/Rev.1.)1 available at https://unfccc.int/files/adaptation/groups_committees/loss_and_damage_executive_committee/application/pdf/ref_8_decision_xcp.21.pdf [Last accessed on 16.10.2016].

Wiessner S. & A.R. Willard, Policy-Oriented Jurisprudence, 44 GERMAN Y.B. INTL'L L. 96, 97, 101, 107-112 (2001)

Williams, Angela, 2008. Turning the Tide: Recognizing Climate Change Refugees in International Law. Law and Policy 502. 517.

ANNEX

Findings from the Pacific Project—This project was a joint initiative of the Pacific Islands Forum Secretariat, The United Nations University (UNU) Institute on Globalization, Culture and Mobility and The United Nations University Institute for Environment and Human Security.

I. Reasons for Migration

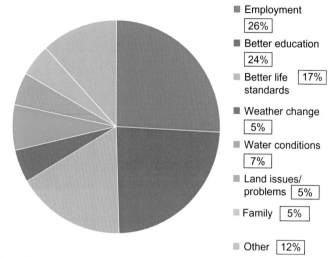

- Employment 26%
- Better education 24%
- Better life standards 17%
- Weather change 5%
- Water conditions 7%
- Land issues/problems 5%
- Family 5%
- Other 12%

II. Main challenges faced by migrants when arriving in new communities

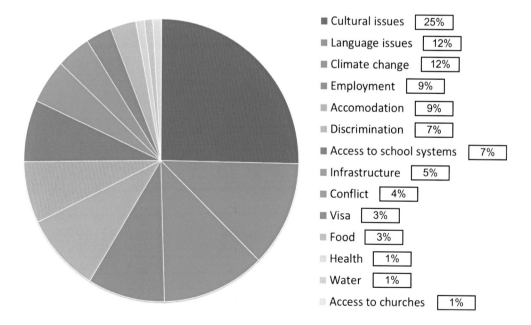

- Cultural issues 25%
- Language issues 12%
- Climate change 12%
- Employment 9%
- Accomodation 9%
- Discrimination 7%
- Access to school systems 7%
- Infrastructure 5%
- Conflict 4%
- Visa 3%
- Food 3%
- Health 1%
- Water 1%
- Access to churches 1%

III. Assistance provided to migrants in receiving countries

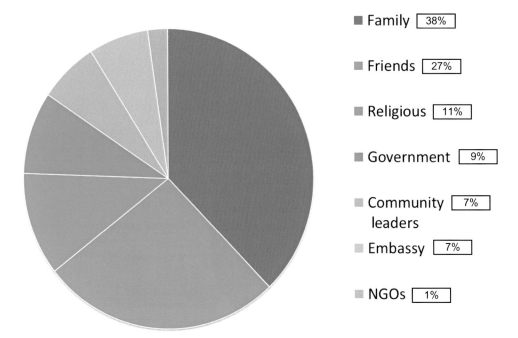

■ Family 38%

■ Friends 27%

■ Religious 11%

■ Government 9%

■ Community 7%
  leaders

■ Embassy 7%

■ NGOs 1%

Integrated Loss and Damage–Climate Change Adaptation–Disaster Risk Reduction Framework: The Case of the Philippines

Ebinezer R. Florano

Center for Policy and Executive Development, National College of Public Administration and Governance, University of the Philippines, Quezon City, Philippines

26.1 INTRODUCTION

In response to the threats of climate change (CC) to small-island nations, the Warsaw International Mechanism (WIM) for Loss and Damage (L&D)[1] was approved during the 19th Conference of the Parties meeting of the United Nations Framework Convention on Climate Change in Warsaw, Poland in November 2013. WIM is meant to address L&D associated with impacts of CC, including extreme events and slow onset events, in developing countries that are particularly vulnerable to its adverse effects. In the preamble of the Warsaw decision, it is recognized that there are limits to adaptation, which must be addressed, i.e., "that loss and damage in some cases involves more than that which can be reduced by adaptation" (Warner, 2013). In 2014, the work plan of the executive committee of WIM was approved. Its foremost action area is to "(e)nhance the understanding of the capacity and coordination needs with regard to preparing for, responding to and building resilience against loss and damage associated with extreme and slow onset events, including through recovery and rehabilitation."

In response to the call of the WIM's action plan, this paper discusses the determination of L&D[2] in the fields of disaster risk reduction (DRR) and climate change adaptation (CCA) as they are being utilized in the Philippines. It aims to answer the following questions: Are there limits to adaptation in the Philippines? Does it have policy and institutional frameworks to determine and address L&D? Are they attuned for CC-induced disasters? How can they be improved? For this purpose, the policy and institutional frameworks for dealing with disasters and climate change problems are briefly discussed, which could shed light on the dichotomy between the two fields. Second, the definitions of L&D

---

1. The UNFCCC defines L&D as "the actual and/or potential manifestation of impacts associated with climate change in developing countries that negatively affect human and natural systems" (UNFCCC, 2012, November 15).
2. L&Ds is used when referring to the plural form "losses and damages."

Resilience. DOI: https://doi.org/10.1016/B978-0-12-811891-7.00026-8

employed by government agencies are explained together with the basic steps involved in determining them in the fields of DRR and CCA. While the two systems have distinct advantages in terms of responding to climate-induced and man-made disasters separately, there are issues when it comes to making inventories of L&Ds. The last part discusses the effectiveness of a more recent integrated L&D−CCA−DRR framework developed to strengthen action planning based on L&D assessments of disasters caused by climate/disaster risks from extreme and slow-onset events.

26.2 CLIMATE CHANGE AND DISASTERS IN THE PHILIPPINES

The Philippines is highly vulnerable to disasters. Typhoons or cyclonic storms are the dominant risk followed by floods, earthquakes, volcanoes, droughts, and landslides (UN-ODRR and WB, 2010). Table 26.1 shows that storms and floods indeed account for the highest frequency and economic significance since the Philippines is ravaged by an average of 20 typhoons annually, aside from the seasonal southwest monsoon during the rainy season commencing in June is expected to occur.

These disasters could be amplified or attenuated by climate change projections made by the country's meteorological agency (Philippine Atmospheric, Geophysical and Astronomical Services Administration or PAGASA):
Temperature

- There has been an increase in the observed mean annual temperature by 0.65°C from 1951−2010 and it is projected to increase by an average 1°C by 2020;
- All areas of the Philippines will get warmer, more so in the relatively warmer summer months; and
- Annual mean temperatures (average of maximum and minimum temperatures) in all areas in the country are expected to rise by 0.9°C to 1.1°C in 2020 and by 1.8°C to 2.2°C in 2050.

Rainfall

- Reduction in rainfall is expected to occur in most provinces during the summer season (MAM) making the usually dry season drier; while
- Rainfall increases are likely in most areas of Luzon and Visayas during the southwest monsoon (JJA) and the SON seasons, making these seasons still wetter, and thus with the likelihood of both droughts and floods in areas where these are projected; and
- Drier season becoming drier and wetter season becoming wetter and greater with time in 2020 and 2050.

Extreme Events
Projections for extreme events in 2020 and 2050 show that:

- Hot temperatures (indicated by the number of days with maximum temperature exceeding 35°C) will continue to become more frequent;
- Number of dry days (days with less than 2.5 mm of rain) will increase in all parts of the country; and
- Heavy daily rainfall (exceeding 300 mm) events will also continue to increase in number in Luzon and Visayas (Cinco et al., 2013).

While the above report focused on three important factors—extreme events and hazards such as storms, flooding correlated with rainfall during monsoon months, and droughts correlated with the El Niño Southern Oscillation (ENSO)—the threat of sea level rise as a climate impact is also very clear. Based on several global warming scenarios, world estimates of sea level rise range from 28 to 43 cm above base level by 2100. The Philippines as an archipelagic

TABLE 26.1 Disaster and Risk Profile of the Philippines (Internationally reported losses 1990−2014)

Hazard	Frequency (%)	Mortality (%)	Economic Issue (%)
Storm	51.3	78.7	79.0
Flood	31.9	5.9	17.3
Landslide	6.4	5.6	−
Volcano	4.6	−	−
Earthquake	4.1	7.9	2.2
Other	1.7	2.0	1.5

Source: Preventionweb (n.d.).

country is especially vulnerable to sea level rise and storm surge since around 60% of its municipalities and 10 of its largest cities are located along the coast (where roughly 60% of the population resides). Rapid urbanization will most likely drive this higher in the future (WB-GFDRR, 2011; Kahn, 2015).

Thus, it should not come as a surprise that from 2011 to 2016, the Philippines ranked high both in the Global Climate Risk Index[3] of Germanwatch, and in the World Risk Index[4] of Alliance Development Works and United Nations University—Institute of Environment and Human Security (UNU-EHS). In the former, the placement of the Philippines varied between 14 and 1 making it one of the high-risk countries in climate change (see Table 26.2). In the latter, the country was consistently ranked number 3, reaching 2nd place in 2014. Again, this meant that the country was among the high-risk countries for climate and nonclimate disasters (see Table 26.3).

TABLE 26.2 Philippines: Global Climate Risk Index, 2011—2016

Year	Overall Rank
2016	4
2015	1
2014	2
2013	14
2012	10
2011	7

Sources: Germanwatch (2011—2016).

TABLE 26.3 Philippines: World Risk Index, 2011—2016

Year	Overall Rank
2016	3
2015	3
2014	2
2013	3
2012	3
2011	3

Sources: BEH-UNU-EHS (2011—2016).

From 2011 to 2016, L&Ds to properties due to natural disasters, environmental hazards, human-induced and hydro-meteorological events was on the rise. Compared to the baseline L&D in 2004—2010 worth USD 371 million, substantial increases occurred every year, topping in 2013 at USD 2.17 billion due to Super Typhoon Haiyan/Yolanda (see Table 26.4). The increasing amount of L&Ds in the Philippines during the 6-year period, not to mention casualties and

TABLE 26.4 Loss and Damage, 2011—2016 vs. Baseline (in USD)

Baseline	Year					
2004—2010 Average	2011	2012	2013	2014	2015	2016
371 m	488 m	977 m	2.17 b	1.075 b	821 m	801 m

Legend: m-million, b-billion
Note: Converted amount in USD from the original Philippine Peso (PHP) value using the exchange rate USD1:PHP0.0196.
Source: PSA (2017).

---

3. The Global Climate Risk Index measures the climate risks of countries based on their average annual fatalities, fatalities per 1,000 inhabitants, losses in USD million (purchasing power parity), and losses per unit GDP in percent.
4. The World Risk Index measures the climate and nonclimate risks of countries based on their exposure, and vulnerability which is a function of susceptibility, coping capacities, and adaptive capacities.

injuries and those related to noneconomic L&D, is proof of the limits to adaptation especially with regard to extreme weather events, i.e., no amount of adaptation could prepare a disaster-prone country to withstand the extraordinary fury of nature.

The wrath of Super Typhoon Haiyan/Yolanda is a testament to the limits of adaptation. The Philippines attracts an average of 20 tropical cyclones every year, hence, its people have somewhat adjusted/adapted to their visits to the extent that it becomes a normal part of their lives. However, there was no way that communities in eight out of the 17 regions in the Philippines, mostly in the central Visayas region, could escape from the typhoon's extraordinary wind power and strength, when it made landfall on 8 November 2013. When the super typhoon made landfall and the threat of storm surge was broadcasted, many thought that they would survive it just like in the past. However, when the typhoon departed, it left in its trail the following casualties: 6,300 deaths, 28,689 injuries, and 1,061 missing people. The damages to the economy were estimated at USD 2.05 billion (NDRRMC, 2014). Such was the power of Haiyan that even Dr. Zhang Qiang, a disaster expert from Beijing Normal University in China, admitted that, indeed, there was no way out of it, no matter the preparations. He said that "(s)ometimes, no matter how much and how carefully you prepare, the disaster is just too big" (AP, 2013, November 11).

26.3 POLICY AND INSTITUTIONAL FRAMEWORKS FOR DETERMINING L&D

In response to the threats of climate change and man-made disasters, two distinct laws were enacted by the Philippine Congress in 2009 and 2010. These laws are relevant for determining L&D in the Philippines. These are the "Climate Change Act of 2009" (Republic Act No. 9729 or RA 9729), and "The Philippine Disaster Risk Reduction and Management Act of 2010" (Republic Act No. 10121 or RA 10121). The former delves on the country's response to climate change (CC) through the implementation of adaptation and mitigation measures in the country. The latter, on the other hand, focuses on DRR through its four main pillars, i.e., prevention and mitigation, preparation, response, and recovery and rehabilitation.

As a consequence of the two laws, two separate agencies are responsible to implement them, i.e., the Climate Change Commission (CCC) under the Office of the President and the National Disaster Risk Reduction and Management (NDRRMC).[5] Moreover, Congress appropriates separate budgets for the CCC and the National Disaster Risk Reduction and Management Fund (NDRRM Fund).

In terms of determining L&D, the climate change law does not directly mention the need to conduct the inventory of L&D. However, it is implied with the use of vulnerability and risk assessment methods. On the other hand, the DRRM law makes it obligatory to conduct Post-Disaster Needs Analysis (PDNA) in the aftermath of disasters. Specifically, its Section 9 (m) states that the Office of Civil Defense (OCD) shall "conduct early recovery and post-disaster needs assessment institutionalizing gender analysis as part of it."

26.4 DETERMINING L&D: THE DISASTER RISK REDUCTION APPROACH

Loss and damage are well-defined in the field of DRR, mostly in terms of economic value and whether or not they can be still be recovered. The manual on PDNA of the OCD operationalizes and differentiates them with the following definitions:

- **Damages**——"are total or partial destruction of capital assets, infrastructure such as animal sheds, storage, ice plants, irrigation, inventory of goods like agricultural inputs; equipment, machinery; and raw materials for production, among others. Damages are valued as the cost of replacement or repair of destroyed structures or equipment at predisaster prices while agricultural products are valued at predisaster farm gate prices" (WB, 2012, October, p. 1).
- **Losses**——"are the values due to the change in economic flow during the period of recovery and reconstruction following the disaster. They are the current value of goods and services that were not and/or will not be produced over a time span due to the disaster until full recovery is attained" (WB, 2012, October, pp. 1−2).

The inventory of L&D in the Philippines is conducted through PDNA, which is specifically defined as "a multi-sectoral and multi-disciplinary structured approach for assessing disaster impacts and prioritizing recovery and reconstruction needs" (Florano et al., 2016, p. 140). The work is led and undertaken by government agencies in collaboration

---

5. Historically speaking, DRR has a longer years of existence in the Philippines than CC because the former started since 1954 (as "disaster management") and has undergone several reorganizations, the most recent of which was the creation of the National Disaster Coordinating Council in 1978 through Presidential Decree No. 1566. On the other hand, the CC law was enacted only in 2009.

with international development partners and the private sector. PDNA is generally intended to estimate: (1) short-term interventions to initiate recovery from the damages and losses; and (2) financial requirements needed to achieve a holistic post-disaster recovery, reconstruction, and risk management.

PDNA usually starts three weeks after a disaster struck and may take two months to finish. The following are the steps involved in conducting a PDNA:

1. **Collect and/or validate the baseline data:** Baseline data cover all the assets of the locality and the related replacement and repair costs. It is important to note that baseline data are compiled by LGUs prior to the occurrence of a disaster.
2. **Estimate damages and losses:** Using the baseline data, the damages and losses are estimated.
3. **Validate information on damages and losses:** The assessments teams meet to validate and reconcile the data collected from the field and prevent double counting.
4. **Analyze the impacts of the damages and losses to affected population:** Assessment teams analyze the potential impacts of the damages and losses in relation to the delays in the socioeconomic activities, the losses in government revenues, and any other disaster-related issue.
5. **Forward the impact assessment to appropriate agencies for aggregate analysis:** The results of the analysis are forwarded to the concerned agencies and technical experts.
6. **Estimate the recovery and reconstruction needs:** Based on the damages and losses of the agency together with their aggregate analyses, the needs are identified.
7. **Provide the National Economic and Development Authority (NEDA) with the lists of identified programs and projects for recovery and reconstruction:** NEDA undertakes macroeconomic impact analysis of the programs and projects for the current and future years as well as prepares a strategic action plan for recovery and reconstruction in consideration of the Philippine Development Plan.
8. **Analyze financing options:** The assessment suggests the possible sources of financing such as Quick Response Fund, NDRRM Fund, existing funds that can be redirected to the present needs, grants from international development partners and the private sector, etc.
9. **Write the draft of the implementation plan of the identified programs and projects:** The identified needs incorporate a rough schedule of implementation outlining timing and budget required and other needed information.
10. **Write the draft of the PDNA Report:** With all the information gathered and analyzed, a PDNA report of the each agency is drafted (Florano et al., 2016, pp. 142−145).

The draft PDNA reports are submitted to the OCD for consolidation and inclusion in the overall PDNA report and the disaster recovery plan (see OCD-NDRRMC, 2014). Based on the PDNA, all government agencies and other development partners can now proceed with the implementation of rehabilitation and recovery projects and activities. An example of a recovery and rehabilitation plan based on a PDNA report is the "Reconstruction Assistance on Yolanda," a.k.a., RAY. The PDNA reported that the losses and damages from the Typhoon Yolanda/Haiyan are worth PhP 104.6 billion (USD 2.2 billion). However, RAY estimates that PhP360.9 billion (USD 7.7 billion) is needed for recovery and rehabilitation (NEDA, n.d., p. 29).

26.5 DETERMINING L&D: THE CLIMATE CHANGE ADAPTATION APPROACH

As stated above, there is no specific provision in the Climate Change Act of 2009 about determining L&D in the context of climate change. However, it is implied in the various vulnerability and risk assessment methods employed for the formulation of climate change adaptation (CCA) plans. In 2008, NEDA, the central planning agency of the Philippines, published the "Guidelines on Mainstreaming Disaster Risk Reduction in Subnational Development and Land Use/Physical Planning." It was issued to help improve regional and provincial planning analysis by recognizing risks posed by natural hazards and the vulnerability of the population, economy, and the environment to these hazards. It was originally conceived purely for DRR. However, later, CC variables have been included such as temperature increase, precipitation change, sea level rise, and so forth. This is illustrated in Fig. 26.1.

First, the guidelines advise LGUs to undertake disaster risk assessment (DRA). Briefly, DRA involves the following: (1) hazard characterization/frequency analysis; (2) consequence analysis; (3) risk estimation; and (4) risk prioritization. Below are the summaries for each of the processes involved in each step of the DRA:

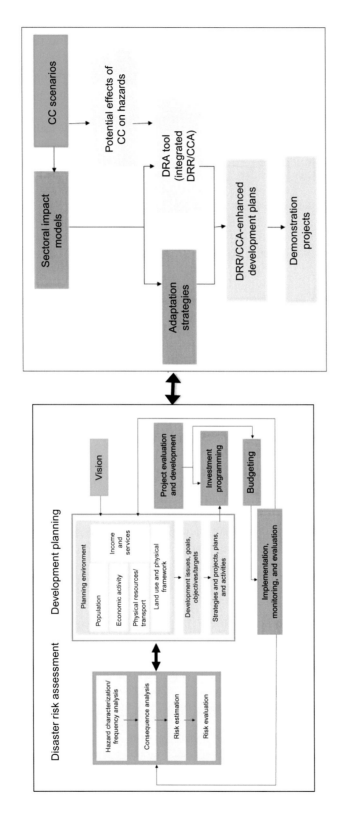

FIGURE 26.1 Integrating CCA in local development planning. Source: *De Guzman (2010)*.

1. **Hazard characterization/frequency analysis**——the in-depth study and monitoring of hazards to determine their potential, origin (e.g., geologic or hydrometeorologic), geographical extent, and hazard impact characteristics including their magnitude-frequency behavior, historical behavior, and initiating (or triggering) factors
2. **Consequence analysis**——determining or defining the elements at risk from a given hazard and defining their vulnerability
3. **Risk estimation**——involves the integration of the results of hazard characterization and frequency analysis (or hazard analysis) with consequence analysis to derive an overall measure of risk
4. **Risk prioritization**——is undertaken to guide the identification of areas needing urgent attention (NEDA-UNDP-EU HA, 2008, pp. 34–45).

The result of the DRA is then incorporated in the making of the medium-term Comprehensive Land Use Plan, and Comprehensive Development Plan of each local government, which will be the basis for policy/program/project formulation and yearly budgeting (NEDA-UNDP-EU HA, 2008, pp. 121–136).

26.6 INTEGRATED L&D–CCA–DRR FRAMEWORK

In a workshop on L&D conducted by a group of researchers of a local consulting firm (Oscar M. Lopez Center or OLMC) in 2015, many issues were identified with the assessment of L&D by the distinct CCA and DRR systems and their possible implications if not addressed. These are summarized in Table 26.5 below.

Policy makers and scientists in the workshop realized the need to plug-in these loopholes in the current L&D system to achieve "a holistic approach in viewing the importance of L&D knowledge for policymaking and effective strategic action to improve adaptation and mitigation strategies" (Gabriel et al., 2015, p. 2). Thus, a framework (see Fig. 26.2) that integrates L&D, CCA, and DRR was formulated in 2015 from a yearlong series of workshops conducted by the OLM which were participated in by several stakeholders from national government agencies, local government units, UN agencies, academics, and scientists. The framework works to "ensure a cyclic process that can enhance the resiliency of the people and reduce their vulnerability to future climate-related disasters" (Gabriel et al., 2015, p. 2).

The color-coded framework[6] shows that L&Ds from current and future stressors (no. 1) that lead to climate/disaster risks (no. 2) which impacts the socioecological system (no. 3). L&D are measured in terms of the potential and actual losses and damages (nos. 4A and 4B) which will serve as a guide and basis in planning (no. 5) and implementation (no. 6). The potential L&D generally refer to projected risks from both extreme and slow-onset events (e.g., sea level rise/coastal erosion, heat waves, droughts), while "actual" ones are those which took place already. The assessments for actual L&Ds are categorized as short term, which includes rapid/early assessment of extreme weather events (no. 4B1) used for relief and recovery planning, and medium- to long-term assessments, which include both extreme and slow-onset events subjected to in-depth evaluation and analysis (no. 4B2) used for participatory integrated CCA-DRRM

TABLE 26.5 Summary of Perceived Issues on L&D in the Philippines

Issues	Results
Lack of awareness of LGUs on the tool being used	Tool not efficiently used
Lack of a standardized process	Data mismatch
Lack of baseline data and projected damages and losses	Over- or underestimation of post-disaster data
Lack of capability of national government to properly distribute resources	Improver distribution of resources; some affected areas do not get the relief they needed
Poor governance, especially for the implementation of institutional arrangements	Misguided actions; wrong prioritization
Slow assessment system	Relief, recovery, and rehabilitation are delayed
Assessors are victims themselves	Delayed assessment of needs
Data loss and absence of integrated data from various sectors	No basis for comparative analysis of data
Tool has temporal limitations; focused on short-term and direct impacts	Difficulty in assigning value to noneconomic and consequential long-term losses

Source: Table 26.1 in Gabriel et al. (2015, June).

---

6. The succeeding discussions about the framework are based on the PowerPoint presentation of Lasco (2016).

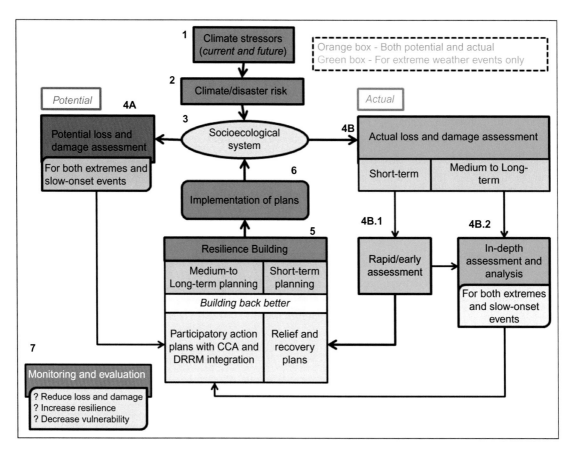

FIGURE 26.2   Integrated L&D−CCA−DRR framework. Source: *Lasco (2016)*.

planning. The plans mentioned above, from short to long terms, shall be anchored on the principle "building back better." The monitoring and evaluation (no. 7) of the implementation of these plans vis-à-vis disasters shall be measured by the following parameters: reduced L&D, increased resilience, and decreased vulnerability.

The framework has already been tested by the researchers of OLMC in 2015 in Ormoc City located in the Leyte province in the Eastern Visayas region of the Philippines. It was among those which were hit by Typhoon Haiyan (a.k. a. Yolanda) on 8 November 2013. The testing aims to know the current practices of the local government in determining losses and damages from the point of view of climate change and disaster risk reduction, in planning for recovery and rehabilitation, and the implementation and monitoring and evaluation of these plans.

It was found out that assessment studies on potential L&D were already conducted the city with the *Deutsche Gessellschaft für Internationale Zusammenarbeit* (GIZ) in the past. They used cost−benefit analysis to project the agricultural land area that could be damaged whenever typhoons visit the city. On the other hand, actual L&D studies were also conducted before by the city, this time, with assistance from UN Habitat and Oxfam. Physical observation or ocular survey of PDNA was conducted to assess the damages to the agricultural sector and to infrastructure and to determine the families that were affected. The information from the in-depth assessment were used to help the communities "build back better." Those from rapid field assessments were used for immediate relief and response to the affected communities. Furthermore, the results of the PDNA were employed for national rehabilitation and recovery planning. In terms of implementation, the city officials showed strong interest in creating programs and policies for resilience building (Lasco, 2016).

However, the researchers of OMLC were able to detect problems in the determination of L&D using CCA and DRR lenses. First, the potential and actual L&D assessments were not comprehensive enough. The potential L&D assessment was discontinued because of the lack of technical knowledge of the city government staff. In the case of the actual assessments for Typhoon Haiyan, the focus was mainly on damages, i.e., those whose predisaster value or form could be still be recovered or rehabilitated like damaged road, bridges, school buildings, etc. Second, the researchers reported that "political dynamics" at the local government level is a hurdle in the implementation of resilience-building projects,

TABLE 26.6 Ways on which L&D Information can be of Relevance to CCA and DRR

Sector	Importance
Budgeting and Accounting	To ensure that climate change expenditures are integrated or included in the annual budget of the city government
Interior and Local Government	To make sure that the planned projects for each barangay, i.e., village, will not be affected again by disasters
Engineering	To build back better infrastructure projects
Planning and Development	For comprehensive recovery and rehabilitation programs
Social Welfare & Development	To measure the extent of damage in the locality
Environment and Natural Resources	To gauge the extent of damaged forest lands and actual/existing forest cover
Tourism	The account for damaged heritage houses and ruins

Source: Lasco (2016).

policies, and programs. Third, CC and DRR actions plans have yet to be mainstreamed in local comprehensive development plans. Thus, L&D mechanisms are still lacking in the majority of local governments (Lasco, 2016).

Having learned the potentials of the integrated L&D–CCA–DRR framework in building resilient communities as illustrated by the Ormoc City study, experts cited its importance to various sectors of local governance and resilience building (Table 26.6). However, they have yet to be adopted and practiced by other local governments.

26.7 CONCLUSIONS

Recent data about losses and damages sustained by the Philippines from natural disasters reveal that no amount of preparations could make its people prepared for the fury of nature. This fact, in a way, supports the contention of the WIM that "residual" L&Ds can still be incurred by some countries, especially developing ones, even with all climate change adaptation plans in place. Unfortunately, the two distinct laws on CC and DRR of the Philippines are not yet harmonized to determine L&D. The DRR law has PDNA that are regularly utilized after the occurrence of disasters. Ironically, the CC law does not have clear guidelines on L&D determination. It can only be assumed that L&D is made part of the vulnerability and risk assessments for climate change action planning. With these two distinct systems, various issues have been raised which affect the capability of the government to effectively assess L&D from climate- and man-induced disasters. In 2015, an integrated framework for L&D–CCA–DRR was formulated by a group of researchers. The framework was designed to assess potential and actual L&Ds from extreme and slow-onset events. The assessments would then be fed into immediate relief and recovery plans and medium- to long-term CCA and DRR plans. However, much needs to be undertaken to institutionalize it and sharpen its methodologies.

REFERENCES

Bündnis Entwicklung Hilft (Alliance Development Works) (BEH) and United Nations University – Institute for Environment and Human Security (UNU-EHS). (2011-2016). *World Risk Index*. Retrieved from http://www.uni-stuttgart.de/ireus/Internationales/WorldRiskIndex/.

Cinco, T.A. et al. (2013). Climate trends and projections in the Philippines. A paper presented at the 12th National Convention on Statistics on 1–2 October 2013 at EDSA Shangri-la Hotel, Mandaluyong City, Metro Manila, Philippines. Retrieved from http://nap.psa.gov.ph/ncs/12thncs/papers/INVITED/IPS-43%20Science%20and%20Technology%20and%20Innovation%20Statistics/IPS-43_1%20Climate%20Trends%20and%20Projections%20in%20the%20Philippines.pdf.

De Guzman, M.V. (2010). Mainstreaming disaster risk reduction and climate change adaptation in development planning in the Philippines. A PowerPoint presentation made on 22 October 2010 at the 2nd Asia-Pacific Climate Change Adaptation Forum held in Bangkok, Thailand.

Florano, E.R., et al., 2016. Building back a better nation: Disaster recovery and rehabilitation in the Philippines. In: Kaneko, Yuka, et al., (Eds.), Asian Law in Disasters: Toward a Human-Centered Recovery. Routledge, Oxon, UK, pp. 131–160.

Gabriel A.V. et al. (2015, June). Assessing the linkages between CCA, DRR, and loss and damage in the Philippines. *Proceedings of the Resilient Cities 2015 Congress*. Paper presented at the 6th Global Forum on Urban Resilience & Adaptation, Bonn, Germany (1–12). Bonn: ICLEI. Retrieved from http://resilient-cities.iclei.org/fileadmin/sites/resilient-cities/files/Resilient_Cities_2015/RC2015_congress_proceedings_Gabriel__Pulhin__Lasco.pdf.

Germanwatch (2011–2016). *Global Climate Risk Index*. Retrieved from https://germanwatch.org/en/cri.

Kahn, B., 2015. Sea level could rise at least 6 meters. Scientific American. Retrieved from: http://www.scientificamerican.com/article/sea-level-could-rise-at-least-6-meters/.

Lasco, R. (2016). *Framework for L&D-CCA-D*RR Integration [PowerPoint slides]. Presented at the Linking Loss & Damage with Climate Change Adaptation and Disaster Risk Reduction in the Philippines: A Science-Policy Forum, Ortigas, Pasig, Philippines.

National Economic and Development Authority (NEDA). (n.d.). *Reconstruction Assistance on Yolanda: Implementation for Results.* Pasig, Metro Manila, Philippines: NEDA.

NEDA, United Nations Development Programme, and European Commission Humanitarian Aid (NEDA-UNDP-EU HA), 2008. Guidelines on Mainstreaming Disaster Risk Reduction in Subnational Development and Land Use/Physical Planning. NEDA-UNDP-EU HA, Pasig City, Philippines.

Office of Civil Defense-National Disaster Risk Reduction and Management Council (OCD-NDRRMC), 2014. Post-Disaster Needs Assessment (PDNA) in TY Yolanda-Affected Areas. OCD-NDRMMC, Quezon City, Philippines.

Philippine Statistics Authority (PSA) (2017). 2016 Statistical indicators on Philippine Development. Retrieved from http://psa.gov.ph/sites/default/files/Environment.pdf.

Preventionweb (n.d.). Philippines — disaster risk & profile. Retrieved from http://www.preventionweb.net/countries/phl/data/.

United Nations Office for Disaster Risk Reduction and World Bank (UN-ODRR and WB), *Synthesis Report on Ten ASEAN Countries Disaster Risks Assessment*, ASEAN Disaster Risk Management Initiative, December 2010. Retrieved from http://www.unisdr.org/files/18872_asean.pdf.

Warner, K. (2013). Significance of Warsaw International Mechanism. Retrieved from https://ehs.unu.edu/blog/articles/significance-of-the-warsaw-international-mechanism.html.

World Bank (WB) (2012, October). Capacity building for post-disaster assessments in the Philippines: Draft PDNA guidance notes.

World Bank Global Facility for Disaster Risk Reduction (WB-GFDRR), 2011. *Vulnerability, Risk Reduction and Adaptation to Climate Change, Philippines.* Available at http://sdwebx.worldbank.org/climateportalb/home.cfm?page = country_profile&CCode = PHL.

Next Steps

Chapter 27

Intelligent Tinkering in Climate Change Adaptation

Zinta Zommers[1] and Keith Alverson[2]

[1]*Food and Agriculture Organization of the United Nations, Freetown, Sierra Leone,* [2]*International Environmental Technology Center, UN Environment, Osaka, Japan*

Chapter Outline

27.1 INTRODUCTION

From sea level rise to heat waves, chapters in this book highlight the current and expected impacts of climate change. At present, the Nationally Determined Contributions in the Paris Agreement provide only one-third of the emissions reductions needed to stay below 2°C of global warming (UNEP, 2017). As the "emissions gap" increases so does the need for adaptation. The Paris Climate Agreement creates a global goal for adaptation and enables countries to include adaptation and resilience goals in their Nationally Determined Contributions. Adaptation is also receiving greater attention in climate finance discussions. Governments are encouraged to, "scale up their level of financial support, with a concrete roadmap to achieve the goal of jointly providing USD 100 billion annually by 2020 for mitigation and adaptation while significantly increasing adaptation finance from current levels" (UNFCCC, 2015).

As adaptation investments increase, there is a concurrent need to ensure activities are correctly scaled up and finance well used. In this vein, the Paris Agreement launched a four-year technical examination process on adaptation, "to identify concrete opportunities for strengthening resilience, reducing vulnerabilities and increasing the understanding and implementation of adaptation actions." This is no small task. Methodologies to validate progress are still nascent and the evidence base for what works is limited (STAP, 2017). Chapters in this book point to the ongoing debate about what constitutes successful adaptation. Florano (Chapter 26) has noted that in the *Philippines Development Plan 2011−2016 Midterm Update* success was measured primarily by focusing on outcomes—"reduction in the amount of damages and losses to properties due to natural disasters, environmental hazards, human-induced and hydro-meteorological events." By contrast, other authors appear to measure success as a process or as strengthened capacities or behavioral change. Regardless of definition, comparative studies with which to evaluate success are scarce.

As a result of these gaps in knowledge, there is a significant risk that investments and programs may lead to maladaptation—adverse climate-related outcomes, increased vulnerability to climate change, or diminished welfare. Lessons can be learnt from medicine. In this field, lack of evidence, inappropriate use of case study data, and spurious correlations have, in the past, resulted in incorrect conclusions with tragic outcomes. Thalidomide is a well-known example. Given to thousands of pregnant women, it resulted in stillbirths and phocomelia. In 1962 U.S. Congress enacted the Kefauver-Harris Amendments to the Food, Drug, and Cosmetic Act requiring drugs be proven efficacious in "adequate and well-controlled investigations" (Bothwell et al., 2016). Medical procedures need to be tested and evaluated in addition to drugs. For example, prior to 1985 it was widely believed that STA−MCA bypass surgery produced benefits to stroke patients. But in 1985, a large randomized controlled trial showed no reduction in the subsequent rate of stroke

Resilience. DOI: https://doi.org/10.1016/B978-0-12-811891-7.00027-X

with the procedure. Some patients were actually worse off as the surgery blunted the natural rate of recovery from the initial stroke (Legovini, 2010).

No matter what the field, understanding causal links between interventions and outcomes is critical for decision-making (Legovini, 2010). Monitoring and Evaluation (M&E) plays an essential role in identifying where to focus investments, what is working and what is not, why this is the case, and how to learn from experience to maximize impact (STAP, 2017). However, there is a lack of well-established standards of "best practice" for adaptation (Christiansen et al., 2016). This chapter therefore focuses on how we evaluate evidence and identify success. It highlights tools from other fields including conservation, development economics and medicine. The chapter discusses evidence-based medicine, and highlights different types of impact evaluations including randomized control trials and systematic reviews. It is hoped that wider application of these tools in climate change adaptation will result in improved effectiveness and durability of adaptation actions.

27.2 LEARNING FROM OTHER FIELDS

Over a decade ago, William Sutherland et al. (2004) wrote that, "Much of current conservation practice is based upon anecdote and myth rather than upon the systematic appraisal of the evidence, including experience of others who have tackled the same problem." The authors report that site managers responsible for managing 2996 ha of fen in Broadland, UK, predominantly used anecdotal evidence to make decisions (Sutherland et al., 2004). Out of 61 management actions, only 2.4% of decisions were based on primary scientific literature. Instead, 32.4% of decisions were made based on perceived common sense, 21.8% were based on personal experience, 20.0% were made after speaking to other managers in the region. These statistics are alarming because individual experience can be biased and lead to spurious conclusions. Further, experience that lies solely in the memory of a single practitioner risks getting lost or going unused.

The evidence-based conservation movement promotes the increased use of scientific information in conservation. In evidence-based conservation, decisions are made using systematically assessed scientific information from published, peer-review texts, as well as from practitioners' experiences, independent expert assessments, and local knowledge. Systematic type reviews are used to accumulate information from case studies. They can offer generalizable lessons when combined into one study (see for example the Baker et al. (2013) study of animal welfare in illegal wildlife trade). Sutherland et al. (2004) originally argued that these reviews should be compiled in an accessible database of information that can be searched by country, site name, animal, etc. Similarly in medicine, the Cochrane Library is a collection of six databases that contain different types of high-quality, independent evidence for decision-making[1].

Indeed, long before conservation, medicine had its own movement to combine generalized evidence with practitioner knowledge. In 1992, "evidence-based medicine" (EBM) was proclaimed as the new paradigm for the profession (Haynes, 2002). EBM is a systematic approach to clinical problem-solving combining the best available research evidence with clinical expertise and patient values (Akobeng, 2005a). EBM advocates have created procedures to objectively identify and summarize emerging evidence, and resources that allow users to find it when and where it is needed (Haynes, 2002). EBM steps include: (1) creating an answerable clinical question; (2) searching for evidence; (3) appraising the evidence; (4) deciding if the evidence applies to the current case; and (5) evaluating whether or not this is improving performance (Akobeng, 2005a). Through these steps EBM hopes to achieve safer, more consistent, and cost-effective care. Indeed, achievements since 1992 have included not only establishing the Cochrane Collaboration but setting methodological and publication standards for primary and secondary research, building systems for developing and updating clinical practice guidelines, and developing resources and courses for teaching critical appraisal.

27.3 HIERARCHY OF EVIDENCE

Difference methods can be used to appraise the evidence base. However, some research designs are more powerful than others in their ability to limit bias and answer questions on the effectiveness of interventions and limit bias (Figure 27.1) (Akobeng, 2005b). This has been referred to as a hierarchy of evidence.

In medicine, randomized control trials (RCTs) are often considered the gold standard with which to measure the effectiveness of a drug or procedure. An RCT is an impact evaluation that uses random assignment to allocate resources, run programs, or apply policies as part of the study design. In other words, individuals or communities are randomly assigned to control and treatment groups and then assessed. This minimizes the risk of confounding factors

---

1. See http://www.cochranelibrary.com/about/about-the-cochrane-library.html

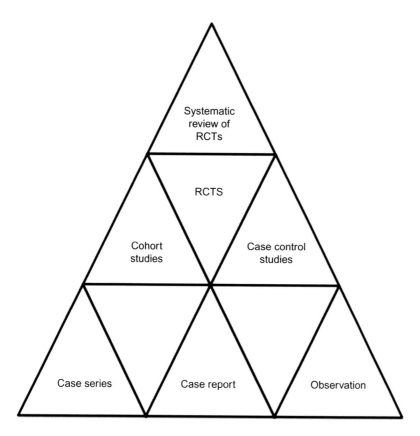

FIGURE 27.1   Hierarchy of evidence for questions about the effectiveness of an intervention or a treatment. *Adapted by permission from BMJ Publishing Group Limited [Arch.Dis.Child, Akobeng, 90, 840–844, 2005b].*

influencing the results (Akobeng, 2005b). To further eliminate bias, the participants and researchers may not even know who is in the treatment or control groups. Like all impact evaluations, the main purpose of randomized evaluations is to determine whether an intervention was successful. But well designed RCTs can also quantify impact, answering questions such as: How effective was it? Were there unintended side-effects? Who benefited most? Who was harmed? Why did it work or not work? What lessons can be applied to other contexts? How cost-effective was the program? Compared to nonrandomized evaluations, results from RCTs are less subject to methodological debates, easier to convey, and more likely to be convincing to program funders or policy-makers (JPAL, 2017).

Randomization was introduced into scientific experiments in the 1920s in agriculture trials (JPAL, 2017). Large-scale randomized clinical trials have been the norm in medicine since the 1960s. Between the 1960s and 1990s, randomized trails were also used to evaluate some government social programs—electricity-pricing schemes, employment programs, and housing allowances (JPAL, 2017). Some development economists claim that, "Randomized trials revolutionized medicine in the 20th century and have the potential to revolutionize social policy during the 21st" (Bothwell et al., 2016). Indeed, the application of RCTs to development and social policy has grown rapidly in the 21st century with the establishment of groups such as the Abdul Latif Jameel Poverty Action Lab (JPAL) and Innovation for Poverty Action (IPA). JPAL was founded in 2003 by three professors at the Massachusetts Institute of Technology with a mission to reduce poverty by ensuring that policy is informed by scientific evidence. The results are already being used to improve development programs (Box 27.1). At the time of writing, JPAL had grown to a network of 145 affiliated professors in 49 universities, with a database of 863 randomized evaluations from over 80 countries.

Systematic reviews (seen at the top of the pyramid in Figure 27.1) are then often conducted on groups of RCTs to critically appraise evidence. Systematic reviews apply scientific strategies to collect, assess, and synthesize relevant studies and other evidence (Akobeng, 2005c). Traditional narrative reviews have a high risk of bias because they use informal methods to select and interpret studies, and may inadvertently reinforce preconceived ideas. Systematic reviews reduce the risk of bias by using clear steps to comprehensively search for, include, and exclude studies (Figure 27.2). By assembling a comprehensive data set and analyzing results, it is possible to determine if findings are consistent across a range of studies, populations, or subgroups.

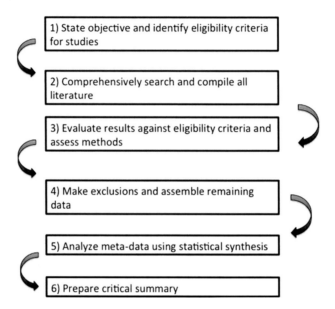

FIGURE 27.2 Methodology for systematic review. *Adapted by permission from BMJ Publishing Group Limited [Arch. Dis. Child, Akobeng, 90, 845–848, 2005c].*

27.4 CHALLENGES AND APPLICATION TO ADAPTATION

Despite the clear utility of RCTs and systematic reviews in helping identify success of interventions across different situations, such formalized scientific approaches also face problems. First, reviews may be so rigorous that they also end up discarding much of the data or ignoring critical information. "Grey" literature, studies not published in English, and older research may be missed. Facts or values specific to unique instances may also be important to patients or to program design. These are not captured by RCTs. Further some interventions are difficult to standardize for clinical trials. For example in surgery, Bothwell et al. (2016) report that each patient has unique pathological findings, each surgeon has different skills, each operation involves countless choices about anesthesia, premedication, surgical approach, instrumentation, and postoperative care. Standardization may be particularly challenging for climate change adaptation, as local context is critically important and circumstances highly variable.

A nuanced and sophisticated understanding of evidence is therefore needed. A realist approach to evaluations has developed as a result of such concerns, and could be considered for adaptation. Realist approaches assume that nothing works everywhere for everyone. Realist evaluations therefore don't ask, "Does this program work?" but instead focus on "What works for whom, in what circumstances, and in what respects, and how?" (Westhorp 2014). In realist evaluations program theories are tested for the purpose of refining them. Conclusions are always tentative and programs can always be refined further.

Another challenge to building an evidence base is the cost of impact evaluations. Over USD 30 million may be needed to run an RCT (Bothwell et al., 2016). However, a movement has developed to reduce costs by using existing data. Regular information systems or general administrative data can be used to evaluate programs, provided data is of reliable quality and there are control and treatment groups. In Senegal routine data from medical centers was used to evaluate different government HIV prevention programs. The government wanted to roll out a new peer-counseling program and evaluate the impacts. A formal RCT was not used. Instead, districts were assigned to one of three "treatments"—the new peer counseling program, the old social mobilization program, or no program. Regular monitoring indicated that peer-counseling resulted in a greater number of people getting tested for HIV but social mobilization encouraged partners of HIV-positive patients to get tested. Legovini (2010) concludes that this learning was done at almost no additional cost above program implementation. The analytical structure alone—a random assignment of health districts to different interventions—enabled the use of routine information for the purpose of a rigorous impact evaluation.

Concerns also exist over the ethics of RCTs, particularly if individuals in the control group are excluded from receiving potential benefits of action (see Lurie and Wolfe, 1997, for example). However this may be ameliorated by using a layered approach to programming and offering the intervention, if successful, to the control group after the trial. Others have argued that credible impact evaluations are international public goods: the benefits of knowing that a program works or does not work extends well beyond the organization or the country implementing the program (Duflo and Kremer, 2005).

Another challenge is that evaluations and systematic reviews can be a poor fit for multiple-morbidity (Greenhalgh et al., 2014). Greenhalgh et al. (2014) predict that this will be a growing problem in medicine, "As the population ages and the prevalence of chronic degenerative diseases increases, the patient with a single condition that maps unproblematically to a single evidence based guideline is becoming a rarity." A similar problem exists for climate change adaptation. Adaptation interventions occur against the background of evolving climate, environmental, and developmental circumstances. It may be difficult to determine the specific impact of the adaptation action when other impacts and interventions are occurring. Christiansen et al. (2016) write, "Ideally, the baseline for adaptation interventions should be 'development as it would have happened in the absence of adaptation investments,' that is, including the effects of any regular development projects or investments made for purposes other than addressing climate change." However, "due to the nature of adaptation as an additional but not easily distinguishable factor in an already dynamic development process, the definition of specific baselines for an isolated adaptation investment is difficult."

Challenges related to shifting baselines are also an issue for conservation. In the 19th century it was believed that nature, freed from human influence, tends towards a state of harmony and balance (Botkin, 1990, Budiansky, 1995). During the 20th century opinion changed. It was now argued that disturbances—such as tree falls, wind, and fire—are inherent in natural systems (Botkin, 1990). This poses a significant dilemma for management. As Dan Botkin (1990) asks, "How do you manage something that is always changing?"

Finally, timing is a problem. The long-term nature of climate change, and the rarity of certain extreme events, makes measuring impact difficult. The results of adaptation efforts may only become apparent over long periods of time. While long-term RCTs have been conducted (e.g., IPA's work examining the impact of community-driven development in Sierra Leone[2]) lengthy timelines increase costs and delay effective implementation of lessons learned. Given the magnitude and immediacy of the climate change impacts, there is an urgent need for quick action. Traditionally, learning and research is conducted before and after implementation. Programs are implemented, validated, and then, if successful, scaled up. But this approach may not even be appropriate for rapidly evolving fields in which action is quickly needed. The results can be outdated by the time they are published. During the AIDS crises activists developed flexible approaches to clinical research, such as use of surrogate end points, conditional approvals, and ways to access drugs outside of trials (Bothwell et al., 2016). Similar flexibility may be needed in adaptation.

27.5 INTELLIGENT TINKERING

Ultimately, it seems clear that we need some sort of flexible approach that balances impact evaluations with action. In other words, climate change adaptation may need intelligent tinkering and iterative risk management. "Intelligent tinkering" includes planning experiments—from small and rapid to large and long-term—into projects (Mills et al., 2015). Iterative risk management is an ongoing process of assessment, action, reassessment, and response (STAP, 2017). The Thames Estuary 2100 project, described in Chapter 6 in this book, provides an example. It used a method of developing

---

2. For further details see: https://www.poverty-action.org/study/community-driven-development-sierra-leone

options together with a process or regular updating in which options and decisions are reviewed based on changing circumstances.

USAID's Refine and Implement Program provides another example in which implementation includes a strong learning or experimental component. USAID's Office of Food for Peace (FFP) works to improve food security of the most vulnerable populations in over 50 countries around the world. FFP recently started sponsoring programs with a 1-year refinement period, called "Refine and Implement" (Whelan, 2016). During this time, organizations that receive funding are expected to carry out extensive research and community consultation, develop more refined partnership engagements, and train staff. At the end of the 1-year period, the recipients can propose changes to the implementation plans (Whelan, 2016). This inception period provides time for partners to understand and adapt to the local context, to try adaptive management. This approach is being piloted in Liberia and the Democratic Republic of Congo, both shock-prone operating environments. As illustrated in Chapter 23, fragile states pose challenges to experimentation and it may be difficult to monitor work or establish success. However, the Refine and Implement approach is also going to be applied to the FFP program for refugees in Uganda along with a randomized trial to test the effectiveness of the intervention. This combined approach of learning and testing may both help adapt the program to local circumstances and measure impacts for future scale-up. The control group will receive the most effective intervention during a second implementation round.

The Scientific and Technical Advisory Panel of the Global Environment Facility has also recently developed a new tool specific to climate change adaptation—RAPTA. RAPTA was created to help project designers embed resilience, adaptation and transformation into projects (O'Connell et al., 2016). The RAPTA guidelines try to provide "a structured approach to learning from the interventions to enable constant improvement and adaptation of our management interventions, while understanding how the systems we are managing are themselves rapidly changing" (O'Connell et al., 2016). The RAPTA process comprises several steps: (1) scoping; (2) multistakeholder engagement and governance; (3) theory of change development; (4) system description; (5) system assessment; (6) identification of intervention options and adaptive implementation pathways; and (7) monitoring and assessment, learning and knowledge management.

27.6 STEPS FORWARD

There is some indication that evidence-based approaches are beginning to penetrate the field of climate change adaptation. As an example, in this book, E. Vogel uses bibliographic network analysis to review 3,480 sources. Berrang-Ford et al. (2015) report that systematic review approaches have been used in a few other climate change adaptation studies. A search of adaptation-related documents on Web of Knowledge, revealed 27 articles using systematic review. Yet, "despite this emerging application of systematic approaches, there is negligible evidence of standardization, guidelines for review documentation, or a methodological baseline for what constitutes a 'systematic review' in the context of adaptation literature," Berrang-Ford et al. (2015) conclude.

So what is the way forward?

The first step is to remember that, just as the patient's well-being is the top priority in medicine, adaptation must benefit those vulnerable to climate change. Through meaningful conversations, studies must be designed with communities and the results must be shared with them. Communities must also demand that evidence is better presented, better explained, and better applied to address their needs.

A second step is to try to develop and promote best-practice guidelines for the "science of adaptation." Ford and Berrang-Ford (2016) suggest the researchers, program managers, and communities should strive to develop 4Cs in adaptation tracking: consistency, comparability, comprehensiveness, and coherence. In other words, practitioners should utilize a **consistent** and operational conceptualization of adaptation; focus on **comparable** units of analysis; develop **comprehensive** datasets on adaptation action; and be **coherent** with our understanding of what constitutes real adaptation. Ongoing efforts by the UNFCCC and other intergovernmental organizations may contribute to this. For example, the International Organization for Standardization (ISO) is working to develop standards to manage and mitigate greenhouse gas emissions as well as to harmonize adaptation efforts. Berrang-Ford et al. (2015) have also developed guidelines for systematic reviews in climate change adaptation.

Third, we must learn from other fields and apply relevant techniques to adaptation. Where possible this should include randomized evaluations. As Duflo and Kremer (2005) wrote about development interventions, "We do not propose that all projects be subject to randomized evaluations. But we argue that there is currently a tremendous imbalance in evaluation methodology and that increasing the share of projects subject to randomized evaluation from near-zero to even a small fraction could have a tremendous impact on knowledge about what works in development." This argument can also be made for adaptation.

Fourth, we must remain flexible. It is clear there is no approach appropriate for all. Experimental and nonexperimental methods are complementary. We will need mixed methods. Fifty percent of the World Bank's Development Impact Evaluation Initiative (DIME) studies use multiple methods. For example, the electrification of rural towns in Ethiopia was evaluated using a nonexperimental geographical matching of towns and distance from the poles. The level of subsidy required to ensure high household connection to the grid was tested experimentally using town-wide lotteries (Torero, 2006 as reported in Legovini, 2010).

In the end, while we work to improve climate change adaptation, we should not get stuck in epistemological debates. While the title of this book refers to adaptation as a science, adaptation also contains elements of art. As chapters in this book show, there are cultural, sociological, psychological, and ethnographic nuances that must be taken into account for adaptation measures to build a resilient society. Adaptation should also be viewed through the lens of history. History can help us see that our current state is not necessarily inevitable or eternal but a historical moment (Snyder, 2017). We need to stop and evaluate where we are, the structures that got us here, and choose to build a better future.

REFERENCES

Akobeng, A.K., 2005a. Principles of evidence based medicine. Arch. Dis. Child. 90, 837–840.

Akobeng, A.K., 2005b. Understanding Randomized Control Trials. Arch. Dis. Child. 90, 840–844.

Akobeng, A.K., 2005c. Understanding Systematic Reviews and Meta-Analysis. Arch. Dis. Child. 90, 845–848.

Baker, S., Caine, R., Van Kesteren, F., Zommers, Z., D'Cruze, N., Macdonald, D., 2013. Rough Trade: Animal Welfare in the Global Wildlife Trade. Bioscience. 63 (12), 928–938.

Berrang-Ford, L., Pearce, T., Ford, J., 2015. Systematic review approaches for climate change adaptation research. Reg Environ Change 15, 755–769.

Bothwell, L., Greene, J., Podolsky, S., Jones, D., 2016. Assessing the Gold Standard — Lessons from the History of RCTs. N. Engl. J. Med. 374 (22), 2175–2181.

Botkin, D., 1990. Discordant Harmonies. Oxford University Press, Oxford.

Budiansky, S., 1995. Nature's Keepers: The New Science of Nature Management. Weidenfeld and Nicolson, London.

Christiansen, L., Schaer, C., Larsen, C., Naswa, P. (2016). *Monitoring & Evaluation for climate change adaptation: A summary of key challenges and emerging practice. Understanding, discussing and ex*emplifying the key challenges of M&E for adaptation. UNEP DTU Partnership Working Papers series, Climate Resilient Development Programme, Working Paper 1.

Duflo, E., Kremer, M., 2005. Use of Randomization in the Evaluation of Development Effectiveness. In: Pitman, G., Feinstein, O., Ingram, G. (Eds.), *Evaluating Development Effectiveness. World Bank Series on Evaluation and Development Volume 7.* Transaction Publishers, London.

Ford, J., Berrang-Ford, L., 2016. The 4Cs of adaptation tracking: consistency, comparability, comprehensiveness, coherency. Mitig Adapt Strateg Glob Change 21, 839–859.

Greenhalgh, T., Howick, J., Maskrey, N., 2014. Evidence based medicine: a movement in crisis? BMJ. 348, g3725. Available from: https://doi.org/10.1136/bmj.g3725.

Haynes, R., 2002. What kind of evidence is it that Evidence-Based Medicine advocates want health care providers and consumers to pay attention to? BMC. Health. Serv. Res Available from: http://www.biomedcentral.com/1472-6963/2/3.

JPAL (2017) Introduction to Evaluations. [online], available at: https://www.povertyactionlab.org/research-resources/introduction-evaluations. [accessed, 5 November 2017].

Legovini, A., 2010. Development impact evaluation initiative: a World Bank-wide strategic approach to enhance developmental effectiveness. World Bank, Washington, DC.

Lurie, P., Wolfe, S.M., 1997. Unethical trials of interventions to reduce perinatal transmission of the human immunodeficiency virus in developing countries. N. Engl. J. Med. 337, 853–856.

Mills, A., Van der Vyver, M., Gordon, I., Patwardhan, A., Marais, C., Blignaut, J., et al., 2015. Prescribing innovation within a large-scale restoration programme in degraded subtropical thicket in South Africa. Forests 6 (11), 4328–4348.

O'Connell, D., Abel, N., Grigg, N., Maru, Y., Butler, J., Cowie, A., et al., 2016. Designing projects in a rapidly changing world: Guidelines for embedding resilience, adaptation and transformation into sustainable development projects (Version 1.0). Global Environment Facility, Washington, D.C.

Snyder, T. (2017) From Inevitability to Eternity: The New Politics of Unfreedom. 9th Isaiah Berlin Memorial Lecture, Riga, Latvia, online video, viewed 5 November 2017, https://www.youtube.com/watch?v = AC0ITS4WaTE.

STAP, 2017. Strengthening Monitoring and Evaluation of Climate Change Adaptation: A STAP Advisory Document. Global Environment Facility, Washington DC.

Sutherland, W., Pullin, A., Dolman, P., Knight, T., 2004. The Need for Evidence-Based Conservation. Trends in Ecology & Evolution 19 (6), 305–308.

Torero, M., 2006. Estimating Impact of Rural Electrification Programs in Ethiopia. Powerpoint presentation. IFPRI. Available from: http://siteresources.worldbank.org/EXTIMPEVA/Resources/Torero.ppt.

UNEP, 2017. Emissions Gap Report 2017 – Executive Summary. United Nations Environment Programme, Nairobi.

UNFCCC (2015) The Paris Agreement, United Nations Framework Convention on Climate Change, 21st Conference of the Parties, Paris: United Nations.

Westhorp, G., 2014. Realist impact evaluation: an introduction. Methods Lab. Overseas Development Institute (ODI), the Australian Department of Foreign Affairs and Trade (DFAT) and Better Evaluation.

Whelan, J., 2016. The Refine and Implement Pilot: USAID Food for Peace's Approach to Adaptable Mechanisms. 2016 CLA Case Competition. USAID, Washington.

Author Index

Subject Index

Note: Page numbers followed by "*b*," "*f*," and "*t*" refer to boxes, figures, and tables, respectively.

PRIME project. *See* Pastoralist Areas
  Resilience Improvement Through
  Market Expansion (PRIME) project
Program for Infrastructure Development in
  Africa (PIDA), 187
Programa Nasional Dezenvolvimentu Suku
  (PNDS), 171
Programme Steering Committee (PSC), 219
Project Delivery Teams (PDTs), 219
Projection of future change, 10
Protection, 22, 22*f*
Prototype ECI, 251
PSC. *See* Programme Steering Committee
  (PSC)
PSP. *See* Participatory Scenario Planning (PSP)
Public awareness of flooding, 82
Public services, 139
Public/institutional attitudes to flood risk, 85
Punta Cana, tourist destination in Dominican
  Republic, 148
Punta Cana Ecological Foundation, 148

Q

Qualitative methods, 272
Quantitative analysis of survey data, 257, 258*t*
Quantitative project monitoring tool, 257
Quinhagak, 121−122

R

Rabi, Banabans in, 310
RADA. *See* Rural and Agricultural
  Development Agency (RADA)
"Radical listening" concept, 199
Rainfall, 231, 283
Randomization, 331
Randomized control trials (RCTs), 330−331,
  332*b*, 333
RAPTA process, 334
RAY. *See* Reconstruction Assistance on
  Yolanda (RAY)
RCCC. *See* Red Cross Red Crescent Climate
  Centre (RCCC)
RCP4.5 scenario, 298
RCTs. *See* Randomized control trials (RCTs)
Rebuilding fisheries, 200
Reconstruction Assistance on Yolanda (RAY),
  321
Red Cross Red Crescent Climate Centre
  (RCCC), 238−239
Red Cross volunteers, 219
Reduced Emissions from Deforestation and
  Degradation (REDD +), 176, 201
  Kasigau corridor, 201
"Refine and Implement" approach, 334
Regional approach, 309−311
  Fiji, 309−311
  Vanuatu, 311
Regional governance, 306
Regional validation team (RVT), 257
Relative sea-level rise (RSLR), 16−19, 16*f*
  relative sea-level change, 16−17
Relocation institutional framework, 122−124
Relocation process, 118
Remote sensing satellite indices, 107

Renewable energy, 156, 187
Renewable production, 181
"Research, Modeling, and Prediction" pillar,
  224
Researcher access, 167
Residual costs, 42
Resilience, 20, 105, 108, 254−255, 267, 287,
  306. *See also* Urban resilience
  approach, 279−280
  characterizes, similarities, and differences of
    CSA practices, 169−175
  data analysis, 168−169
  dividends and changing root, 287−288
  framing, 280*b*
  implications for theory and practice,
    175−176
  mixed-crop farming in Mindanao, 171*f*
  risk and uneven resilience, 268
  sampling and data collection, 167−168, 168*t*
  social capital, 269−272
    measurement, 272
Resilient behavior, 194
Resource-poor farmers, 254
Retroreflective materials, 96
"Rich get richer, poor get poorer" paradigm,
  296
Risk
  analysis, 267
  assessment, 9
    for adaptation, 7−9
    and adaptation analysis, 146−147, 147*f*
    impact observation and, 65−66
  estimation, 323
  financial options and risk management, 41
  prioritization, 323
  and uneven resilience, 268
Risk Management Guidelines for Climate
  Change Adaptation Decision-Making,
  147
Risk society theory, 268
"Roadmaps", 221
RSLR. *See* Relative sea-level rise (RSLR)
Rubble, 160−161
Rural and Agricultural Development Agency
  (RADA), 227, 230−231
RVT. *See* Regional validation team (RVT)

S

SA. *See* South Africa (SA)
Sacramento Valley, 127−128
"Safe Drinking Water" program,
  139
Sahel, past climate shifts analysis in, 246*b*
SAI. *See* Stratospheric aerosol injection (SAI)
San Joaquin River Restoration Program, 131
San Joaquin Valley, 127−128, 128*f*, 131, 133*f*
SASSCAL. *See* Southern African Science
  Service Center for Climate Change and
  Adaptive Land Management
  (SASSCAL)
SB 88. *See* Senate Bill 88 (SB 88)
SCCF. *See* Special Climate Change Fund
  (SCCF)
SCF. *See* Standing Committee on Finance
  (SCF)

"Science of adaptation", 334
Scientific and Technical Advisory Panel of the
  Global Environment Facility, 334
SDGs. *See* Sustainable Development Goals
  (SDGs)
Sea level indicators, 85
Sea-level rise (SLR), 13, 39, 117, 120
  adaptation to, 19−25
    choosing between adaptation measures/
      options, 24
    experience, 25
    processes and frameworks, 23−24
    strategies and options, 20−23
  coastal systems, 14−16, 14*f*
  global-mean and relative sea-level change,
    16−17
  impacts, 17−19
    on socioeconomic sectors in coastal zones,
      18*t*
  natural system effects, 15*t*
Seasonal forecasting and observations, 41
Sectoral and cross-cutting approach, 55−56
Senate Bill 88 (SB 88), 136
Sendai Framework for Disaster Risk
  Reduction, 57−59, 69−70, 156−157,
  207−208, 281
"Service failure", 282
Setbacks, 120
SGMA. *See* Sustainable Groundwater
  Management Act (SGMA)
Shishmaref, 121
Shoreline management planning (SMP), 24
Short-term adaptations, 41
Sierra Nevadan rivers, 129
"Sigma". *See* Sustainable Groundwater
  Management Act (SGMA)
Siltation, 311
Simazine, 129, 132
"Single track" massive water infrastructure,
  134
Sipi River in Uganda, 211
SLA. *See* Sustainable Livelihoods and
  Emergencies Approach (SLA)
"Slash-and-burn" region, 199−200
SLR. *See* Sea-level rise (SLR)
Small island coastal communities in Caribbean,
  144
Smallholder, 165
  farmers, 165, 167, 173, 174*t*, 256
  resilience, 174−175
  systems, 175
SMP. *See* Shoreline management planning
  (SMP)
Social capital, 196−197, 200, 262*b*, 268−269
  expanding levels of moral responsibility,
    196*f*
  measurement, 272
  and resilience, 269−272
  risk and uneven resilience, 268
  vulnerability, 110
Social Capital Community Benchmark Survey,
  272
Social connectivity, 43
Social ecological indicators, 124
Social isolation, 92
Social learning, 43